Selected Titles in This Series

163 **A. Ya. Dorogovtsev, D. S. Silvestrov, A. V. Skorokhod, and M. I. Yadrenko,** Probability theory: Collection of problems, 1997

162 **M. V. Boldin, G. I. Simonova, and Yu. N. Tyurin,** Sign-based methods in linear statistical models, 1997

161 **Michael Blank,** Discreteness and continuity in problems of chaotic dynamics, 1997

160 **V. G. Osmolovskiĭ,** Linear and nonlinear perturbations of the operator div, 1997

159 **S. Ya. Khavinson,** Best approximation by linear superpositions (approximate nomography), 1997

158 **Hideki Omori,** Infinite-dimensional Lie groups, 1997

157 **V. B. Kolmanovskiĭ and L. E. Shaĭkhet,** Control of systems with aftereffect, 1996

156 **V. N. Shevchenko,** Qualitative topics in integer linear programming, 1997

155 **Yu. Safarov and D. Vassiliev,** The asymptotic distribution of eigenvalues of partial differential operators, 1997

154 **V. V. Prasolov and A. B. Sossinsky,** Knots, links, braids and 3-manifolds. An introduction to the new invariants in low-dimensional topology, 1997

153 **S. Kh. Aranson, G. R. Belitsky, and E. V. Zhuzhoma,** Introduction to the qualitative theory of dynamical systems on surfaces, 1996

152 **R. S. Ismagilov,** Representations of infinite-dimensional groups, 1996

151 **S. Yu. Slavyanov,** Asymptotic solutions of the one-dimensional Schrödinger equation, 1996

150 **B. Ya. Levin,** Lectures on entire functions, 1996

149 **Takashi Sakai,** Riemannian geometry, 1996

148 **Vladimir I. Piterbarg,** Asymptotic methods in the theory of Gaussian processes and fields, 1996

147 **S. G. Gindikin and L. R. Volevich,** Mixed problem for partial differential equations with quasihomogeneous principal part, 1996

146 **L. Ya. Adrianova,** Introduction to linear systems of differential equations, 1995

145 **A. N. Andrianov and V. G. Zhuravlev,** Modular forms and Hecke operators, 1995

144 **O. V. Troshkin,** Nontraditional methods in mathematical hydrodynamics, 1995

143 **V. A. Malyshev and R. A. Minlos,** Linear infinite-particle operators, 1995

142 **N. V. Krylov,** Introduction to the theory of diffusion processes, 1995

141 **A. A. Davydov,** Qualitative theory of control systems, 1994

140 **Aizik I. Volpert, Vitaly A. Volpert, and Vladimir A. Volpert,** Traveling wave solutions of parabolic systems, 1994

139 **I. V. Skrypnik,** Methods for analysis of nonlinear elliptic boundary value problems, 1994

138 **Yu. P. Razmyslov,** Identities of algebras and their representations, 1994

137 **F. I. Karpelevich and A. Ya. Kreinin,** Heavy traffic limits for multiphase queues, 1994

136 **Masayoshi Miyanishi,** Algebraic geometry, 1994

135 **Masaru Takeuchi,** Modern spherical functions, 1994

134 **V. V. Prasolov,** Problems and theorems in linear algebra, 1994

133 **P. I. Naumkin and I. A. Shishmarev,** Nonlinear nonlocal equations in the theory of waves, 1994

132 **Hajime Urakawa,** Calculus of variations and harmonic maps, 1993

131 **V. V. Sharko,** Functions on manifolds: Algebraic and topological aspects, 1993

130 **V. V. Vershinin,** Cobordisms and spectral sequences, 1993

129 **Mitsuo Morimoto,** An introduction to Sato's hyperfunctions, 1993

128 **V. P. Orevkov,** Complexity of proofs and their transformations in axiomatic theories, 1993

127 **F. L. Zak,** Tangents and secants of algebraic varieties, 1993

126 **M. L. Agranovskiĭ,** Invariant function spaces on homogeneous manifolds of Lie groups and applications, 1993

(*Continued in the back of this publication*)

Probability Theory: Collection of Problems

Translations of
MATHEMATICAL
MONOGRAPHS

Volume 163

Probability Theory: Collection of Problems

A. Ya. Dorogovtsev
D. S. Silvestrov
A. V. Skorokhod
M. I. Yadrenko

American Mathematical Society
Providence, Rhode Island

А. Я. Дороговцев, Д. С. Сільвестров, А. В. Скороход, М. Й. Ядренко

ТЕОРІЯ ЙМОВІРНОСТЕЙ. ЗБІРНИК ЗАДАЧ

«Вища школа», Київ, 1976

Translated by O. I. Klesov and V. A. Kotov

1991 *Mathematics Subject Classification.* Primary 60–01.

ABSTRACT. The book contains problems on principal topics of probability theory and selected topics of random processes and mathematical statistics. In addition to standard problems, the topics that are widely used in applications are discussed. Answers to most problems are given. For more complicated problems, hints and solutions are presented.

The book will be useful for professors and students involved in teaching or learning probability theory.

Library of Congress Cataloging-in-Publication Data

Teoriiā ĭmovirnosteĭ, English.

Probability theory : collection of problems / A. Ya. Dorogovtsev ; [translated by O. I. Klesov and V. A. Kotov].

p. cm. — (Translations of mathematical monographs ; v. 163)

Includes bibliographical references.

ISBN 0-8218-0372-7

1. Probabilities—Problems, exercises, etc. I. Dorogovt͡sev, A. I͡A. (Anatoliĭ I͡Akovlevich) II. Title. III. Series.

QA273.25.T4613 1997

519.2—dc21 97-5939 CIP

∞ The paper used in this book is acid-free and falls within the guidelines
established to ensure permanence and durability.

10 9 8 7 6 5 4 3 2 1 02 01 00 99 98 97

Contents

Preface to the English Edition

This book contains problems in probability theory. The topics covered include classical probability and combinatorics, geometric probabilities, axiomatics of probability, conditional probability, random variables, random walks, series of random variables, the law of large numbers and the strong law of large numbers, limit theorems for sums of independent random variables and infinitely divisible distributions, compound Poisson processes and Wiener processes, martingales, and Markov chains. In addition, there are many problems devoted to such techniques of probability as moment generating functions and characteristic functions, probability inequalities (Chebyshev, Kolmogorov), renewal models, and so forth.

The problems are arranged in consecutive order so that the reader who solves them may become proficient in probability methods. The book is self-contained; each section has all the necessary theoretical results to allow the reader to use the collection of problems without recourse to other sources.

The book is addressed primarily to researches who wish to use the methods of probability in further work. They will benefit greatly from the second part of the book, which contains complete solutions for the most difficult problems and hints to the solutions of other problems.

Because of the diverse problems it contains, the book can be used to teach both beginning and advanced probability. A teacher will find in it problems that correspond to any goals of the class.

The book may be of interest also to those who study analysis, as the authors have tried to emphasize the intimate relationship between probability theory and analysis. For example, in Chapter III, the reader will find many problems that first appeared in analysis and later became part of probability theory.

Finally, the book may appeal to readers interested in probability, as it contains some "historical" problems unearthed from older books.

A. V. Skorokhod

Preface

This book of problems is intended for students of pure or applied mathematics. In addition to problems in traditional areas of probability theory, it also includes problems in the theory of stochastic processes, which finds wide application in the theory of automatic control, queueing theory, reliability theory, and many other areas of modern science and engineering.

The book has been complied in such a way as to be useful also to students majoring in pedagogical and engineering areas. For this reason much attention is given in Chapter I to problems in combinatorics.

The problems included in this book were chosen on the basis of the authors' broad experience in teaching probability at Kyiv National University, and present a modern approach to this subject. In particular, heavy emphasis is placed on the standard axiomatics of probability theory due to A. N. Kolmogorov.

The authors will appreciate receiving suggestions and remarks.

A. Ya. Dorogovtsev
D. S. Silvestrov
A. V. Skorokhod
M. I. Yadrenko

CHAPTER I

Random Events

§I.1. Operations on sets. Algebras and σ-algebras of sets

Consider a set (space) $\Omega = \{\omega\}$ and its subsets, which are denoted by capital letters. The notation $A \subset B$ (to be read: "A is a subset of B") means that every element of the set A belongs to B. Sets A and B are called *equal* ($A = B$) if $A \subset B$ and $B \subset A$.

The notation $\omega \in A$ (to be read: "ω belongs to A") means that the element ω belongs to A. The notation $\omega \notin A$ (to be read: "ω does not belong to A") means that the element ω does not belong to the set A. The set that contains no elements is called *the empty* set and is denoted by the symbol $\varnothing$; the empty set is considered to be a subset of any set.

The sum (*union*) $A \cup B$ of sets A and B is the set of those and only those elements that belong at least to one of the sets A or B.

The product (*intersection*) $A \cap B$ of sets A and B is the set of those and only those elements that belong to both A and B.

The difference $A \setminus B$ of sets A and B is the set of those and only those elements that belong to A and do not belong to B.

The complement $\overline{A}$ of a set A is the set of those and only those elements of Ω that do not belong to A, $\overline{A} = \Omega \setminus A$.

The symmetric difference $A \triangle B$ of sets A and B is the set $(A \setminus B) \cup (B \setminus A)$.

Sets A and B are said to be *disjoint* if $A \cap B = \varnothing$.

An algebra of sets. A class $\mathfrak{M}$ of subsets of Ω is called *an algebra* if the following conditions are fulfilled:

1) if $A \in \mathfrak{M}$ and $B \in \mathfrak{M}$, then $A \cup B \in \mathfrak{M}$;
2) if $A \in \mathfrak{M}$, then also $\overline{A} = \Omega \setminus A \in \mathfrak{M}$.

A σ-algebra of sets. A class F of subsets of Ω is called *a σ-algebra* if the following conditions are fulfilled:

1) if $A_n \in F$, $n = 1, 2, \ldots$, then also $\bigcup_{n=1}^{\infty} A_n \in \mathfrak{M}$;
2) if $A \in \mathfrak{M}$, then also $\overline{A} = \Omega \setminus A \in \mathfrak{M}$.

Problems

I.1.1. Prove the identities:

a) $A \setminus B = A \setminus (A \cap B) = (A \cup B) \setminus B$;
b) $A \cap (B \setminus C) = (A \cap B) \setminus (A \cap C)$;
c) $(A \setminus C) \cap (B \setminus C) = (A \cap B) \setminus C$.

I.1.2. Let A and B be subsets of the plane $\Omega = \mathbf{R}^2$ defined as follows:

$$A = \{(x, y) \colon x + y \leq 1\} \quad \text{and} \quad B = \{(x, y) \colon y \leq 2x + 2\}.$$

Depict geometrically the sets $A \cap B$, $A \cup B$, $A \setminus B$, $\overline{A}$, $\overline{B}$, and $\overline{A} \cap \overline{B}$.

I.1.3. Let $A_n = (\frac{1}{2n}, \frac{1}{n})$. Describe the sets

$$A = \bigcap_{n=1}^{\infty} A_n \quad \text{and} \quad B = \bigcup_{n=1}^{\infty} A_n.$$

I.1.4. Let A_n, $n \geq 1$, be a sequence of subsets of the plane defined as $A_n = \{(x, y): x^2 + y^2 \leq n^{-2}\}$. Describe the sets

$$A = \bigcup_{n=1}^{\infty} A_n \quad \text{and} \quad B = \bigcap_{n=1}^{\infty} A_n.$$

I.1.5. Prove that
a) $A \triangle B = (A \cup B) \setminus (A \cap B)$;
b) $A \triangle (B \triangle C) = (A \triangle B) \triangle C$;
c) $A \triangle \varnothing = A$, $A \triangle \Omega = \overline{A}$;
d) $A \triangle A = \varnothing$, $A \triangle \overline{A} = \Omega$.

I.1.6. Let $\{A_n, n \geq 1\}$ be a sequence of sets. Prove that $\bigcup_{n=1}^{\infty} A_n = \bigcup_{n=1}^{\infty} B_n$, where $B_1 = A_1$ and

$$B_n = A_n \setminus \bigcup_{i=1}^{n-1} A_i.$$

Show that the sets B_i are pairwise disjoint.

I.1.7. Let I be a set of indices. Prove that

$$\overline{\bigcup_{\alpha \in I} A_\alpha} = \bigcap_{\alpha \in I} \overline{A}_\alpha, \qquad \overline{\bigcap_{\alpha \in I} A_\alpha} = \bigcup_{\alpha \in I} \overline{A}_\alpha.$$

I.1.8. Let

$$\chi_A(\omega) = \begin{cases} 1, & \omega \in A, \\ 0, & \omega \notin A. \end{cases}$$

(The function $\chi_A(\omega)$ is called *the indicator of a set A*.) Prove that
a) $\chi_{A \cap B}(\omega) = \chi_A(\omega)\chi_B(\omega)$;
b) $\chi_{A \cup B}(\omega) = \chi_A(\omega) + \chi_B(\omega) - \chi_A(\omega)\chi_B(\omega)$;
c) $\chi_{\overline{A}}(\omega) = 1 - \chi_A(\omega)$;
d) $\chi_{A \setminus B}(\omega) = \chi_A(\omega)[1 - \chi_B(\omega)]$;
e) $\chi_{A \triangle B}(\omega) = |\chi_A(\omega) - \chi_B(\omega)|$;
f) $\chi_{A \triangle B}(\omega) = \chi_A(\omega) + \chi_B(\omega) \mod 2$.

I.1.9. Using the results of Problem I.1.8, prove the following equalities:
a) $(A \cap B) \cap C = A \cap (B \cap C)$;
b) $(A \cap B) \cup C = (A \cup C) \cap (B \cup C)$;
c) $\overline{A \cap B} = \overline{A} \cup \overline{B}$.

I.1.10. Let $\{A_n, n \geq 1\}$ be a sequence of subsets of Ω, A^* the set of those elements ω that belong to infinitely many sets A_n, and A_* the set of those elements ω that belong to all but a finite number of sets A_n.

Prove that a) $A_* \subset A^*$; b) $A_* = \bigcup_{n=1}^{\infty} \bigcap_{m=n}^{\infty} A_m$; c) $A^* = \bigcap_{n=1}^{\infty} \bigcup_{m=n}^{\infty} A_m$.

REMARK. Sometimes the sets A^* and A_* are denoted as follows:

$$A^* = \limsup A_n, \qquad A_* = \liminf A_n.$$

In the case $\limsup A_n = \liminf A_n$, it is customary to say that there exists $\lim_{n\to\infty} A_n$ defined as

$$\lim_{n\to\infty} A_n = \liminf A_n = \limsup A_n.$$

I.1.11. Let

$$A_n = \begin{cases} A & \text{if } n \text{ is even,} \\ B & \text{if } n \text{ is odd.} \end{cases}$$

Prove that

$$\limsup A_n = A \cup B, \qquad \liminf A_n = A \cap B.$$

I.1.12. Let $\{A_n, n \geq 1\}$ be a sequence of subsets of Ω such that $A_n \supset A_{n+1}$ for all $n \geq 1$. Prove that

$$\limsup A_n = \liminf A_n = \bigcap_{n=1}^{\infty} A_n.$$

I.1.13. Let $\{A_n, n \geq 1\}$ be a sequence of subsets of Ω such that $A_n \subset A_{n+1}$ for all $n \geq 1$. Prove that

$$\limsup A_n = \liminf A_n = \bigcup_{n=1}^{\infty} A_n.$$

I.1.14. Let $\{A_n, n \geq 1\}$ be a sequence of pairwise disjoint sets. Prove that

$$\limsup A_n = \liminf A_n = \varnothing.$$

I.1.15. Let $\mathfrak{M}$ be an algebra of sets. Prove that
a) if $A_k \in \mathfrak{M}$, $k = 1, 2, \ldots, n$, then $\bigcup_{k=1}^{n} A_n \in \mathfrak{M}$;
b) if $A_k \in \mathfrak{M}$, $k = 1, 2, \ldots, n$, then $\bigcap_{k=1}^{n} A_n \in \mathfrak{M}$;
c) $\Omega \in \mathfrak{M}$ and $\varnothing \in \mathfrak{M}$.

I.1.16. Let $I = \{\alpha\}$ be a set of indices, and let $\mathfrak{M}_\alpha$ be a class of algebras enumerated with indices α, $\alpha \in I$. Prove that $\bigcap_{\alpha\in I} \mathfrak{M}_\alpha$ is an algebra. (The intersection of any number of algebras is an algebra.)

I.1.17. Prove that the set of all subsets of a set Ω is an algebra.

I.1.18. Let K be a class of subsets of Ω. An algebra $\mathfrak{M}_0(K)$ is called *the minimal algebra containing the class* K if
1) every set in K belongs to $\mathfrak{M}_0(K)$;
2) whatever an algebra $\mathfrak{M}$ that contains the class K may be, the inclusion $\mathfrak{M}_0(K) \subset \mathfrak{M}$ holds.

Prove that for every class K the minimal algebra containing the class K exists.

I.1.19. Let a class K consist of a single set $A \subset \Omega$. Describe the minimal algebra $\mathfrak{M}_0(K)$ that contains the class K.

I.1.20. Let $\Omega = \{1, 2, 3, 4, 5, 6\}$. Describe all the algebras of sets that contain the sets $A = \{2, 3, 4\}$ and $B = \{4, 6\}$. Also, describe the minimal algebra that contains the sets A and B.

I.1.21. A class of sets $A_1, \ldots, A_n$ is said to be *a partition of* Ω if

1) $A_1, \ldots, A_n$ are pairwise disjoint, and
2) $\bigcup_{i=1}^n A_i = \Omega$.

Prove that the set of all possible unions of the sets $A_1, \ldots, A_n$ forms an algebra. Describe the minimal algebra $\mathfrak{M}_0(K)$ that contains the class $K = \{A_1, \ldots, A_n\}$.

I.1.22. A nonempty set A is called *an atom of an algebra* $\mathfrak{M}$ if

1) $A \in \mathfrak{M}$;
2) $B \in \mathfrak{M}$ and $B \neq \varnothing$ imply $B = A$.

If $\mathfrak{M}$ is a finite algebra, then the set of all atoms of $\mathfrak{M}$ forms a partition $A_1, \ldots, A_n$ of Ω. The minimal algebra that contains the sets $A_1, \ldots, A_n$ coincides with $\mathfrak{M}$. Prove this assertion.

I.1.23. Let K be an arbitrary class of subsets of Ω. Let us form successively the following classes:

1) the class K_1 consisting of $\{\varnothing, \Omega\}$ and sets A such that either $A \in K$ or $\overline{A} \in K$;
2) the class K_2 of all finite intersections of subsets of the class K_1;
3) the class K_3 of all finite sums of disjoint subsets of the class K_2.

Prove that the class K_3 coincides with the minimal algebra $\mathfrak{M}_0(K)$ that contains the class K. Construct the classes K_1, K_2, and K_3 for $K = \{\{1,2,3\}, \{3,4\}\}$.

I.1.24. Prove that if F is a σ-algebra of sets and $A_n \in F$ for all $n \geq 1$, then

$$\bigcap_{n=1}^{\infty} A_n \in F.$$

I.1.25. Prove that the intersection of any number of σ-algebras is a σ-algebra.

I.1.26. Let K be a class of subsets of Ω. *The minimal σ-algebra* that contains the class K (*the σ-algebra generated by the class* K) is defined as the σ-algebra $F_0(K)$ such that

1) every set in the class K belongs to $F_0(K)$;
2) the inclusion $F_0(K) \subset K$ holds, whatever a σ-algebra F containing the class K may be.

Prove that for every class K the minimal σ-algebra $F_0(K)$ containing the class K exists.

I.1.27. Let K_1 and K_2 be two classes of subsets of Ω such that $K_1 \subset K_2$. Prove that $F_0(K_1) \subset F_0(K_2)$.

I.1.28. Let K be the class of intervals of the form $[a, b)$ on the real line. The minimal σ-algebra $\mathfrak{B} = F_0(K)$ that contains the class K is called *the σ-algebra of Borel sets*; the subsets of $\mathfrak{B}$ are called *Borel sets*.

Prove that

a) for every a the set $\{a\}$ is a Borel set;
b) every countable set on the real line is a Borel set;
c) an interval (a, b) is a Borel set;
d) an open set on the real line is a Borel set;
e) a closed set on the real line is a Borel set.

I.1.29. Let K be the class of open intervals on the real line. Prove that the minimal σ-algebra that contains the class K is the σ-algebra $\mathfrak{B}$ of Borel sets.

I.1.30. Let K be the class of open sets of the real line. Prove that the minimal σ-algebra that contains the class K is the σ-algebra of Borel sets.

I.1.31. Let K be the class of closed sets of the real line. Prove that the minimal σ-algebra that contains the class K is the σ-algebra of Borel sets.

I.1.32. Let K be the real line. Prove that the collection of sets of the form $F \cap G$, where F, $G \in \mathbf{R}$, F being a closed and G an open set, forms an algebra, but it does not form a σ-algebra. What sets should be added to the collection in order to get a σ-algebra?

I.1.33. A class M of subsets of Ω_0 is called a *monotone* class if

$$\lim A_n \subset M$$

for any monotone sequence of sets $\{A_n, n \geq 1\}$ in M.

In order that an algebra $\mathfrak{M}$ be a σ-algebra, it is necessary and sufficient that it be a monotone class. Prove this assertion.

I.1.34. Prove that the intersection of any number of monotone classes is a monotone class.

I.1.35. Let K be a class of subsets of Ω. A monotone class $M_0(K)$ is called *the minimal monotone class containing the class* K if the following conditions are fulfilled:

1) every set in K belongs to $M_0(K)$;
2) whatever a monotone class M containing K may be, the following inclusion holds:

$$M_0(K) \subset M.$$

Prove that for any class K the minimal monotone class $M_0(K)$ containing the class K exists.

I.1.36. Let K be a class of sets, $\Gamma_0(K)$ the minimal σ-algebra that contains K, and $M_0(K)$ the minimal monotone class containing K. Prove that

$$M_0(K) \subset F_0(K).$$

I.1.37. Let $\mathfrak{M}$ be an algebra of sets, $F_0(\mathfrak{M})$ the minimal σ-algebra that contains $\mathfrak{M}$, and $M_0(\mathfrak{M}) = M$ the minimal monotone class that contains $\mathfrak{M}$. Prove the following assertions:

a) the class of sets $\widetilde{M} = \{B \colon B \in M \text{ and } \overline{B} \in M\}$ coincides with M;
b) if $A \in M$, then the class of sets $M_A = \{B \colon B \in M, A \cap B \in M\}$ coincides with M;
c) M is a σ-algebra of sets;
d) $F_0(\mathfrak{M}) = M_0(\mathfrak{M})$, i.e., the minimal σ-algebra that contains the algebra $\mathfrak{M}$ coincides with the minimal monotone class that contains $\mathfrak{M}$.

§I.2. Combinatorics

The fundamental principle of combinatorics (the multiplication rule). Suppose it is necessary to perform k consecutive operations. If the first operation can be performed in n_1 ways, the second in n_2 ways, the third in n_3 ways, and so on, up to the kth operation inclusive, which can be performed in n_k ways, then all k operations can be performed in $n_1 n_2 \cdots n_k$ ways.

Combinations of n elements taken k at a time. Let a set A contain n elements. Then the number of subsets of the set A consisting of k elements is equal to

$$\binom{n}{k} = \frac{n!}{k!\,(n-k)!},$$

where $n! = 1 \cdot 2 \cdot \cdots \cdot n$.

k-element subsets of the set $A = \{a_1, a_2, \ldots, a_n\}$ are called *combinations of n elements $\{a_1, a_2, \ldots, a_n\}$ taken k at a time.*

Ordered sets. A set of n elements is called *an ordered set* if to every element of the set a certain number from 1 to n is assigned so that to different elements there correspond different numbers. Ordered sets are regarded as different if they differ in their elements or in their order.

Permutations of elements of a given set. Ordered sets that differ only in the order of their elements (that is, the ordered sets that can be formed from the same set) are called *permutations* of the elements of this set. The number of permutations of n elements is equal to $P_n = n!$.

Arrangements of n elements taken k at a time. Ordered k-element subsets of a set containing n elements are called *arrangements of n elements taken k at a time*. The number of arrangements of n elements taken k at a time is equal to

$$A_n^k = k!\binom{n}{k} = n(n-1)\cdots(n-k+1).$$

The number of partitions of an n-element set into m groups. Let $k_1, k_2, \ldots, k_m$ be nonnegative integers such that $k_1 + k_2 + \cdots + k_m = n$. The number of ways in which a set A consisting of n elements can be represented as the union of its subsets $B_1, B_2, \ldots, B_m$ consisting of $k_1, k_2, \ldots, k_m$ elements, respectively, is equal to

$$C_n(k_1, k_2, \ldots, k_m) = \frac{n!}{k_1!\,k_2!\cdots k_m!}.$$

Permutations with repetitions. The number of different permutations that can be obtained out of n elements, among which k_1 elements are of the first type, k_2 elements are of the second type, ..., k_m elements are of the mth type, is equal to

$$C_n(k_1, k_2, \ldots, k_m) = \frac{n!}{k_1!\,k_2!\cdots k_m!}.$$

The (Cartesian) product of sets. Given sets $A_1, A_2, \ldots, A_k$, the set of all k-tuples of the form $(a_1, a_2, \ldots, a_k)$, where $a_1 \in A_1, a_2 \in A_2, \ldots, a_k \in A_k$, is called *the product of the sets* $A_1, A_2, \ldots, A_k$ and is denoted by $A_1 \times A_2 \times \cdots \times A_k$.

EXAMPLE. If $A = \{a, b\}$ and $B = \{c, d, e\}$, then

$$A \times B = \{(a, c), (a, d), (a, e), (b, c), (b, d), (b, e)\}.$$

The number of elements in the product of sets. Let $N(A)$ be the number of elements of a set A. Then

$$N(A_1 \times A_2 \times \cdots \times A_k) = N(A_1)N(A_2)\cdots N(A_k).$$

Problems

I.2.1. There are n roads joining a town A to a town B, and m roads from B to a town C. How many roads join the town A to the town C?

I.2.2. There are seven trails to the top of a mountain. In how many ways can a hiker climb the mountain and come down? Answer the same question under the condition that the hiker uses different trails for going up and down.

I.2.3. Seventeen teams take part in the soccer championship of a country. In how many ways can the gold, silver, and bronze medals be distributed among the teams?

I.2.4. How many three-digit numbers can be written using the digits $0, 1, 2, 3$, and 4?

I.2.5. How many three-digit numbers can be written using the digits $0, 1, 2, 3$, and 4 if every digit must be used only once?

I.2.6. In how many ways can seven persons form a line to a cashier?

I.2.7. A group of students signed for ten courses. There are six classes on Monday, all classes being different. In how many ways can the schedule for Monday be set up?

I.2.8. How many five-digit numbers are divisible by 5?

I.2.9. Suppose n and m points, respectively, are taken on two sides of a triangle, and each vertex of the third side is joined to the points on the opposite side by straight lines. In how many parts is the triangle subdivided by these straight lines?

I.2.10. n players participate in a chess tournament. How many games will be played in the tournament if every participant plays one game with any other participant?

I.2.11. How many diagonals can be drawn in a convex n-sided polygon?

I.2.12. n straight lines are drawn on the plane, no two being parallel and no three intersecting at one point. How many intersection points are there in this configuration?

I.2.13. What is the largest number of parts into which n straight lines can subdivide the plane?

I.2.14. What is the largest number of parts into which n planes can subdivide the space?

I.2.15. What is the largest number of parts into which n circles can subdivide the plane?

I.2.16. What is the largest number of parts into which n spheres can subdivide the space?

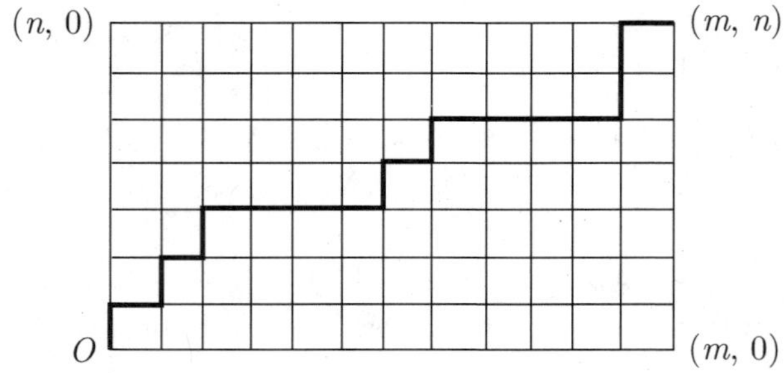

FIGURE 1

I.2.17. In how many ways can two castles of different colors be placed on an $n \times n$ chessboard so that they cannot take each other?

I.2.18. A car plate consists of two letters and four digits. How many different plates can be made up using the 26 letters of the English alphabet?

I.2.19. If a page with some numbers on it is turned upside down, then the digits 0, 1, and 8 do not change, the digits 6 and 9 transpose, and the other digits lose their meaning. How many seven-digit numbers do not change?

I.2.20. Sixteen teams participate in the soccer championship of a country. The teams that finish in the first, second, and third places are rewarded with the gold, silver, and bronze medal, respectively, while the teams that finish in the last two places leave the field. How many different outcomes can the championship have?

I.2.21. a) How many different divisors does the number $3^5 \cdot 5^4$ have?

b) Let $p_1, p_2, \ldots, p_n$ be different primes. How many divisors does the number $p_1^{\alpha_1} p_2^{\alpha_2} \cdots p_n^{\alpha_n}$ have?

I.2.22. A passenger put his luggage in a locker. When he comes back to take his luggage, it turns out that he has forgotten the combination. The passenger remembers only that it contained 23 and 37. To open the locker, he must dial a five-digit combination. What is the largest number of combinations he might need to dial to open the locker?

I.2.23. A rectangular table of m rows and n columns with entries $+1$ and -1 must be written in such a way that the product of the entries of each row and each column is 1. In how many ways can this be done?

I.2.24. In how many ways can a committee of four members be formed out of nine persons?

I.2.25. In how many ways can a reader choose three books out of six?

I.2.26. In how many points do the diagonals of a convex n-gon intersect if no three of them intersect at one point?

I.2.27. Consider a rectangular net of squares (a "chess-town" consisting of mn squares separated by $n-1$ "horizontal" and $m-1$ "vertical" streets (Figure 1)). How many different shortest paths lead from the bottom left corner (the point $(0,0)$) to the top right corner (the point (m,n))?

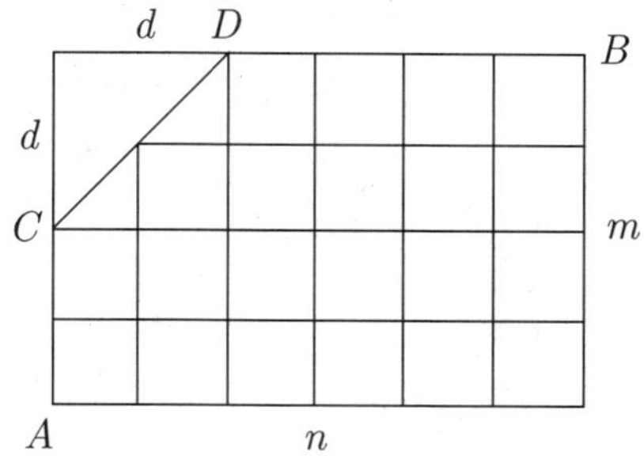

FIGURE 2

I.2.28. Using the previous problem, prove geometrically the following equalities:

a) $\binom{n}{k} = \binom{n-1}{k-1} + \binom{n-1}{k}$;

b) $\binom{n}{0}^2 + \binom{n}{1}^2 + \binom{n}{2}^2 + \cdots + \binom{n}{n}^2 = \binom{2n}{n}$;

c) $\binom{n}{m}\binom{k}{0} + \binom{n-1}{m-1}\binom{k+1}{1} + \cdots + \binom{n-m}{0}\binom{k+m}{m} = \binom{n+k+1}{m}$;

d) $\sum_{k=0}^{n-r} \binom{n-k-1}{r-1} = \binom{n}{r}$.

I.2.29. Find the number of shortest paths from the point A to the point B on the chessboard depicted in Figure 2.

I.2.30. An international committee consists of nine members. Documents of the committee are kept in a safe. What is the minimum number of locks that the safe must have, and the minimum number of keys that must be made, and how must the keys be distributed among the members of the committee in order that the access to the documents be possible only when at least six members of the committee are present? Consider the case where the committee consists of n members and the safe can be opened only in the presence of at least m members.

I.2.31. All diagonals are drawn in a convex n-sided polygon. It is known that no three of them intersect at one point. Into how many parts is the polygon thereby subdivided?

I.2.32. Prove that

$$(a+b)^n = \sum_{k=0}^{n} \binom{n}{k} a^{n-k} b^k.$$

I.2.33. a) Prove that

$$\binom{n}{k+1} > \binom{n}{k} \qquad \text{for } k < \frac{n-1}{2},$$
$$\binom{n}{k+1} < \binom{n}{k} \qquad \text{for } k > \frac{n-1}{2}.$$

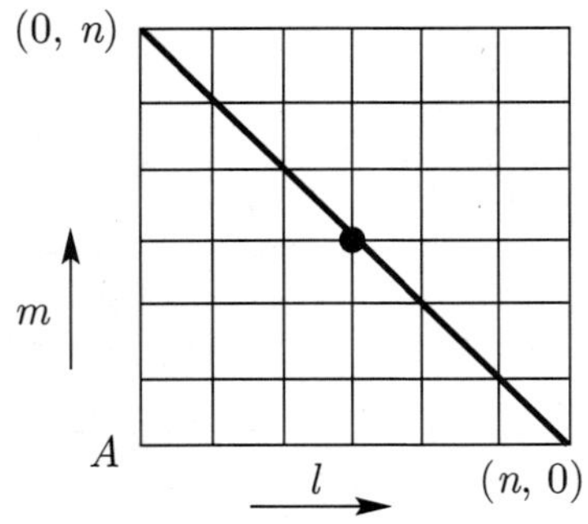

FIGURE 3

b) Find the largest among the numbers $\binom{n}{k}$, where $k = 0, 1, \ldots, n$.

I.2.34. Let p be a prime number. Prove that the numbers

$$\binom{p}{1}, \binom{p}{2}, \ldots, \binom{p}{p-1}$$

are divisible by p.

I.2.35. Fermat's Theorem. Let p be a prime number. Prove that for any integer a the difference $a^p - a$ is divisible by p.

I.2.36. Prove that the difference $\left[(2 + \sqrt{5})^p \right] - 2^{p+1}$ is divisible by p if p, $p > 2$, is a prime number. (The symbol $[x]$ denotes the integer part of x.)

I.2.37. How many subsets does an n-element set have? (The empty set is a subset of every set.)

I.2.38. There are n lamps in a room. In how many ways can the room be lit?

I.2.39. There is a network of roads (Figure 3). 2^n walkers start at the point A; half of them go in the direction l and the other half in the direction m. Upon reaching the first crossroad, every group divides into two groups of equal size; half go in the direction l, and the other half in the direction m. Such a division occurs at every crossroad. How many walkers will be there at every crossroad after they cover n segments?

I.2.40. Let A be a set of n points on the plane such that m points lie on one straight line and no other three points lie on one straight line. How many triangles can be formed on the plane with vertices in A?

I.2.41. In how many ways can four books be arranged on a bookshelf (denote them by A, B, C, and D)?

I.2.42. In how many ways can a set $\{1, 2, \ldots, 2n\}$ be ordered so that every even number occupy an even position?

I.2.43. How many permutations of n elements are such that two given elements are not next to each other?

I.2.44. In how many ways can eight castles be placed on a chessboard so that they cannot take one another?

I.2.45. In how many ways can four students be seated in a clasroom that seats 25 people?

I.2.46. A student must take four exams in eight days. In how many ways can this be accomplished?

I.2.47. In how many ways can the numbers $1, 2, \ldots, n$ be ordered so that 1, 2, and 3 go one after another in ascending order?

I.2.48. How many positive integers not exceeding 10^9 have digits that are all different and arranged in ascending order?

I.2.49. How many permutations of n elements have exactly r elements between given two elements?

I.2.50. Four persons, A, B, C, and D, are to address a meeting. In how many ways can they be entered in the list of speakers if B cannot speak before A?

I.2.51. In how many ways can n guests be seated at a round table?

I.2.52. In how many ways can a set $\{1, 2, \ldots, n\}$ be ordered so that every multiple of 2 and every multiple of 3 occupy a position divisible by 2 and 3, respectively?

I.2.53. How many different "words" can be formed by permuting letters in the word "mathematics"?

I.2.54. Let there be k_1 letters a_1, k_2 letters a_2, ... , k_m letters a_m $(k_1 + k_2 + \cdots + k_m = n)$. How many "words" can be formed of these letters?

I.2.55. How many different "words" can be formed of the letters a, b, and c provided that the letter a occurs in a "word" no more than twice, the letter b no more than once, and the letter c no more than three times?

I.2.56. How many different "words" can be formed by permuting letters in the word "combinatorics"?

I.2.57. In how many ways can $m + n + s$ objects be divided into three groups so that there will be m objects in the first group, n objects in the second, and s in the third?

I.2.58. In how many ways can $3n$ different objects be distributed among three persons so that every person will receive n objects?

I.2.59. The polynomial theorem. Prove that $(a_1 + a_2 + \cdots + a_k)^n$ is equal to the sum of all possible terms of the form

$$\frac{n!}{r_1!\, r_2! \cdots r_k!} a_1^{r_1} a_2^{r_2} \cdots a_k^{r_k},$$

where $r_1 + r_2 + \cdots + r_k = n$, i.e.,

$$(a_1 + a_2 + \cdots + a_k)^n = \sum_{\substack{r_1 \geq 0, r_2 \geq 0, \ldots, r_k \geq 0, \\ r_1 + r_2 + \cdots + r_k = n}} \frac{n!}{r_1!\, r_2! \cdots r_k!} a_1^{r_1} a_2^{r_2} \cdots a_k^{r_k}.$$

I.2.60. n identical balls are placed into N urns. Prove that

a) the number of different distributions of the balls among the urns equals $\binom{N+n-1}{n} = \binom{N+n-1}{N-1}$;

b) the number of those distributions when every urn contains at least one ball equals $\binom{n-1}{N-1}$.

I.2.61. a) How many nonnegative integral solutions does the equation $x_1 + x_2 + \cdots + x_N = n$ have?

b) How many positive integral solutions does the equation $x_1+x_2+\cdots+x_N = n$ have?

I.2.62. Let $f(x_1, \ldots, x_N)$ be an analytic function of N variables. How many partial derivatives of order n has this function?

I.2.63. In how many ways can n similar presents be distributed among N children? In how many cases every child gets at least one present?

I.2.64. Combinations of m elements taken n at a time with repetitions are groups of n elements, each being of one of m different kinds (the order of elements in a group is not taken into consideration).

a) Write all combinations of three elements taken two at a time with repetitions.

b) Prove that the number of combinations of m elements taken n at a time with repetitions is equal to

$$f_m^n = \binom{m+n-1}{m-1}.$$

I.2.65. In how many ways can one choose six cakes, similar or different, in a bakery that carries 11 different kinds of cakes?

I.2.66. How many domino tiles can be formed using numbers $0, 1, \ldots, r$?

I.2.67. How many nonnegative integral solutions does the inequality $x_1 + x_2 + \cdots + x_m \leq n$ have?

I.2.68. How many nonnegative integral solutions does the system of inequalities

$$\begin{aligned} x_1 &\leq r+1, \\ x_1 + x_2 &\leq r+2, \\ &\cdots\cdots\cdots \\ x_1 + x_2 + \cdots + x_n &\leq r+n, \end{aligned}$$

have, where n is a nonnegative integer?

I.2.69. Let there be given a set $A = \{a_1, a_2, \ldots, a_n\}$ of n elements, which we shall call a population, and let r elements $a_{i_1}, \ldots, a_{i_r}$ be taken one by one from the population A. The set $\{a_{i_1}, \ldots, a_{i_r}\}$ is called *a sample of size r from the population A*. Consider two kinds of choice: the choice with replacement (at each step, the chosen element is returned to the population A) and the choice without replacement (at each step, the chosen element is not returned to the population A).

Prove that the number of different samples of size r from a population containing n elements equals n^r if the choice with replacement is exercised, and $A_n^r = n(n-1)\cdots(n-r+1)$ for the choice without replacement.

I.2.70. Suppose a set X contains k elements, and a set Y contains n elements. How many different functions are there such that X is their domain of definition and Y is their range of values?

I.2.71. Let $N(A)$ be the number of elements in a set A. Prove that

a) $N(A \cup B) = N(A) + N(B) - N(A \cap B)$;

b) $N(A \cup B \cup C) = N(A) + N(B) + N(C)$
$$- [N(A \cap B) + N(A \cap C) + N(B \cap C)] - N(A \cap B \cap C);$$

c) $N(A_1 \cup \cdots \cup A_n)$
$$\begin{aligned}&= N(A_1) + \cdots + N(A_n)\\ &- [N(A_1 \cap A_2) + N(A_1 \cap A_3) + \cdots + N(A_{n-1} \cap A_n)]\\ &+ [N(A_1 \cap A_2 \cap A_3) + N(A_1 \cap A_2 \cap A_4) + \cdots + N(A_{n-2} \cap A_{n-1} \cap A_n)]\\ &- \cdots + (-1)^{n-1} N(A_1 \cap \cdots \cap A_n).\end{aligned}$$

The right-hand side of this equality is the sum of n terms, the kth term being of the form
$$(-1)^{k-1} S_k(A_1, \dots, A_n),$$
where $S_k(A_1, \dots, A_n)$ is the sum of numbers $N(A_{i_1} \cap \cdots \cap A_{i_k})$ over all possible intersections of k different sets from the sets $A_1, \dots, A_n$.

I.2.72. Of 35 students in a class, 20 joined a mathematics club, 11 joined a physics club, and 10 students did not join any of the clubs. How many students joined both the mathematics and physics clubs? How many students joined only the mathematics club?

I.2.73. Of 100 students, 28 know English, 30 know German, 42 know French, 8 students know English and German, 10 know English and French, and 3 students know all the three languages. How many students do not know any of the three languages?

I.2.74. (Lewis Carroll.) In a furious fight, at least 70% of warriors lost one eye, at least 75% lost one ear, at least 80% lost one hand, and at least 85% lost one leg. What is the minimum number of warriors who lost simultaneously an eye, an ear, a hand, and a leg?

I.2.75. All permutations of n numbers $1, 2, \dots, n$ are considered. Find the number of those permutations in which at least one number stands in its place.

I.2.76. Let $a_1, a_2, \dots, a_n$ be relatively prime natural numbers, and N a natural number. How many positive natural numbers do not exceed N and are not divisible by any of the numbers $a_1, a_2, \dots, a_n$.

I.2.77. Let a natural number n factorize as follows:
$$n = p_1^{\alpha_1} p_2^{\alpha_2} \cdots p_k^{\alpha_k},$$

where $p_1, p_2, \dots, p_k$ are prime numbers. Let $\varphi(n)$ be the number of positive integers that do not exceed n and are relatively prime with n. Prove that

$$\varphi(n) = n\left(1 - \frac{1}{p_1}\right)\left(1 - \frac{1}{p_2}\right)\cdots\left(1 - \frac{1}{p_k}\right)$$

(the function $\varphi(n)$ is called *the Euler function*).

I.2.78. How many terms in the expansion of an nth order determinant contain at least one diagonal element?

I.2.79. Let $N_{[m]}(A_1 \cup \dots \cup A_n)$ be the number of elements that belong to exactly m sets from $A_1, A_2, \dots, A_n$. Prove that

$$\begin{aligned} N_{[m]}(A_1 \cup \dots \cup A_n) = \binom{m}{m} & \sum_{1 \le i_1 < i_2 < \dots < i_m \le n} N(A_{i_1} \cap \dots \cap A_{i_m}) \\ & - \binom{m+1}{m} \sum_{1 \le i_1 < i_2 < \dots < i_{m+1} \le n} N(A_{i_1} \cap \dots \cap A_{i_{m+1}}) \\ & + \dots + (-1)^{n-m}\binom{n}{m} \sum_{1 \le i_1 < i_2 < \dots < i_n \le n} N(A_{i_1} \cap \dots \cap A_{i_n}). \end{aligned}$$

I.2.80. Find the number of all permutations of $1, 2, \dots, n$ in which exactly m numbers stand in their places.

I.2.81. Sperner's theorem. Let A be a set of n elements, and $A_1, A_2, \dots, A_k$ a family of subsets in A none of which is a subset of another subset. Prove that $k \le \binom{n}{l}$, where l is the integer part of $n/2$.

I.2.82. Let A be a set of n elements, $A_1, \dots, A_k$ subsets of A none of which is a subset of another subset, and $i_1, \dots, i_k$ the number of elements in the subsets $A_1, \dots, A_k$, respectively. Prove that

$$\sum_{r=1}^{k} \binom{n}{i_r}^{-1} \le 1.$$

I.2.83. Let $x_1, \dots, x_n$ be real numbers such that $|x_i| \ge 1$, $1 \le i \le n$. Then in any interval of length 2 there are at most $\binom{n}{l}$ sums of the form $\sum_{k=1}^{n} \varepsilon_k x_k$, where $\varepsilon_k = \pm 1$ and l is the integer part of $n/2$. Prove this assertion.

I.2.84. Let $A_1, \dots, A_s$ be subsets of a set of n elements such that there are no subsets A_i and A_j such that $A_i \supset A_j$ and $A_i \setminus A_j$ contains more than $r - 1$ elements. Then s does not exceed the sum of r maximal binomial coefficients among $\binom{n}{k}$, $0 \le k \le n$. Prove this assertion.

I.2.85. Let r be an arbitrary positive integer, and let $x_1, \dots, x_n$ be real numbers such that $|x_i| \ge 1$, $1 \le i \le n$. Prove that for any interval of length $2r$ the number of sums of the form $\sum_{k=1}^{n} \varepsilon_k x_k$, where $\varepsilon_k = \pm 1$, lying in this interval does not exceed the sum of r maximal binomial coefficients among $\binom{n}{k}$, $0 \le k \le n$.

TABLE 1

Set theory	Probability theory
Set Ω	Ω is the space of elementary events
Set Ω	Ω is a certain event that occurs in every run of the experiment
$\varnothing$ (the empty set)	$\varnothing$ is an impossible event, i.e., an event that does not occur in any run of the experiment
$A \subset B$	The event A implies the event B
$A \cup B$ (the sum (union) of sets)	$A \cup B$ is the event consisting in that at least one of the events A or B occurs
$A \cap B$ (the intersection of sets)	$A \cap B$ is the event consisting in that both A and B occur
$\overline{A} = \Omega \setminus A$ (the complement of a set)	$\overline{A}$ is the negation of the event A consisting in that A does not occur
$A \cap B = \varnothing$ (A and B are disjoint sets)	$A \cap B = \varnothing$: A and B are disjoint events
$A \setminus B$ (the difference of sets)	$A \setminus B$ is the difference of events A and B: A occurs, but B does not

§I.3. A stochastic experiment. The space of elementary events

Probability theory deals with stochastic experiments, which can be repeated any number of times, but the results of which cannot be predicted with certainty. With every experiment one associates the space of elementary events Ω, i.e., the set of all possible results of the experiment. Random events associated with a given stochastic experiment are subsets of Ω.

The set $\mathfrak{A}$ of all subsets that are interpreted as events associated with a given experiment is assumed to be a σ-algebra.

Basic notions of probability theory can be described in terms of set theory as shown in Table 1.

A space of elementary events Ω is called *discrete* if it is finite or countable.

Probabilities of elementary events in discrete spaces. Let the space of elementary events $\Omega = \{\omega_1, \dots, \omega_n, \dots\}$ be discrete. Assume that with every elementary event ω_k a positive number p_k (the probability of ω_k) is associated, so that $\sum_{k=1}^{\infty} p_k = 1$. If A is a random event, $A \subset \Omega$, then

$$\mathsf{P}(A) = \sum_{\omega_k \in A} p_k$$

is called *the probability of the random event A*.

The following relations are valid: a) $\mathsf{P}(A) \geq 0$, b) $\mathsf{P}(A \cup B) = \mathsf{P}(A) + \mathsf{P}(B)$ if random events A and B are disjoint, and c) $\mathsf{P}(\Omega) = 1$.

Problems

I.3.1. A coin is tossed twice. Describe the space of elementary events. Describe the events A: a head will appear at least once, and B: a head will appear at the second toss.

I.3.2. A die is rolled once. Describe the space of elementary events. Describe the event A: the die will show a number divisible by 3.

I.3.3. A die is rolled twice. Describe the space of elementary events. Describe the events A: the sum of the points shown equals 8, and B: a 6 will appear at least once.

I.3.4. A coin is flipped, and then a die is rolled. Describe the space of elementary events.

I.3.5. A coin is flipped until a head occurs. Describe the space of elementary events.

I.3.6. Construct the space of elementary events in the following experiment: a coin is flipped, until a head occurs twice.

I.3.7. An experiment consists in measuring two random events that take values in the interval $[0, 1]$. Describe the space of elementary events.

I.3.8. Construct the space of elementary events for an experiment in which n random variables $\xi_1, \ldots, \xi_n$ are measured, each taking arbitrary real values.

I.3.9. Find the probabilities of the events A and B in Problem I.3.1, under the condition that all elementary events are equiprobable.

I.3.10. Find the probabilities of the events A and B in Problem I.3.3, under the condition that all elementary events are equiprobable.

I.3.11. a) Find the probability of the event A in Problem I.3.2, under the condition that all elementary events are equiprobable.

b) A loaded die that shows a face with the probability proportional to its number is rolled. Describe the space of elementary events and find the probability of every elementary event. Calculate the probability of the event A in Problem I.3.2.

I.3.12. From a batch of N items, among which m items are defective, n items are taken. Describe the space of elementary events. Describe the event A: among the n items sold, l items are defective, $n < N$, $l \leq m$. Calculate the probability of A, under the condition that all elementary events are equally probable.

I.3.13. Construct the set of elementary events in the experiment that consists in choosing k balls from an urn containing m white and n black balls ($k < n + m$). What is the number of elementary events?

Solve this problem if balls are drawn consecutively, one after another.

I.3.14. A batch of ready-made items is checked for size and weight. The items whose size and weight are less than the standard ones are considered defective; the items whose size and weight are greater than the standard ones are returned for remaking; and the other items are supplied to a customer. Describe the set of elementary events in the case where a batch of n items is checked. What is the total number of elementary events?

I.3.15. From the numbers 1, 2, 3, 4, and 5 one number is chosen; then from the remaining four numbers a second number is chosen. Describe the set of elementary events. Assuming that all elementary events are equally probable, calculate the probabilities of the following events:

a) the first number is even;
b) the second number is odd;
c) both the first and the second numbers are odd.

I.3.16. A coin is tossed until it shows the same side in succession. Describe the space of elementary events. Every elementary event that needs n tosses is assumed to occur with probability 2^{-n}. Calculate the probabilities of the following events:

a) the experiment will be over before the sixth toss;
b) the coin will be tossed an even number of times;
c) the experiment will last infinitely long.

I.3.17. Three players a, b, and c hold a chess tournament as follows: the first game is played by a and b, while the player c is free; the second game is played by the winner of the first round and the player c, while the player who lost the first game is free, and so on. The tournament continues until one of the players wins two games in succession (he/she is considered to be the winner of the tournament). No game can end in a draw. Describe the space of elementary events. Every elementary event consisting of k games is assumed to occur with probability 2^{-k}. Calculate the probabilities of the following events:

a) a wins;
b) b wins;
c) c wins;
d) the winner is not known till the nth game;
e) the tournament will never end.

I.3.18. Tom and Jerry agreed to meet in the interval between 9 and 10 a.m. Each of them arrives to a meeting place during this interval. If x is the moment of Tom's arrival and y of Jerry's, then the space of elementary events is a square of points (x, y) such that $9 \leq x \leq 10$ and $9 \leq y \leq 10$. Let the probability of the event that a point (x, y) belongs to a certain domain in this square be equal to the area of this domain.

a) Assume that the person who comes first waits for the second person during time τ and then goes away. What is the probability that Tom and Jerry will meet?
b) What is the probability of the event that Tom will come to the meeting place before Jerry?
c) What is the probability that Tom will come q hours ($q < 1$) earlier than Jerry?

I.3.19. An experiment consists in measuring three quantities that take values between 0 and 1. The space of elementary events is a cube with side 1, and an elementary event is a point of this cube with coordinates (ξ, η, ζ). The probability of the event that a point with coordinates (ξ, η, ζ) belongs to a certain part K_1 of the cube is equal to the volume of K_1.

a) What is the probability of the existence of a triangle with sides ξ, η, ζ?
b) What is the probability of the event that consists in the fulfillment of the inequality $\xi + \eta + \zeta < \frac{3}{2}$?
c) Find the probability of the event $\max(\xi, \eta, \zeta) < t,\ 0 < t < 1$.
d) Find the probability of the event that the distance of a point (ξ, η, ζ) from the origin is at least t, $0 < t < 1$.
e) What is the probability of the event $\xi\eta\zeta < t$, $0 < t < 1$?

I.3.20. Two different balls are put into two empty urns. Describe the space of elementary events. Assuming all elementary events to be equally probable, calculate the probability that there will be an empty urn.

I.3.21. Two identical balls are put into two empty urns. Describe the space of elementary events. Assuming all elementary events equally probable, calculate the probability that there will be an empty urn.

I.3.22. A target consists of 10 disks bounded by concentric circles of radii r_k, $k = 1, 2, \ldots, 10$, where $r_1 < r_2 < \cdots < r_{10}$. The event A_k consists in that the bullet hits the disk of radius r_k. What is the meaning of the events

$$B = \bigcup_{k=1}^{5} A_k \quad \text{and} \quad C = \bigcap_{k=1}^{10} A_k\,?$$

I.3.23. Describe the negations of the following events
a) A: the occurrence of a head in two tosses of a coin;
b) B: three hits in three shots;
c) C: at least one hit in three shots.

I.3.24. Three shots are fired at a target. Let A_i be the event consisting in that there is a hit at the ith $(i = 1, 2, 3)$ shot. Express the following events in terms of the events A_i:
a) there are three hits;
b) there are no hits;
c) there is only one hit;
d) there are at least two hits.

I.3.25. Two dice are rolled. Let A be the event consisting in that the sum of the numbers shown is odd, and let B be the event that at least one 1 is obtained. Describe the events $A \cap B$, $A \cup B$, and $\overline{A} \cap B$. Find their probabilities.

I.3.26. A space of elementary events consists of all permutations of the digits 1, 2, 3, and 4. All elementary events are equally probable. Denote by A_i the event consisting in that the number i $(i = 1, 2, 3, 4)$ stands in the ith place. Prove that
a) $A_1 \cap A_2 \cap A_3 \subset A_4$;
b) $A_1 \cap A_2 \cap \overline{A}_3 \subset \overline{A}_4$.
Compute the probabilities of the events $A_1 \cap A_2$, $A_1 \cap A_2 \cap A_3$, $A_1 \cup A_2 \cup A_3 \cup A_4$.

I.3.27. Let A, B, and C be three random events. Describe the events consisting in that out of A, B, and C
a) only the event A has occurred;
b) the events A and B have occurred but the event C has not;
c) all three events have occurred;
d) at least one event has occurred;

e) one and only one event has occurred;
f) at most two events have occurred;
g) none of the events has occurred.

I.3.28. Prove the equalities $\overline{A \cup B} = \overline{A} \cap \overline{B}$ and $\overline{A \cap B} = \overline{A} \cup \overline{B}$.

I.3.29. Prove that if $A \supset B$, then a) $\overline{B} \supset \overline{A}$; b) $A \cap B = B$; c) $A \cup B = A$.

I.3.30. Give simpler expressions for the following events:
a) $(A \cup B) \cup (A \cup \overline{B})$;
b) $(A \cup B) \cap (\overline{A} \cup B) \cap (A \cup \overline{B})$;
c) $(A \cup B) \cap (B \cup C)$.

I.3.31. Prove the equalities:
a) $(A \cup B) \setminus B = A \setminus (A \cap B) = A \cap \overline{B}$;
b) $A \cap A = A \cup A = A$;
c) $A \cup B = (A \setminus A \cap B) \cup B$;
d) $(A \cup B) \setminus (A \cap B) = (A \cap \overline{B}) \cup (\overline{A} \cap B)$;
e) $(A \cup B) \cap C = (A \cap C) \cup (B \cap C)$.

I.3.32. Prove the following equalities:
a) $(A \cap B) \cup (C \cap D) = (\overline{A} \cup \overline{B}) \cap (\overline{C} \cup \overline{D})$;
b) $(A \cup B) \cap (A \cup \overline{B}) \cup (\overline{A} \cup B) \cap (\overline{A} \cup \overline{B}) = \Omega$;
c) $(A \cup B) \cap (A \cup \overline{B}) \cap (\overline{A} \cup B) \cap (\overline{A} \cup \overline{B}) = \varnothing$;
d) $(A \cup B) \cap (A \cup C) \cap (B \cup C) = (A \cap B) \cup (A \cap C) \cup (B \cap C)$;
e) $A \setminus (B \cap C) = (A \setminus B) \cup (A \setminus C)$;
f) $A \setminus (B \cup C) = (A \setminus B) \setminus C$.

I.3.33. The sum of two events $A \cup B$ can be expressed as the sum of two disjoint events, namely, $A \cup B = (A \setminus A \cap B) \cup B$. Represent the sum of three events in a similar fashion.

I.3.34. Let $\{A_n, n \geq 1\}$ be a sequence of random events, and let B_m be the event that of the events $A_1, A_2, \dots$ the event A_m will occur first. Express the event B_m in terms of the events $A_1, \dots, A_m$. Prove that the events B_n, $n \geq 1$, are disjoint. Express the event $\bigcup_{n=1}^{\infty} B_n$ in terms of the events $A_1, A_2, \dots$.

I.3.35. Let Ω be a discrete space of elementary events, $\Omega = \{\omega_1, \dots, \omega_n, \dots\}$. Assume that to every elementary event ω there corresponds a probability p_k, $p_k \geq 0$, $\sum_{k=1}^{\infty} p_k = 1$, and the probability of an event A is $\mathsf{P}(A) = \sum_{\omega_k \in A} p_k$. Prove that
a) $\mathsf{P}(A \cup B) = \mathsf{P}(A) + \mathsf{P}(B) - \mathsf{P}(A \cap B)$;
b) $\mathsf{P}(A \cup B \cup C) = \mathsf{P}(A) + \mathsf{P}(B) + \mathsf{P}(C) - \mathsf{P}(A \cap B) - \mathsf{P}(A \cap C) - \mathsf{P}(B \cap C) + \mathsf{P}(A \cap B \cap C)$;
c) $\mathsf{P}(A_1 \cup A_2 \cup \dots \cup A_n) = \sum_{k=1}^{n} (-1)^{k-1} S_k^{(n)}$, where

$$S_k^{(n)} = \sum_{1 \leq i_1 < \dots < i_k \leq n} \mathsf{P}(A_{i_1} \cap A_{i_2} \cap \dots \cap A_{i_k}).$$

I.3.36. Let $P_{[m]}$ be the probability that out of random events $A_1, \dots, A_n$ exactly m events will occur. Prove that

$$P_{[m]} = \sum_{k=0}^{n-m} (-1)^k \binom{m+k}{k} S_{m+k}^{(n)}.$$

I.3.37. Let P_m be the probability that at least m events out of given events $A_1, \dots, A_n$ occur. Prove that

$$P_m = \sum_{k=0}^{n-m} (-1)^k \binom{m-1+k}{m-1} S_{m+k}^{(n)}.$$

§I.4. Classical definition of probability

Let a space Ω consist of n elementary equiprobable events, and let a random event A consist of m events from Ω. Then

$$\mathsf{P}(A) = \frac{m}{n}.$$

Problems

I.4.1. Find the probability that among k randomly selected digits
a) there is no digit 0;
b) there is no digit 1;
c) there are no digits 0 and 1;
d) there is no digit 0 or there is no digit 1.

I.4.2. There are r students in a group. What is the probability that the birthdays of at least two of them coincide?

I.4.3. There are r students in a group. What is the probability that at least two of them were born in the same month?

I.4.4. A die is rolled six times. Calculate the probability that all six faces will be shown.

I.4.5. There are seven persons in an elevator. The elevator stops on ten floors. What is the probability that no two persons will exit on the same floor?

I.4.6. De Mere's paradox. Prove that to obtain at least one 1 at a throw of four dice is more probable than to obtain at least once two 1's at 24 throws of two dice. (A court cavalier Chevalier de Mere, a contemporary of Blaise Pascal, considered these probabilities equal.)

I.4.7. Calculate the probability that the birthdays of 12 persons present are in different months.

I.4.8. Given a group of 30 people, calculate the probability that two birthdays fall on each of six months of the year, and three birthdays on each of the remaining six months.

I.4.9. n persons, A and B among them, form a rank in arbitrary order. What is the probability that there are exactly r persons between A and B?

I.4.10. a) From an urn that contains n white and m black balls, k balls are drawn at random. What is the probability that there are r $(r \le n)$ white balls among them?

b) Prove the identity

$$\sum_{r=\max\{0,k-m\}}^{\min\{n,k\}} \binom{n}{r}\binom{m}{k-r} = \binom{n+m}{k}.$$

I.4.11. From N items, M of which are defective, n items are randomly selected. What is the probability that m $(m < M)$ of them are defective? What is the probability that more than m of them are defective?

I.4.12. There are n lottery tickets, among which m tickets are winning. What is the probability for a holder of r tickets to win?

I.4.13. Am exam consists of N questions. A student knows answers to n questions. Professor asks the student k questions. In order to pass the exam, the student must answer at least r, $r < k$, questions. What is the probability that the student will pass the exam?

I.4.14. A participant of the "Sportloto" lottery must choose six sports out of 49 (numbered from 1 to 49). The highest prize is awarded to the participant who guesses correctly all six sports. Prizes are given also to those who guess at least three sports. Find the probability of receiving the highest prize in "Sportloto". Calculate the probability that a participant in "Sportloto" will guess 5, 4, and 3 sports. What is the probability to receive a prize in "Sportloto"?

I.4.15. In order to decrease the total number of games, $2n$ teams are divided into two subgroups of n teams each. What is the probability that two strongest teams will end up a) in different subgroups? b) in the same subgroup?

c) What is the probability that of four strongest teams two will be in one subgroup and two in the other?

I.4.16. k different numbers are chosen from $1, 2, \ldots, N$. What is the probability that

a) each of the numbers chosen is a multiple of a given number q?
b) each of the numbers chosen is a multiple of at least one of two given relatively prime numbers q_1 and q_2?
c) among the numbers chosen there is at least one multiple of q?

I.4.17. From a set $\{1, 2, \ldots, N\}$, k numbers are chosen one after another, with replacement (each of the numbers $1, 2, \ldots, N$ can be chosen $0, 1, \ldots, k$ times). What is the probability that

a) each of the numbers chosen is a multiple of a given number q?
b) each of the numbers chosen is a multiple of at least one of two given relatively prime numbers q_1 and q_2?
c) among the numbers chosen there is at least one multiple of q?

I.4.18. k numbers are randomly chosen one after another, without replacement, from a set $\{1, 2, \ldots, N\}$. What is the probability that

a) a group of consecutive numbers will be chosen?
b) a group of consecutive numbers arranged in ascending order will be chosen?
c) a group of different numbers arranged in ascending order will be chosen?

I.4.19. k numbers are randomly chosen one after another, with replacement, from a set $\{1, 2, \dots, N\}$ (each number can be taken $0, 1, \dots, k$ times). What is the probability that

a) a group of consecutive numbers will be chosen?
b) a group of consecutive numbers arranged in ascending order will be chosen?
c) all chosen numbers will be different?
d) a group of different numbers arranged in ascending order will be chosen?

I.4.20. Twelve dice are rolled. What is the probability to obtain twice each of the numbers $1, 2, \dots, 6$?

I.4.21. n dice are rolled. What is the probability to obtain n_1 1's, n_2 2's, ..., and n_6 6's, $n_1 + n_2 + \cdots + n_6 = n$?

I.4.22. Five dice are rolled one after another. What is the probability that

a) the sum of points shown by the first two dice will be greater than that shown by the last three dice?
b) there are two dice such that the sum of points shown by these dice is greater than that shown by the other three dice?

I.4.23. In the office, n letters are distributed at random among n envelopes with correct addresses. What is the probability of

a) at least one correct mailing?
b) m correct mailings?

I.4.24. Numbers from 1 to n are arranged at random. What is the probability that at least one number will be on its place? What is the limit of this probability as $n \to \infty$?

I.4.25. A term of a determinant of the nth order is randomly selected. What is the probability that it does not contain elements of the main diagonal?

I.4.26. The Maxwell–Bolzmann statistic. Each of n distinguishable particles is thrown at random into one of N cells.

a) What is the probability to find $n_1, n_2, \dots, n_N$ particles in the first, second, ..., nth cell, respectively?
b) Find the probability p_k for a given cell to contain k particles. For which value of k the probability p_k is maximal?
c) Prove that

$$p_k \to e^{-\lambda} \frac{\lambda^k}{k!}$$

if n and N infinitely increase so that the average number of particles per one cell, n/N, tends to λ.

d) What is the probability that there is at least one particle in each cell ($n \geq N$)?
e) What is the probability that exactly r cells are not empty?

I.4.27. A flux of n cosmic-ray particles is captured by a system of N counters. Every particle is captured by one of the counters with equal probability. What is the probability that the presence of particles will be registered by exactly r counters?

I.4.28. A train consists of N cars. Each of n passengers chooses a car at random. What is the probability that

a) there will be at least one passenger in each car?

b) exactly r cars will be occupied?

I.4.29. Nine passengers get into three cars. What is the probability that

a) exactly three passengers will get into a given car?

b) exactly three passengers will get into each of the cars?

c) four passengers will get into one of the cars, three into another, and two into the third?

I.4.30. The Bose–Einstein statistic. Each of n indistinguishable particles gets into one of N cells.

a) What is the probability that $n_1, n_2, \ldots, n_N$ particles will get into the first, second, ... , Nth cell, respectively, if all arrangements that differ in the number of particles in the cells are assumed to be equally probable?

b) Prove that the probability q_k that there will be k particles in a given cell is

$$q_k = \binom{N+n-k-2}{n-k} \Big/ \binom{N+n-1}{n}.$$

c) Prove that $q_0 > q_1 > q_2 > \cdots$ for $N > 2$, that is, 0 is the most probable number of particles in a given cell.

d) Prove that

$$q_k \to \frac{\lambda^k}{(1+\lambda)^{k+1}}$$

as n and N increase so that $n/N \to \lambda$.

e) What is the probability that exactly m cells will be empty?

I.4.31. The Dirac–Fermi statistic. Each of n indistinguishable particles gets into one of N cells, $N > n$. What is the probability that $n_1, n_2, \ldots, n_N$ particles will get into the first, second, ... , Nth cell, respectively, if only arrangements that satisfy the Pauli exclusion principle (in every cell there is at most one particle) are allowed and all these arrangements are equally probable?

I.4.32. At the ticket office of a movie theater there are $m+n$ people, n of them with 50 copeck coins and the other m ($m \leq n$) people with only one rouble bill each (a rouble is equal to 100 copecks). The price of a ticket is 50 copecks. What is the probability that none of the people in the line will have to wait for a change if at the beginning

a) the cashier had no money at all?

b) the cashier had p 50 copeck coins?

I.4.33. Bertrand's vote problem. In elections, a candidate A received a votes, and a candidate B received b votes, $a > b$. Voters voted one after another. What is the probability that the candidate A was always in the lead during the elections?

I.4.34. If during elections a candidate A receives a votes and a candidate B receives b votes, where $a \geq \mu b$ and μ is a nonnegative integer, then the probability that during the consecutive counting of votes the number of votes for A will always be at least μ times the number of votes for B equals

$$\frac{a-\mu b}{a+b}.$$

Prove this assertion.

I.4.35. There are a cards labeled by the number 0, and b cards labeled by a number $\mu+1$ in an urn. Cards are drawn from the urn one by one, without replacement. The probability that for all r, $r=1,2,\dots,a+b$, the sum of the first r numbers drawn is less than r equals

$$\frac{a-\mu b}{a+b}.$$

Prove this assertion.

I.4.36. Someone drinks in random order n glasses of wine and n glasses of water (all glasses are of the same volume). Calculate the probability that after each glass the quantity of wine consumed will not exceed the quantity of water consumed. What is the probability that after exactly $2r$ glasses the quantity of wine consumed will not exceed the quantity of water consumed?

I.4.37. There were n guests at a party, each wearing a pair of galoshes over the shoes. Upon leaving, the guests chose galoshes at random. What is the probability that each of the guests took a left and a right galosh?

I.4.38. In a town with population of $n+1$ people, someone learns a news. He communicates it to the first person he comes across, that person communicates it to one more person, and so on. At every step, a person who learns the news can communicate it, with equal probabilities, to any of n other people. What is the probability that after r steps

a) the news will not return to the man who was the first to learn it?

b) the news will not be repeated by anyone?

Solve this problem under the assumption that at every step the news is communicated to a group of N randomly selected people.

I.4.39. A set K consists of $n+1$ people. A person A writes two letters to two addressees randomly chosen from K; these addressees form the "first generation" set K_1. Each person from K_1 does the same to form the "second generation" set, and so on. Find the probability that the person A does not belong to any of the "generations" $K_1, K_2, \dots, K_r$.

§I.5. Geometric probabilities

In the problems formulated below the following scheme is considered: in a certain region Ω of the n-dimensional space a point is chosen at random. By the expression "a point is chosen at random" we mean the following: the probability $\mathsf{P}(A)$ of the event that a point is taken in a region A, $A\subset\Omega$, equals

$$\mathsf{P}(A)=\frac{m(A)}{m(\Omega)},$$

where $m(\cdot)$ is Lebesgue measure on Ω.

Problems

I.5.1. On a segment of length l two points are taken at random. What is the probability that the distance between these points does not exceed kl, where $0 < k < 1$?

I.5.2. Two vessels have to arrive at the same wharf. Arrivals of vessels are independent random events, equiprobable over a twenty-four-hour period. Find the probability that one of the vessels will have to wait until the wharf is vacant, if the berthing time for the first vessel is one hour and for the second two hours.

I.5.3. A coin of diameter d is thrown to a parqueted floor. The parquet has the form of squares with side a, $a > d$. What is the probability that the coin will not intersect any of the sides of parquet squares?

I.5.4. A regular n-gon is inscribed into the circle of radius R. A point is thrown into the circle. What is the probability that the point will fall inside the n-gon?

I.5.5. What is the probability that a triangle can be constructed from three randomly selected segments of length at most a?

I.5.6. Parallel straight lines are drawn on the plane at a distance of $2a$ from one another. A circle of radius r, $r < a$, is thrown onto the plane. What is the probability that the circle does not intersect any of the straight lines?

I.5.7. Parallel straight lines are drawn on the plane, the distances between consecutive lines alternating between 1.5 cm and 8 cm. A circle of radius 2.5 cm is thrown onto the plane. What is the probability that the circle does not intersect any of the straight lines?

I.5.8. A point is chosen at random on the circle of unit radius with center at the origin. What is the probability that

a) the projection of the point on the Ox axis is at a distance at most r, $0 < r < 1$, from the origin?

b) the distance from the point chosen to the point with coordinates $(1, 0)$ does not exceed r?

I.5.9. Two points are taken at random on a circle of radius R. What is the probability that the distance between these points does not exceed r, $r \leq 2R$?

I.5.10. A point is thrown at random into a disk of radius R. What is the probability that the distance from this point to the center of the disk does not exceed r?

I.5.11. Three points A, B, and C are taken at random on a circle. What is the probability that the triangle ABC is acute?

I.5.12. A rod of length l is broken at random into three parts. What is the probability that the parts obtained can form a triangle?

I.5.13. A rod of length l is broken at random into two parts. What is the probability that the length of the shortest part does not exceed $5l/6$?

I.5.14. N points are chosen at random inside a sphere of radius R. What is the probability that the distance from the center to the nearest point chosen is at least r? Find the limit of this probability as

$$R \to \infty \quad \text{and} \quad \frac{N}{R^3} \to \frac{4}{3}\pi\lambda$$

for some λ.

I.5.15. Buffon's problem. Parallel straight lines are drawn on the plane at a distance $2a$ from one another.

a) A needle of length $2l$ ($l < a$) is thrown at random to the plane. What is the probability that the needle will intersect one of these straight lines?
b) A convex contour of diameter less than $2a$ is thrown onto the plane. What is the probability that the contour will intersect one of the straight lines?

I.5.16. How thick must a coin be in order that the probability that it falls on the rim be $\frac{1}{3}$?

§I.6. Axioms of probability theory

Let Ω be a space of elementary events and $\mathfrak{A}$ a σ-algebra of subsets of Ω (a σ-algebra of random events), that is,

A_1) $\Omega \in \mathfrak{A}$;
A_2) if $A \in \mathfrak{A}$, then $\overline{A} \in \mathfrak{A}$;
A_3) if $A_i \in \mathfrak{A}$, $i \geq 1$, then $\bigcup_{i=1}^{\infty} A_i \in \mathfrak{A}$.

Assume that with every random event A a number $\mathsf{P}(A)$ is associated so that the following conditions are satisfied:

P_1) $\mathsf{P}(A) \geq 0$ for every $A \in \mathfrak{A}$;
P_2) $\mathsf{P}(\Omega) = 1$;
P_3) if a sequence $\{A_n, n \geq 1\}$ of random events is such that $A_i \cap A_j = \varnothing$ for $i \neq j$, then

$$\mathsf{P}\left(\bigcup_{k=1}^{\infty} A_k\right) = \sum_{k=1}^{\infty} \mathsf{P}(A_k).$$

The assertions A_1), A_2), A_3), P_1), P_2), P_3) constitute *the system of axioms of probability theory*. The measure space $(\Omega, \mathfrak{A}, \mathsf{P})$, the space Ω with measure P, is called *a probability space*.

Problems

I.6.1. Using the system of axioms of probability theory, prove that

a) $\mathsf{P}(\varnothing) = 0$;
b) $\mathsf{P}\left(\overline{A}\right) = 1 - \mathsf{P}(A)$;
c) if $A \subset B$, then $\mathsf{P}(A) \leq \mathsf{P}(B)$;
d) $\mathsf{P}(A) \leq 1$ for every random event $A \in \mathfrak{A}$.

I.6.2. Let A and B be random events. Prove that

$$\mathsf{P}(A \cup B) = \mathsf{P}(A) + \mathsf{P}(B) - \mathsf{P}(A \cap B).$$

I.6.3. Let A, B, and C be random events. Prove that

$$\mathsf{P}(A \cup B \cup C) = \mathsf{P}(A) + \mathsf{P}(B) + \mathsf{P}(C) - \mathsf{P}(A \cap B) - \mathsf{P}(A \cap C) - \mathsf{P}(B \cap C) + \mathsf{P}(A \cap B \cap C).$$

I.6.4. Prove that

$$\mathsf{P}(A \setminus B) = \mathsf{P}(A) - \mathsf{P}(A \cap B);$$
$$\mathsf{P}(A \triangle B) = \mathsf{P}(A) + \mathsf{P}(B) - 2\mathsf{P}(A \cap B).$$

I.6.5. Prove that for any two random events A and B we have

$$\mathsf{P}(A \cap B) \le \mathsf{P}(A) \le \mathsf{P}(A \cup B) \le \mathsf{P}(A) + \mathsf{P}(B);$$
$$\max[\mathsf{P}(A), \mathsf{P}(B)] \le \mathsf{P}(A \cup B) \le 2\max[\mathsf{P}(A), \mathsf{P}(B)].$$

I.6.6. The probabilities of events A, B, and $A \cap B$ are known. Find the probabilities of the events a) $\overline{A} \cup \overline{B}$; b) $\overline{A} \cap \overline{B}$; c) $\overline{A} \cap B$; d) $\overline{\overline{A} \cap B}$; e) $\overline{\overline{A} \cup \overline{B}}$; f) $\overline{A \cap B}$; g) $A \cup B$; and h) $A \cup (\overline{A} \cap B)$.

I.6.7. Let $\Omega = (-\infty, +\infty)$, and let $\mathfrak{A}$ be the σ-algebra of all subsets of Ω. For every $A \subset \mathfrak{A}$ put

$$\mathsf{P}(A) = \sum_{n \in A} \frac{1}{2^n},$$

where n runs over positive integers in A. Is the triple $(\Omega, \mathfrak{A}, \mathsf{P})$ a probability space?

I.6.8. Let $\Omega = \{\omega_1, \omega_2, \dots\}$ be a countable set, $\{P_n, n \ge 1\}$ a sequence of nonnegative numbers such that $\sum_{n=1}^{\infty} P_n = 1$, and $\mathfrak{A}$ the system of all subsets of Ω. For every $A \in \mathfrak{A}$ put

$$\mathsf{P}(A) = \sum_{\omega_i \in A} P_i.$$

Prove that the triple $(\Omega, \mathfrak{A}, \mathsf{P})$ is a probability space.

I.6.9. Let Ω be a region of finite Lebesgue measure in the n-dimensional Euclidean space $\mathbf{R}^n$, $\mathfrak{M}$ a system of Lebesgue measurable subsets of $\mathbf{R}^n$, and $m(\cdot)$ Lebesgue measure on $\mathbf{R}^n$. Let the relation

$$\mathsf{P}(A) = \frac{m(A)}{m(\Omega)}$$

hold for every set $A \in \mathfrak{M}$. Prove that $(\Omega, \mathfrak{M}, \mathsf{P})$ is a probability space.

I.6.10. Let A and B be random events, p_0 the probability that neither of these events will occur, p_1 the probability that only one event will occur, and p_2 the probability that both events will occur. Express p_0, p_1, p_2 in terms of $\mathsf{P}(A)$, $\mathsf{P}(B)$, and $\mathsf{P}(A \cap B)$.

I.6.11. Let p_n be the probability that n out of the events A, B, and C will occur, $n = 0, 1, 2, 3$. Express the probabilities p_0, p_1, p_2, and p_3 in terms of $\mathsf{P}(A)$, $\mathsf{P}(B)$, $\mathsf{P}(C)$, $\mathsf{P}(A \cap B)$, $\mathsf{P}(A \cap C)$, $\mathsf{P}(B \cap C)$, and $\mathsf{P}(A \cap B \cap C)$.

I.6.12. Let $\mathsf{P}(A) \ge 0.8$ and $\mathsf{P}(B) \ge 0.8$. Prove that $\mathsf{P}(A \cap B) \ge 0.6$.

I.6.13. Let $A_1, \dots, A_n$ be random events. Prove that

a) $$\mathsf{P}\left(\bigcup_{i=1}^{n} A_i\right) = 1 - \mathsf{P}\left(\bigcap_{i=1}^{n} \overline{A}_i\right) \le \sum_{i=1}^{n} \mathsf{P}(A_i);$$

b) $$\mathsf{P}\left(\bigcap_{i=1}^{n} A_i\right) \ge 1 - \sum_{i=1}^{n} \mathsf{P}\left(\overline{A}_i\right).$$

I.6.14. Prove that

$$\mathsf{P}(A_1 \cap \cdots \cap A_n) \geq \mathsf{P}(A_1) + \cdots + \mathsf{P}(A_n) - n + 1$$

for any n random events $A_1, \ldots, A_n$.

I.6.15. Prove that if $A_1 \cap A_2 \cap \cdots \cap A_n \subset A$, then

$$\mathsf{P}(A) \geq \mathsf{P}(A_1) + \cdots + \mathsf{P}(A_n) - n + 1.$$

I.6.16. Prove that for any two random events A and B the following relation holds:

$$|\mathsf{P}(A \cap B) - \mathsf{P}(A)\mathsf{P}(B)| \leq \frac{1}{4}.$$

I.6.17. Prove that for any three random events A, B, and C the following relation holds:

$$|\mathsf{P}(A \cap B) - \mathsf{P}(A \cap C)| \leq \mathsf{P}(B \triangle C).$$

I.6.18. Prove that

$$\mathsf{P}^2(A \cap B) + \mathsf{P}^2(\overline{A} \cap B) + \mathsf{P}^2(A \cap \overline{B}) + \mathsf{P}^2(\overline{A} \cap \overline{B}) \geq \frac{1}{4},$$

with the inequality becoming the equality if and only if

$$\mathsf{P}(A) = \mathsf{P}(B) = \frac{1}{2} \quad \text{and} \quad \mathsf{P}(A \cap B) = \frac{1}{4}.$$

I.6.19. Prove the inequality

$$\mathsf{P}(A \triangle B) \leq \mathsf{P}(A \triangle C) + \mathsf{P}(C \triangle B).$$

REMARK. The meaning of this problem is the following: if the quantity

$$\rho(A, B) = \mathsf{P}(A \triangle B)$$

is regarded as the distance between random events A and B, then the relation to be proved is the triangle inequality

$$\rho(A, B) \leq \rho(A, C) + \rho(C, B).$$

I.6.20. Define the distance between random events A and B as

$$\rho^*(A, B) = \begin{cases} \mathsf{P}(A \triangle B)/\mathsf{P}(A \cup B) & \text{if } \mathsf{P}(A \cup B) > 0, \\ 0 & \text{if } \mathsf{P}(A \cup B) = 0. \end{cases}$$

Prove that $\rho^*(A, B)$ satisfies the triangle inequality.

I.6.21. Let p_1, p_2, and p_{21} be real numbers. Prove that in order that in some probability space there exist random events A and B such that $p_1 = \mathsf{P}(A)$, $p_2 = \mathsf{P}(B)$, and $p_{12} = \mathsf{P}(A \cap B)$, it is necessary and sufficient that the following inequalities be fulfilled:

$$1 - p_1 - p_2 - p_{12} \geq 0,$$
$$p_1 - p_{12} \geq 0, \qquad p_2 - p_{12} \geq 0, \qquad p_{12} \geq 0.$$

I.6.22. Let $A_1, \dots, A_n$ be random events, and let

$$S_k^{(n)} = \sum_{1 \le i_1 < i_2 < \dots < i_k \le n} \mathsf{P}(A_{i_1} \cap A_{i_2} \cap \dots \cap A_{i_k}).$$

Prove by induction that

$$\mathsf{P}\left(\bigcup_{i=1}^{n} A_i\right) = \sum_{k=1}^{n} (-1)^{k+1} S_k^{(n)}.$$

I.6.23. Let A_1 and A_2 be random events and $\mathfrak{M}$ the minimal algebra that contains them. Prove that every random event B belonging to $\mathfrak{M}$ is a sum of several random events of the form

$$A_1 \cap A_2, \quad A_1 \cap \overline{A}_2, \quad \overline{A}_1 \cap A_2, \quad \overline{A}_1 \cap \overline{A}_2.$$

How many different random events does the algebra $\mathfrak{M}$ contain?

I.6.24. Let $A_1, \dots, A_n$ be random events and $\mathfrak{M}$ the minimal algebra containing them. Prove that every random event B belonging to $\mathfrak{M}$ is the sum of at most 2^n "fundamental" random events of the form

$$\gamma = A_{i_1} \cap A_{i_2} \cap \dots \cap A_{i_k} \cap \overline{A}_{j_1} \cap \overline{A}_{j_2} \cap \dots \cap \overline{A}_{j_{n-k}},$$

where $(i_1, \dots, i_k)$ is a subset of the set $\{1, 2, \dots, n\}$ and $(j_1, j_2, \dots, j_{n-k})$ is the complementary set. How many sets does the algebra $\mathfrak{M}$ contain?

I.6.25. Let $A_1, \dots, A_n$ be random events, $\mathfrak{M}$ the minimal algebra containing them, $c_1, \dots, c_m$ real numbers, and $B_1, \dots, B_m$ arbitrary random events from $\mathfrak{M}$. Prove that the inequality $\sum_{k=1}^{m} c_k \mathsf{P}(B_k) \ge 0$ holds for all $A_1, \dots, A_n$ if and only if this inequality holds for any system of random events $A_1, \dots, A_n$ in which $\mathsf{P}(A_k)$, $1 \le k \le n$, is either zero or one.

REMARK. Problem I.6.25 can be used for proving various inequalities and equalities for probabilities of random events (Problems I.6.26–32). For example, in order to prove the relation

$$\sum_{k=1}^{m} c_k \mathsf{P}(B_k) \ge 0$$

and, in particular,

$$\sum_{k=1}^{m} c_k \mathsf{P}(B_k) = 0$$

for arbitrary random events $A_1, \dots, A_m$, it is sufficient to check these relations in the case where each of the events $A_1, \dots, A_m$ has the probability 0 or 1.

I.6.26. Using Problem I.6.25, solve Problem I.6.22.

I.6.27. Let $p_{[r]}$ be the probability that exactly r events out of $A_1, \dots, A_n$ occur. Prove that

$$p_{[r]} = \sum_{k=0}^{n-r} (-1)^k \binom{r+k}{r} S_{r+k}^{(n)}, \qquad r = 0, 1, \dots, n,$$

where $S_0^{(n)} = 1$ and

$$S_k^{(n)} = \sum_{1 \le i_1 < i_2 < \cdots < i_k \le n} \mathsf{P}(A_{i_1} \cap A_{i_2} \cap \cdots \cap A_{i_k}).$$

I.6.28. Let p_r be the probability that r or more events out of $A_1, \ldots, A_n$ occur. Prove that

$$p_r = \sum_{k=0}^{n-r} (-1)^k \binom{r+k-1}{r-1} S_{r+k}^{(n)}.$$

I.6.29. Fréchet's inequality. Prove that

$$\frac{S_{r+1}^{(n)}}{\binom{n}{r+1}} \le \frac{S_r^{(n)}}{\binom{n}{r}}, \qquad r = 0, 1, \ldots, n-1.$$

I.6.30. Prove that

$$\frac{\binom{n}{r+1} - S_{r+1}^{(n)}}{\binom{n-1}{r}} \le \frac{\binom{n}{r} - S_r^{(n)}}{\binom{n-1}{r-1}}, \qquad r = 1, 2, \ldots, n-1.$$

I.6.31. Prove that

$$\mathsf{P}(A_1 \cap A_2 \cap \cdots \cap A_n) = \sum_{k=1}^{n} (-1)^{k-1} R_k^{(n)},$$

where

$$R_k^{(n)} = \sum_{1 \le i_1 < i_2 < \cdots < i_k \le n} \mathsf{P}(A_{i_1} \cup A_{i_2} \cup \cdots \cup A_{i_k}).$$

I.6.32. Prove that

$$S_r^{(n)} - (r+1) S_{r+1}^{(n)} \le p_{[r]} \le S_r^{(n)}.$$

I.6.33. Let $p_1, \ldots, p_r$ be nonnegative numbers such that $\sum_{i=1}^n p_i = 1$. Denote by $S_r(p_1, \ldots, p_r)$ a probability space $(\Omega, \mathfrak{A}, \mathsf{P})$, where Ω is a set of r elements $\omega_1, \ldots, \omega_r$, $\mathfrak{A}$ the family of all subsets of Ω, and $\mathsf{P}(A) = \sum_{\omega_i \in A} p_i$. Let $A_1, \ldots, A_n$ be arbitrary random events, and let $\mathfrak{M}$ be the minimal algebra that contains $A_1, \ldots, A_n$. Let $B_1, \ldots, B_m$ be arbitrary random events in $\mathfrak{M}$. Consider the inequality

$$\text{(1)} \qquad \sum_{j=1}^{m} \sum_{i=1}^{m} c_{ji} \mathsf{P}(B_j) \mathsf{P}(B_i) \ge 0.$$

Prove the following assertion: *if the inequality* (1) *holds*

a) *for any system of events $A_1, \ldots, A_n$, each belonging to $S_1(1)$,*

b) *for any system of events $A_1, \ldots, A_n$, each belonging to $S_2(\frac{1}{2}, \frac{1}{2})$,*

then the inequality (1) *holds identically for any system of events $A_1, \ldots, A_n$.*

REMARK. The assertion of this problem can be used for proving some inequalities and equalities for random events (Problems I.6.34–37).

I.6.34. Prove that for any random events A_1 and A_2 the inequality

$$\mathsf{P}(A_1 \cup A_2) \mathsf{P}(A_1 \cap A_2) \le \mathsf{P}(A_1) \mathsf{P}(A_2)$$

holds.

I.6.35. Prove that for any random events A_1 and A_2 the relation

$$\mathsf{P}^2(A_1 \cup B) + \mathsf{P}^2(A_1 \cap B) = \mathsf{P}^2(A_1) + \mathsf{P}^2(A_2) + 2\mathsf{P}(A_1 \cap \overline{A}_2)\mathsf{P}(\overline{A}_1 \cap A_2)$$

holds.

I.6.36. Prove that for any random events $A_1, \dots, A_n$ the inequality

$$\mathsf{P}\left(\bigcup_{i=1}^{n} A_i\right) \geq \frac{\left[\sum_{i=1}^{n} \mathsf{P}(A_i)\right]^2}{2\sum_{1 \leq i \leq j \leq n} \mathsf{P}(A_i \cap A_j) + \sum_{i=1}^{n} \mathsf{P}(A_i)}$$

holds.

I.6.37. Prove that

$$kS_k^{(n)} \geq S_{k-1}^{(n)} \left(S_1^{(n)} - k + 1\right).$$

I.6.38. Let $A_1, \dots, A_n$ be random events and

$$C_k^{(n)} = \bigcup_{1 \leq i_1 < i_2 < \cdots < i_k \leq n} A_{i_1} \cap A_{i_2} \cap \cdots \cap A_{i_k}.$$

Prove that

a) $\displaystyle\sum_{k=1}^{n} \mathsf{P}\left(C_k^{(n)}\right) = \sum_{k=1}^{n} \mathsf{P}(A_k)$;

b) $\displaystyle\prod_{k=1}^{n} \mathsf{P}\left(C_k^{(n)}\right) \leq \prod_{k=1}^{n} \mathsf{P}(A_k)$.

I.6.39. Let Ω be a countable set $\{\omega_1, \dots, \omega_n, \dots\}$, let $\mathfrak{A}$ be the family of all subsets of Ω, and let a measure $\mathsf{P}(\cdot)$ be given by the relations

$$\mathsf{P}(\{\omega_n\}) = p_n,$$

where $p_n \geq p_{n+1} \geq 0$ and $\sum_{n=1}^{\infty} p_n = 1$. Prove that

a) the set of those x for which there exists an event $A \subset \mathfrak{A}$ such that $\mathsf{P}(A) = x$ is a perfect set;

b) the set $\{\mathsf{P}(A), A \in \mathfrak{A}\}$ coincides with the segment $[0, 1]$ if and only if

$$p_n \leq \sum_{k=n+1}^{\infty} p_k, \qquad n = 1, 2, \dots.$$

I.6.40. A probability space $(\Omega, \mathfrak{A}, \mathsf{P})$ is called *a nonatomic probability space* if for every event $A \in \mathfrak{A}$ there exists an event $B \subseteq A$ such that $0 < \mathsf{P}(B) < \mathsf{P}(A)$. Prove that for a nonatomic probability space $(\Omega, \mathfrak{A}, \mathsf{P})$ the set of values $\mathsf{P}(A)$, $A \in \mathfrak{A}$, coincides with the whole interval $[0, 1]$.

I.6.41. Prove that in the case of an arbitrary probability space $(\Omega, \mathfrak{A}, \mathsf{P})$ the set of values $\mathsf{P}(A)$, $A \in \mathfrak{A}$, is a closed set.

I.6.42. Let $(\Omega, \mathfrak{A}, \mathsf{P})$ be a probability space. A random event $A \in \mathfrak{A}$ is called *an atom* if

a) $\mathsf{P}(A) > 0$;

b) for every random event $B \in \mathfrak{A}$, $B \subset A$, either $\mathsf{P}(B) = 0$ or $\mathsf{P}(A \setminus B) = 0$.

Prove that

a) the set of all atoms of the probability space $(\Omega, \mathfrak{A}, \mathsf{P})$ is at most countable;

b) Ω can be decomposed into at most a countable union of atoms and the "nonatomic" part;

c) for any $\varepsilon > 0$ there exists a finite partition of Ω into sets of $\mathfrak{A}$ such that every set in this partition either has a probability that does not exceed ε or is an atom with probability greater than ε.

§I.7. Conditional probability. Independent random events

Conditional probability. Let $(\Omega, \mathfrak{A}, \mathsf{P})$ be a probability space. *The conditional probability of an event A, $A \in \mathfrak{A}$,* given an event B such that $\mathsf{P}(B) > 0$, is defined by the relation

$$\mathsf{P}(A/B) = \frac{\mathsf{P}(A \cap B)}{\mathsf{P}(B)}.$$

The probability multiplication formula. If $\mathsf{P}(B) > 0$, then $\mathsf{P}(A \cap B) = \mathsf{P}(B)\mathsf{P}(A/B)$.

Independent random events. Random events A and B $(A, B \in \mathfrak{A})$ are called *independent* if

$$\mathsf{P}(A/B) = \mathsf{P}(A)\mathsf{P}(B).$$

Jointly independent random events. Random events $A_1, \ldots, A_n$ $(A_i \in \mathfrak{A}, 1 \leq i \leq n)$ are called *jointly independent* if

$$\mathsf{P}(A_{i_1} \cap \cdots \cap A_{i_k}) = \mathsf{P}(A_{i_1}) \cdots \mathsf{P}(A_{i_k})$$

for any $k = 1, 2, \ldots, n$ and $1 \leq i_1 < i_2 < \cdots < i_k \leq n$.

A complete group of events. Random events $H_1, \ldots, H_n$ $(H_i \in \mathfrak{A}, 1 \leq i \leq n)$ form *a complete group of events* if

1) any two events are disjoint: $H_i \cap H_j = \varnothing$;

2) $\bigcup_{i=1}^{n} H_i = \Omega$.

The law of total probability. Let $H_1, \ldots, H_n$ be a complete group of events and $\mathsf{P}(H_i) > 0$, $1 \leq i \leq n$. Then for any event A, $A \in \mathfrak{A}$, the following relation holds:

$$\mathsf{P}(A) = \sum_{i=1}^{n} \mathsf{P}(H_i)\mathsf{P}\left(A/H_i\right).$$

The total probability formula holds also for a countable number of events: if $\{H_n, n \geq 1\}$ is a sequence of random events such that

1) $H_i \cap H_k = \varnothing$, $i \neq k$, and

2) $\bigcup_{i=1}^{\infty} H_i = \Omega$,

then for any event $A \in \mathfrak{A}$ the following relation holds:

$$\mathsf{P}(A) = \sum_{i=1}^{\infty} \mathsf{P}(H_i)\mathsf{P}\left(A/H_i\right).$$

The Bayes formula. If $H_1, \dots, H_n$ is a complete group of events, $\mathsf{P}(H_i) > 0$, $1 \le i \le n$, and $B \in \mathfrak{A}$ is an arbitrary event such that $\mathsf{P}(B) > 0$, then

$$\mathsf{P}\left(H_i/B\right) = \frac{\mathsf{P}(H_i)\mathsf{P}\left(B/H_i\right)}{\sum_{k=1}^{n} \mathsf{P}(H_k)\mathsf{P}\left(B/H_k\right)}.$$

Problems

I.7.1. A coin is tossed twice.

a) Describe the space of elementary events.

b) Describe the random event A: at the first throw of the coin a head is obtained, and the event B: at the second throw of the coin a head is obtained.

c) Calculate the probabilities $\mathsf{P}(A)$, $\mathsf{P}(B)$, $\mathsf{P}(A \cap B)$, and $\mathsf{P}\left(B/A\right)$.

I.7.2. A coin is tossed three times.

a) Describe the space of elementary events.

b) Describe the event A: a head is obtained twice and the event B: a head is obtained at least once.

c) Calculate $\mathsf{P}(A \cap B)$, $\mathsf{P}(B)$, and $\mathsf{P}\left(A/B\right)$.

I.7.3. From a number of families that have two children one family is selected. All elementary events are assumed to be equally probable. What is the probability that

a) there are two boys in the family if it is known that one of the siblings is a boy?

b) there are two boys in the family if it is known that the elder child is a boy?

I.7.4. Two dice are rolled. What is the probability of getting at least one 6 if it is known that the sum of points shown is 8?

I.7.5. Three dice are rolled. What is the probability of obtaining at least one 6 if it is known that all the three dice showed different faces?

I.7.6. From an urn that contains m white and n black balls two balls are drawn successively. It is known that the first ball is white. What is the probability for the second ball to be also white?

I.7.7. It is known that in a tossing of ten dice at least one 1 was obtained. What is the probability of obtaining two or more 1's?

I.7.8. It is known that 5% of men and 0.25% of women are color-blind. A person chosen at random appeared to be color-blind. What is the probability that the person is a man? (The number of men and women is assumed to be the same.)

I.7.9. Prove that if A and B are disjoint and $\mathsf{P}(A \cup B) \neq 0$, then

$$\mathsf{P}\left(A/A \cup B\right) = \frac{\mathsf{P}(A)}{\mathsf{P}(A) + \mathsf{P}(B)}.$$

I.7.10. Given $\mathsf{P}(A/B) = 0.7$, $\mathsf{P}\left(A/\overline{B}\right) = 0.3$, and $\mathsf{P}\left(B/A\right) = 0.6$, calculate $\mathsf{P}(A)$.

I.7.11. Let $\mathsf{P}(A) = p$ and $\mathsf{P}(B) = 1 - \varepsilon$, where ε is a small number. Find an upper and a lower bound for $\mathsf{P}(A/B)$.

I.7.12. Prove that

$$\mathsf{P}(A_1 \cap A_2 \cap \cdots \cap A_n) = \mathsf{P}(A_1) \prod_{k=1}^{n-1} \mathsf{P}\left(A_{k+1}/A_1 \cap \cdots \cap A_k\right).$$

I.7.13. Prove that the events A and B in Problem I.7.1 are independent.

I.7.14. From a group of families with two children, one family is selected. Describe the space of elementary events. Assuming all elementary events equally probable, describe the random event A: there are a boy and a girl in that family, and the random event B: there is no more than one girl in the family. Calculate $\mathsf{P}(A)$, $\mathsf{P}(B)$, and $\mathsf{P}(A \cap B)$ and prove that the events A and B are dependent.

I.7.15. From a group of families with three children, one family is selected. Describe the space of elementary events and the random events A and B defined in Problem I.7.14. All elementary events are equiprobable. Calculate $\mathsf{P}(A)$, $\mathsf{P}(B)$, and $\mathsf{P}(A \cap B)$ and prove that the events A and B are independent.

I.7.16. Let $\mathsf{P}(B) > 0$, and let the equality $\mathsf{P}(A/B) + \mathsf{P}(\overline{A}) = 1$ be fulfilled. What can be said about the events A and B?

I.7.17. Events A and B are disjoint, and $\mathsf{P}(B) > 0$. Calculate $\mathsf{P}(A/B)$.

I.7.18. Prove that if A and B are disjoint events with positive probabilities, then they are dependent.

I.7.19. If random events A and B are independent, then $\overline{A}$ and B, A and $\overline{B}$, and $\overline{A}$ and $\overline{B}$ are also independent. Prove this assertion.

I.7.20. Let $\mathsf{P}(A) > 0$ and $\mathsf{P}\left(B/\overline{A}\right) = \mathsf{P}(A/B)$. Prove that A and B are independent.

I.7.21. Random events A and B_1 and A and B_2 are independent, the events B_1 and B_2 being disjoint. Prove that the events A and $B_1 \cup B_2$ are independent.

I.7.22. Two dice are rolled. Consider the following random events:
A_1: the number of points shown by the first die is even;
A_2: the number of points shown by the second die is odd;
A_3: the sum of the points shown by all the dice is odd.
Prove that any two of the events A_1, A_2, and A_3 are independent, but the events A_1, A_2, and A_3 are not jointly independent.

I.7.23. A regular tetrahedron, three faces of which are painted red, yellow and blue, respectively, and the fourth with all the three colors, is thrown onto the plane. The events A, B, and C consist in that the tetrahedron lands so that the red, yellow, or blue color, respectively, touches the plane. Prove that any two events among A, B, and C are independent, but A, B, and C are not jointly independent.

I.7.24. Let events A and B be independent, and $A \subset B$. Prove that either $\mathsf{P}(A) = 0$ or $\mathsf{P}(B) = 1$.

I.7.25. If an event does not depend on itself, then either $\mathsf{P}(A) = 0$ or $\mathsf{P}(A) = 1$. Prove this assertion.

I.7.26. Prove that if an event A does not depend on $B \cap C$ and $B \cup C$, B does not depend on $A \cap C$, C does not depend on $A \cap B$, and $\mathsf{P}(A)$, $\mathsf{P}(B)$, and $\mathsf{P}(C)$ are positive, then A, B, and C are jointly independent.

I.7.27. If events A, B, and C are jointly independent, then the events A and $B \cup C$ and also A and $B \setminus C$ are independent. Prove this assertion.

I.7.28. Let A, B and A, C be independent, and let either $B \supset C$ or $B \cap C = \varnothing$. Then either the events A and $B \setminus C$ or the events A and $B \cup C$ are independent. Prove this assertion.

I.7.29. Let $\Omega = \{\omega_1, \dots, \omega_r\}$ be a set of r elements, $\mathfrak{A}$ the collection of all subsets of Ω, and $\mathsf{P}(A) = d(A)/r$, where $d(A)$ is the number of elements in the set A. Prove that $(\Omega, \mathfrak{A}, \mathsf{P})$ is a probability space. Describe all pairs of independent events A and B in the probability space $(\Omega, \mathfrak{A}, \mathsf{P})$ in the case when r is a prime number.

I.7.30. Assume that an event A is independent of every event of a sequence $\{B_n, n \geq 1\}$ of random events such that $B_k \cap B_j = \varnothing$ for $k \neq j$. Prove that the events A and $\bigcup_{k=1}^{\infty} B_k$ are independent.

I.7.31. Events $A_1, \dots, A_n$ are jointly independent and $\mathsf{P}(A_k) = p_k$. What is the probability that

a) none of the events $A_1, \dots, A_n$ occurs?
b) at least one of the events $A_1, \dots, A_n$ occurs?
c) exactly one of the events $A_1, \dots, A_n$ occurs?

I.7.32. In every scanning cycle a radar tracking a space object detects the object with constant probability p. What is the probability of detecting the object in n cycles?

I.7.33. There are m radars, each detecting an object in one scanning cycle, with probability p (independent of other cycles and other radars). Each radar performs n cycles in time T. Find the probabilities of the following events: A: in time T the object will be detected by at least one radar, and B: in time T the object will be detected by all radars.

I.7.34. A radar keeps k objects under surveillance. During the surveillance time the ith object can be lost with probability p_i. Find the probability of the following events: A: none of the objects will be lost; B: that at least one object will be lost; and C: no more than one object will be lost.

I.7.35. Missiles are fired at a target. The probability of each missile hitting the target is p, hits of different missiles being independent events. Each missile that hits the target destroys it with probability p_1. Shots are fired until the target is destroyed or the whole supply of missiles is exhausted. There are n missiles. What is the probability that

a) some missiles will remain unused?
b) after the destruction of the target there will be at least two missiles left?
c) at most two missiles will be needed to destroy the target?

I.7.36. Chevalier de Mere's problem. How many times should two dice be rolled in order that the probability of occurrence of at least one 6 be greater than $\frac{1}{2}$?

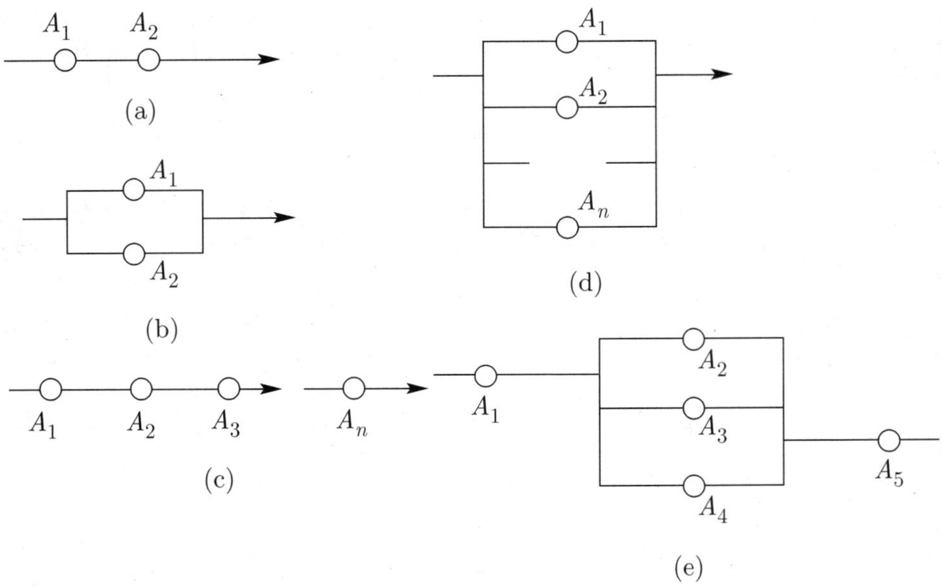

FIGURE 4

I.7.37. Let A_1 and A_2 be independent events, and p_i the probability that exactly i events of the events A_1 and A_2 will occur, $i = 0, 1, 2$. Express the probabilities p_0, p_1, and p_2 in terms of $\mathsf{P}(A_1)$ and $\mathsf{P}(A_2)$.

I.7.38. Let A_1, A_2, and A_3 be independent events, and let p_i be the probability that exactly i, $i = 0, 1, 2, 3$, events of the events A_1, A_2, and A_3 will occur. Express the probabilities p_0, p_1, p_2, and p_3 in terms of $\mathsf{P}(A_1)$, $\mathsf{P}(A_2)$, and $\mathsf{P}(A_3)$.

I.7.39. An electric circuit is set up as shown in Figure 4. Let A_i be the event that over time T the ith element of the circuit will get out of order, and let $\mathsf{P}(A_i) = p_i$. Different elements of the circuit get out of order independently of one another. Let A be the event that the circuit will fail in time T. For each of the schemes (a)–(e), express the event A in terms of the events A_i and calculate the probability of A.

REMARK. In Problems I.7.40–44, the reliability of a device or its separate units is understood as the probability of their trouble-free operation over a certain time interval.

I.7.40. A device consisting of n units fails if at least one of its units fails. The units fail independently of one another. The reliability of each unit is p. Calculate the reliability of the device.

I.7.41. To improve the reliability of a device, it is duplicated by a similar device. The reliability of each device is p. If the first device breaks down, the system is immediately switched to the second. Find the reliability of

a) this system of devices;
b) the system in which the switching operates with reliability p_1.

I.7.42. To improve the reliability of a device, it is duplicated by $n - 1$ similar devices, each having reliability p.

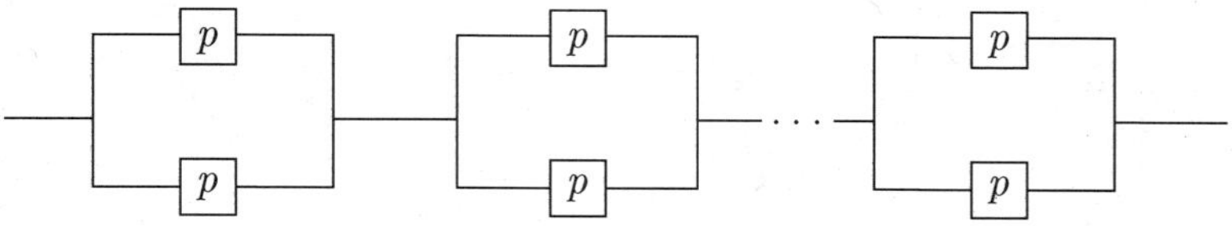

(a)

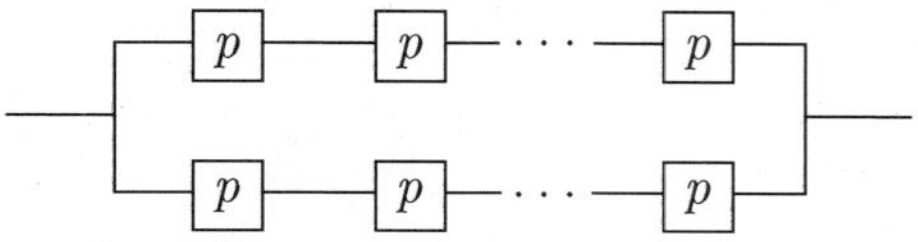

(b)

FIGURE 5

a) Find the reliability of this system. How many devices must be installed in order that the reliability of the system be at least P?
b) Find the reliability of the system of devices if the switch to a duplicate has reliability p_1. How many duplicates must be taken to ensure the reliability of at least P?

I.7.43. A device consists of three units. The first unit contains n_1 elements, the second n_2 elements, and the third n_3 elements. The device fails to operate if the first unit becomes disabled; the second and the third units duplicate each other. The failure of an element results in the failure of the unit that contains it. The reliability of each element is p. The elements fail independently of one another. Find the reliability of the device.

I.7.44. A technological system consists of n units, the reliability of each unit being p. The failure of at least one unit results in the failure of the entire system. To increase the reliability of the system, it is duplicated by n similar units. Which way of duplication provides higher reliability:
a) the duplication of every unit (Figure 5(a))?
b) the duplication of the whole system (Figure 5(b))?

I.7.45. Players A and B play a chess match. The winner of a game gets one point. The probability that A wins a game is α, while for B it equals β, $\alpha > \beta$, $\alpha + \beta = 1$. The player who is 2 points ahead of the opponent wins the match.
a) What is the probability that A will win the match?
b) What is the probability that B will win the match?
c) What is more advantageous for A: to play one game or to play the whole match?
d) What is the probability that the match will never terminate?

I.7.46. Solve Problem I.7.45 under the condition that the winner is the player who wins two consecutive games.

I.7.47. Two contestants, A and B, shoot at a target in turn. The contestant who hits first, wins. The probabilities of hitting the target are p_1 and p_2 for A and B, respectively. The contestant A is the first to shoot. Calculate the probability

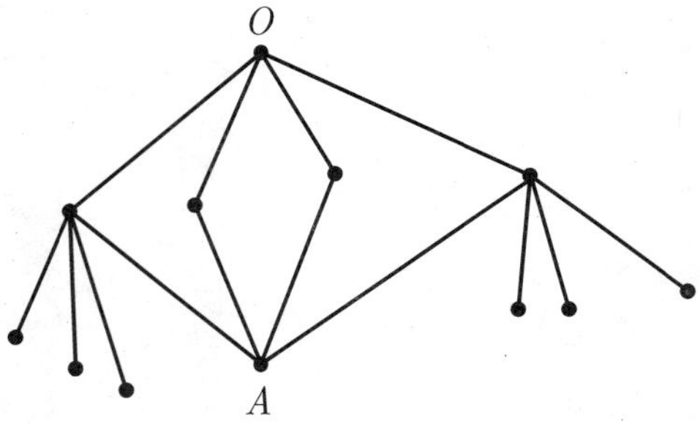

FIGURE 6

of winning for each contestant. What is probability that the contest will continue infinitely long?

I.7.48. Two players flip a coin in turn. The player who obtains a head first, wins. Find the probability of winning for each of the players.

I.7.49. There are n white and m black balls in an urn. Two players draw balls from the urn in turn, returning the balls drawn back to the urn. The player who draws a white ball first, wins. Find the probability of winning for each player.

I.7.50. Among N question cards, n cards are "lucky". Students draw the cards in turn, one after another. Who is more probable to draw a "lucky" card: the student who draws first or the second student?

I.7.51. There are n balls in an urn. All possible assumptions on the number of white balls in the urn are equally probable. A ball is drawn from the urn at random. What is the probability that this ball is white?

I.7.52. In N urns there are $n_1, \dots, n_N$ balls, respectively, of which $m_1, \dots, m_N$ balls are white. From a randomly selected urn a ball is drawn. What is the probability that the ball is white?

I.7.53. A traveler starts from point O (Figure 6) and at every cross-road chooses at random one of all possible roads. What is the probability for the traveler to get to the point A?

I.7.54. Two urns contain n_1 and n_2 balls, respectively, of which m_1 and m_2 balls are white. A ball whose color is unknown is transferred from the first urn into the second. After that a ball is drawn from the second urn. What is the probability that the ball is white?

I.7.55. (Lewis Carroll.) An urn contains a single ball, which is known to be either white or black. A white ball is put into the urn and then, after thorough mixing, one ball is drawn from the urn. The ball turns out to be white. What is the probability that the second ball is also white?

I.7.56. A radar station tracks an object that can jam. If the object does not jam, then the radar detects it in one scanning cycle, with probability p_0; if it does, then the object is detected with probability p_1, $p_1 < p_0$. The probability of jamming during one scanning cycle is p, and it does not depend on jammings in previous cycles. Find the probability of detecting the object at least once in n scanning cycles.

I.7.57. A device consists of n units that can duplicate one another, and it can operate in either favorable or unfavorable mode. In the favorable mode the reliability of operation of each unit is p_1, and in the unfavorable p_2. The probability that the device will operate in the favorable mode is p, and in the unfavorable $1-p$. Calculate the reliability of the device.

I.7.58. Let the probability p that there are n children in a family equal αp^n for $n \geq 1$, and let $p_0 = 1 - \alpha p(1 + p + p^2 + \cdots)$. Assume that the probabilities of the birth of a boy or a girl are equal.

a) Prove that for $k \geq 1$ the probability that there are k boys in a family equals $2\alpha p^k/(2-p)^k$.
b) Let a family be known to have at least one boy. What is the probability that there are two or more boys in the family?

I.7.59. Each of $N+1$ urns contains N balls. The kth urn contains k red and $N-k$ white balls, $k = 0, 1, \ldots, N$. From a randomly selected urn a ball is drawn, with replacement, n times. Calculate

a) the probability that all n balls are red;
b) the conditional probability that the next $(n+1)$th ball is red, provided that all previous balls were red.

I.7.60. A white ball is put into an urn that contains n balls. What is the probability that the ball drawn from the urn is white if all assumptions on the initial composition of the urn are equally probable?

I.7.61. Each of n urns contains m white and k black balls. A ball is randomly transferred from the first into the second urn, then from the second to the third, and so on. Compute the probability to draw a white ball from the last urn.

I.7.62. An urn contains n balls. All assumptions on the number of white balls in the urn are equally probable. A randomly drawn ball appears to be white. Calculate the probabilities of all assumptions on the composition of balls in the urn. Which assumption is most probable?

I.7.63. From an urn that contains n balls of unknown colors, one ball is drawn and it appears to be white. After that, one more ball is drawn. What is the probability that this ball is also white? (All assumptions concerning the initial composition of the urn are equiprobable.)

I.7.64. Each of k_1 urns contains m_1 white and n_1 black balls, and each of k_2 urns contains m_2 white and n_2 black balls. From a randomly selected urn a ball is drawn and it appears to be white. What is the probability that the ball is drawn from the first group of urns?

I.7.65. The first gunman hits a target with probability $p_1 = 0.6$, the second with probability $p_2 = 0.5$, and the third with probability $p_3 = 0.4$. All three gunmen fired at the target. Two hits were registered. What is more probable: the third gunman hit the target or he did not?

I.7.66. The probabilities of random events A, B, and C are p_1, p_2, and p_3, respectively. Given that two events have occurred and one has not, prove that the probability that the event C has occurred is greater than $\frac{1}{2}$ if

$$\frac{1}{p_1} + \frac{1}{p_2} - \frac{1}{p_3} > 1.$$

I.7.67. At the input of a radar receiver there can arrive a noisy signal, with probability p, or only a noise, with probability $1-p$. In the case of a noisy signal the receiver registers the presence of a signal with probability p_1; if there comes only noise, then the receiver registers the presence of a signal with probability p_2. It is known that the receiver has registered a signal. What is the probability that there was a signal at the input of the radar receiver?

I.7.68. Three hunters took three shots, one shot each, at a bear. The bear was killed with one bullet. What is the probability that the bear was killed by the first, the second, or the third hunter if their probabilities of hit are $p_1 = 0.2$, $p_2 = 0.4$, and $p_3 = 0.6$, respectively?

I.7.69. From an urn that contains m ($m > 3$) white and n black balls one ball has been lost. In order to determine the composition of balls in the urn, two balls are drawn and they appear to be white. Calculate the probability that the lost ball was white.

I.7.70. From an urn that contains 3 white and 2 black balls, 2 balls are transferred into another urn that contains 4 white and 4 black balls. What is the probability now to draw a white ball from the second urn?

I.7.71. Parts are made at two plants. The overall production of the second plant is n times larger than that of the first plant. The proportion of defective parts at the first plant is p_1 and at the second p_2. A randomly selected part appears to be detective. What is the probability that it was made at the second plant?

I.7.72. A probabilistic formulation of Fermat's Last Theorem. Two urns contain one and the same number of balls, several white and several black balls each. From these urns n balls are drawn with replacement, $n \geq 3$. Find the number n and composition of each urn if the probability that all balls taken from the first urn are white is equal to the probability that the balls drawn from the second urn are either all white or all black.

I.7.73. Consider a sequence of independent trials, each consisting in a throw of an unbiased die. Find the probability that three consecutive 6's will occur earlier than two consecutive 1's.

I.7.74. Consider a sequence of independent trials with probability of "success" p and that of "failure" $q = 1-p$. Find the probability that a consecutive "successes" will occur earlier than b consecutive "failures".

I.7.75. Consider a quadratic equation $Ax^2 + Bx + C = 0$, whose coefficients A, B, and C are determined as the results of three consecutive throws of a die. Find the probability that

a) the equation has real roots;
b) the equation has rational roots.

I.7.76. A coastal artillery gets sight of an enemy cruiser at a distance of 1 km from the coast and starts shelling it, making one shot every minute. After the first shot the cruiser begins to go away to sea at 60 km/h. Let the probability of hitting the cruiser be inversely proportional to the square of the distance from the coastal gun to the cruiser. Namely: if the cruiser is at a distance x, then the probability of hit is $0.75x^{-2}$. The probability that the cruiser will withstand n hits and will not go down is 4^{-n}. Calculate the probability that the cruiser will manage to go away.

CHAPTER II

Random Variables

§II.1. Discrete random variables

Discrete random variables. Let $(\Omega, \mathfrak{A}, \mathsf{P})$ be a probability space. A function $\xi(\omega)$ on Ω taking a finite or countable number of values $x_1, x_2, \dots, x_n, \dots$ and measurable with respect to the σ-algebra $\mathfrak{A}$ is called *a discrete random variable*. For every x_i,

$$\{\omega: \xi(\omega) = x_i\} \in \mathfrak{A},$$

and therefore the probability $\mathsf{P}(\xi(\omega) = x_i) = p_i$ is meaningful.

The distribution of a discrete random variable. Let $\xi(\omega)$ be a discrete random variable taking values $x_1, \dots, x_i, \dots$. The set of numbers $\mathsf{P}(\xi(\omega) = x_i) = p_i$, $i = 1, 2, \dots$, is called *the distribution of the discrete random variable* ξ. Clearly, $p_i \geq 0$ and $\sum_i p_i = 1$.

Every so often the distribution of a discrete random variable is given in the form of a table in which the values of the random variable and the corresponding probabilities are listed:

ξ	x_1	$\dots$	x_i	$\dots$
P	p_1	$\dots$	p_i	$\dots$

The distribution function of a random variable $\xi(\omega)$ is defined by the equality

$$\mathsf{P}\{\omega: \xi(\omega) < x\} = \sum_{i: x_i < x} p_i.$$

The expectation of a random variable. Let $\xi(\omega)$ be a discrete random variable taking values x_i with probabilities p_i, $i = 1, 2, \dots$. Assume that the series $\sum_i |x_i| p_i$ converges. Then the sum of the series

$$\mathsf{E}\,\xi(\omega) = \sum_i x_i p_i$$

is called *the expectation of the random variable* $\xi(\omega)$.

If $\sum_i |x_i| p_i = +\infty$, then the random variable $\xi(\omega)$ is said *to have no expectation. The expectation of a sum of random variables is equal to the sum of their expectations.*

The variance of the random variable $\xi(\omega)$ is defined by the equality

$$\operatorname{Var} \xi = \mathsf{E}[\xi - \mathsf{E}\,\xi]^2 = \sum_i (x_i - \mathsf{E}\,\xi)^2 p_i.$$

The joint distribution of random variables $\xi(\omega)$ and $\eta(\omega)$. Let $\xi(\omega)$ be a discrete random variable that takes values $x_1, x_2, \dots, x_i, \dots$, and let $\eta(\omega)$ be a discrete random variable that takes values $y_1, y_2, \dots, y_j, \dots$. The set of numbers

$$\mathsf{P}\{\omega: \xi(\omega) = x_i, \eta(\omega) = y_j\} = p_{ij}, \qquad i, j \geq 1,$$

is called *the joint distribution of the random variables* ξ and η (or *the joint distribution of the random vector* (ξ, η)).

The following assertions are valid:

a) $p_{ij} \geq 0$, $\sum_i \sum_j p_{ij} = 1$;

b) $\sum_j p_{ij} = p_i$, $\sum_i p_{ij} = q_j$,

where $\{p_i\}$ is the distribution of $\xi(\omega)$, and $\{q_j\}$ is the distribution of $\eta(\omega)$.

Independent random variables. Random variables ξ and η are called *independent* if for any i and j

$$\mathsf{P}(\xi(\omega) = x_i, \eta(\omega) = y_j) = \mathsf{P}(\xi(\omega) = x_i)\mathsf{P}(\eta(\omega) = y_j).$$

The correlation coefficient. Given random variables ξ and η, the quantity

$$r(\xi, \eta) = \frac{\mathsf{E}(\xi - \mathsf{E}\,\xi)(\eta - \mathsf{E}\,\eta)}{\sqrt{\operatorname{Var}\xi}\sqrt{\operatorname{Var}\eta}}$$

is called *the correlation coefficient of the random variables* ξ and η.

The following assertions are valid:

a) $|r(\xi, \eta)| \leq 1$;

b) if ξ and η are independent, then $r(\xi, \eta) = 0$;

c) if $|r(\xi, \eta)| = 1$, then $\eta = a\xi + b$ with probability 1, where a and b are constants.

The covariance of random variables ξ *and* η is the quantity

$$\operatorname{cov}(\xi, \eta) = \mathsf{E}(\xi - \mathsf{E}\,\xi)(\eta - \mathsf{E}\,\eta).$$

The binomial distribution. Suppose n independent trials are carried out. Each trial can have two outcomes: a "success", with probability p, and a "failure", with probability $q = 1 - p$. Denote by ξ the number of "successes". Then

$$\mathsf{P}(\xi = k) = \binom{n}{k} p^k q^{n-k}, \qquad k = 0, 1, \dots, n.$$

This distribution is called *the binomial distribution*, or *the Bernoulli distribution* (the above scheme of trials is called *the scheme of independent trials*, or *the Bernoulli scheme*).

The geometric distribution. A random variable ξ taking values $0, 1, \dots, k, \dots$ has *the geometric distribution with parameter* p if

$$\mathsf{P}(\xi = k) = (1 - p)^k p.$$

The random variable ξ can be interpreted as the number of trials until the first "success" in the scheme of independent trials, with probability of "success" equal to p.

The Poisson distribution. A random variable ξ taking values $0, 1, \dots, k, \dots$ has *the Poisson distribution with parameter* λ if

$$\mathsf{P}(\xi = k) = e^{-\lambda} \frac{\lambda^k}{k!}.$$

The polynomial distribution. Suppose n independent trials are carried out, each with r possible outcomes: E_1 with probability p_1; E_2 with probability p_2; ... ; E_r with probability p_r. Let ν_i be the number of occurrences of E_i. Then the random vector $(\nu_1, \nu_2, \dots, \nu_r)$ has the so-called *geometric distribution*

$$\mathsf{P}(\nu_1 = m_1, \dots, \nu_r = m_r) = p_n(m_1, \dots, m_r) = \frac{n!}{m_1! \cdots m_r!} p_1^{m_1} \cdots p_r^{m_r},$$

where $m_1 + m_2 + \cdots + m_r = n$.

Problems

II.1.1. A coin is tossed twice. Describe the space of elementary events Ω. Let $\xi(\omega)$ be the number of occurrences of a head. Find the distribution of the random variable ξ, the expectation $\mathsf{E}\,\xi$, and the variance $\operatorname{Var}\xi$.

II.1.2. A die is rolled twice. Describe the space of elementary events Ω. Let $\xi(\omega)$ be the sum of points shown. Find the distribution of the random variable $\xi(\omega)$ and $\mathsf{E}\,\xi$.

II.1.3. A coin is tossed until a head is obtained. Describe the space of elementary events Ω. Let $\xi(\omega)$ be the number of tosses. Calculate
a) the distribution of the random variable ξ;
b) $\mathsf{P}(\xi > 1)$ and $\mathsf{P}(\xi \le n)$.

II.1.4. Shots are fired at a target until the first hit. Hits are independent events, each occurring with probability p. Describe the space of elementary events Ω. Let $\xi(\omega)$ be the number of shots fired. Calculate the distribution of the random variable $\xi(\omega)$.

II.1.5. Indicate which of the sequences given below is the distribution of a random variable:
a) $p^k q^2$, $q = 1 - p$, $0 < p \le 1$, $k = 1, 2, \dots$;
b) $p^{k-n} q$, $q = 1 - p$, $0 \le p \le 1$, $n > 0$, $k = n, n+1, \dots$;
c) $(k(k+1))^{-1}$, $k \ge 1$;
d) $\int_k^{k+1} f(x)\,dx$, $k = 0, 1, \dots$, where $\int_0^\infty f(x)\,dx = 1$;
e) $2^k e^{-2}/k!$, $k \ge 0$.

II.1.6. Let ξ be a random variable taking values $0, \pm 1, \dots, \pm n$ with probabilities $\mathsf{P}(\xi = i) = (2n+1)^{-1}$. Calculate $\mathsf{E}\,\xi$ and $\operatorname{Var}\xi$.

II.1.7. A random variable ξ has the distribution

ξ	1	-1
P	$\frac{1}{2}$	$\frac{1}{2}$

Calculate $\mathsf{E}\,\xi^n$ and $\mathsf{E}\,e^{it\xi}$.

II.1.8. A random variable ξ has the distribution

ξ	-1	0	1
p_i	$\frac{1}{3}$	$\frac{1}{3}$	$\frac{1}{3}$

Find

a) the distribution of the random variable $\eta = |\xi|$;
b) $\mathsf{E}\,\eta$ and $\operatorname{Var}\eta$.

II.1.9. A random variable ξ has the distribution

x_i	-1	0	1	2
p_i	0.2	0.1	0.3	0.4

Find the expectation and variance of the random variable $\eta = 2^\xi$.

II.1.10. Let ξ be a random variable with the distribution

ξ	1	-1
P	$\frac{1}{2}$	$\frac{1}{2}$

Find the distribution of the random variable $\eta = \sin \xi\pi$.

II.1.11. A random variable ξ has the distribution

ξ	-1	-0.5	-0.1	0	0.1	0.2	0.5	1.0	1.5	2.0
P	0.005	0.012	0.074	0.102	0.148	0.231	0.171	0.160	0.081	0.016

Calculate

a) the expectation and variance of ξ;
b) $\mathsf{P}(|\xi| \leq 0.5)$.

II.1.12. Let a random variable ξ take a finite number of nonnegative values $x_1, \dots, x_r$. Prove that

a) $\displaystyle\lim_{n\to\infty} \frac{\mathsf{E}\,\xi^{n+1}}{\mathsf{E}\,\xi^n} = \max_{1\leq i\leq r} x_i,$

b) $\displaystyle\lim_{n\to\infty} \sqrt[n]{\mathsf{E}\,\xi^n} = \max_{1\leq i\leq r} x_i.$

II.1.13. Let ξ be a random variable taking nonnegative integral values, and let $\mathsf{E}\,\xi < +\infty$. Prove that

$$\mathsf{E}\,\xi = \sum_{i=1}^{\infty} \mathsf{P}(\xi \geq i).$$

II.1.14. Two dice are rolled one after another. Describe the space of elementary events. Let $\xi(\omega)$ be the number of 6's shown by the first die, and let $\eta(\omega)$ be the number of 6's shown by the second die. Find the joint distribution of $\xi(\omega)$ and $\eta(\omega)$. Prove that the random variables $\xi(\omega)$ and $\eta(\omega)$ are independent.

II.1.15. Two dice are rolled one after another. Describe the space of elementary events Ω. Let $\xi(\omega)$ be the number shown by the first die, and $\eta(\omega)$ the number shown by the second die.

a) Find the joint distribution of $\xi(\omega)$ and $\eta(\omega)$.

b) Prove that the random variables $\xi(\omega)$ and $\eta(\omega)$ are independent.

II.1.16. Two dice are rolled. Let ξ be the number shown by the first die, and let η be the greater of the two numbers.

a) Find the joint distribution of ξ and η;

b) Calculate $\mathsf{E}\,\xi$, $\mathsf{E}\,\eta$, $\operatorname{Var}\xi$, $\operatorname{Var}\eta$, and the covariance of ξ and η.

II.1.17. Let $\xi_1, \dots, \xi_n$ be random variables that have finite variances, and let $S_n = \xi_1 + \dots + \xi_n$. Prove that

a) $\operatorname{Var} S_n = \sum_{k=1}^n \operatorname{Var}\xi_k + 2\sum_{i<j} \operatorname{cov}(\xi_i, \xi_j)$;

b) if $\xi_1, \dots, \xi_n$ are pairwise independent, then

$$\operatorname{Var} S_n = \sum_{k=1}^n \operatorname{Var}\xi_k.$$

II.1.18. Let γ_1 and γ_2 be independent identically distributed random variables, and let $\xi = \gamma_1 + \gamma_2$ and $\eta = \gamma_1 - \gamma_2$. Prove that $r(\xi, \eta) = 0$.

II.1.19. Let ξ and η be the sum and the difference of the numbers shown by two dice. Prove that $r(\xi, \eta) = 0$. Prove that the random variables ξ and η are independent.

II.1.20. Let ξ take values ± 1 and ± 2, each with probability $\frac{1}{4}$, and let $\eta = \xi^2$.

a) Find the joint distribution of ξ and η.

b) Prove that $r(\xi, \eta) = 0$.

c) Prove that ξ and η are dependent.

II.1.21. If each of random variables ξ and η takes only two values and $r(\xi, \eta) = 0$, then ξ and η are independent. Prove this assertion.

II.1.22. Let ξ and η be discrete random variables, each taking a finite number of values: ξ takes values x_i, $i = 1, \dots, m$, and η takes values y_j, $j = 1, \dots, n$. Assume that the random variables ξ^h and η^k are uncorrelated for $h = 1, \dots, m-1$ and $k = 1, \dots, n-1$, i.e.,

$$\mathsf{E}\,\xi^h \eta^k = \mathsf{E}\,\xi^h\,\mathsf{E}\,\eta^k.$$

Prove that the random variables ξ and η are independent.

II.1.23. In an office, n letters are distributed randomly among n envelopes with correct addresses. Let ξ_n be the number of correct mailings. Calculate $\mathsf{E}\,\xi_n$ and $\operatorname{Var}\xi_n$.

II.1.24. Let ξ and η be discrete independent random variables taking values $x_1, \dots, x_n, \dots$ with probabilities

$$\mathsf{P}(\xi = x_n) = a_n, \qquad \mathsf{P}(\eta = x_n) = b_n.$$

Calculate $\mathsf{P}(\xi = \eta)$.

II.1.25. Let ξ and η be independent nonnegative random variables taking integral values, and let $\mathsf{E}\,\xi < +\infty$. Prove that

$$\mathsf{E}\min(\xi, \eta) = \sum_{i=1}^{\infty} \mathsf{P}(\xi \geq i)\mathsf{P}(\eta \geq i).$$

II.1.26. Let ξ and η be independent random variables taking values $0, 1, \ldots, n$, and let

$$\mathsf{P}(\xi = i) = \mathsf{P}(\eta = i) = \frac{1}{n+1}.$$

Find the distribution of the random variable $\gamma = \xi + \eta$.

II.1.27. The Bernoulli scheme with variable probability of a success. In a sequences of independent trials, the probability of success in the kth trial is p_k and that of failure is $q_k = 1 - p_k$. Let ξ_n be the number of successes in n trials.

a) Calculate $\mathsf{E}\,\xi_n$ and $\operatorname{Var}\xi_n$.

b) Prove that for a given value of

$$a = \frac{1}{n}\sum_{k=1}^{n} p_k, \qquad 0 < a < 1,$$

the maximum of $\operatorname{Var}\xi_n$ is attained for $p_1 = p_2 = \cdots = p_n = a$.

c) Calculate

$$\mathsf{E}\left(\xi_n - \sum_{k=1}^{n} p_k\right)^3 \quad \text{and} \quad \mathsf{E}\left(\xi_n - \sum_{k=1}^{n} p_k\right)^4.$$

II.1.28. Let ξ_n be the number of successes in the Bernoulli scheme. Calculate $\mathsf{E}\,\xi_n$, $\operatorname{Var}\xi_n$, and $\mathsf{E}\,|\xi_n - np|$.

II.1.29. Let ξ be a random variable having the binomial distribution with parameters n and p. Given $\mathsf{E}\,\xi = 12$ and $\operatorname{Var}\xi = 4$, find n and p.

II.1.30. For a sequence of n Bernoulli trials in which the probability of success is p, let m_0 be the most probable number of successes, i.e., the value of m for which the probability is at its maximum. Prove that $np - q \leq m_0 \leq np + p$.

II.1.31. A gun is fired at a target n times. Hits are independent events, each occurring with probability p. Let ξ be the total number of hits. Find

a) the distribution of ξ;

b) $\mathsf{E}\,\xi$ and $\operatorname{Var}\xi$;

c) the most probable number of hits if $p = 0.4$ and $n = 20$.

II.1.32. A die is rolled 5 times. Find the probability that the die shows twice a number that is a multiple of 3.

II.1.33. A gun is fired at a target 20 times, the probability of hit being 0.7 every time. Calculate

a) the probability of at least one hit;

b) the probability that there are no more than two hits;

c) most probable number of hits;

d) the expectation and variance of the number of hits.

II.1.34. A battery fires 14 shots at a target, the probability of hitting at one shot being 0.2. Calculate

a) the most probable number of hits and its probability;

b) the probability of destroying the target if at least 4 hits are needed to achieve that.

II.1.35. Find the probability of getting

a) at least one 6, if six dice are rolled;

b) at least two 6's, if 12 dice are rolled;

c) at least three 6's, if 18 dice are rolled.

II.1.36. A die is rolled n times. Let ξ be the number of occurrences of a 6. Calculate

a) the distribution of ξ;

b) $\mathsf{E}\,\xi$.

II.1.37. a) What is the probability of hitting a target at least twice if the probability of one hit is $1/5$ and the number of independent shots is 10?

b) Find the probability of at least two hits under the condition that the target has been hit once already.

II.1.38. Each of two persons tosses a coin n times. Find the probability that they will get the same number of heads.

II.1.39. Banach's problem. A certain mathematician always carries two match boxes. Every time he needs a match, he selects a box at random. Initially each box contains N matches.

a) Find the probability that the second box contains r $(N \geq r)$ matches.

b) Find the probability that when he finds a box empty for the first time, the other box contains exactly r matches.

c) Find the probability that the box first found empty is different from the one that became empty first.

II.1.40. Playing with a partner of equal skill, what is more probable: to win 4 games out of 8 or 3 games out of 5?

II.1.41. n Bernoulli trials are carried out, each with probability of a "success" equal to p. Find the probability to get the "success" at the kth trial, under the condition that in the first n trials the "success" has occurred once.

II.1.42. Let ξ_n be the number of "successes" in n Bernoulli trials, and let

$$\eta = \begin{cases} 1, & \xi_n \text{ is an even number,} \\ 0, & \xi_n \text{ is an odd number.} \end{cases}$$

Calculate $\mathsf{E}\,\eta$.

II.1.43. Let $\xi_1, \ldots, \xi_n$ be a sequence of independent random variables taking the values 0 or 1, and let

$$\mathsf{P}(\xi_i = 1) = p, \qquad \mathsf{P}(\xi_i = 0) = 1 - p.$$

Put $\eta_i = 0$ if $\xi_i + \xi_{i+1}$ is an even number, $\eta_i = 1$ if $\xi_i + \xi_{i+1} = 1$, and $\gamma = \sum_{i=1}^{n} \eta_i$. Calculate $\mathsf{E}\,\gamma$ and $\operatorname{Var}\gamma$.

II.1.44. Find the probability that the number of "successes" in n Bernoulli trials is divisible by r.

II.1.45. Let ξ be a random variable having the geometric distribution with parameter p. Prove that $\mathsf{E}\,\xi = qp^{-1}$ and $\operatorname{Var}\xi = qp^{-2}$.

II.1.46. A die is rolled till the first occurrence of a 6. Let ξ be the number of rolls.

a) Find the distribution of the random variable ξ.
b) Find the expectation and variance of the random variable ξ.
c) What is the probability that the die is rolled at most three times?

II.1.47. a) **The characteristic property of the geometric distribution.** Prove that the random variable ξ taking values $0, 1, 2, \ldots$ has the geometric distribution if and only if

$$\mathsf{P}\left(\xi = k + r / \xi \geq k\right) = \mathsf{P}(\xi = r), \qquad r \geq 0.$$

b) The duration of a long distance telephone conversation is measured in minutes and it is a random variable with geometric distribution. What is the probability that a conversation will continue another 3 minutes if it already goes on for 10 minutes. (The parameter of the geometric distribution is p.)

II.1.48. Random variables ξ_1 and ξ_2 are independent and have the same geometric distribution $\{q^k p, k \geq 0\}$. Prove that

$$\mathsf{P}\left(\xi_1 = k / \xi_1 + \xi_2 = n\right) = \frac{1}{n+1}, \qquad k = 0, 1, \ldots, n.$$

Is the converse valid?

II.1.49. Random variables ξ_1 and ξ_2 are independent and have the same geometric distribution $\{q^k p, k \geq 0\}$. Let $\eta = \max(\xi_1, \xi_2)$. Find the distribution of the random variable η and the joint distribution of the random variables η and ξ_1.

II.1.50. Let ξ be the number of Bernoulli trials with constant probability of success p needed to achieve the rth success. Prove that

$$\mathsf{P}(\xi = n) = \binom{n}{r-1} p^r q^{n-r+1}, \qquad n \geq r.$$

REMARK. The above distribution is called *the negative binomial distribution with parameters* (r, p); for $r = 1$ we get the geometric distribution.

II.1.51. Let a random variable ξ have the negative binomial distribution with parameters (r, p). Prove that

$$\mathsf{E}\,\xi = r\frac{p}{q} \quad \text{and} \quad \operatorname{Var}\xi = r\frac{q}{p^2}.$$

II.1.52. Shots are fired at a target until the rth hit. The probability of hitting at one shot is equal to p. Let ξ be the number of shots taken. Find the distribution of the random variable ξ. Calculate $\mathsf{E}\,\xi$ and $\operatorname{Var}\xi$.

II.1.53. A die is rolled until the rth appearance of a 6. Find the expectation of the number of rolls.

II.1.54. Tickets in boxes are numbered with digits from 1 to n. In order to win, one needs to take a complete set of tickets, that is, a set that contains all the tickets with numbers 1 through n. Only one ticket is taken from each box. Find the expectation of the number of boxes to be tried in order to take a complete set of tickets.

II.1.55. a) Let ξ be the length of the first series (of successes or failures) in a sequence of Bernoulli trials. Find the distribution of ξ, $\mathsf{E}\,\xi$, and $\operatorname{Var}\xi$.

b) Let η be the length of the second series. Find the distribution of η, $\mathsf{E}\,\eta$, $\operatorname{Var}\eta$, and the joint distribution of ξ and η.

II.1.56. Let ξ be a random variable having the Poisson distribution with parameter λ. Prove that $\mathsf{E}\,\xi = \lambda$ and $\operatorname{Var}\xi = \lambda$.

II.1.57. Let ξ_1 and ξ_2 be independent random variables having the Poisson distributions with parameters λ_1 and λ_2, respectively.

a) Prove that the random variable $\eta = \xi_1 + \xi_2$ has the Poisson distribution with parameter $\lambda_1 + \lambda_2$.

b) Prove that the conditional distribution of the random variable ξ, given $\xi_1 + \xi_2 = n$, is the binomial distribution with parameters n and $p = \lambda_1/(\lambda_1 + \lambda_2)$, i.e.,

$$\mathsf{P}\left(\xi_1 = k/\xi_1 + \xi_2 = n\right) = \binom{n}{k}\left(\frac{\lambda_1}{\lambda_1+\lambda_2}\right)^k \left(1 - \frac{\lambda_1}{\lambda_1+\lambda_2}\right)^{n-k}, \qquad 0 \le k \le n.$$

II.1.58. A random variable ξ has the Poisson distribution with parameter λ. Calculate

$$\mathsf{E}\,\frac{1}{1+\xi}.$$

II.1.59. Let η be a random variable taking nonnegative integral values. Each of η balls is put either in urn A (with probability p) or in urn B (with probability $q = 1 - p$). Let η_A be the number of balls in the urn A, and η_B the number of balls in the urn B.

Prove that the random variables η_A and η_B are independent if and only if η has the Poisson distribution.

II.1.60. Random variables ξ and η are independent and have the Poisson distributions with parameters λ_1 and λ_2, respectively. Prove that

$$\mathsf{P}(\xi - \eta = k) = e^{-\lambda_1 - \lambda_2}\sqrt{\left(\frac{\lambda_1}{\lambda_2}\right)^k}\, I_{|k|}(2\sqrt{\lambda_1\lambda_2}),$$

where

$$I_m(x) = \sum_{k=0}^{\infty} \frac{1}{k!\,\Gamma(k+m+1)}\left(\frac{x}{2}\right)^{2k+m}$$

($I_m(x)$ is the Bessel function of the second kind).

II.1.61. The Pascal distribution. A random variable ξ takes nonnegative integral values with probabilities

$$\mathsf{P}(\xi = n) = \frac{a^n}{(1+a)^{n+1}},$$

where $a > 0$. Calculate the expectation and variance of ξ.

II.1.62. An urn contains N balls labeled by numbers from 1 to N. From the urn, n balls are drawn one after another, each ball being immediately replaced after drawing. Let ξ be the largest digit obtained. Find the distribution of the random variable ξ and its expectation.

II.1.63. A die is rolled n times. Calculate
a) the probability to obtain a 1 n_1 times, a 2 n_2 times, ... , a 6 n_6 times;
b) the probability of the event that a 6 will not appear.

II.1.64. A regular triangle is inscribed into a circle.
a) What is the probability that a point thrown at random into the disk falls inside the triangle?
b) Thirteen points are thrown into the disk. What is the probability that there are three points in each circular segment and four points inside the triangle?

II.1.65. Twelve dice are rolled. What is the probability that each face will be shown twice?

II.1.66. Prove that for the maximal probability $p_n(k_1, \ldots, k_r)$ of the binomial distribution the following inequality holds:

$$np_i - 1 < k_i \leq (n + r - 1)p_i, \qquad i = 1, 2, \ldots, r.$$

II.1.67. A volume V containing N molecules of a gas is divided into n compartments. The probability for a molecule to get into a compartment is the same for all molecules and all compartments and does not depend on the distribution of other molecules among the compartments.
a) What is the probability that there are m_1 molecules in the first compartment, m_2 molecules in the second compartment, ... , and m_n molecules in the nth compartment ($m_1 + m_2 + \cdots + m_n = N$)?
b) What is the probability that there are exactly l molecules in the first compartment?
c) Let $N = kn$, where k is a positive integer. Which distribution of molecules among the compartments is most probable?

II.1.68. Let a vector $(\nu_1, \ldots, \nu_r)$ have the polynomial distribution, i.e.,

$$\mathsf{P}(\nu_1 = m_1, \ldots, \nu_r = m_r) = \frac{n!}{m_1! \cdots m_r!} p_1^{m_1} \cdots p_r^{m_r}.$$

Prove that

a) $\mathsf{E}\,\nu_i = np_i, \quad \operatorname{Var}\nu_i = \sqrt{np_i(1 - p_i)};$

b) $r(\nu_i, \nu_j) = -\sqrt{\dfrac{p_i p_j}{(1 - p_i)(1 - p_j)}}, \qquad i \neq j.$

II.1.69. Find the correlation coefficient between the number of occurrences of a 1 and the number of occurrences of a 6 in n throws of a die.

II.1.70. An urn contains N balls, of which n balls are white. From the urn, m ($m \leq n$) balls are drawn. Describe the space of elementary events. Let there be $\xi(\omega)$ white balls among those drawn. Find
a) the distribution of the random variable ξ (this distribution is called *the hypergeometric distribution*);
b) $\mathsf{E}\,\xi$ and $\operatorname{Var}\xi$.

§II.2. General notion of a random variable

Let $(\Omega, \mathfrak{A}, \mathsf{P})$ be a probability space. A function $\xi(\omega)$ on Ω measurable with respect to the σ-algebra $\mathfrak{A}$, i.e., a function such that for every real x

$$\{\omega \colon \xi(\omega) < x\} \in \mathfrak{A},$$

is called *a random variable.*

The function

$$F(x) = \mathsf{P}\{\omega \colon \xi(\omega) < x\}$$

is called *the distribution function of the random variable* ξ.

Problems

II.2.1. Consider the following stochastic experiment: a point is thrown at random into the interval $[0, T]$. Let ω denote the coordinate of this point, $\Omega = \{\omega \colon 0 \le \omega \le T\}$, and $\mathfrak{A}$ the σ-algebra of Lebesgue measurable subsets of the interval $[0, T]$. For $A \subset \mathfrak{A}$ put

$$\mathsf{P}(A) = \frac{m(A)}{T},$$

where $m(A)$ is the Lebesgue measure of the set A.

a) Prove that $(\Omega, \mathfrak{A}, \mathsf{P})$ is a probability space.

b) Prove that the distance $\xi(\omega)$ from the point ω to the point 0 is a random variable and find its distribution function.

II.2.2. Consider the following stochastic experiment: a point P is thrown at random into the disk of radius R with center at the origin. Let

$$\Omega = \{(u, v) \colon u^2 + v^2 \le R^2\},$$

and let $\mathfrak{A}$ be the σ-algebra of Lebesgue measurable subsets of Ω. For $A \in \mathfrak{A}$ put

$$\mathsf{P}(A) = \frac{m(A)}{m(\Omega)},$$

where $m(\cdot)$ is Lebesgue measure on the plane.

a) Prove that $(\Omega, \mathfrak{A}, \mathsf{P})$ is a probability space.

b) Prove that the distance from the point P to the origin is a random variable.

c) Find the distribution function of this random variable.

II.2.3. Let Ω be the square on the plane XOY with vertices $(0, 0)$, $(0, 1)$, $(1, 0)$, and $(1, 1)$. Denote the coordinates of points of the square by (x, y). As $\mathfrak{A}$ we take the minimal σ-algebra containing all sets formed by the points of polygons lying inside the square. Every set A in $\mathfrak{A}$ is squarable. The probability of a set $A \in \mathfrak{A}$ is equal to the area of A. Prove that the following functions are random variables defined on the probability space $(\Omega, \mathfrak{A}, \mathsf{P})$ and find their distribution functions and distribution densities:

a) $\xi = x + y$; b) $\xi = x - y$; c) $\xi = xy$; d) $\xi = x^2 + y^2$; e) $\xi = x/y$; f) $\xi = \min(x, y)$; g) $\xi = \max(x, y)$; h) $\xi = |x - y|$; i) $\xi =$ the distance of (x, y) to the diagonal that joins the points $(0, 0)$ and $(1, 1)$.

II.2.4. Let $\xi(\omega)$ be a random variable on a probability space $(\Omega, \mathfrak{A}, \mathsf{P})$, and let $F(x)$ be its distribution function.

a) Prove that the sets

$$\{\omega: \xi(\omega) \leq x\}, \qquad \{\omega: \xi(\omega) = x\},$$
$$\{\omega: a \leq \xi(\omega) < b\}, \qquad \{\omega: a < \xi(\omega) < b\}$$

are random events.

b) Prove that

$$\mathsf{P}\{\omega: \xi(\omega) \leq x\} = F(x+0),$$
$$\mathsf{P}\{\omega: \xi(\omega) = x\} = F(x+0) - F(x),$$
$$\mathsf{P}\{\omega: a \leq \xi(\omega) < b\} = F(b) - F(a),$$
$$\mathsf{P}\{\omega: a < \xi(\omega) < b\} = F(b) - F(a+0).$$

c) Let $F(x) = \mathsf{P}\{\omega: \xi(\omega) \leq x\}$. Prove that $\mathsf{P}\{\omega: \xi(\omega) < x\} = F(x-0)$.

II.2.5. Let $\xi(\omega)$ be a random variable on a probability space $(\Omega, \mathfrak{A}, \mathsf{P})$ with the distribution function $F(x)$.

a) Prove that for every c the sets $\{\omega: \xi(\omega) \geq c\}$ and $\{\omega: \xi(\omega) > c\}$ are random events.

b) Express the probabilities of these events in terms of the distribution function.

II.2.6. Let A be a random event. Prove that

$$\chi_A(\omega) = \begin{cases} 1, & \omega \in A, \\ 0, & \omega \notin A, \end{cases}$$

is a random variable.

II.2.7. Consider a probability space $(\Omega, \mathfrak{A}, \mathsf{P})$, where $\Omega = [0,1]$, $\mathfrak{A}$ is the σ-algebra of Lebesgue measurable sets, and $\mathsf{P}(A) = m(A)$ for $A \in \mathfrak{A}$ ($m(\cdot)$ is Lebesgue measure on $[0,1]$).

Let E be a Lebesgue nonmeasurable subset of $[0,1]$, and let

$$\xi(\omega) = \begin{cases} 1, & \omega \in E, \\ -1, & \omega \notin E. \end{cases}$$

Is the function $\xi(\omega)$ a random variable on $(\Omega, \mathfrak{A}, \mathsf{P})$?

II.2.8. Let $\xi(\omega)$ be a random variable. Prove that the functions

a) $\eta_1(\omega) = a\xi(\omega)$;

b) $\eta_2(\omega) = |\xi(\omega)|$;

c) $\eta_3(\omega) = \xi^2(\omega)$

are also random variables.

II.2.9. Let $\eta(\omega) = \xi^2(\omega)$ be a random variable. Is it true that

a) $\xi(\omega)$ is a random variable?

b) $|\xi(\omega)|$ is a random variable?

II.2.10. Let $\Omega = [0,1]$, $\mathfrak{A}$ the σ-algebra of Lebesgue measurable sets, and $\mathsf{P}(\cdot)$ Lebesgue measure on $[0,1]$. Consider the probability space $(\Omega, \mathfrak{A}, \mathsf{P})$. Let E be a Lebesgue nonmeasurable subset of $[0,1]$, and

$$\xi(\omega) = \begin{cases} \omega, & \omega \in E, \\ -\omega, & \omega \notin E. \end{cases}$$

Is $\xi(\omega)$ a random variable?

II.2.11. If $\xi(\omega)$ is a random variable, then $\{\omega\colon \xi(\omega) = c\}$ is a random event for every c. Is the converse true?

II.2.12. Let $\xi_1(\omega), \xi_2(\omega), \ldots, \xi_n(\omega)$ be random variables. Prove that

$$\max_{1 \le i \le n} \xi_i(\omega) \quad \text{and} \quad \min_{1 \le i \le n} \xi_i(\omega)$$

are also random variables.

II.2.13. Let I be a countable set and $\xi_\alpha(\omega)$ a random variable for every $\alpha \in I$. Prove that

$$\sup_{\alpha \in I} \xi_\alpha(\omega) \quad \text{and} \quad \inf_{\alpha \in I} \xi_\alpha(\omega)$$

are random variables. Is this assertion true if the set I is not countable?

II.2.14. Prove that a random variable $\xi(\omega)$ assumes a value x with positive probability if and only if its distribution function has a discontinuity at the point x. Formulate a condition for discreteness of a random variable in terms of its distribution function.

II.2.15. A number x is called *a point of increase of a distribution function* $F(x)$ if for every $\varepsilon > 0$ the inequality

$$F(x+\varepsilon) - F(x-\varepsilon) > 0$$

holds. Construct an example of a discrete distribution for which every real number is a point of increase.

II.2.16. Let $(\Omega, \mathfrak{A}, \mathsf{P})$ be a probability space. A random event $A \in \mathfrak{A}$ such that $\mathsf{P}(A) > 0$ is called *an atom* if for every random event $B \in \mathfrak{A}$ the inclusion $B \subset A$ implies that either $\mathsf{P}(B) = 0$ or $\mathsf{P}(B) = \mathsf{P}(A)$. Prove that a random variable $\xi(\omega)$ is constant with probability 1 on every atom.

II.2.17. Let $\Omega = [0,1]$, $\mathfrak{A}$ be the σ-algebra of Borel subsets of Ω, and $\mathsf{P}(\cdot)$ Lebesgue measure on $[0,1]$. Prove that

a) $(\Omega, \mathfrak{A}, \mathsf{P})$ is a probability space;

b) every Borel function on $[0,1]$ is a random variable on the probability space $(\Omega, \mathfrak{A}, \mathsf{P})$.

II.2.18. Let F_0 be a class of subsets of Ω such that the class of finite sums of disjoint sets in F_0 is an algebra of sets, and let $\mathcal{K}$ be a class of functions of ω that have the following properties:

1) if $A \in F_0$, then

$$\chi_A(\omega) = \begin{cases} 1, & \omega \in A, \\ 0, & \omega \notin A, \end{cases}$$

belongs to $\mathcal{K}$;

2) if $\xi_1(\omega), \dots, \xi_n(\omega)$ are functions in $\mathcal{K}$, then a linear combination of these functions also belongs to $\mathcal{K}$;

3) if $\xi_n(\omega) \in \mathcal{K}$ and $\lim_{n\to\infty} \xi_n(\omega) = \xi(\omega)$ for all ω, then $\xi(\omega) \in \mathcal{K}$.

Also, let $\mathfrak{B}(F_0)$ be the minimal σ-algebra that contains the class F_0. Prove that $\mathcal{K}$ contains all functions measurable with respect to $\mathfrak{B}(F_0)$.

II.2.19. Let $\Omega = \mathbf{R}^n$, and let F_0 be the class of parallelepipeds $\{x: a_i \le x_i < b_i, i = 1, 2, \dots, n\}$. Prove that $\mathcal{K}$ contains all Borel functions.

II.2.20. Let $(\Omega, \mathfrak{A}, \mathsf{P})$ be a probability space, $\xi_1(\omega), \dots, \xi_n(\omega)$ random variables on this space, and $f(x_1, \dots, x_n)$ a Borel function of n variables. Prove that $f(\xi_1(\omega), \dots, \xi_n(\omega))$ is a random variable.

II.2.21. Let A be an n-dimensional Borel set. Prove that

$$\{\omega: (\xi_1(\omega), \dots, \xi_n(\omega)) \in A\} \in \mathfrak{A}.$$

II.2.22. Let $\{\xi_t, t \in T\}$ be a collection of random variables. The minimal σ-algebra containing all sets of the form $\{\omega: \xi_t(\omega) \in A\}$, where $t \in T$ and A is an arbitrary Borel set on the real line, is the minimal σ-algebra with respect to which all functions $\xi_t(\omega)$, $t \in T$, are measurable. This σ-algebra is denoted $\mathfrak{B}(\xi_t(\omega), t \in T)$. Prove that a function $\eta(\omega)$ is measurable with respect to $\mathfrak{B}(\xi_1(\omega), \dots, \xi_n(\omega))$ if and only if

$$\eta(\omega) = f(\xi_1(\omega), \dots, \xi_n(\omega)),$$

where $f(x_1, \dots, x_n)$ is a Borel function.

II.2.23. Let $\{\xi_t(\omega), t \in T\}$ be a set of random variables, $\mathfrak{B}_S = \mathfrak{B}\{\xi_t: t \in S\}$, where $S \subset T$, and T an uncountable set. Prove that

a) if $\Lambda \in \mathfrak{B}_T$, then there exists a countable subset S (depending on Λ) of the set T such that $\Lambda \in \mathfrak{B}_S$;

b) if $\eta(\omega)$ is a random variable measurable with respect to $\mathfrak{B}_T$, then there exists a countable subset S (depending on $\eta(\omega)$) of the set T such that $\eta(\omega)$ is measurable with respect to $\mathfrak{B}_S$.

§II.3. Random variables and random vectors

The distribution function $F(x)$ of a random variable ξ is the probability

$$F(x) = \mathsf{P}(\xi < x).$$

The distribution function $F(x)$ has the properties:

a) $F(x)$ is left continuous;

b) $F(x)$ is nondecreasing on $(-\infty, +\infty)$;

c) $F(-\infty) = 0$ and $F(+\infty) = 1$.

For every function $F(x)$ with these properties it is possible to construct a probability space $(\Omega, \mathfrak{A}, \mathsf{P})$ and a random variable $\xi(\omega)$ on this space that has the distribution function $F(x)$.

If $F(x)$ is the distribution function of a random variable ξ, then

$$\mathsf{P}(a \le \xi < b) = F(b) - F(a), \qquad a < b.$$

The distribution density of a random variable ξ. If the distribution function $F(x)$ of a random variable ξ can be represented in the form

$$F(x) = \int_{-\infty}^{x} p(u)\,du,$$

then the random variable ξ is said to have *the distribution density* $p(x)$. For almost all x the equality $F'(x) = p(x)$ holds. The distribution density $p(x)$ is a nonnegative function, and $\int_{-\infty}^{+\infty} p(u)\,du = 1$.

The following inequality is valid:

$$\mathsf{P}(a \le \xi < b) = \int_a^b p(u)\,du.$$

If $p(x)$ is a continuous function, then

$$\mathsf{P}(x \le \xi < x + \Delta x) = p(x)\Delta x + o(\Delta x).$$

The uniform distribution. A random variable ξ has *the uniform distribution on a segment* $[a, b]$ if the distribution density of ξ is equal to

$$p(x) = \begin{cases} 1/(b-a), & x \in (a,b), \\ 0, & x \notin (a,b). \end{cases}$$

The normal distribution $N(a, \sigma^2)$. A random variable has *the normal distribution $N(a, \sigma^2)$* if its distribution density is equal to

$$p(x) = \frac{1}{\sqrt{2\pi}\sigma} \exp\left\{-\frac{1}{2}\frac{(x-a)^2}{\sigma^2}\right\}.$$

The exponential distribution. A random variable ξ has *the exponential distribution with parameter* λ if its distribution density is equal to

$$p(x) = \begin{cases} \lambda e^{-\lambda x}, & x > 0, \\ 0, & x < 0. \end{cases}$$

The distribution function of a random vector $(\xi_1, \dots, \xi_n)$ is the probability

$$F(x_1, \dots, x_n) = \mathsf{P}(\xi_1 < x_1, \dots, \xi_n < x_n).$$

The distribution density of a random vector. If the distribution function $F(x_1, \dots, x_n)$ of a random vector $(\xi_1, \dots, \xi_n)$ can be represented as

$$F(x_1, \dots, x_n) = \int_{-\infty}^{x_1} \cdots \int_{-\infty}^{x_n} p(u_1, \dots, u_n)\,du_1 \cdots du_n,$$

then the random vector $(\xi_1, \dots, \xi_n)$ is said *to have the distribution density* (or *the probability distribution density*) $p(x_1, \dots, x_n)$.

The distribution density $p(x_1, \dots, x_n)$ of the random vector $(\xi_1, \dots, \xi_n)$ is a nonnegative function and

$$\int_{-\infty}^{\infty} \cdots \int_{-\infty}^{\infty} p(x_1, \dots, x_n)\,dx_1 \cdots dx_n = 1.$$

If A is a Borel set in $\mathbf{R}^n$, and a vector $(\xi_1, \ldots, \xi_n)$ has a distribution density $p(x_1, \ldots, x_n)$, then

$$\mathsf{P}((\xi_1, \ldots, \xi_n) \in A) = \int \overset{n}{\cdots} \int_A p(x_1, \ldots, x_n)\, dx_1 \cdots dx_n.$$

Given the distribution density of a random vector, the distribution density of each of its components can be found. For instance,

$$p_{\xi_1}(x_1) = \int_{-\infty}^{\infty} p_{\xi_1, \xi_2}(x_1, x_2)\, dx_2.$$

If a distribution density $p(x_1, \ldots, x_n)$ is a continuous function, then

$$\begin{aligned} &\mathsf{P}(x_1 < \xi_1 < x_1 + \Delta x_1, \ldots, x_n < \xi_n < x_n + \Delta x_n) \\ &\quad = p(x_1, \ldots, x_n)\Delta x_1 \cdots \Delta x_n + o(\Delta x_1 \cdots \Delta x_n). \end{aligned}$$

Independent random variables. Random variables $\xi_1, \ldots, \xi_n$ are independent if

$$\mathsf{P}(\xi_1 < x_1, \ldots, \xi_n < x_n) = \mathsf{P}(\xi_1 < x_1) \cdots \mathsf{P}(\xi_n < x_n).$$

The expectation of a random variable. Let $\xi(\omega)$ be a random variable on a probability space $(\Omega, \mathfrak{A}, \mathsf{P})$. The random variable $\xi(\omega)$ has *an expectation* if the integral

$$\mathsf{E}\,\xi(\omega) = \int_\Omega \xi(\omega)\, d\mathsf{P} = \mathsf{E}\,\xi$$

exists. If $F(x)$ is the distribution function of ξ, then

$$\mathsf{E}\,\xi = \int_{-\infty}^{\infty} x\, dF(x).$$

If ξ has a distribution density, then

$$\mathsf{E}\,\xi = \int_{-\infty}^{\infty} xp(x)\, dx.$$

If $g(x)$ is a Borel function and

$$\int_{-\infty}^{\infty} |g(x)|\, dF(x) < +\infty,$$

then

$$\mathsf{E}\, g(\xi) = \int_{-\infty}^{\infty} g(x)\, dF(x).$$

The expectation of the sum of random variables is equal to the sum of their expectations.

The expectation of the product of independent random variables is equal to the product of their expectations.

Problems

II.3.1. Which of the functions below are distribution functions:

a) $F(x) = \dfrac{3}{4} + \dfrac{1}{2\pi} \arctan x;$

b) $F(x) = \begin{cases} 0, & x \le 0, \\ [x]/2, & 0 < x \le 2, \\ 1, & x > 2; \end{cases}$

c) $F(x) = \begin{cases} 0, & x \le 0, \\ x/(1+x), & x > 0; \end{cases}$

d) $F(x) = e^{-e^{-x}};$

e) $F(x) = \begin{cases} 0, & x < 0, \\ 1 - (1 - e^{-x})/x, & x > 0? \end{cases}$

II.3.2. A random variable ξ has the distribution density $p(x) = ae^{-\lambda|x|}$, $\lambda > 0$.
a) Compute the coefficient a.
b) Find the distribution function of ξ.
c) Compute $\mathsf{E}\,\xi$ and $\operatorname{Var}\xi$.
d) Construct the graphs of the distribution density and distribution function.

II.3.3. The functions that follow depend on certain parameters. Determine the values of the parameters for which these functions are distribution densities.

a) $f(x) = \begin{cases} c, & x \in [a,b], \\ 0, & x \notin [a,b]; \end{cases}$

b) $f(x) = \begin{cases} k|x-a|, & c \le x \le d, \\ 0, & x \notin [c,d]; \end{cases}$

c) $f(x) = \begin{cases} c/(ax+b), & d \le x < \infty, \\ 0, & x < d; \end{cases}$

d) $f(x) = \begin{cases} ax^2+bx+c, & ax^2+bx+c \ge 0, \\ 0, & ax^2+bx+c < 0; \end{cases}$

e) $f(x) = \begin{cases} cx^\alpha e^{-\lambda x}, & x > 0, \\ 0, & x \le 0; \end{cases}$

f) $f(x) = ce^{\alpha(x-b)^2};$

g) $f(x) = \dfrac{d}{a+bx+cx^2}.$

II.3.4. A random variable ξ has the distribution density $p(x) = \pi^{-1}(1+x^2)^{-1}$ (the Cauchy distribution). Compute the probabilities:
a) $\mathsf{P}(\xi \ge 1)$;
b) $\mathsf{P}(|\xi| \ge 1)$.

II.3.5. A random variable ξ has the exponential distribution with parameter $\lambda = \frac{1}{3}$. Compute the probabilities
a) $\mathsf{P}(\xi > 3)$;
b) $\mathsf{P}(\xi > 6/\xi > 3)$;
c) $\mathsf{P}(\xi > t+3/\xi > t)$.

II.3.6. Let $F(x)$ be the distribution function of a random variable ξ. Find the distribution function and distribution density of the random variable $\eta = e^\xi$.

II.3.7. Let $F(x)$ be the distribution function of a random variable ξ. Find the distribution function of the random variable $\eta = -\xi$.

II.3.8. A random variable ξ is called *a symmetric random variable* if the distribution functions of the random variables ξ and $-\xi$ coincide. Formulate a condition for symmetry of a random variable

a) in terms of its distribution function;
b) in terms of its distribution density.

II.3.9. Let $F(x)$ be the distribution function of a random variable ξ. Find the distribution function of the random variable $\eta = \operatorname{sgn} \xi$ and compute $\mathsf{E}\,\eta$.

II.3.10. A random variable ξ has the distribution function $F(x)$. Find the distribution functions of the following random variables:

a) $a\xi + b$, where a and b are arbitrary real numbers;
b) ξ^2;
c) $g(\xi)$, where $g(x)$ is a monotone function;
d) $|\xi|$;
e) $\sin \xi$;
f) $\tan \xi$.

II.3.11. A random variable ξ has a distribution density $p(x)$. Find the distribution densities of the following random variables:

a) $a\xi + b$, where $a \neq 0$;
b) $|\xi|$;
c) ξ^2;
d) $g(\xi)$, where $g(x)$ is a monotone differentiable function;
e) $\sin \xi$;
f) $g(\xi)$ if $g(\xi)$ is a piecewise monotone differentiable function.

II.3.12. Let ξ be a random variable uniformly distributed on the segment $[-1, 1]$. Find the distribution of the random variable $\eta = |\xi|$.

II.3.13. A random variable ξ is uniformly distributed on the segment $[0, 1]$. Find the distribution densities of the random variables

a) $\eta = \xi^2$;
b) $\eta = 1/\xi$;
c) $\eta = e^\xi$;

and construct their graphs.

II.3.14. Suppose the distribution density of a random variable ξ is $p(x) = \pi^{-1}(1+x^2)^{-1}$. Find the distribution function of the random variable $\eta = \arctan \xi$.

II.3.15. A random variable ξ is uniformly distributed on the segment $[0, 2]$. Find the distribution function of the random variable $\eta = |\xi - 1|$.

II.3.16. A random variable ξ is uniformly distributed on the segment $[-1, 2]$. Find the distribution function of the random variable $\eta = |\xi|$.

II.3.17. A random variable ξ has the exponential distribution with parameter λ. Find the distribution density of the random variable $\eta = \xi^2$.

II.3.18. A random variable ξ has the distribution density

$$p(x) = \frac{1}{\sqrt{2\pi}} \frac{1}{x^2} e^{-x^2/2}, \qquad -\infty < x < +\infty.$$

Find the distribution of the random variable $\eta = 1/\xi$.

II.3.19. A random variable ξ has the distribution density

$$p(x) = \frac{1}{\pi} \frac{1}{1+x^2}.$$

Find the distribution density of the random variable $\eta = \xi^2$.

II.3.20. A random variable ξ is uniformly distributed on the interval $\left[-\frac{\pi}{2}, \frac{\pi}{2}\right]$. Find the distribution density of the random variable $\eta = \sin \xi$.

II.3.21. Let ξ be uniformly distributed on $[0, 4\pi]$. Find the distribution function and distribution density of the random variable $\eta = \sin \xi$.

II.3.22. A random variable ξ is uniformly distributed on the interval $\left[-\frac{\pi}{2}, \frac{\pi}{2}\right]$. Find the distribution density of the random variable $\eta = |\sin \xi|$.

II.3.23. A random variable ξ has the Cauchy distribution with density

$$p(x) = \frac{1}{\pi} \frac{1}{1+x^2}, \qquad -\infty < x < +\infty.$$

Find the distribution density of the random variable $\eta = 1/\xi$.

II.3.24. The diameter of a disk is a random variable ξ uniformly distributed on the interval $[a, b]$. Find the distribution function of the area of the disk.

II.3.25. Let ξ be a random variable with increasing continuous distribution function $F(x)$, and let $\eta = F(\xi)$. Find the distribution function of the random variable η.

II.3.26. Let $F(x)$ be a distribution function such that $F(0) = 0$. Prove that the function

$$G(x) = \begin{cases} F(x) - F(1/x), & x \geq 1, \\ 0, & x < 1, \end{cases}$$

is a distribution function.

II.3.27. Let $F(x)$ be a distribution function. Prove that the functions

$$G_1(x) = \frac{1}{h} \int_x^{x+h} F(u)\, du,$$

$$G_2(x) = \frac{1}{2h} \int_{x-h}^{x+h} F(u)\, du$$

are distribution functions for every $h > 0$.

II.3.28. A point P is uniformly distributed in a disk of radius R. Let η be the distance from P to the center of the disk. Find the distribution function $F_\eta(x)$ and distribution density $p_\eta(x)$ of the random variable η. Construct the graphs of the functions $F_\eta(x)$ and $p_\eta(x)$. Compute $\mathsf{E}\,\eta$ and $\operatorname{Var} \eta$.

II.3.29. Let O be the origin on the plane, P a random point on the Ox axis, and Q the point with coordinates $(0,1)$. The angle OQP is known to be uniformly distributed on the segment $[-\pi/2, \pi/2]$. Find the distribution function and distribution density of the abscissa of the point P.

II.3.30. A point A is uniformly distributed on the unit circle. Let ξ be the x-coordinate of A. Find

a) the distribution function of ξ;

b) the distribution density of ξ.

Compute c) $\mathsf{E}\,\xi$; d) $\mathsf{P}\left(|\xi| \geq \frac{1}{2}\right)$.

II.3.31. A van wheel of unit radius has a crack on its rim. Let ξ be the height of the crack above ground after an accidental stop of the van. Find the distribution function of ξ.

II.3.32. On a circle of radius R two points are taken at random according to a uniform distribution. Find the distribution function of the distance γ between them and compute $\mathsf{E}\,\gamma$.

II.3.33. A point P is uniformly distributed on the circle of radius R with center at the origin. Find the distribution function and distribution density of the length γ of the segment PQ of the tangent at P, where Q is the point of its intersection with the Ox axis. Does $\mathsf{E}\,\gamma$ exist?

II.3.34. A point is thrown at random into the interval of the Oy axis with ends $(0,0)$ and $(0,R)$ (the y-coordinate of the point is uniformly distributed in $(0,R)$). Through this point the chord is drawn across the disk $x^2+y^2=R^2$ perpendicular to the Oy axis. Find the distribution function of the length of this chord.

II.3.35. A rod of length l is broken at random into two parts. Find the distribution function of the smaller part.

II.3.36. Two points are thrown into a segment $[0,T]$. Let γ be the distance between these points. Find the distribution function of γ and compute $\mathsf{E}\,\gamma$, $\operatorname{Var}\gamma$, and $\mathsf{E}\,\gamma^n$, $n>2$.

II.3.37. Two persons agreed to meet over a time interval $[0,T]$. Let γ be the time one of them will have to wait till the moment of their meeting. Find the distribution function of γ and compute $\mathsf{E}\,\gamma$.

II.3.38. On a segment $[0,T]$, two points are taken at random according to the uniform distribution. These points divide the segment $[0,T]$ into three parts. Find the distribution functions of the length of each of the three parts.

II.3.39. On a segment $[0,T]$, n points are chosen at random with the uniform distribution. These points divide the segment into $n+1$ parts. Prove that the length of every part has the same distribution function. Find this distribution function.

II.3.40. Suppose a device starts operating at the zero moment of time and gets out of order at a random time ξ. Suppose that the conditional probability that the device does not break down during the time interval $(x, x+\Delta x)$, provided that this has not happened before the moment x, is equal to $\lambda\Delta x + o(\Delta x)$. Prove that for $x>0$

$$\mathsf{P}(\xi < x) = 1 - e^{-\lambda x},$$

i.e., the random variable ξ has the exponential distribution with parameter λ.

II.3.41. Let ξ be the time of trouble-free operation of some device, which starts working at the zero moment of time. Suppose that the conditional probability that the device gets out at order during the time interval $(x, x + \Delta x)$, provided that it has not failed before the moment x, is equal to $\lambda(x)\Delta x + o(\Delta x)$. Prove that the distribution function of ξ is of the form

$$F(x) = 1 - \exp\left\{ - \int_0^x \lambda(u)\, du \right\}.$$

II.3.42. The Weibull–Gnedenko distribution. Prove that if in the previous problem $\lambda(x) = c\alpha x^{\alpha-1}$, $\alpha > 1$, $c > 0$, then the distribution function ξ is of the form

$$F(x) = 1 - \exp\{-cx^\alpha\}.$$

II.3.43. Let ξ be a random variable with the exponential distribution, and let $t > 0$ be a fixed real number. Find the distribution of $\xi - t$ given $\xi \geq t$.

II.3.44. Let $F(x)$ be the distribution function of a positive random variable ξ with the property $\mathsf{P}\left(\xi < t + x / \xi > t\right) = \mathsf{P}(\xi < x)$. Prove that

$$F(x) = 1 - e^{-\lambda x}, \qquad x > 0.$$

II.3.45. Let ξ have the exponential distribution with parameter λ. Compute a) $\mathsf{E}\,\xi^k$; b) $\operatorname{Var}\xi$; and c) $\mathsf{P}(\xi \geq 1)$.

II.3.46. An electric bulb burns out after some time of service. This time is a random variable exponentially distributed with parameter $\lambda = 0.003$. In a year the bulb is replaced, even if it has not burned out. Find the expectation of the lifetime of the bulb.

II.3.47. Let ξ be a random variable uniformly distributed on $(0, 1)$. Find the distribution function of the random variable $\eta = \ln \frac{1}{\xi}$. Calculate $\mathsf{E}\,\eta$.

II.3.48. Let ξ be a random variable uniformly distributed on $[0, 1]$. Find the distribution of the random variable $\eta = -\lambda^{-1} \ln(1 - \xi)$.

II.3.49. Let ξ be a random variable exponentially distributed with parameter λ. Find the distribution of the random variable $\eta = [\xi]$, where $[\cdot]$ stands for the integer part. Calculate $\mathsf{E}\,\eta$.

II.3.50. Let ξ have the normal distribution $N(0, 1)$. Find the distribution of the random variable $\eta = \xi^{-2}$. Does $\mathsf{E}\,\eta$ exist?

II.3.51. A random variable ξ has the normal distribution $N(0, \sigma^2)$. For which value of σ is the probability of its falling into an interval (a, b) maximal?

II.3.52. Prove the following properties of a distribution function:

a) $\lim\limits_{x\to+0} x \int_x^\infty \frac{1}{u}\, dF(u) = 0;$

b) $\lim\limits_{x\to+\infty} x \int_x^\infty \frac{1}{u}\, dF(u) = 0;$

c) $\lim\limits_{x\to-0} x \int_{-\infty}^x \frac{1}{u}\, dF(u) = 0;$

d) $\lim\limits_{x\to-\infty} x \int_{-\infty}^x \frac{1}{u}\, dF(u) = 0.$

II.3.53. A random variable ξ has the distribution density $f(x)$. Calculate its expectation if

a) $f(x) = \begin{cases} (b-a)^{-1}, & a \le x \le b, \\ 0, & x < a \text{ or } x > b; \end{cases}$

b) $f(x) = \begin{cases} 1-|x-1|, & x \in [0,2], \\ 0, & x < 0 \text{ or } x > 2; \end{cases}$

c) $f(x) = \begin{cases} ke^{-k|x-a|}, & x \ge a, \\ 0, & x < a; \end{cases}$

d) $f(x) = \frac{1}{\sqrt{2\pi b}} \exp\left\{-\frac{(x-a)^2}{2b}\right\};$

e) $f(x)$ is an even function.

II.3.54. Find $\operatorname{Var}\xi$ if ξ has the density $f(x)$ given by the formulas a), b), and c) of Problem II.3.53.

II.3.55. Let ξ be a random variable uniformly distributed on a segment $[-a, a]$. Calculate a) $\mathsf{E}\,\xi$; b) $\operatorname{Var}\xi$; and c) $\mathsf{P}\,(|\xi| > a/2)$.

II.3.56. A random variable ξ is uniformly distributed on $[0, 1]$. Compute a) $\mathsf{E}\sin^2 \pi\xi$; b) $\mathsf{E}\, e^{\xi}$.

II.3.57. Let $F(x)$ be a continuous distribution function. Prove that

a) $\int_{-\infty}^{\infty} F(x)\, dF(x) = \frac{1}{2};$

b) $\int_{-\infty}^{\infty} F^k(x)\, dG(x) = \frac{n}{n+k},$

where $G(x) = F^n(x)$.

II.3.58. a) Find $\mathsf{E}\,|\xi|$, if a random variable ξ is normally distributed with parameters (a, σ^2).

b) Let ξ be normally distributed with parameters (a, σ^2). Compute $\mathsf{E}\,|\xi - a|$.

II.3.59. Let ξ be a random variable with the distribution density

$$p_\xi(x) = \frac{1}{\pi} \frac{1}{(1+x^2)}.$$

Calculate $\mathsf{E}\min(|\xi|, 1)$.

II.3.60. In textile manufacturing, the following quantity is used:

$$t = \frac{a'' - a'}{2a},$$

where a is the average length of a cotton fiber, a'' is the expectation of the length of cotton fibers longer than the average, and a' is the expectation of the length of cotton fibers shorter than the average. Assuming that the length of a cotton fiber has the normal distribution $N(a, \sigma^2)$, express t in terms of a and σ^2.

II.3.61. Let ξ be a random variable taking values in the interval $[a, b]$. Prove that $a \le \mathsf{E}\,\xi \le b$ and $\operatorname{Var}\xi \le (b-a)^2/4$.

II.3.62. Prove that the expectation of a random variable ξ is finite if and only if the series

$$\sum_{n=0}^{\infty} \mathsf{P}(|\xi| \ge n)$$

converges.

II.3.63. Prove that if $F(x)$ is the distribution function of a random variable ξ and $\mathsf{E}\,\xi$ exists, then the following relations hold:

$$\lim_{x\to+\infty} x(1 - F(x)) = 0,$$
$$\lim_{x\to-\infty} xF(x) = 0.$$

II.3.64. Prove that if $\mathsf{E}\,\xi$ exists, then the equality

$$\mathsf{E}\,\xi = \int_0^\infty (1 - F(y))\,dy - \int_{-\infty}^0 F(y)\,dy = \int_0^\infty (1 - F(+y) + F(-y))\,dy$$

holds, and conversely, the existence of the integrals on the right-hand side implies the existence of $\mathsf{E}\,\xi$.

II.3.65. Let ξ be a random variable with the distribution function $F(x)$ that has an expectation. Prove that

$$\mathsf{E}\,|\xi| = \int_{-\infty}^0 F(x)\,dx + \int_0^\infty [1 - F(x)]\,dx.$$

II.3.66. If $F(x)$ is the distribution function of a random variable ξ and $\mathsf{E}\,\xi^2$ exists, then

$$\lim_{x\to\infty} x^2[1 - F(x) + F(-x)] = 0$$

and

$$\mathsf{E}\,\xi^2 = 2\int_0^\infty x[1 - F(x) + F(-x)]\,dx.$$

II.3.67. Let $F(x)$ be a distribution function. Prove that for every $s > 0$

$$\int_0^\infty x^s\,dF(x) = s\int_0^\infty x^{s-1}[1 - F(x)]\,dx,$$

the convergence of one of these integrals implying the convergence of the other.

II.3.68. Let $F(x)$ be the distribution function of a random variable ξ. Prove that $\mathsf{E}\,|\xi|^s$, $s > 0$, exists if and only if the function $|x|^{s-1}(1 - F(x) + F(-x))$ is integrable on $(-\infty, \infty)$.

II.3.69. On a sphere of radius R two points, A and B, are taken. Find the distribution function and expectation of the length of the chord AB.

II.3.70. On a sphere of radius R in the n-dimensional space two points, A and B, are taken. Find the distribution density and expectation of the length of the chord AB.

II.3.71. A point P is uniformly distributed on the circle $x^2 + y^2 = 1$. Let η be the length of the projection of the vector $\overrightarrow{OP}$ to a fixed straight line (say, the Ox axis). Find the distribution density and expectation of the random variable η.

II.3.72. Let a point P be uniformly distributed on the sphere $x^2+y^2+z^2 = 1$. Let η be the length of the projection of the vector $\overrightarrow{OP}$ to a fixed straight line (say, the Ox axis). Find the distribution function and expectation of the random variable η.

II.3.73. A point P is uniformly distributed on the sphere $x^2 + y^2 + z^2 = 1$. Let γ be the length of the projection of the vector $\overrightarrow{OP}$ to a fixed plane (say, the xOy plane). Find the distribution density and expectation of γ.

REMARK. In Problems II.3.74–76 that follow the words "a random direction is chosen in the n-dimensional space" or "randomly oriented" must be understood as "a point is chosen at random on the unit sphere".

II.3.74. Let N be a fixed point on a sphere of radius R. A straight line is drawn through the point N in a random direction, which intersects the sphere at a point P. Prove that the length of the chord NP is a random variable uniformly distributed on $[0, 2R]$.

II.3.75. An isosceles triangle is formed by the unit vector on the Ox axis and a randomly oriented unit vector. Find the distribution function of the length of the third side of this triangle

a) in the case of the plane;

b) in the case of the three-dimensional space.

II.3.76. Let N be a fixed point on a circle of radius R. A randomly oriented straight line is drawn through the point N, which interests the circle at a point P. Find the distribution function of the length of the chord NP.

II.3.77. The distribution density of a random vector (ξ, η) is of the form

$$p(x,y) = \begin{cases} [(1+ax)(1+ay) - a]e^{-x-y-axy}, & x > 0\ y > 0, \\ 0, & \text{otherwise,} \end{cases}$$

where $0 < a < 1$. Find

a) the distribution density of ξ and η;

b) the distribution function of (ξ, η).

II.3.78. The distribution density of a random vector (ξ, η) is

$$p(x,y) = \begin{cases} 24x^2y(1-x), & 0 < x < 1,\ 0 < y < 1, \\ 0, & \text{otherwise.} \end{cases}$$

Find
  a) the distribution density of ξ;
  b) the distribution density of η.
Prove that ξ and η are independent.

II.3.79. Let $p_1(x)$ and $p_2(x)$ be distribution densities, $F_1(x)$ and $F_2(x)$ the corresponding distribution functions, and let $|a| < 1$. Prove that
  a) $f(x,y) = p_1(x)p_2(y)[1+a(2F_1(x)-1)(2F_2(x)-1)]$ is the distribution density of some random vector (ξ, η);
  b) the distribution densities of ξ and η are equal to $p_1(x)$ and $p_2(x)$, respectively.

II.3.80. Let ξ_1 and ξ_2 be independent random variables, both having the normal distribution $N(0,1)$. Compute

$$\mathsf{P}(\xi_1^2 + \xi_2^2 \leq R^2).$$

II.3.81. The distribution density of a random vector (ξ, η) is equal to

$$p(x,y) = \begin{cases} 24xy(1-x^2), & 0 \leq x \leq 1,\ 0 \leq y \leq 1, \\ 0, & \text{otherwise.} \end{cases}$$

Prove that ξ and η are independent.

II.3.82. The distribution density of a random vector (ξ, η) is equal to

$$p(x,y) = \begin{cases} xe^{-x(1+y)}, & x > 0,\ y > 0, \\ 0, & \text{otherwise.} \end{cases}$$

Compute
  a) the distribution density of ξ;
  b) the distribution density of η.

II.3.83. The distribution density of a random vector (ξ, η) is equal to

$$p(x,y) = \begin{cases} x^{p-1}(y-x)^{q-1}e^{-y}/(\Gamma(p)\Gamma(q)), & 0 < x < y, \\ 0, & \text{otherwise.} \end{cases}$$

Find the distribution densities of ξ and η.

II.3.84. Let ξ and η be independent random variables having the exponential distribution with parameter λ. Find the distribution function of the random variable $\xi/(\xi+\eta)$.

II.3.85. Let ξ and η be independent random variables having the exponential distribution with parameter λ. Find the distribution function of the random variable $(\xi+\eta)/\xi$.

II.3.86. Random variables ξ_1 and ξ_2 are independent and exponentially distributed with parameters λ_1 and λ_2, respectively. Find the distribution function of the random variable

$$\eta = \frac{\xi_1}{\xi_1 + \xi_2}.$$

II.3.87. Random variables ξ_1 and ξ_2 are independent and uniformly distributed on the segment $[0, 1]$. Find the distribution function of the random variable

$$\eta = \frac{\xi_1}{\xi_1 + \xi_2}.$$

II.3.88. Runners A and B start simultaneously and finish at times T_A and T_B, which are independent identically distributed random variables having the exponential distribution with parameter λ. Let $T_B > T_A$. Find the probability that the time of the winner exceeds the difference between the time of the loser and the time of the winner ($T_A > T_B - T_A$).

II.3.89. The distribution density of a random vector (ξ, η) is equal to

$$f(x, y) = \frac{1}{2\pi\sigma^2} \exp\left\{-\frac{x^2 + y^2}{2\sigma^2}\right\}.$$

Find the distribution density of the random variable $\gamma = \max(|\xi|, |\eta|)$.

II.3.90. Let ξ_1, ξ_2, and θ be independent random variables, ξ_1 and ξ_2 being identically normally distributed with parameters 0 and 1 and θ being uniformly distributed on the segment $[0, 2\pi]$. Find the distribution function of the random variable $\xi_1 \cos\theta + \xi_2 \sin\theta$.

REMARK. In Problems II.3.91–94 the distribution functions of random variables ξ_1 and ξ_2 are assumed to be continuous.

II.3.91. Let ξ_1 and ξ_2 be independent random variables, and let $\eta = \xi_1 + \xi_2$. Prove that

a) the distribution function of η is given by

$$F_\eta(x) = \int_{-\infty}^{\infty} F_{\xi_2}(x - u)\, dF_{\xi_1}(u) = \int_{-\infty}^{\infty} F_{\xi_1}(x - u)\, dF_{\xi_2}(u);$$

b) if at least one of the random variables ξ_1 or ξ_2 has the distribution density, then the random variable η has the distribution density

$$p_\eta(x) = \int_{-\infty}^{\infty} p_{\xi_1}(x - v)\, dF_{\xi_2}(v),$$

or

$$p_\eta(x) = \int_{-\infty}^{\infty} p_{\xi_2}(x - u)\, dF_{\xi_1}(u);$$

c) if the random variables ξ_1 and ξ_2 have the distribution densities, then

$$p_\eta(x) = \int_{-\infty}^{\infty} p_{\xi_1}(x - v) p_{\xi_2}(v)\, dv = \int_{-\infty}^{\infty} p_{\xi_2}(x - u) p_{\xi_1}(u)\, du.$$

II.3.92. Let ξ_1 and ξ_2 be independent random variables, and let $\eta = \xi_1 - \xi_2$. Prove that

a) the distribution function of η is given by

$$F_\eta(x) = \int_{-\infty}^{\infty} F_{\xi_1}(x + v)\, dF_{\xi_2}(v) = \int_{-\infty}^{\infty} [1 - F_{\xi_2}(u - x)]\, dF_{\xi_1}(u);$$

b) if at least one of the random variables, ξ_1 or ξ_2, has the distribution density, then η has the distribution density and

$$p_\eta(x) = \int_{-\infty}^{\infty} p_{\xi_1}(x+v)\, dF_{\xi_2}(v),$$

or

$$p_\eta(x) = \int_{-\infty}^{\infty} p_{\xi_2}(u-x)\, dF_{\xi_1}(u);$$

c) if the distribution densities of ξ_1 and ξ_2 exist, then

$$p_\eta(x) = \int_{-\infty}^{\infty} p_{\xi_1}(x+v) p_{\xi_2}(v)\, dv = \int_{-\infty}^{\infty} p_{\xi_2}(u-x) p_{\xi_1}(u)\, du.$$

II.3.93. Let ξ_1 and ξ_2 be independent random variables, and let $\eta = \xi_1 \xi_2$. Prove that

a) the distribution function of η is given by

$$F_\eta(x) = \int_{-\infty}^{0} \left[1 - F_{\xi_2}\left(\frac{x}{u}\right)\right] dF_{\xi_1}(u) = \int_{0}^{\infty} F_{\xi_2}\left(\frac{x}{u}\right) dF_{\xi_1}(u);$$

b) if at least one of the random variables, ξ_1 or ξ_2, has the distribution density, then η has the distribution density and

$$p_\eta(x) = \int_{-\infty}^{\infty} \frac{1}{|v|} p_{\xi_1}\left(\frac{x}{v}\right) dF_{\xi_2}(v),$$

or

$$p_\eta(x) = \int_{-\infty}^{\infty} \frac{1}{|u|} p_{\xi_2}\left(\frac{x}{u}\right) dF_{\xi_1}(u);$$

c) if the random variables ξ_1 and ξ_2 have the distribution densities, then

$$p_\eta(x) = \int_{-\infty}^{\infty} \frac{1}{|u|} p_{\xi_1}\left(\frac{x}{u}\right) p_{\xi_2}(u)\, du.$$

II.3.94. Let ξ_1 and ξ_2 be independent random variables, and let $\eta = \xi_1/\xi_2$. Prove that

a) the distribution function of η is given by

$$F_\eta(x) = \int_{-\infty}^{0} [1 - F_{\xi_1}(vx)]\, dF_{\xi_2}(v) + \int_{0}^{\infty} F_{\xi_1}(vx)\, dF_{\xi_2}(v);$$

b) if the random variable ξ_1 has the distribution density, then the random variable η also has the distribution density and

$$p_\eta(x) = \int_{-\infty}^{\infty} |v| p_{\xi_1}(vx)\, dF_{\xi_2}(v);$$

c) if ξ_1 and ξ_2 have the distribution densities, then

$$p_\eta(x) = \int_{-\infty}^{\infty} |v| p_{\xi_1}(vx) p_{\xi_2}(v)\, dv.$$

II.3.95. Let ξ_1 and ξ_2 be independent random variables uniformly distributed on $\left[-\frac{1}{2}, \frac{1}{2}\right]$. Find the distribution density of the random variable $\eta = \xi_1 + \xi_2$.

II.3.96. Random variables ξ_1 and ξ_2 are independent and uniformly distributed on the segment $[0,1]$. Find the distribution densities of the random variables

a) $\xi_1\xi_2$;

b) $\xi_1-\xi_2$;

c) $|\xi_1-\xi_2|$.

II.3.97. Let ξ_1 and ξ_2 be independent random variables uniformly distributed on a segment $[-a/2,a/2]$. Find the distribution density of the random variable $\eta=\xi_1-\xi_2$.

II.3.98. Random variables ξ_1 and ξ_2 are independent and uniformly distributed on the segment $[0,1]$. Find the distribution function of the random variable $\eta=\xi_1+\xi_2$.

II.3.99. Let ξ_1 and ξ_2 be independent identically distributed random variables with the distribution density $p(x)=\frac{1}{2}e^{-|x|}$. Find the distribution density of the random variable $\eta=\xi_1+\xi_2$.

II.3.100. Let ξ and η be independent random variables having the exponential distribution with parameter λ. Calculate

a) the distribution density of $\xi-\eta$;

b) the distribution density of $|\xi-\eta|$.

II.3.101. Random variables ξ and η are independent and have the distribution densities $p_\xi(x)=12x^2(1-x)$, $x\in(0,1)$ and $p_\eta(y)=y$, $y\in(0,1)$. Find the distribution density of the product $\xi\eta$.

II.3.102. Random variables ξ and η are independent and have the distribution densities

$$p_\xi(x)=\begin{cases}(\pi\sqrt{1-x^2})^{-1}, & |x|<1,\\ 0, & |x|\geq 1,\end{cases}\qquad p_\eta(x)=\begin{cases}0, & x\leq 0,\\ xe^{-x^2/2}, & x>0.\end{cases}$$

Prove that the random variable $\xi\eta$ has the normal distribution.

II.3.103. Random variables ξ and η are independent and have the normal distribution $N(0,\sigma^2)$. Prove that the ratio $\gamma=\xi/\eta$ has the Cauchy distribution.

II.3.104. Random variables ξ and η are independent and identically distributed with the distribution density $p_\xi(x)=p_\eta(x)=c/(1+x^4)$.

a) Compute the constant c.

b) Find the distribution density of $\gamma=\xi/\eta$.

II.3.105. Let ξ and η be independent random variables having the exponential distribution with parameter λ. Find

a) the distribution function of ξ/η;

b) the distribution density of ξ/η;

c) the expectation of ξ/η.

II.3.106. Let ξ and η be independent random variables having the exponential distributions with parameters λ_1 and λ_2, respectively. Prove that the distribution density of the random variable $\xi+\eta$ is equal to

$$\lambda_1\lambda_2\frac{e^{-\lambda_1 x}-e^{-\lambda_2 x}}{\lambda_1-\lambda_2},\qquad x>0.$$

II.3.107. Let $\xi_0, \xi_1, \dots, \xi_n$ be independent random variables having the exponential distributions with parameters $\lambda_0, \lambda_1, \dots, \lambda_n$, respectively, $\lambda_k \neq \lambda_j$, $k \neq j$. Prove that the distribution density of the random variable $S_n = \xi_0 + \xi_1 + \cdots + \xi_n$ is of the form

$$\lambda_0 \lambda_1 \cdots \lambda_n \left[\psi_{0n} e^{-\lambda_0 x} + \cdots + \psi_{nn} e^{-\lambda_n x}\right], \qquad x > 0,$$

where

$$\frac{1}{\psi_{kn}} = (\lambda_0 - \lambda_k) \cdots (\lambda_{k-1} - \lambda_k)(\lambda_{k+1} - \lambda_k) \cdots (\lambda_n - \lambda_k).$$

II.3.108. Suppose that a random vector (ξ_1, ξ_2) has the distribution density $p_{(\xi_1,\xi_2)}(x, y)$. Prove that the random variables $\xi_1 + \xi_2$ and $\xi_1 - \xi_2$ also have the distribution densities and

a) $p_{\xi_1+\xi_2}(z) = \int_{-\infty}^{\infty} p_{(\xi_1,\xi_2)}(x, z - x)\, dx = \int_{-\infty}^{\infty} p_{(\xi_1,\xi_2)}(z - y, y)\, dy,$

b) $p_{\xi_1-\xi_2}(z) = \int_{-\infty}^{\infty} p_{(\xi_1,\xi_2)}(z + y, y)\, dy = \int_{-\infty}^{\infty} p_{(\xi_1,\xi_2)}(x, x - z)\, dx.$

II.3.109. Suppose that a random vector (ξ_1, ξ_2) has the distribution density $p_{(\xi_1,\xi_2)}(x, y)$. Prove that the random variables $\xi_1\xi_2$ and ξ_1/ξ_2 also have the distribution densities and

a) $p_{\xi_1\xi_2}(z) = \int_{-\infty}^{\infty} p_{(\xi_1,\xi_2)}\left(\frac{z}{y}, y\right) \frac{dy}{|y|} = \int_{-\infty}^{\infty} p_{(\xi_1,\xi_2)}\left(x, \frac{z}{x}\right) \frac{dx}{|x|},$

b) $p_{\xi_1/\xi_2}(z) = \int_{-\infty}^{\infty} p_{(\xi_1,\xi_2)}(zy, y)|y|\, dy.$

II.3.110. A random variable (ξ_1, ξ_2) has the distribution density

$$p(x, y) = \begin{cases} x + y, & 0 \le x \le 1,\ 0 \le y \le 1, \\ 0, & \text{otherwise.} \end{cases}$$

Find the distribution density of the random variable $\eta = \xi_1 + \xi_2$.

II.3.111. The distribution density of a random vector (ξ_1, ξ_2) is

$$f(x, y) = \begin{cases} \lambda^2 e^{-2x}, & 0 < y < x, \\ 0, & \text{otherwise.} \end{cases}$$

Find the distribution density of the random variable $\eta = \xi_1 + \xi_2$.

II.3.112. A random vector (ξ_1, ξ_2) has the distribution density

$$p(x, y) = \begin{cases} \Gamma(p+q+r)/\big(\Gamma(p)\Gamma(q)\Gamma(r)\big) x^{p-1} y^{q-1} (1 - x - y) & \\ & \text{if } x > 0,\ y > 0,\ x + y < 1, \\ 0 & \text{otherwise.} \end{cases}$$

Calculate the distribution density of $\eta = \xi_1 + \xi_2$.

II.3.113. Suppose that a random vector (ξ_1, ξ_2) has the distribution density $p_{(\xi_1,\xi_2)}(x,y)$. Prove that the random variable $\rho = \sqrt{\xi_1^2 + \xi_2^2}$ also has the distribution density and

$$p_\rho(z) = \begin{cases} z \int_0^{2\pi} p_{(\xi_1,\xi_2)}(z\cos\theta, z\sin\theta)\, d\theta, & z \ge 0, \\ 0, & z < 0. \end{cases}$$

II.3.114. The Rayleigh distribution. Let ξ_1 and ξ_2 be independent normally distributed random variables, $\mathsf{E}\,\xi_1 = \mathsf{E}\,\xi_2 = 0$, $\operatorname{Var}\xi_1 = \operatorname{Var}\xi_2 = \sigma^2$. Put $\gamma = \sqrt{\xi_1^2 + \xi_2^2}$. Prove that for $z > 0$

$$p_\gamma(z) = \frac{z}{\sigma^2} \exp\left\{ -\frac{z^2}{2\sigma^2} \right\}.$$

II.3.115. A random vector (ξ, η) has the distribution density

$$p(x,y) = \begin{cases} 24xy(1-x^2), & 0 < x < 1,\ 0 < y < 1, \\ 0, & \text{otherwise}. \end{cases}$$

Calculate the distribution density of $\gamma = \xi\eta$.

II.3.116. The distribution density of a random vector (ξ, η) is equal to

$$p(x,y) = \begin{cases} xe^{-x(1+y)}, & x > 0,\ y > 0, \\ 0, & \text{otherwise}. \end{cases}$$

Calculate the distribution density of $\xi\eta$.

II.3.117. A random variable ξ has the gamma distribution with parameters (α, β), i.e., the distribution density of ξ is equal to

$$p(x) = \begin{cases} \beta^\alpha \Gamma^{-1}(\alpha) x^{\alpha-1} e^{-\beta x}, & x > 0, \\ 0, & x < 0. \end{cases}$$

Calculate a) $\mathsf{E}\,\xi$; b) $\operatorname{Var}\xi$; and c) $\mathsf{E}\,\xi^n$.

II.3.118. Let ξ have the normal distribution $N(0,1)$. Prove that the random variable $\eta = \xi^2$ has the gamma distribution with parameters $\left(\frac{1}{2}, \frac{1}{2}\right)$.

II.3.119. Let ξ_1 and ξ_2 be independent random variables having the gamma distributions with parameters (α_1, β) and (α_2, β), respectively. Prove that the random variable $\eta = \xi_1 + \xi_2$ has the gamma distribution with parameters $(\alpha_1 + \alpha_2, \beta)$.

II.3.120. The Erlang distribution. Let $\{\xi_k, 1 \le k \le n\}$ be independent identically distributed random variables each having the exponential distribution with parameter $\lambda > 0$. Prove that the random variable $S_n = \xi_1 + \cdots + \xi_n$ has the following distribution density:

$$p_{S_n} = \begin{cases} \frac{(\lambda x)^{n-1}}{(n-1)!} \lambda e^{-\lambda x}, & x \ge 0, \\ 0, & x < 0. \end{cases}$$

II.3.121. The χ^2 distribution with n degrees of freedom. Let random variables $\xi_1, \dots, \xi_n$ be independent and all have the normal distribution $N(0,1)$. Prove that the distribution density of the random variable $\chi^2 = \xi_1^2 + \dots + \xi_n^2$ is equal to

$$f_{\chi^2}(x) = \begin{cases} x^{n/2-1}e^{-x/2}/(2^{n/2}\Gamma(n/2)), & x > 0, \\ 0, & x < 0. \end{cases}$$

II.3.122. Let $\xi_1, \dots, \xi_m$, $\eta_1, \dots, \eta_n$ be independent random variables having the normal distribution $N(0,1)$. Find the distribution density of the random variable

$$\frac{\xi_1^2 + \dots + \xi_m^2}{\eta_1^2 + \dots + \eta_n^2}.$$

II.3.123. The Student distribution. Let $\xi, \xi_1, \dots, \xi_n$ be independent random variables having the normal distribution $N(0,1)$. Prove that the distribution density of the random variable

$$\eta = \frac{\xi}{\sqrt{\xi_1^2 + \dots + \xi_n^2}}$$

is equal to

$$f_\eta(x) = \frac{1}{\sqrt{\pi}} \frac{\Gamma((n+1)/2)}{\Gamma(n/2)} \frac{1}{(1+x^2)^{(n+1)/2}}, \qquad x > 0.$$

II.3.124. Let $\xi_1, \dots, \xi_n$ be independent random variables uniformly distributed on $[0,1]$. Find the distribution function of $\eta = \xi_1 \xi_2 \cdots \xi_n$. Calculate $\mathsf{E}\,\eta$.

II.3.125. Let $\{\xi_k, 1 \le k \le n\}$ be independent random variables uniformly distributed on $[0,1]$. Prove that the random variable $-2\ln(\xi_1 \cdots \xi_n)$ has the χ^2 distribution with $2n$ degrees of freedom.

II.3.126. Let ξ_1 and ξ_2 be independent normally distributed random variables with zero expectation and variance σ_1^2 and σ_2^2, respectively. Prove that the random variable $\eta = \xi_1\xi_2/\sqrt{\xi_1^2 + \xi_2^2}$ is normally distributed with parameters $(0, \sigma^2)$, where

$$\frac{1}{\sigma} = \frac{1}{\sigma_1} + \frac{1}{\sigma_2}.$$

II.3.127. Random variables ξ_1 and ξ_2 are independent and have the normal distributions $N(a_1, \sigma_1^2)$ and $N(a_2, \sigma_2^2)$, respectively. Prove that the random variable $\eta = \xi_1 + \xi_2$ has the normal distribution $N(a_1 + a_2, \sigma_1^2 + \sigma_2^2)$.

II.3.128. Let ξ_1 and ξ_2 be independent random variables, each having the normal distribution with parameters 0 and 1. Prove that the random variables $\xi_1 - \xi_2$ and $\xi_1 + \xi_2$ are independent.

II.3.129. Let ξ_1 and ξ_2 be independent random variables having exponential distributions with parameters λ_1 and λ_2, respectively. Prove that the random variables $\xi_1 - \xi_2$ and $\min(\xi_1, \xi_2)$ are independent.

II.3.130. Let ξ_1 and ξ_2 be independent random variables, each having the exponential distribution with parameter $\lambda > 0$. Denote by $\xi_{(1)}$ and $\xi_{(2)}$ the greater and the smaller, respectively, of the ξ_1 and ξ_2. Prove that the random variables $\xi_{(1)}$ and $\xi_{(2)} - \xi_{(1)}$ are independent and find the distribution of each of them.

II.3.131. Random variables ξ and η are independent and have the exponential distribution with parameter λ. Prove that the random variables $\xi + \eta$ and ξ/η are also independent.

II.3.132. Let ξ and η be independent random variables having the normal distribution $N(0, \sigma^2)$. Prove that the random variables $\xi^2 + \eta^2$ and ξ/η are independent.

II.3.133. Let $\{\xi_k, 1 \le k \le n\}$ be independent random variables having the normal distribution $N(0,1)$. Prove that the random variables

$$\frac{\xi_k^2}{\sum_{i=1}^n \xi_i^2} \quad \text{and} \quad \sum_{i=1}^n \xi_i^2$$

are independent.

II.3.134. Let $\xi_1, \dots, \xi_n, \eta$ be independent random variables, each of the random variables ξ_k, $1 \le k \le n$, taking the values 0 or 1 with probability $\frac{1}{2}$. The random variable η is uniformly distributed on $[0,1]$. Prove that the random variable

$$\frac{\eta}{2^n} + \sum_{k=1}^n \frac{\xi_k}{2^k}$$

is uniformly distributed on $[0,1]$.

II.3.135. A random variable ξ has the Cauchy distribution. Prove that the random variables

$$\text{a) } \frac{1}{\xi}; \qquad \text{b) } \frac{2\xi}{1-\xi^2}; \qquad \text{c) } \frac{3\xi - \xi^3}{1 - 3\xi^2}$$

have the Cauchy distribution.

II.3.136. Let ξ_1 and ξ_2 be random variables having the Cauchy distribution. Prove that the random variable

$$\eta = \frac{\xi_1 + \xi_2}{1 - \xi_1 \xi_2}$$

also has the Cauchy distribution.

II.3.137. Let ξ_1 and ξ_2 be independent random variables with the Cauchy distribution. Find the distribution density of the random variable $\eta = \xi_1 \xi_2$.

II.3.138. A random variable ξ is uniformly distributed on the segment $[-1,1]$. Find the correlation coefficients of the following pairs of random variables:

(a) ξ and ξ^2;
(b) ξ and $\sin \pi\xi/2$;
(c) ξ and ξ^3;
(d) $\sin \pi\xi/2$ and $\cos 2\pi\xi/2$.

II.3.139. A point (ξ, η) is uniformly distributed inside the triangle with vertices $(0,0)$, $(1,0)$, and $(0,1)$. Find the correlation coefficient between ξ and η.

II.3.140. Prove that if a random variable ξ is uncorrelated with each of random variables $\eta_1, \dots, \eta_k$, then it is also uncorrelated with a random variable $\alpha_1\eta_1 + \dots + \alpha_k\eta_k$, for arbitrary real numbers $\alpha_1, \dots, \alpha_k$.

II.3.141. Prove that if random variables $\xi_1, \dots, \xi_n$ are pairwise uncorrelated, then $\operatorname{Var}(\xi_1 + \dots + \xi_n) = \operatorname{Var}\xi_1 + \dots + \operatorname{Var}\xi_n$.

II.3.142. Random variables ξ and η are independent and normally distributed with the same parameters a and σ^2. Find the correlation coefficient of the random variables $\gamma_1 = \alpha\xi + \beta\eta$ and $\gamma_2 = \alpha\xi - \beta\eta$.

II.3.143. Random variables ξ and η are independent and have the normal distribution $N(0,1)$. Compute $\mathsf{E}(\operatorname{sgn}\xi + \operatorname{sgn}\eta)^2$.

II.3.144. A point $P = (\xi, \eta)$ is uniformly distributed in the region $x^2/a^2 + y^2/b^2 = 1$. Find
  a) the distribution density of ξ;
  b) the distribution density of η;
  c) $\mathsf{E}\,\xi$ and $\mathsf{E}\,\eta$;
  d) the covariance of ξ and η.
Notice that the random variables ξ and η are uncorrelated but dependent.

II.3.145. A point $P = (\xi, \eta)$ is uniformly distributed in the region $|x| + |y| \le a$. Find
  a) the distribution densities of ξ and η;
  b) $\mathsf{E}\,\xi$ and $\mathsf{E}\,\eta$;
  c) the covariance of ξ and η.

II.3.146. Let $f(x)$ be a distribution density concentrated on $(0, +\infty)$. Put

$$p(x,y) = \begin{cases} \frac{f(x+y)}{x+y}, & x > 0,\ y > 0, \\ 0, & \text{otherwise.} \end{cases}$$

Prove that $p(x, y)$ is the distribution density of some random vector (ξ_1, ξ_2). Calculate the covariance matrix of (ξ_1, ξ_2).

II.3.147. Let ξ_i, $i = 1, 2, \dots, n$, be independent identically distributed random variables, and let $\mathsf{E}(\xi_i - \mathsf{E}\,\xi_i)^3 = 0$. Prove that the random variables

$$\xi = \sum_{i=1}^{n} \xi_i \quad \text{and} \quad s^2 = \sum_{i=1}^{n} (\xi_i - \xi)^2$$

are uncorrelated.

II.3.148. A point a is fixed on the interval $[0,1]$. A random variable ξ is uniformly distributed on the interval $[0,1]$. Let η be the distance between ξ and a. Find the correlation coefficient between ξ and η. For which values of a are the random variables ξ and η uncorrelated?

II.3.149. Let

$$\begin{aligned} y_1 &= f_1(x_1, \dots, x_n), \\ &\cdots\cdots\cdots\cdots \\ y_n &= f_n(x_1, \dots, x_n) \end{aligned} \tag{1}$$

be a one-to-one mapping of $\mathbf{R}^n$ onto $\mathbf{R}^n$, and let

$$x_1 = g_1(y_1, \dots, y_n),$$
(2) $$\dots\dots\dots\dots\dots\dots$$
$$x_n = g_n(y_1, \dots, y_n)$$

be the inverse mapping. Also, let $(\xi_1, \dots, \xi_n)$ be a random vector with distribution density $p_{(\xi_1,\dots,\xi_n)}(x_1, \dots, x_n)$, and let

$$\eta_1 = f_1(\xi_1, \dots, \xi_n),$$
$$\dots\dots\dots\dots\dots\dots$$
$$\eta_n = f_n(\xi_1, \dots, \xi_n).$$

Prove that the distribution density $p_{(\eta_1,\dots,\eta_n)}(y_1, \dots, y_n)$ of the random vector $(\eta_1, \dots, \eta_n)$ is equal to

$$p_{(\eta_1,\dots,\eta_n)}(y_1, \dots, y_n) = p_{(\xi_1,\dots,\xi_n)}\big(g_1(y_1, \dots, y_n), \dots, g_n(y_1, \dots, y_n)\big) \left| \frac{D(x_1, \dots, x_n)}{D(y_1, \dots, y_n)} \right|,$$

where $\frac{D(x_1,\dots,x_n)}{D(y_1,\dots,y_n)}$ is the Jacobian of the transformation (2).

II.3.150. The distribution density of a random vector (ξ, η) is equal to

$$p(x, y) = \frac{1}{2\pi\sigma^2} \exp\left\{-\frac{x^2 + y^2}{2\sigma^2}\right\}.$$

Let (R, Φ) be the polar coordinates of the point. Calculate
 a) the joint distribution density of (R, Φ);
 b) the distribution density of R;
 c) the distribution density of Φ.
Prove that R and Φ are independent random variables.

II.3.151. Let ξ_1, ξ_2, and ξ_3 be independent random variables having the normal distribution $N(0, 1)$, and let

$$\eta_1 = \frac{1}{\sqrt{2}}(\xi_1 - \xi_2),$$
$$\eta_2 = \frac{1}{\sqrt{6}}(\xi_1 + \xi_2 - 2\xi_3),$$
$$\eta_3 = \frac{1}{\sqrt{3}}(\xi_1 + \xi_2 + \xi_3).$$

Prove that the random variables η_1, η_2, and η_3 are independent and each of them has the normal distribution $N(0, 1)$.

II.3.152. Let ξ_1 and ξ_2 be independent identically distributed random variables having the normal distribution $N(0, 1)$. Prove that the random vector (η_1, η_2), where

$$\eta_1 = \exp\left\{-\frac{1}{2}(\xi_1^2 + \xi_2^2)\right\},$$
$$\eta_2 = \frac{1}{\pi}\arctan\frac{\xi_1}{\xi_2} + \frac{1}{2},$$

has the uniform distribution in the rectangle $[0,1] \times [0,1]$.

II.3.153. The distribution density of a random vector (ξ, η) has the form

$$p(x,y) = \frac{2}{\pi}(1 - x^2 - y^2), \qquad 0 \le x^2 + y^2 \le 1.$$

Calculate the distribution density of (R, Φ), where (R, Φ) are the polar coordinates of (ξ, η).

II.3.154. Random variables ξ and η are independent, ξ having the exponential distribution with parameter λ and η being uniformly distributed on the interval $[0, 2\pi]$. Let $\gamma_1 = \sqrt{\xi}\cos\eta$ and $\gamma_2 = \sqrt{\xi}\sin\eta$. Prove that γ_1 and γ_2 are independent identically distributed random variables with the distribution density

$$p_{\gamma_1}(x) = p_{\gamma_2}(x) = \sqrt{\frac{\lambda}{\pi}}\exp\{-\lambda^2 x\}.$$

II.3.155. A random variable (ξ_1, ξ_2) has a distribution density

$$p(x,y) = g\left(\sqrt{x^2 + y^2}\right).$$

Find the distribution density of the random variable (R, Φ), where R and Φ are the polar coordinates of the point (ξ_1, ξ_2).

II.3.156. Let

$$p(x_1, x_2) = \frac{1}{\sqrt{2\pi}}\frac{1}{\sqrt{(1 + x_1^2 + x_2^2)^3}}.$$

a) Prove that $p(x_1, x_2)$ is the distribution density of a random vector (ξ_1, ξ_2).
b) Find the distribution density of ξ_1 and the distribution density of ξ_2.
c) Find the joint distribution density of (R, Φ), where R and Φ are the polar coordinates of (ξ_1, ξ_2).

II.3.157. Let a random vector (ξ_1, ξ_2, ξ_3) have the distribution density

$$p(x_1, x_2, x_3) = g\left(\sqrt{x_1^2 + x_2^2 + x_3^2}\right)$$

for some function g. Find the joint distribution density of (R, Θ, Φ), where R, Θ, and Φ are the spherical coordinates of the point (ξ_1, ξ_2, ξ_3).

II.3.158. Let ξ_1, ξ_2, and ξ_3 be independent random variables having the normal distribution $N(0, \sigma^2)$. Find the joint distribution density of (R, Θ, Φ), where R, Θ, and Φ are the spherical coordinates of the point (ξ_1, ξ_2, ξ_3).

II.3.159. Let

$$p(x_1, x_2, x_3) = \frac{1}{\pi^2} \frac{1}{(1 + x_1^2 + x_2^2 + x_3^2)^2}.$$

a) Prove that $p(x_1, x_2, x_3)$ is the distribution density of some random vector (ξ_1, ξ_2, ξ_3).
b) Find the distribution densities of ξ_1, ξ_2, and ξ_3.
c) Find the distribution density of (R, Θ, Φ), where R, Θ, Φ are the spherical coordinates of the point (ξ_1, ξ_2, ξ_3).

II.3.160. A random vector whose direction is taken to be random and whose length is a random variable independent of its direction will be called *a spherically symmetric random vector*. Let $V(x)$ be the distribution function of a spherically symmetric random vector in $\mathbf{R}^3$, $F(x)$ the distribution function of the length of the random vector on the Ox axis, and $v(x)$ and $f(x)$ the corresponding distribution densities. Prove that

a) $F(t) = \int_0^t V\left(\frac{t}{x}\right)\, dx, \qquad t > 0;$

b) $v(t) = -tf'(t), \qquad t > 0.$

II.3.161. The Maxwell distribution. Let the projections of a velocity vector on the coordinate axes be independent identically distributed random variables with the normal distribution $N(0, 1)$. Prove that the distribution density of velocity is equal to

$$v(t) = \sqrt{\frac{2}{\pi}} t^2 e^{-t^2/2}, \qquad t > 0.$$

II.3.162. Let $V(x)$ be the distribution function of a spherically symmetric vector in $\mathbf{R}^2$ and $F(x)$ the distribution function of the length of the projection of this vector on the Ox axis. Prove that

$$F(x) = \frac{2}{\pi} \int_0^{\pi/2} V\left(\frac{x}{\sin\varphi}\right) d\varphi.$$

II.3.163. Let $\xi_1, \dots, \xi_n$ be independent identically distributed random variables having the exponential distribution with parameter λ. Prove that the random variable $\xi_{(1)} = \min(\xi_1, \dots, \xi_n)$ has the exponential distribution with parameter $n\lambda$.

II.3.164. A system consists of n units. Let ξ_i be the time of trouble-free operation of the ith unit. The random variables $\xi_1, \dots, \xi_n$ are independent and have the exponential distribution with parameter λ. The system fails if at least one of its units fails. Find the distribution function of the time of trouble-free operation of the system and the expectation of the time of its trouble-free operation.

II.3.165. Let $\xi_1, \dots, \xi_n$ be independent identically distributed random variables having the exponential distribution with parameter λ. Find the distribution function of the random variable $\xi_{(n)} = \max(\xi_1, \dots, \xi_n)$.

II.3.166. A system consists of n units. Let ξ_i be the time of trouble-free operation of the ith unit. The random variables $\xi_1, \dots, \xi_n$ are independent and have the exponential distribution with parameter λ. The system operates until at least one of its units is in operative condition. Find the distribution function of the time of trouble-free operation of the system and the expectation of the time of its trouble-free operation.

II.3.167. Let $\xi_1, \dots, \xi_n$ be independent identically distributed random variables with continuous distribution function $F(x)$. Let us order them by magnitude to obtain a nondecreasing sequence $\xi_{(1)} \leq \xi_{(2)} \leq \cdots \leq \xi_{(n)}$. Find

a) the distribution function of $\xi_{(1)} = \min(\xi_1, \dots, \xi_n)$;
b) the distribution function of $\xi_{(n)} = \max(\xi_1, \dots, \xi_n)$;
c) the distribution function of $\xi_{(m)}$, $1 < m < n$.

II.3.168. The distribution function of a random variable ξ_i is assumed to have a distribution density $p(x)$. Find the distribution density of the random variable $\xi_{(m)}$, $1 \leq m \leq n$.

II.3.169. Find the joint distribution function and joint distribution density of the random variables $\xi_{(k)}$ and $\xi_{(m)}$, $k < m$.

II.3.170. Find the distribution function of the random variable $\eta_n = \xi_{(n)} - \xi_{(1)}$.

II.3.171. A system consists of n units. Let ξ_i be the time of trouble-free operation of the ith unit. The random variables $\xi_1, \dots, \xi_n$ are independent and exponentially distributed with parameter λ. The system operates until k units fail. Denote by γ the time of trouble-free operation of the system. Find the distribution function and distribution density of γ.

II.3.172. Three customers, A, B, and C, come to the post office and find two vacant windows. The service of A and B starts immediately, whereas the service of C starts on completion of servicing either A or B. The service times are independent identically distributed random variables with the exponential distribution.

a) What is the probability that C will not be the last to leave the post office?
b) What is the distribution function and expectation of the time spent by C in the post office?

II.3.173. Let $\xi_1, \dots, \xi_n$ be independent random variables uniformly distributed on the segment $[0, T]$. Let us order them by magnitude to obtain a nondecreasing sequence $\xi_{(1)} \leq \xi_{(2)} \leq \cdots \leq \xi_{(n)}$.

a) Find the joint distribution function of the random vector $(\xi_{(1)}, \dots, \xi_{(n)})$.
b) Find the distribution function and distribution density of the random variable $\xi_{(k)}$.

II.3.174. The Liouville integral. Usinig probabilistic considerations, establish the following formula for multiple integral transformation:

$$\underset{\substack{x_1 + \cdots + x_n \leq 1 \\ x_i \geq 0}}{\int \overset{n}{\cdots} \int} f(x_1 + \cdots + x_n) x_1^{\alpha_1 - 1} \cdots x_n^{\alpha_n - 1} \, dx_1 \cdots dx_n$$

$$= \frac{\Gamma(\alpha_1) \cdots \Gamma(\alpha_n)}{\Gamma(\alpha_1 + \cdots + \alpha_n)} \int_0^1 f(u) u^{\alpha_1 + \cdots + \alpha_n - 1} \, du.$$

REMARK. Let us introduce the following notation, which will be used in Problems II.3.175–178:

$$x_+ = \frac{x + |x|}{2} = \begin{cases} x, & x \geq 0, \\ 0, & x < 0. \end{cases}$$

II.3.175. Let $\xi_1, \dots, \xi_n$ be independent random variables uniformly distributed on $[0, T]$. The points $\xi_1, \dots, \xi_n$ divide the segment $[0, T]$ into $n+1$ parts. Denote the lengths of these parts by $\eta_1, \dots, \eta_n, \eta_{n+1}$.

a) Find the joint distribution of $\eta_1, \dots, \eta_n$.

b) Prove that the random variables $\eta_1, \dots, \eta_n$ all have the same distribution function. Find it.

c) Prove that for any $x_1 \geq 0, \dots, x_{n+1} \geq 0$

$$\mathsf{P}(\eta_1 > x_1, \dots, \eta_{n+1} > x_{n+1}) = \frac{[T - x_1 - \dots - x_{n+1}]_+^n}{T^n}.$$

d) Let $p_n(T)$ be the probability that the length of each of the parts $\eta_1, \dots, \eta_{n+1}$ is greater than h. Prove that

$$p_n(T) = \frac{[T - (n+1)h]_+^n}{T^n}.$$

e) Calculate

$$\mathsf{E}\, \eta_1^{\alpha_1 - 1} \eta_2^{\alpha_2 - 1} \cdots \eta_n^{\alpha_n - 1}, \qquad \mathsf{E}(\eta_1^2 + \dots + \eta_n^2).$$

II.3.176. Let $\xi_1, \dots, \xi_n$ be independent random variables uniformly distributed on the segment $[0, 1]$. Find the distribution function of the random variable $\eta_n = \xi_{(n)} - \xi_{(1)}$ and $\mathsf{E}\, \eta_n$. Calculate the probability that all n points $\xi_1, \dots, \xi_n$ lie in an interval of length at most t.

II.3.177. Let $\xi_1, \dots, \xi_n$ be independent identically distributed random variables having the exponential distribution with parameter λ, and let $\eta_n = \xi_1 + \dots + \xi_n$. Put

$$\gamma_k = \frac{\xi_k}{\eta_n}, \quad k = 1, 2, \dots, n-1, \qquad \gamma_n = \eta_n.$$

1) Find
   a) the joint distribution density of $(\gamma_1, \dots, \gamma_n)$;
   b) the joint distribution density of $(\gamma_1, \dots, \gamma_{n-1})$.
2) Prove that $(\gamma_1, \dots, \gamma_{n-1})$ have the same distribution density as in the case where γ_k is the length of the kth interval resulting from a random partition of the interval $(0, 1)$ by $n - 1$ uniformly distributed points.

II.3.178. (Feller.) 1) Let $\xi_1, \dots, \xi_n$ be independent random variables uniformly distributed on $[0, a]$ and let $s_n = \xi_1 + \dots + \xi_n$. Denote by $U_n(x)$ and $u_n(x)$ the distribution function and distribution density, respectively, of s_n.

Prove that

a) $u_{n+1}(x) = \dfrac{1}{a} \displaystyle\int_0^a u_n(x - y)\, dy = \dfrac{U_n(x) - U_n(x - a)}{a}$;

b) $U_n(x) = \dfrac{1}{n} \dfrac{1}{a^n n!} \displaystyle\sum_{k=0}^{n} (-1)^k \binom{n}{k} (x - ka)_+^n$;

c) $u_n(x) = \dfrac{1}{a^n (n-1)!} \displaystyle\sum_{k=0}^{n} (-1)^k \binom{n}{k} (x - ka)_+^{n-1}$.

2) Let $\xi_1, \dots, \xi_n$ be independent random variables uniformly distributed on $[-b, b]$, and let $s_n = \xi_1 + \dots + \xi_n$. Calculate the distribution density of s_n.

3) There are given n arcs of length a on a circle of length t, whose midpoints are chosen independently and at random. Prove that the probability $\varphi_n(t)$ that these n arcs cover the whole circle is equal to

$$\varphi_n(t) = a^n (n-1)! \, \frac{u_n(t)}{t^{n-1}},$$

i.e.,

$$\varphi_n(t) = \sum_{k=0}^{n} (-1)^k \binom{n}{k} \left(1 - k\frac{a}{t}\right)_+^{n-1}.$$

4) Let the interval $(0,t)$ be divided into n parts by choosing independently $n-1$ points $\xi_1, \dots, \xi_{n-1}$ uniformly distributed on $(0,t)$. Prove that the probability that the length of none of these parts exceeds a is equal to $\varphi_n(t)$.

5) Consider n independent unit random vectors in $\mathbf{R}^3$ whose directions are chosen at random. Let L_n be the length of the sum of these vectors. Find the distribution density of L_n.

II.3.179. Let D be a cube in the n-dimensional space, and let $V(D)$ be its volume (Lebesgue measure).

a) Using probabilistic considerations, deduce the following formula that expresses a $2n$-fold multiple integral in terms of an ordinary integral:

$$\int_D \int_D f(|x-y|) \, dx \, dy = V^2(D) \int_{-\infty}^{\infty} f(z) \, dF_D(z),$$

where $|x-y|$ is the distance between the points x and y and $F_D(z)$ is the distribution function of the distance between two points taken in D independently with the uniform distribution.

b) Prove that

$$\int_0^T \int_0^T f(|u-v|) \, du \, dv = 2 \int_0^T (T-z) f(z) \, dz.$$

c) Let $S_n(R)$ be the sphere of radius R in the n-dimensional space, and let $m(\cdot)$ be Lebesgue measure on $S_n(R)$. Prove that

$$\int_{S_n(R)} \int_{S_n(R)} f(|x-y|) \, m_n(dx) \, m_n(dy) = \frac{2^n (\pi R)^{n-1}}{(n-2)!} \int_0^{2R} z^{n-2} \left(1 - \frac{z^2}{R^2}\right)^{(n-3)/2} f(z) \, dz.$$

II.3.180. Let ξ be a nonnegative random variable with a distribution density, η a random variable uniformly distributed on $(0,\xi)$, and $\gamma = \xi - \eta$. Prove that the random variables γ and η are independent if and only if

$$p_\xi(x) = \begin{cases} a^2 x e^{-ax}, & x > 0, \\ 0, & x < 0. \end{cases}$$

II.3.181. Let ξ_{11}, ξ_{12}, ξ_{21}, and ξ_{22} be independent random variables having the normal distribution $N(0,1)$. Find the distribution of the random variable

$$\Delta = \begin{vmatrix} \xi_{11} & \xi_{12} \\ \xi_{21} & \xi_{22} \end{vmatrix}.$$

II.3.182. Let $\xi_1, \dots, \xi_n$ be positive independent identically distributed random variables. Calculate

$$\mathsf{E}\,\frac{\xi_1+\xi_2+\cdots+\xi_k}{\xi_1+\xi_2+\cdots+\xi_n}.$$

II.3.183. Let ξ be a random variable. The function

$$Q_\xi(l) = \sup_{x\in\mathbf{R}^1} \mathsf{P}(x \le \xi \le x+l), \qquad 0 \le l \le \infty,$$

is called *the concentration function of* ξ.

a) Find $Q_\xi(+\infty)$ and $Q_\xi(0)$.

b) Prove that $Q_\xi(l)$ does not decrease.

c) Express $Q_\xi(l)$ in terms of the distribution function of ξ.

d) Let ξ and η be independent random variables. Prove that

$$Q_{\xi+\eta}(l) \le Q_\xi(l) \quad \text{and} \quad Q_{\xi+\eta}(l) \le Q_\eta(l).$$

e) Prove that if the distribution function of one of the random variables ξ or η is continuous, then the distribution function of $\xi+\eta$ is also continuous.

II.3.184. Let a random variable ξ have the Cauchy distribution with the distribution density

$$p(x) = \frac{1}{\pi}\frac{a}{x^2+a^2}, \qquad x \in \mathbf{R}^1.$$

Find $Q_\xi(l)$.

II.3.185. Let ξ_1 and ξ_2 be independent random variables. Prove that

a) for all nonnegative x_1 and x_2,

$$Q_{\xi_1}(x_1)Q_{\xi_2}(x_2) \le Q_{\xi_1+\xi_2}(x_1+x_2);$$

b) for any $x \ge 0$,

$$Q_{\xi_1+\xi_2}(x) \le Q_{\xi_1}(x)Q_{\xi_2}(x) + (1-Q_{\xi_1}(x))(1-Q_{\xi_2}(x)).$$

II.3.186. Suppose ξ and η are independent random variables and ξ has the probability density $p(x)$. Denote by $q(x)$ the probability density of the sum $\xi+\eta$. Prove that

$$\sup_{x\in\mathbf{R}} q(x) \le \sup_{x\in\mathbf{R}} p(x).$$

§II.4. The normal distribution on the plane

A random vector (ξ_1, ξ_2) has the normal distribution $N(a_1, \sigma_1, a_2, \sigma_2, \rho)$ if its distribution density is of the form

$$p_{(\xi_1,\xi_2)}(x,y) = \frac{1}{2\pi\sigma_1\sigma_2\sqrt{1-\rho^2}} \exp\{-K(x,y)\},$$

where

$$K(x,y) = \frac{1}{2(1-\rho^2)}\left[\frac{(x-a_1)^2}{\sigma_1^2} - 2\rho\frac{(x-a_1)(y-a_2)}{\sigma_1\sigma_2} + \frac{(y-a_1)^2}{\sigma_2^2}\right].$$

In this case

$$\begin{aligned} \mathsf{E}\,\xi_1 &= a_1, & \mathsf{E}\,\xi_2 &= a_2, \\ \operatorname{Var}\xi_1 &= \sigma_1^2, & \operatorname{Var}\xi_2 &= \sigma_2^2, \\ \mathsf{E}(\xi_1 - a_1)(\xi_2 - a_2) &= \rho\sigma_1\sigma_2. \end{aligned}$$

Problems

II.4.1. Suppose that a random vector (ξ_1, ξ_2) has the normal distribution $N(a_1, \sigma_1, a_2, \sigma_2, \rho)$. Prove that the random variables ξ_1 and ξ_2 have the normal distributions $N(a_1, \sigma_1^2)$ and $N(a_2, \sigma_2^2)$, respectively.

II.4.2. Let $\varphi_1(x, y)$ and $\varphi_2(x, y)$ be two densities of bivariate normal distributions on the plane, with zero expectations, unit variances, and different correlation coefficients. Prove that

a) the function $\frac{1}{2}(\varphi_1(x, y) + \varphi_2(x, y))$ is the distribution density of some random vector (ξ_1, ξ_2);
b) the random vector (ξ_1, ξ_2) is not normal;
c) each of the random variables ξ_1 and ξ_2 has the normal distribution $N(0, 1)$.

II.4.3. Let $u(x)$ be an odd continuous function on $(-\infty, +\infty)$ equal to zero outside the interval $(-1, 1)$, and let

$$|u(x)| \le \frac{1}{\sqrt{2\pi e}}.$$

Further, let

$$\varphi(x) = \frac{1}{\sqrt{2\pi}} e^{-x^2/2}.$$

Prove that

a) the function $\varphi(x)\varphi(y) + u(x)u(y)$ is the distribution density of some random vector (ξ_1, ξ_2);
b) the random vector (ξ_1, ξ_2) is not normally distributed;
c) each of the random variables ξ_1 and ξ_2 has the normal distribution.

II.4.4. Let

$$\varphi(x) = \frac{1}{\sqrt{2\pi}} e^{-x^2/2},$$

and let

$$f(x, y) = \begin{cases} 0, & (x, y) \in E_1 \cup E_3 \cup E_5 \cup E_7, \\ 2\varphi(x)\varphi(y), & (x, y) \in E_2 \cup E_4 \cup E_6 \cup E_8, \end{cases}$$

where the regions $E_1, \dots, E_8$ are defined as shown in Figure 7. Prove that $f(x, y)$ is the distribution density of some random vector (ξ_1, ξ_2) and the distribution density of each of the random variables ξ_1 and ξ_2 is normal.

REMARK. The results of Problems II.4.2–6 show that while each component of the random vector (ξ_1, ξ_2) is normally distributed, the vector itself can have a distribution other than normal.

II.4.5. Prove that any linear combination of the components of a normally distributed random vector has the normal distribution.

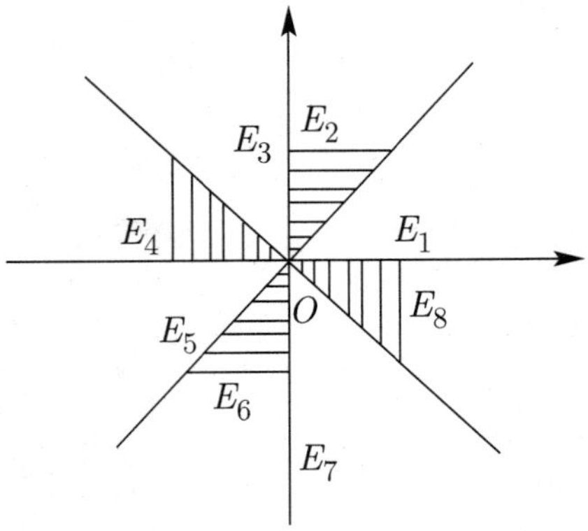

FIGURE 7

II.4.6. A random vector (ξ_1, ξ_2) is normally distributed on the plane. Prove that the random variables ξ_1 and ξ_2 are independent if and only if the correlation coefficient of ξ_1 and ξ_2 is zero.

II.4.7. Let $\xi_1, \dots, \xi_n$ be independent identically distributed random variables having the normal distribution $N(0,1)$. Also, let

$$\eta_1 = \sum_{k=1}^{n} a_k \xi_k \quad \text{and} \quad \eta_2 = \sum_{k=1}^{n} b_k \xi_k.$$

Prove that η_1 and η_2 are independent if and only if $\sum_{k=1}^{n} a_k b_k = 0$.

II.4.8. Let (ξ_1, ξ_2) be a normal random vector with $a_1 = a_2 = 0$. Put $\eta_1 = \xi_1 \cos\alpha + \xi_2 \sin\alpha$ and $\eta_2 = -\xi_1 \sin\alpha + \xi_2 \cos\alpha$. Prove that

a) the distribution density of the random vector (η_1, η_2) is equal to

$$p_{(\eta_1,\eta_2)}(y_1, y_2) = \frac{1}{2\pi\sigma_1\sigma_2\sqrt{1-\rho^2}} \exp\left\{-\frac{1}{2(1-\rho^2)}\left[Ay_1^2 - 2By_1y_2 + Cy_2^2\right]\right\},$$

where

$$A = \frac{\cos^2\alpha}{\sigma_1^2} - 2\rho\frac{\cos\alpha\sin\alpha}{\sigma_1\sigma_2} + \frac{\sin^2\alpha}{\sigma_2^2},$$

$$B = \frac{\cos\alpha\sin\alpha}{\sigma_1^2} - \rho\frac{\sin^2\alpha - \cos^2\alpha}{\sigma_1\sigma_2} - \frac{\cos\alpha\sin\alpha}{\sigma_2^2},$$

$$C = \frac{\sin^2\alpha}{\sigma_1^2} + 2\rho\frac{\sin\alpha\cos\alpha}{\sigma_1\sigma_2} + \frac{\cos^2\alpha}{\sigma_2^2}.$$

b) If $\tan 2\alpha = 2\rho\sigma_1\sigma_2/(\sigma_1^2 - \sigma_2^2)$, then η_1 and η_2 are independent.

II.4.9. Let (ξ_1, ξ_2) be a normally distributed random vector, and $a_1 = a_2 = 0$. Prove that the distribution density of the random variable $\eta = \xi_1/\xi_2$ is equal to

$$p_\eta(x) = \frac{\sigma_1\sigma_2\sqrt{1-\rho^2}}{\pi\left(\sigma_2^2x^2 - 2\rho\sigma_1\sigma_2x + \sigma_1^2\right)}.$$

II.4.10. Let (ξ_1, ξ_2) be a normally distributed random vector with $\mathsf{E}\,\xi_1 = a_1$, $\mathsf{E}\,\xi_2 = a_2$, $\operatorname{Var}\xi_1 = \sigma_1^2$, $\operatorname{Var}\xi_2 = \sigma_2^2$ (the correlation coefficient of ξ_1 and ξ_2 is equal to ρ). Put

$$Q(x_1, x_2) = \frac{1}{2(1-\rho^2)}\left[\left(\frac{x_1 - a_1}{\sigma_1}\right)^2 - 2\rho\frac{(x_1 - a_1)(x_2 - a_2)}{\sigma_1\sigma_2} + \left(\frac{x_2 - a_2}{\sigma_2}\right)^2\right].$$

Find the distribution function of the random variable $\eta = Q(\xi_1, \xi_2)$.

II.4.11. Prove that there is only one distribution of a random vector (ξ_1, ξ_2) with the following properties:

a) ξ_1 and ξ_2 are independent identically distributed random variables;

b) the distribution of (ξ_1, ξ_2) is spherically symmetric.

Show that the only distribution with these properties is the normal distribution on the plane with density

$$\frac{1}{2\pi\sigma_1\sigma_2}\exp\left\{-\frac{1}{2}\frac{x^2+y^2}{\sigma^2}\right\}.$$

II.4.12. Let (ξ_1, ξ_2) be a normally distributed random vector on the plane ($\mathsf{E}\,\xi_1 = \mathsf{E}\,\xi_2 = 0$), and let (R, Φ) be the polar coordinates of the point (ξ_1, ξ_2). Find

a) the joint distribution density of (R, Φ);

b) the distribution density of R;

c) the distribution density of Φ.

II.4.13. Let a random vector (ξ, η) have the normal distribution, and let $\mathsf{E}\,\xi = \mathsf{E}\,\eta = 0$, $\operatorname{Var}\xi = \operatorname{Var}\eta = 1$, $\mathsf{E}\,\xi\eta = \rho$. Prove that

$$\mathsf{P}(\xi\eta > 0) = \frac{1}{2} + \frac{1}{\pi}\arcsin\rho,$$

$$\mathsf{P}(\xi\eta < 0) = \frac{1}{\pi}\arccos\rho.$$

II.4.14. Let a random vector (ξ_1, ξ_2) have the normal distribution, and let $\mathsf{E}\,\xi_1 = \mathsf{E}\,\xi_2 = 0$, $\operatorname{Var}\xi_1 = \operatorname{Var}\xi_2 = 1$, $\mathsf{E}\,\xi\eta = \rho$. Prove that

$$\mathsf{E}\max(\xi_1, \xi_2) = \sqrt{\frac{1-\rho}{\pi}}.$$

II.4.15. Let (ξ_1, ξ_2) be a normally distributed random vector such that $\mathsf{E}\,\xi_1 = \mathsf{E}\,\xi_2 = 0$ and $\mathsf{E}\,\xi_1\xi_2 < 0$. Prove that for $a > 0$, $b > 0$,

$$\mathsf{P}(\xi_1 \geq a, \xi_2 \geq b) \leq \mathsf{P}(\xi_1 \geq a)\mathsf{P}(\xi_2 \geq b).$$

II.4.16. Let (ξ_1, ξ_2) and (η_1, η_2) be two normally distributed random vectors such that $\mathsf{E}\,\xi_k = \mathsf{E}\,\eta_k = 0$, $\mathsf{E}\,\xi_k^2 = \mathsf{E}\,\eta_k^2$, $k = 1, 2$, and $\mathsf{E}\,\xi_1\xi_2 \leq \mathsf{E}\,\eta_1\eta_2$. Prove that

$$\mathsf{P}(\max(\xi_1, \xi_2) > a) \geq \mathsf{P}(\max(\eta_1, \eta_2) > a).$$

II.4.17. (A. N. Shiryaev.) Let ξ, η, and ζ be independent random variables with the Gaussian distribution $N(0, 1)$. Prove that the random variable

$$\gamma = \frac{\xi + \eta\zeta}{\sqrt{1+\xi^2}}$$

has the Gaussian distribution.

§II.5. The Chebyshev inequality and other inequalities

Problems

II.5.1. Prove that if $\mathsf{E}\,\xi^2$ exists, then

$$\mathsf{P}(|\xi| > \varepsilon) \leq \frac{1}{\varepsilon^2}\,\mathsf{E}\,\xi^2.$$

II.5.2. The Chebyshev inequality. Prove that if $\mathsf{E}\,\xi^2$ exists and $\mathsf{E}\,\xi = a$, then

$$\mathsf{P}(|\xi - a| > \varepsilon) \leq \frac{\operatorname{Var}\xi}{\varepsilon^2}.$$

II.5.3. Let $\mathsf{E}\,\xi = 1$ and $\operatorname{Var}\xi = 0.04$. Using the Chebyshev inequality, estimate the probability that $0.5 < \xi < 1.5$.

II.5.4. Let $\mathsf{E}\,\xi = a$ and $\operatorname{Var}\xi = \sigma^2$. Estimate the probability $\mathsf{P}(|\xi - a| \leq 3\sigma)$.

II.5.5. Let $f(x)$ be a nonnegative nondecreasing function defined on the range of values of a random variable ξ, and let $\mathsf{E}\,f(\xi)$ exist. Prove that for any $\varepsilon > 0$,

$$\mathsf{P}(\xi > \varepsilon) \leq \frac{\mathsf{E}\,f(\xi)}{f(\varepsilon)}.$$

II.5.6. Let A be a Borel set on the real line, and let $f(x)$ be a Borel function such that

a) $f(x) \geq 0$ for $x \in (-\infty, +\infty)$;
b) $f(x) \geq \varepsilon > 0$ for $x \in A$;
c) $\mathsf{E}\,f(\xi)$ exists.

Prove that

$$\mathsf{P}(\xi \in A) \leq \frac{\mathsf{E}\,f(\xi)}{\varepsilon}.$$

II.5.7. Let A be a Borel set on the real line, and let $f(x)$ be a Borel function such that

a) $f(x) \leq 0$ if $x \notin A$;
b) $f(x) \leq 1$ if $x \in A$;
c) $\mathsf{E}\,f(\xi)$ exists.

Prove that

$$\mathsf{P}(\xi \in A) \geq \mathsf{E}\,f(\xi).$$

II.5.8. Assuming that $\mathsf{E}\,f(\xi)$ exists, prove that for $\mathsf{P}(|\xi| > \varepsilon)$ the following upper and lower bounds are valid:

a) $$\mathsf{P}(|\xi| > \varepsilon) \leq \frac{\mathsf{E}\,f(\xi)}{f(\varepsilon)}$$

if $f(x)$ is a nonnegative nondecreasing even function on (ε, ∞);

b) $$\mathsf{P}(|\xi| > \varepsilon) \geq \frac{\mathsf{E}\,f(\xi) - f(\varepsilon)}{k}$$

if $f(x)$ is a nonnegative nondecreasing bounded ($|f(x)| \leq k$) even function on $(0, \infty)$.

II.5.9. The Markov inequality. Prove that if $\mathsf{E}\,|\xi|^k$ exists, then

$$\mathsf{P}(|\xi| > \varepsilon) \le \frac{\mathsf{E}\,|\xi|^k}{\varepsilon^k}.$$

II.5.10. Prove that if $\mathsf{E}\,e^{c\xi}$ exists for some $c > 0$, then $\mathsf{P}(\xi > \varepsilon) \le e^{-c\xi}\,\mathsf{E}\,e^{c\xi}$.

II.5.11. Let $\mathsf{E}\,\xi = a$ and $\mathsf{E}\,|\xi - a|^r = \mu_r$. Prove that

a) $\mathsf{P}(|\xi - a| \ge \varepsilon) \le \dfrac{\mu_r}{\varepsilon^r}$;

b) $\mathsf{P}\left(\dfrac{|\xi - a|}{\mu_r^{1/r}} \ge \varepsilon\right) \le \dfrac{1}{\varepsilon^r}$.

II.5.12. Let $\mathsf{E}\,\xi = 0$ and $\mathsf{E}\,\xi^2 = \sigma^2$. Prove that

a) $\mathsf{P}(\xi > \varepsilon) \le \dfrac{\sigma^2}{\sigma^2 + \varepsilon^2}$ for $\varepsilon > 0$;

b) $\mathsf{P}(\xi < \varepsilon) \le \dfrac{\sigma^2}{\sigma^2 + \varepsilon^2}$ for $\varepsilon < 0$.

II.5.13. The Cantelli inequality. Let $\mathsf{E}\,\xi = a$ and $\mathsf{E}[\xi - a]^2 = \sigma^2$. Prove that

a) $\mathsf{P}(\xi - a > \varepsilon) \le \dfrac{\sigma^2}{\sigma^2 + \varepsilon^2}$ for $\varepsilon > 0$;

b) $\mathsf{P}(\xi - a < \varepsilon) \le \dfrac{\sigma^2}{\sigma^2 + \varepsilon^2}$ for $\varepsilon < 0$.

II.5.14. Let (ξ_1, ξ_2) be a random vector such that $\mathsf{E}\,\xi_i = a_i$, $\operatorname{Var}\xi_i = \sigma_i^2$, $i = 1, 2$, and let the correlation coefficient of ξ_1 and ξ_2 be equal to r. Prove that

$$\mathsf{P}\left(\max\left(\frac{|\xi_1 - a_1|}{\sigma_1}, \frac{|\xi_2 - a_2|}{\sigma_2}\right) \ge \varepsilon\right) \le \frac{1 - \sqrt{1 - r^2}}{\varepsilon^2}.$$

II.5.15. Prove that

a) for $x > 0$,

$$\int_x^\infty e^{-y^2/2}\,dy \le \frac{1}{x} e^{-x^2/2};$$

b) if ξ is a normally distributed random variable with $\mathsf{E}\,\xi = a$, $\operatorname{Var}\xi = \sigma^2$, then

$$\mathsf{P}(|\xi - a| > \varepsilon\sigma) \le \sqrt{\frac{2}{\pi}}\frac{1}{\varepsilon} e^{-\varepsilon^2/2}.$$

II.5.16. Let a random variable ξ have the normal distribution $N(a, \sigma^2)$. Estimate the probability $\mathsf{P}(|\xi - a| > 3\sigma)$.

II.5.17. Prove that

a) $\mathsf{P}(|\xi_1 + \xi_2| > \varepsilon) \le \mathsf{P}\left(|\xi_1| > \dfrac{\varepsilon}{2}\right) + \mathsf{P}\left(|\xi_2| > \dfrac{\varepsilon}{2}\right)$;

b) $\mathsf{P}\left(\left|\displaystyle\sum_{k=1}^m \xi_k\right| > \varepsilon\right) \le \displaystyle\sum_{k=1}^m \mathsf{P}\left(|\xi_k| > \frac{\varepsilon}{m}\right)$.

II.5.18. The Jensen inequality. *If g is a convex function and* $\mathsf{E}\, g(\xi)$ *exists, then* $\mathsf{E}\, g(\xi) \le g(\mathsf{E}\,\xi)$. *If g is a concave function, then* $g(\mathsf{E}\,\xi) \le \mathsf{E}\, g(\xi)$. Prove these assertions.

II.5.19. The Cauchy–Schwarz–Bunyakovskii inequality. *If ξ and η are arbitrary random variables such that* $\mathsf{E}\,\xi^2 < \infty$ *and* $\mathsf{E}\,\eta^2 < \infty$, *then* $\mathsf{E}\,\xi\eta$ *exists and* $(\mathsf{E}\,\xi\eta)^2 \le \mathsf{E}\,\xi^2\, \mathsf{E}\,\eta^2$. Prove this assertion.

II.5.20. The Hölder inequality. If ξ and η are arbitrary random variables and numbers $p > 1$ and $q > 1$ are such that

$$\frac{1}{p} + \frac{1}{q} = 1,$$

then

$$\mathsf{E}\,|\xi\eta| \le \left(\mathsf{E}\,|\xi|^p\right)^{1/p} \left(\mathsf{E}\,|\eta|^q\right)^{1/q}.$$

Prove this assertion.

II.5.21. Prove that for any random variable ξ the function $u(t) = \ln \mathsf{E}\,|\xi|^t$, $t \ge 0$, is convex with respect to t in any interval where it is finite.

II.5.22. (Yu. V. Kozachenko.) Let ξ be a random variable such that $\mathsf{E}\,\xi = 0$. The quantity

$$\tau(\xi) = \sup_{y>0} \sqrt{\frac{1}{2y^2} \ln \mathsf{E} \exp\{2y\xi\}}$$

is called *the positive normal bias of the random variable* ξ. The quantity

$$\tau^*(\xi) = \sup_{y>0} \sqrt{\frac{1}{2y^2} \ln \mathsf{E} \cosh\{2y\xi\}}$$

is called *the normal bias of the random variable* ξ.

If $\tau^*(\xi) < \infty$, then ξ is called *a sub-Gaussian random variable*. Prove that

a) for symmetric random variables, $\tau^*(\xi) = \tau(\xi)$;

b) if ξ has the normal distribution $N(0, \sigma^2)$, then $\tau(\xi) = \sigma$;

c) if $\lambda > 0$, then $\tau(\lambda\xi) = \lambda\tau(\xi)$;

d) for $x \ge 0$,

$$\mathsf{P}(\xi > x) \le \exp\left\{-\frac{x^2}{2\tau^2(\xi)}\right\}.$$

II.5.23. Let ξ_1 and ξ_2 be sub-Gaussian symmetric random variables. Prove that $\eta = \xi_1 + \xi_2$ is a sub-Gaussian random variable.

§II.6. Conditional probabilities and conditional expectations

Let $(\Omega, \mathfrak{A}, \mathsf{P})$ be a probability space, and let $\mathfrak{B}$ be an arbitrary σ-algebra contained in $\mathfrak{A}$.

The *conditional probability of an event A, given a σ-algebra* $\mathfrak{B}$, is the random variable $\mathsf{P}\left(A/\mathfrak{B}\right)$ with the following properties:

1) $\mathsf{P}\left(A/\mathfrak{B}\right)$ is a $\mathfrak{B}$-measurable function on Ω;

2) for every event $B \in \mathfrak{B}$,

$$\mathsf{P}(A \cap B) = \int_B \mathsf{P}\left(A/\mathfrak{B}\right)\, \mathsf{P}(d\omega). \tag{1}$$

By the Radon–Nikodym theorem, for a given set A there exists a class of random variables satisfying 1) and 2). However, if random variables $\mathsf{P}'\left(A/\mathfrak{B}\right)$ and $\mathsf{P}''\left(A/\mathfrak{B}\right)$ satisfy 1) and 2), then

$$\mathsf{P}(\mathsf{P}'\left(A/\mathfrak{B}\right) = \mathsf{P}''\left(A/\mathfrak{B}\right)) = 1.$$

Thus the definition of conditional probability gives rise to classes of random variables where any two variables of the same class coincide with probability 1. In what follows, by $\mathsf{P}\left(A/\mathfrak{B}\right)$ we mean a representative of the class. This remark applies equally to the definition of conditional expectation given below.

All random variables considered in this section are real.

The conditional probability of an event $A \in \mathfrak{A}$, given an event $B \in \mathfrak{A}$ such that $\mathsf{P}(B) > 0$, is the number

$$\widetilde{\mathsf{P}}(A) = \mathsf{P}\left(A/B\right) = \frac{\mathsf{P}(A \cap B)}{\mathsf{P}(A)}.$$

Note that

$$\mathsf{P}(\mathsf{P}\left(A/B\right) = \mathsf{P}\left(A/\mathfrak{B}\right)) = 1$$

if $\mathfrak{B} = \{\varnothing, B, \overline{B}, \Omega\}$. The function $\widetilde{\mathsf{P}}(A)$ defines a probability measure on $\mathfrak{A}$. Thus we have the new probability space $(\Omega, \mathfrak{A}, \widetilde{\mathsf{P}})$. Let ξ be a random variable such that $\mathsf{E}\,|\xi| < \infty$. By definition, the function

$$F_B(x) = \widetilde{\mathsf{P}}(\xi < x) = \mathsf{P}\left(\xi < x/B\right),$$

defined for every real number x, is *the conditional distribution function of the random variable ξ, given B.*

The random variable

$$\widetilde{\mathsf{E}}\,\xi = \mathsf{E}\left(\xi/B\right) = \int_{-\infty}^{\infty} x\, dF_B(x)$$

is called *the conditional expectation of the random variable ξ, given B*. For a Borel function $f(x)$, $x \in \mathbf{R}$, such that

$$\int_{-\infty}^{\infty} |f(x)|\, dF_B(x) < \infty,$$

we have

$$\widetilde{\mathsf{E}}(f(\xi)) = \mathsf{E}\left(f(\xi)/B\right) = \int_{-\infty}^{\infty} f(x)\, dF_B(x).$$

A more general definition of conditional expectation covers also certain events of probability 0. It goes as follows. *The conditional expectation of a random variable ξ, given a σ-algebra $\mathfrak{B}$*, is a random variable $\mathsf{E}\left(\xi/\mathfrak{B}\right)$ with the following properties:

1) $\mathsf{E}\left(\xi/\mathfrak{B}\right)$ is a $\mathfrak{B}$-measurable function on Ω;
2) for every set $B \in \mathfrak{B}$,

$$\int_B \xi(\omega)\, \mathsf{P}(d\omega) = \int_B \mathsf{E}\left(\xi/\mathfrak{B}\right)\, \mathsf{P}(d\omega). \tag{2}$$

The conditional probability $\mathsf{P}\left(A/\mathfrak{B}\right)$ is a particular case of the corresponding conditional expectation, namely:

$$\mathsf{P}\left(A/\mathfrak{B}\right) = \mathsf{E}\left(\chi_A/\mathfrak{B}\right),$$

where χ_A is the indicator of the set A.

The conditional probability of an event A and conditional expectation of a random variable η, given a random variable ξ, are defined, respectively, as the conditional probability of the event A and conditional expectation of the random variable η, given the σ-algebra generated by the random variable ξ.

Let $\{\eta_t, t \in T\}$ be a family of random variables, and let $\mathfrak{B}$ be the σ-algebra generated by this family. The conditional expectation of a random variable ξ, given the family $\{\eta_t, t \in T\}$, is defined as $\mathsf{E}\left(\xi/\mathfrak{B}\right)$ and is denoted by the symbol

$$\mathsf{E}\left(\xi/\eta_t, t \in T\right).$$

Problems

II.6.1. Find $\mathsf{E}\left(\xi/\mathfrak{B}\right)$ in the following cases:

a) $\mathfrak{B} = \mathfrak{A}$;

b) $\mathfrak{B} = \{\varnothing, \Omega\}$;

c) $\mathfrak{B} = \{\varnothing, A, \bar{A}, \Omega\}$, $0 < \mathsf{P}(A) < 1$.

Prove that for a $\mathfrak{B}$-measurable random variable ξ the equality $\mathsf{E}\left(\xi/\mathfrak{B}\right) = \xi$ holds with probability 1; in particular, for a constant c, $\mathsf{E}\left(c/\mathfrak{B}\right) = c$ with probability 1.

II.6.2. A random variable η takes finitely many values $x_1, x_2, \ldots, x_n$ and $\mathsf{P}(\eta = x_k) > 0$, $1 \le k \le n$. Find $\mathsf{P}\left(A/\eta\right)$ and $\mathsf{E}\left(\xi/\eta\right)$ for a random variable ξ such that $\mathsf{E}\,|\xi| < \infty$.

II.6.3. Let $\Omega = [0,1]$, $\mathfrak{A}$ the σ-algebra of Lebesgue measurable subsets of $[0,1]$, and P Lebesgue measure on Ω. Let $B_1 = [0, a_1]$, $B_2 = (a_1, a_2], \ldots, B_n = (a_{n-1}, 1]$, where $0 < a_1 < a_2 < \cdots < a_{n-1} < 1$ are fixed numbers, $\mathfrak{B}$ the minimal σ-algebra containing the sets $B_1, B_2, \ldots, B_n$, and η a random variable such that $\eta = c_1\chi_{B_1} + \cdots + c_n\chi_{B_n}$, where $c_1, c_2, \ldots, c_n$ are fixed numbers. Find $\mathsf{P}\left(A/B_k\right)$, $1 \le k \le n$, $\mathsf{P}\left(A/\mathfrak{B}\right)$, and $\mathsf{E}\left(\xi/\eta\right)$ for a random variable ξ such that $\mathsf{E}\,|\xi| < \infty$.

II.6.4. Let a random variable ξ have a positive distribution density f and $\mathsf{E}\,\xi$ exist. Let $\mathfrak{B}_1$ be the σ-algebra of Borel sets A such that the inclusion $x \in A$ implies the inclusion $x + k \in A$ for every integer k, and let $\mathfrak{B}_2$ be the σ-algebra of Borel sets symmetric about the point 0. Find the conditional expectations $\mathsf{E}\left(\xi/\mathfrak{B}_1\right)$ and $\mathsf{E}\left(\xi/\mathfrak{B}_2\right)$.

II.6.5. Prove that the following relations hold with probability 1:

a) $\mathsf{E}\left(\xi/\mathfrak{B}\right) \ge 0$ if $\mathsf{P}(\xi \ge 0) = 1$;

b) $\mathsf{E}\left(\xi/\mathfrak{B}\right) \ge \mathsf{E}\left(\eta/\mathfrak{B}\right)$ if $\mathsf{P}(\xi \ge \eta) = 1$ and $\mathsf{E}\,\xi$, $\mathsf{E}\,\eta$ exist;

c) $\left|\mathsf{E}\left(\xi/\mathfrak{B}\right)\right| \le \mathsf{E}\left(|\xi|/\mathfrak{B}\right)$ if $\mathsf{E}\,|\xi| < \infty$;

d) $\mathsf{E}\left(a\xi + b\eta/\mathfrak{B}\right) = a\,\mathsf{E}\left(\xi/\mathfrak{B}\right) + b\,\mathsf{E}\left(\xi/\mathfrak{B}\right)$, where a and b are real numbers.

II.6.6. Prove that $\mathsf{E}\left[\mathsf{E}\left(\xi/\mathfrak{B}\right)\right] = \mathsf{E}\,\xi$.

II.6.7. Let a random variable ξ and a σ-algebra $\mathfrak{B}$ be independent, i.e.,

$$\mathsf{P}(\{\omega : \xi < c\} \cap B) = \mathsf{P}(\xi < c)\mathsf{P}(B)$$

for any real c and any event $B \in \mathfrak{B}$. Prove that $\mathsf{E}\left(\xi/\mathfrak{B}\right) = \mathsf{E}\,\xi$ with probability 1.

II.6.8. Prove that two σ-algebras $\mathfrak{B}_1$ and $\mathfrak{B}_2$ are independent if and only if for every $\mathfrak{B}_1$-measurable random variable ξ such that $\mathsf{P}(\xi \geq 0) = 1$, the equality $\mathsf{E}\left(\xi/\mathfrak{B}_2\right) = \mathsf{E}\,\xi$ holds with probability 1.

II.6.9. Prove that if $\mathfrak{B}_1 \subset \mathfrak{B}_2$, then $\mathsf{E}\left[\mathsf{E}\left(\xi/\mathfrak{B}_2\right)/\mathfrak{B}_1\right] = \mathsf{E}\left(\xi/\mathfrak{B}_1\right)$ with probability 1.

II.6.10. Random variables ξ and η are such that $\mathsf{E}\,|\xi\eta| < \infty$ and $\mathsf{E}\,|\eta| < \infty$, and the random variable ξ is $\mathfrak{B}$-measurable. Prove that $\mathsf{E}\left(\xi\eta/\mathfrak{B}\right) = \xi\,\mathsf{E}\left(\eta/\mathfrak{B}\right)$ with probability 1.

II.6.11. Let $\mathfrak{B}_1 \subset \mathfrak{B}_2$, and let ξ and η be random variables such that $\mathsf{E}\,\xi\eta$ exists and ξ is $\mathfrak{B}_2$-measurable. Prove that $\mathsf{E}\left(\xi\eta/\mathfrak{B}_1\right) = \mathsf{E}\left[\xi\,\mathsf{E}\left(\eta/\mathfrak{B}_2\right)/\mathfrak{B}_1\right]$ with probability 1.

II.6.12. Let ξ be a random variable such that $\mathsf{E}\,|\xi| < \infty$. Prove that the conditional expectation $\mathsf{E}\left(\xi/\mathfrak{B}\right)$ can be defined as a $\mathfrak{B}$-measurable random variable ζ such that the equality $\mathsf{E}\,\xi\eta = \mathsf{E}\,\zeta\eta$ holds with probability 1 for any bounded $\mathfrak{B}$-measurable random variable η.

II.6.13. Let $\{\xi_n, n \geq 1\}$ be a sequence of random variables such that

$$\mathsf{P}(\xi_1 \geq 0) = 1 \qquad \text{and} \qquad \mathsf{P}(\xi_{n+1} \geq \xi_n) = 1, \quad n \geq 1.$$

Prove that $\lim_{n\to\infty} \mathsf{E}\left(\xi_n/\mathfrak{B}\right) = \mathsf{E}\left(\lim_{n\to\infty} \xi_n/\mathfrak{B}\right)$ with probability 1.

II.6.14. Let $\{\xi_n, n \geq 1\}$ be a sequence of random variables such that

$$\mathsf{P}(|\xi_n| \leq \eta) = 1, \qquad n \geq 1,$$

for some random variable η, $\mathsf{E}\,|\eta| < \infty$, and let $\mathsf{P}(\xi_n \to \xi, n \to \infty) = 1$ for some random variable ξ. Prove that $\lim_{n\to\infty} \mathsf{E}\left(\xi_n/\mathfrak{B}\right) = \mathsf{E}\left(\xi/B\right)$ with probability 1.

II.6.15. Let $\{\xi_n, n \geq 1\}$ be a sequences of random variables such that

$$\mathsf{P}(\xi_n \leq \xi) = 1, \qquad n \geq 1,$$

where ξ is a random variable such that $\mathsf{E}\,|\xi| < \infty$. Prove that

$$\mathsf{E}\left(\limsup_{n\to\infty} \xi_n/\mathfrak{B}\right) \geq \limsup_{n\to\infty} \mathsf{E}\left(\xi_n/\mathfrak{B}\right)$$

with probability 1.

Similarly, if $\mathsf{P}(\xi_n \geq \eta) = 1$, $n \geq 1$, where η is a random variable such that $\mathsf{E}\,|\eta| < \infty$, then

$$\mathsf{E}\left(\liminf_{n\to\infty} \xi_n/\mathfrak{B}\right) \leq \liminf_{n\to\infty} \mathsf{E}\left(\xi_n/\mathfrak{B}\right)$$

with probability 1.

II.6.16. Let $\{\xi_n, n \geq 1\}$ be a sequence of random variables such that

$$\mathsf{P}(\xi_n \geq 0) = 1, \qquad n \geq 1.$$

Prove that $\mathsf{E}\left(\sum_{n=1}^{\infty} \xi_n/\mathfrak{B}\right) = \sum_{n=1}^{\infty} \mathsf{E}\left(\xi_n/\mathfrak{B}\right)$ with probability 1.

II.6.17. Let f be a real Borel function on the plane, and let ξ and η be random variables, where ξ is measurable with respect to a σ-algebra $\mathfrak{B}$ and $\mathsf{E}\,|f(\xi,\eta)| < \infty$. Prove that

$$\mathsf{E}\left(f(\xi,\eta)/\mathfrak{B}\right) = \mathsf{E}\left(f(x,\eta)/\mathfrak{B}\right)\Big|_{x=\xi}$$

with probability 1.

II.6.18. Let ξ and η be independent random variables, $\mathfrak{B}$ the σ-algebra generated by η, and f a real Borel function on the plane such that $\mathsf{E}\,|f(\xi,\eta)| < \infty$. Prove that

$$\mathsf{E}\left(f(\xi,\eta)/\mathfrak{B}\right) = \left[\mathsf{E}\, f(\xi,y)\right]\Big|_{y=\eta}$$

with probability 1.

II.6.19. Prove the total probability formula

$$\mathsf{P}(A) = \int_\Omega \mathsf{P}\left(A/\mathfrak{B}\right)\,\mathsf{P}(d\omega) = \mathsf{E}\,\mathsf{P}\left(A/\mathfrak{B}\right).$$

II.6.20. Prove that the relations

a) $0 \le \mathsf{P}\left(A/\mathfrak{B}\right) \le 1$;

b) $\mathsf{P}\left(A/\mathfrak{B}\right) = 0$ if $\mathsf{P}(A) = 0$;

c) $\mathsf{P}\left(A/\mathfrak{B}\right) = 1$ if $\mathsf{P}(A) = 1$;

d) $\lim_{n\to\infty} \mathsf{P}\left(A_n/\mathfrak{B}\right) = \mathsf{P}\left(A/\mathfrak{B}\right)$ if $A_n \subset A_{n+1}$, $n \ge 1$, $A = \bigcup_{n=1}^\infty A_n$ or $A_n \supset A_{n+1}$, $n \ge 1$, $A = \bigcap_{n=1}^\infty A_n$;

e) $\mathsf{P}\left(\bigcup_{n=1}^\infty A_n/\mathfrak{B}\right) = \sum_{n=1}^\infty \mathsf{P}\left(A_n/\mathfrak{B}\right)$ if $A_k \cap A_j = \varnothing$, $k \ne j$,

hold with probability 1.

The conditional probability $\mathsf{P}\left(A/\mathfrak{B}\right) = \mathsf{P}\left(A/\mathfrak{B}\right)(\omega)$ is a function of an event $A \in \mathfrak{A}$ and an elementary event $\omega \in \Omega$. Prove that the equality e) does not imply that $\mathsf{P}\left(A/\mathfrak{B}\right)$ is σ-additive with respect to A for a fixed ω.

II.6.21. An event A and a σ-algebra $\mathfrak{B}$ are independent, i.e., $\mathsf{P}(A\cap B) = \mathsf{P}(A)\mathsf{P}(B)$ for $B \in \mathfrak{B}$. Prove that $\mathsf{P}\left(A/\mathfrak{B}\right) = \mathsf{P}(A)$ with probability 1.

II.6.22. The space $(\Omega, \mathfrak{A}, \mathsf{P})$ is the real line $\mathbf{R}$ with the σ-algebra $\mathfrak{A}$ of Borel sets, and $\mathsf{P}(A) = \int_A f(x)\,dx$, where f is a continuous positive function such that $\int_{-\infty}^\infty f(x)\,dx = 1$. Let $\mathfrak{B}_1$ be the σ-algebra of sets of the form

$$\{x\colon \sin x \in B\}, \qquad B \in \mathfrak{A},$$

and let $\mathfrak{B}_2$ be the σ-algebra of sets of the form

$$\{x\colon g(x) \in B\}, \qquad B \in \mathfrak{A},$$

where g is a piecewise monotone function. Find $\mathsf{P}\left(A/\mathfrak{B}_1\right)$ and $\mathsf{P}\left(A/\mathfrak{B}_2\right)$, $A \in \mathfrak{A}$.

II.6.23. Let $\mathfrak{B}_1$ and $\mathfrak{B}_2$ be two σ-algebras such that $\mathfrak{B}_1 \subset \mathfrak{B}_2$. Prove that

$$\mathsf{E}\left[\mathsf{P}\left(A/\mathfrak{B}_2\right)/\mathfrak{B}_1\right] = \mathsf{P}\left(A/\mathfrak{B}_1\right)$$

with probability 1.

II.6.24. A random variable ξ and an event A such that $\mathsf{P}(A) > 0$ are independent. Prove that

a) $F_A(x) = F(x) = \mathsf{P}(\xi < x)$, where $x \in \mathbf{R}$ and F_A is the conditional distribution function of the random variable ξ, given A;

b) $\mathsf{E}(\xi/A) = \mathsf{E}\,\xi$ if additionally $\mathsf{E}\,|\xi| < \infty$.

II.6.25. For a random variable ξ such that $\mathsf{E}\,|\xi| < \infty$ and an event A such that $\mathsf{P}(A) > 0$, prove that

$$\int_{-\infty}^{\infty} |x|\, dF_A(x) < +\infty.$$

II.6.26. Under the conditions of the previous problem prove the equality

$$\mathsf{E}(\xi/A) = \frac{1}{\mathsf{P}(A)} \int_A \xi(\omega)\, \mathsf{P}(d\omega).$$

II.6.27. Under the conditions of Problem II.6.25, prove the inequalities

$$|\mathsf{E}(\xi/A)| \le \mathsf{E}(|\xi|/A) \le \frac{\mathsf{E}\,|\xi|}{\mathsf{P}(A)}.$$

II.6.28. A random variable ξ has the positive distribution density f, and $\mathsf{E}\,|\xi| < \infty$. Find the expectations

$$\mathsf{E}(\xi/\xi > 0) \quad \text{and} \quad \mathsf{E}(\xi/a \le \xi \le b).$$

II.6.29. Let $\mathfrak{B}$ be the σ-algebra generated by a complete group of events B_n, $n \ge 1$: $B_j \cap B_k = \varnothing$, $j \ne k$; $\bigcup_{n=1}^{\infty} B_n = \Omega$; $\mathsf{P}(B_n) > 0$, $n \ge 1$. Prove that for a random variable ξ such that $\mathsf{E}\,|\xi| < \infty$ the equalities

$$\mathsf{E}(\xi/\mathfrak{B}) = \sum_{n=1}^{\infty} \mathsf{E}(\xi/B_n)\, \chi_{B_n},$$

$$\mathsf{E}\,\xi = \sum_{n=1}^{\infty} \mathsf{E}(\xi/B_n)\, \mathsf{P}(B_n)$$

hold with probability 1.

II.6.30. Let $\nu, \xi_1, \xi_2, \ldots, \xi_n, \ldots$ be independent random variables such that ν takes only positive integral values and $\mathsf{E}\,\nu < \infty$. Let the random variables $\{\xi_n, n \ge 1\}$ be identically distributed with $\mathsf{E}\,|\xi_1| < \infty$. Prove the equality

$$\mathsf{E}(\xi_1 + \cdots + \xi_\nu) = \mathsf{E}\,\nu\, \mathsf{E}\,\xi_1.$$

II.6.31. Let $\{\tau_n, n \ge 1\}$ and $\{\xi_n, n \ge 1\}$ be independent sequences of independent and identically distributed, within the sequence, random variables, τ_1 having the exponential distribution with parameter $\lambda > 0$, $\mathsf{E}\,\xi_1 = m$, $\mathsf{E}(\xi_1 - m)^2 = \sigma^2 < \infty$. For every $t \ge 0$ define the random variable $\xi(t)$ by putting $\xi(t) = \xi_k$ for

$$\sum_{j=0}^{k-1} \tau_j \le t < \sum_{j=0}^{k} \tau_j, \qquad \tau_0 = 0.$$

Calculate

$$m(t) = \mathsf{E}\,\xi(t),$$
$$r(s,t) = \mathsf{E}[\xi(t) - m(t)][\xi(s) - m(s)], \qquad s,t \geq 0.$$

II.6.32. Let ξ be a random variable such that $\mathsf{E}\,|\xi| < \infty$, η a random variable taking values y_n, $n \geq 1$, and $\mathsf{P}(\eta = y_n) > 0$, $n \geq 1$. Calculate $\mathsf{E}\left(\xi/\eta\right)$.

II.6.33. Let $\mathfrak{B}$ be the σ-algebra generated by a complete group of events B_n, $n \geq 1$ (see Problem II.6.29). Prove that

$$\mathsf{P}\left(A/\mathfrak{B}\right) = \sum_{n=1}^{\infty} \mathsf{P}\left(A/B_n\right)\chi_{B_n}, \qquad A \in \mathfrak{A},$$

with probability 1.

II.6.34. The Jensen inequality. Let f be a continuous convex real function on the real axis. Let $\mathsf{E}\,|\xi| < \infty$ and $\mathsf{E}\,|f(\xi)| < \infty$. Prove that $f\left(\mathsf{E}\left(\xi/\mathfrak{B}\right)\right) \leq \mathsf{E}\left(f(\xi)/\mathfrak{B}\right)$ with probability 1.

II.6.35. Let ξ be a random variable such that $\mathsf{E}\,|\xi|^r < \infty$ for some $r \geq 1$. Prove the inequality $\mathsf{E}\left|\mathsf{E}\left(\xi/\mathfrak{B}\right)\right|^r \leq \mathsf{E}\,|\xi|^r$.

II.6.36. Let ξ be a random variable such that $\mathsf{E}\,|\xi|^r < \infty$ for some $r \geq 1$, and let $\mathfrak{M}$ be the family of all σ-algebras $\mathfrak{B}$ such that $\mathfrak{B} \subset \mathfrak{A}$. Prove that the family of random variables $\{|\,\mathsf{E}\left(\xi/\mathfrak{B}\right)|^r, \mathfrak{B} \in \mathfrak{M}\}$ is uniformly integrable, i.e.,

$$\lim_{a\to\infty} \sup_{\mathfrak{B}\in\mathfrak{M}} \int_{\{|\,\mathsf{E}(\xi/\mathfrak{B})|^r > a\}} \left|\mathsf{E}\left(\xi/\mathfrak{B}\right)\right|^r \mathsf{P}(d\omega) = 0.$$

II.6.37. For a random variable ξ such that $\mathsf{E}\,\xi^2 < \infty$, put

$$\operatorname{Var}\left(\xi/\mathfrak{B}\right) = \mathsf{E}\left(\left[\xi - \mathsf{E}\left(\xi/\mathfrak{B}\right)\right]^2/\mathfrak{B}\right).$$

Prove the equality

$$\operatorname{Var}\xi = \mathsf{E}\operatorname{Var}\left(\xi/\mathfrak{B}\right) + \operatorname{Var}\left(\mathsf{E}\left(\xi/B\right)\right).$$

II.6.38. For random variables ξ and η such that $\mathsf{E}\,\xi^2 < \infty$ and $\mathsf{E}\,\eta^2 < \infty$, prove the equality

$$\mathsf{E}\left[\xi\,\mathsf{E}\left(\eta/\mathfrak{B}\right)\right] = \mathsf{E}\left[\eta\,\mathsf{E}\left(\xi/\mathfrak{B}\right)\right].$$

II.6.39. Let ξ be a random variable such that $\mathsf{E}\,\xi^2 < \infty$, $\mathfrak{B}$ a σ-algebra, and $\mathfrak{M}$ the family of all $\mathfrak{B}$-measurable random variables η such that $\mathsf{E}\,\eta^2 < \infty$. Prove that

$$\inf_{\eta\in\mathfrak{M}} \mathsf{E}(\xi - \eta)^2 = \mathsf{E}\left[\xi - \mathsf{E}\left(\xi/\mathfrak{B}\right)\right]^2.$$

II.6.40. Let ξ and η be random variables, and let $\mathsf{E}\,\xi^2 < \infty$. Prove that for any real Borel function f on the real line,

$$\mathsf{E}[\xi - f(\eta)]^2 \geq \mathsf{E}\left[\xi - \mathsf{E}\left(\xi/\eta\right)\right]^2.$$

II.6.41. A sequence of random variables $\{\xi_n, n \geq 1\}$ converges in mean of order $r \geq 1$ to a random variable ξ, i.e., $\mathsf{E}\,|\xi_n|^r < \infty$, $n \geq 1$, $\mathsf{E}\,|\xi|^r < \infty$; $\mathsf{E}\,|\xi_n - \xi|^r \to 0$ as $n \to \infty$. Prove that the sequence $\{\mathsf{E}\,(\xi_n/\mathfrak{B})\,, n \geq 1\}$ converges in mean of order r to $\mathsf{E}\,(\xi/\mathfrak{B})$.

II.6.42. Assume that $\Omega = \mathbf{R}^m$, $\mathfrak{A}$ is the σ-algebra of Borel sets on $\mathbf{R}^m$, and a probability P on $\mathfrak{A}$ is given by a distribution function $F(x_1, \ldots, x_m)$ symmetric with respect to its arguments. A set $A \in \mathfrak{A}$ is called *symmetric* if, along with a point $(x_1, x_2, \ldots, x_m) \in A$, the point $(x_{k_1}, x_{k_2}, \ldots, x_{k_m})$ also belongs to A for an arbitrary permutation $k_1, k_2, \ldots, k_m$ of the numbers $1, 2, \ldots, m$.

Prove that the set $\mathfrak{B}$ of all symmetric sets is a σ-algebra.

Let φ be a real Borel function on $\mathbf{R}^m$ such that $\mathsf{E}\,|\varphi| < \infty$. Calculate $\mathsf{E}\,(\varphi/\mathfrak{B})$. Consider the examples:

a) $\varphi(x_1, \ldots, x_m) = x_1$;

b) $\varphi(x_1, \ldots, x_m) = x_1^2 - x_1 x_2$.

II.6.43. In addition to the conditions of Problem II.6.42, assume that $\mathsf{E}\,\varphi^2 < \infty$. Prove the inequality $\mathsf{E}\left[\mathsf{E}\,(\varphi/\mathfrak{B})\right]^2 \leq \mathsf{E}\,\varphi^2$.

II.6.44. Let ξ and η be random variables, $\mathsf{E}\,|\xi| < \infty$, and let $\mathfrak{B}$ be the σ-algebra generated by the random variable η. Prove that for a $\mathfrak{B}$-measurable random variable ξ (for instance, for functions of η) the equality $\mathsf{E}\,(\xi/\eta) = \xi$ holds with probability 1.

II.6.45. If random variables ξ and η are independent and $\mathsf{E}\,|\xi| < \infty$, then the equality $\mathsf{E}\,(\xi/\eta) = \mathsf{E}\,\xi$ holds with probability 1.

II.6.46. Prove that for any sequences of random variables $\{\eta_n, n \geq 1\}$ and Borel functions $\{g_n\}$, and any random variable ξ such that $\mathsf{E}\,|\xi| < \infty$, the equality

$$\mathsf{E}\left[\mathsf{E}\,(\xi/\eta_n, n \geq 1)\,/g_n(\eta_1, \ldots, \eta_n), n \geq 1\right] = \mathsf{E}\,(\xi/g_n(\eta_1, \ldots, \eta_n), n \geq 1)$$

holds with probability 1.

II.6.47. Random variables ξ_k, $1 \leq k \leq n$, are independent, identically distributed, and $\mathsf{E}\,|\xi_1| < \infty$. Let

$$\bar{\xi} = \frac{1}{n}\sum_{k=1}^{n} \xi_k.$$

Prove that for every k, $1 \leq k \leq n$, the equality

$$\mathsf{E}\,(\xi_k/\bar{\xi}) = \bar{\xi}$$

holds with probability 1.

II.6.48. Let ξ and η be random variables, and let $\mathsf{E}\,|\xi| < \infty$. For a Borel set $A \subset \mathbf{R}$ define the function

$$\nu(A) = \int_{\{\eta \in A\}} \xi\,\mathsf{P}(d\omega),$$

where the integral is taken over the elementary events ω with $\eta(\omega) \in A$. Prove that ν is a signed measure, i.e., the difference of two finite measures on the σ-algebra S of Borel sets in $\mathbf{R}$.

Let $F(A), A \in S$, be Lebesgue–Stieltjes measure generated by the distribution function $F(x) = \mathsf{P}(\eta < x)$ of the random variable η. Prove that $\nu \ll F$, i.e., there exists a Borel function g on $\mathbf{R}$ such that

$$\nu(A) = \int_A g(y)\, dF(y), \qquad A \in S.$$

The random variable $g(y)$ is called *the conditional expectation of the random variable ξ, given $\eta = y$*, and is denoted by the symbol $\mathsf{E}\left(\xi/\eta = y\right)$.

II.6.49. Prove that
a) $\mathsf{E}\,\xi = \int_{-\infty}^{\infty} \mathsf{E}\left(\xi/\eta = y\right)\, dF(y)$;
b) $g(\eta) = \mathsf{E}\left(\xi/\eta\right)$ with probability 1.

II.6.50. For an event $A \in \mathfrak{A}$ we put $\mathsf{P}\left(A/\eta = y\right) = \mathsf{E}\left(\chi_A/\eta = y\right)$ (see Problem II.6.48). Let the random variables ξ and η have a continuous joint distribution density f such that

$$f_2(y) = \int_{-\infty}^{\infty} f(x, y)\, dx > 0, \qquad y \in \mathbf{R}.$$

Also let $\mathsf{E}\,|\xi| < \infty$. Prove that

$$\text{a) } \mathsf{E}\left(\xi/\eta = y\right) = \frac{1}{f_2(y)} \int_{-\infty}^{\infty} x f(x, y)\, dx,$$

$$\text{b) } \mathsf{P}\left(\xi < x/\eta = y\right) = \frac{1}{f_2(y)} \int_{-\infty}^{x} f(u, y)\, du.$$

The function

$$\mathsf{P}\left(\xi < x/\eta = y\right) = \frac{1}{f_2(y)} \int_{-\infty}^{x} f(u, y)\, du, \qquad x \in \mathbf{R},$$

is called *the conditional distribution function of the random variable ξ, given $\eta = y$*.

The function

$$f(x/y) = \frac{1}{f_2(y)} f(x, y), \qquad x \in \mathbf{R},$$

is called *the conditional distribution density of the random variable ξ, given $\eta = y$*.

II.6.51. Under the condition of Problem II.6.50 for $\Delta y > 0$, calculate
a) $\mathsf{E}\left(\xi/y - \Delta y \le \eta < y + \Delta y\right)$;
b) $\mathsf{P}\left(\xi < x/y - \Delta y \le \eta < y + \Delta y\right)$
and prove that
c) $\lim_{\Delta y \to 0} \mathsf{E}\left(\xi/y - \Delta y \le \eta < y + \Delta y\right) = \mathsf{E}\left(\xi/\eta = y\right)$;
d) $\lim_{\Delta y \to 0} \mathsf{P}\left(\xi < x/y - \Delta y \le \eta < y + \Delta y\right) = \mathsf{P}\left(\xi < x/\eta = y\right)$.

II.6.52. Under the condition of Problem II.6.50 prove the equality

$$\mathsf{P}(\xi \in A, \eta \in B) = \int_A \left[\int_B f(u/v) f_2(v)\, dv\right] du.$$

II.6.53. Under the condition of Problem II.6.50 find $\mathsf{E}\left(\xi/\xi + \eta \in B\right)$, where B is a Borel set on $\mathbf{R}$.

II.6.54. Random variables ξ and η are independent and each of them has the normal distribution with mean 0 and variance 1. Let ρ and φ be the polar coordinates of the point (ξ, η), $\rho \geq 0$, $0 \leq \varphi < 2\pi$. Prove that

$$\mathsf{P}\left(\varphi < \tau / \rho = r\right) = \frac{\tau}{2\pi}, \qquad 0 \leq \tau \leq 2\pi,\ r > 0.$$

II.6.55. A random variable ξ_1 has the uniform distribution on $[0,1]$. If $\xi_1 = x_1$, then ξ_2 is a random variable uniformly distributed on $[x_1, x_1+1]$. If $\xi_2 = x_2$, then ξ_3 is a random variable uniformly distributed on $[x_2, x_2+1]$. Random variables $\xi_4, \xi_5, \dots, \xi_n$ are similarly defined for all $n \geq 4$. Calculate $\mathsf{E}\,\xi_n$.

II.6.56. A random variable ξ_1 is uniformly distributed on $[0,1]$. If $\xi_1 = x_1$, then ξ_2 is a random variable uniformly distributed on $[x_1, 1]$. If $\xi_2 = x_2$, then ξ_3 is a random variable uniformly distributed on $[x_2, 1]$. Random variables $\xi_4, \xi_5, \dots, \xi_n$ are similarly defined for all $n \geq 4$. Calculate $\mathsf{E}\,\xi_n$.

II.6.57. Random variables $\xi_1, \xi_2, \dots, \xi_n$ are independent, and each of them has the uniform distribution on the segment $[0, 2a]$, $a > 0$. Let

$$\bar{\xi} = \frac{1}{n} \sum_{k=1}^{n} \xi_k.$$

Prove that

a) $\mathsf{E}\,\xi_k = \mathsf{E}\,\bar{\xi} = a$;

b) $\mathsf{E}\left(\bar{\xi} / \max_{1 \leq k \leq n} \xi_k\right) = \frac{1}{2}(n+1)n^{-1} \max_{1 \leq k \leq n} \xi_k$.

II.6.58. Random variables $\xi_1, \xi_2, \dots, \xi_n$ are independent, and each of them has the uniform distribution on the segment $[0,1]$; $\xi_1^* \leq \xi_2^* \leq \dots \leq \xi_n^*$ are these random variables arranged in nondecreasing order. Prove that the conditional distribution density of the random variables $\xi_1^*, \xi_2^*, \dots, \xi_{n-1}^*$, given $\xi_n^* = y$, is equal to

$$f(x_1, x_2, \dots, x_{n-1}/y) = \begin{cases} \dfrac{(n-1)!}{y^{n-1}}, & 0 < x_1 < x_2 < \dots < x_{n-1} < y, \\ 0, & \text{otherwise.} \end{cases}$$

II.6.59. Random variables ξ and η have the joint bivariate normal distribution with density

$$f(x,y) = \frac{1}{2\pi} \frac{1}{\sigma_1 \sigma_2 \sqrt{1-\rho^2}} \exp\left\{-\frac{K(x,y)}{2(1-\rho^2)}\right\},$$

where

$$K(x,y) = \left(\frac{x-m_1}{\sigma_1}\right)^2 - 2\rho \frac{x-m_1}{\sigma_1} \frac{y-m_2}{\sigma_2} + \left(\frac{y-m_2}{\sigma_2}\right)^2$$

and $m_1, m_2 \in \mathbf{R}$, $\sigma_1 > 0$, $\sigma_2 > 0$, $|\rho| \leq 1$. Prove that

a) $f(x/y) = \dfrac{1}{\sigma_1 \sqrt{2\pi(1-\rho^2)}} \exp\left\{-\dfrac{1}{2\sigma_1^2(1-\rho^2)} \left[x - m_1 - \rho \dfrac{\sigma_1}{\sigma_2}(y - m_2)\right]^2\right\}$;

b) $\mathsf{E}\left(\xi / \eta = y\right) = m_1 + \rho \dfrac{\sigma_1}{\sigma_2}(y - m_2)$.

II.6.60. Let ξ and η be random variables as in the previous problem but with $m_1 = m_2 = 0$ and $\sigma_1 = \sigma_2 = 1$. Prove the equalities:

a) $\mathsf{E}\left(\xi^2/\eta^2 = y^2\right) = \frac{1}{2}\left[\mathsf{E}\left(\xi^2/\eta = y\right) + \mathsf{E}\left(\xi^2/\eta = -y\right)\right]$, $y \geq 0$,

b) $\mathsf{E}\left(\xi^2/\eta^2 = y^2\right) = 1 + \rho^2(y^2 - 1)$.

II.6.61. Let $\{\xi_n, n \geq 1\}$ be a sequence of independent random variables, and let $\eta_n = \xi_1 + \cdots + \xi_n$. Fix $n > 1$. Prove that under the condition $\eta_n = y$ the random variables η_k and η_m with fixed indices $k < n < m$ are independent, i.e.,

$$\mathsf{P}\left(\eta_k < x_1, \eta_m < x_2/\eta_n = y\right) = \mathsf{P}\left(\eta_k < x_1/\eta_n = y\right)\mathsf{P}\left(\eta_m < x_2/\eta_n = y\right).$$

II.6.62. A sequence of random variables $\{\xi_n, n \geq 1\}$ and a σ-algebra $\mathfrak{B}$ are such that $\mathsf{E}\left(|\xi_n|/\mathfrak{B}\right) \to 0$ in probability as $n \to \infty$. Prove that $\xi_n \to 0$ in probability as $n \to \infty$.

II.6.63. Let ξ be a random variable such that $\mathsf{E}\,|\xi| < \infty$, and let $\{\mathfrak{B}_n, n \geq 1\}$ be a sequences of σ-algebras such that $\mathfrak{B}_n \subset \mathfrak{B}_{n+1} \subset \mathfrak{A}$, $n \geq 1$. Assume that $\alpha > 0$, and $m \geq 1$ is an integer. Using the relation

$$\left\{\omega\colon \max_{1\leq k\leq m} \mathsf{E}\left(|\xi|/\mathfrak{B}_k\right) \geq \alpha\right\}$$
$$= \bigcup_{k=1}^{m} \left\{\omega\colon \mathsf{E}\left(|\xi|/\mathfrak{B}_j\right) < \alpha, 1 \leq j < k; \mathsf{E}\left(|\xi|/\mathfrak{B}_k\right) \geq \alpha\right\},$$

prove the inequality

$$\mathsf{P}\Big(\max_{1\leq k\leq m} \mathsf{E}\left(|\xi|/\mathfrak{B}_k\right) \geq \alpha\Big) \leq \frac{\mathsf{E}\,|\xi|}{\alpha}.$$

II.6.64. A transformation $T\colon \Omega \to \Omega$ is called *a one-to-one probability preserving transformation* if T is a one-to-one transformation of Ω onto Ω that transfers the sets $A \in \mathfrak{A}$ to $TA \in \mathfrak{A}$ such that $\mathsf{P}(TA) = \mathsf{P}(A)$. A set $A \in \mathfrak{A}$ is called *invariant with respect the transformation* T if

$$\mathsf{P}((A \setminus TA) \cup (TA \setminus A)) = 0.$$

a) Prove that the family of all invariant sets is a σ-algebra.

b) For a random variable ξ define $T\xi$ as $T\xi(\omega) = \xi(T^{-1}\omega)$. Verify that $\mathsf{E}(T\xi) = \mathsf{E}\,\xi$, provided that $\mathsf{E}\,|\xi| < \infty$.

c) Put

$$\eta_n = \frac{1}{n}\sum_{k=0}^{n-1} T^k\xi, \qquad n \geq 1.$$

Assume that $\mathsf{E}\,|\xi| < \infty$ and $\mathsf{E}\,|\eta_n - \eta| \to 0$ as $n \to \infty$ for some random variable η such that $\mathsf{E}\,|\eta| < \infty$. Prove that $\eta = \mathsf{E}\left(\xi/\mathfrak{B}\right)$ with probability 1.

II.6.65. Let ξ be a random variable such that $\mathsf{E}\,|\xi| < \infty$, and let $\{\mathfrak{B}_n, n \geq 1\}$ be a sequence of σ-algebras such that $\mathfrak{B}_n \subset \mathfrak{B}_{n+1} \subset \mathfrak{A}$, $n \geq 1$. Prove that $\lim_{n\to\infty} \mathsf{E}\left(\xi/\mathfrak{B}_n\right) = \mathsf{E}\left(\xi/\mathfrak{B}\right)$ with probability 1, where $\mathfrak{B}$ is the minimal σ-algebra containing all $\mathfrak{B}_n$, $n \geq 1$.

CHAPTER III

Sequences of Random Events and Sequences of Random Variables

§III.1. Sequences of independent events. The Borel–Cantelli lemma. 0-1 law

Let $(\Omega, \mathfrak{A}, \mathsf{P})$ be a probability space. Throughout this chapter, all events and families of events are assumed to belong to $\mathfrak{A}$.

Events $A_1, \ldots, A_n$ are said to be *independent* if

$$\mathsf{P}\Big(\bigcap_{k=1}^{m} A_{i_k}\Big) = \prod_{k=1}^{m} \mathsf{P}(A_{i_k})$$

for every $1 \le m \le n$ and $1 \le i_1 < i_2 < \cdots < i_m \le n$.

A sequence of events $\{A_n, n \ge 1\}$ is said to be *a sequence of independent events* if every finite subset of this sequence is a set of independent events.

A sequence of events $\{A_n, n \ge 1\}$ is said to be *a sequence of pairwise independent events* if A_k and A_m are independent for $m \ne k$, $k \ge 1$, $m \ge 1$.

Let $\{\mathfrak{B}_n, n \ge 1\}$ be a sequence of σ-algebras such that $\mathfrak{B}_n \subset \mathfrak{A}$, $n \ge 1$. A sequence of σ-algebras $\{\mathfrak{B}_n, n \ge 1\}$ is said to be *a sequence of independent σ-algebras* if every sequence of events $\{A_n, n \ge 1\}$ is a sequence of independent events, where $A_n \in \mathfrak{B}_n$, $n \ge 1$.

Let $\mathfrak{A}_n$ be the minimal σ-algebra containing all the σ-algebras $\mathfrak{B}_m$ for $m \ge n$. The σ-algebra $\bigcap_{n=1}^{\infty} \mathfrak{A}_n$ is said to be the *tail* (or *asymptotic*) *σ-algebra with respect to the sequence of σ-algebras* $\{\mathfrak{B}_n, n \ge 1\}$. Events in the tail σ-algebra are called *tail* (or *asymptotic*) events.

The upper limit of a sequence of events $\{A_n, n \ge 1\}$ is the event defined by

$$\limsup_{n\to\infty} A_n = \bigcap_{n=1}^{\infty} \bigcup_{m=n}^{\infty} A_m.$$

The event $\limsup_{n\to\infty} A_n$ occurs if and only if the events A_n, $n \ge 1$, occur infinitely often.

THE BOREL–CANTELLI LEMMA. *Let $\{A_n, n \ge 1\}$ be a sequence of events such that*

$$\sum_{n=1}^{\infty} \mathsf{P}(A_n) < \infty.$$

Then

$$\mathsf{P}\Big(\limsup_{n\to\infty} A_n\Big) = 0.$$

On the other hand, if

$$\sum_{n=1}^{\infty} \mathsf{P}(A_n) = \infty$$

for a sequence of independent events $\{A_n, n \geq 1\}$, *then*

$$\mathsf{P}\Big(\limsup_{n\to\infty} A_n\Big) = 1.$$

Note that the Borel–Cantelli lemma consists of two parts. The first part concerns an arbitrary sequence of events with convergent series of probabilities.

THE KOLMOGOROV 0-1 LAW. *Let* $\{\mathfrak{A}_n, n \geq 1\}$ *be a sequence of independent* σ*-algebras and let* $\mathfrak{B}$ *be the tail* σ*-algebra with respect to this sequence. Then the probability* $\mathsf{P}(A)$ *of an arbitrary event* $A \in \mathfrak{B}$ *is equal to* 0 *or* 1.

Problems

III.1.1. Under what conditions are two events A and B independent if
a) $A \subset B$;
b) $A \cap B = \varnothing$;
c) $B = \bar{A}$?

III.1.2. Let $\Omega = [0,1]$, $\mathfrak{A}$ the σ-algebra of Lebesgue measurable subsets of $[0,1]$, and $\mathsf{P}(A)$ the Lebesgue measure of a set A. Define the following families of events:
a) $A_1 = [0, \frac{1}{2}]$, $A_2 = [\frac{1}{4}, \frac{3}{4}]$;
b) $A_1 = [0, \frac{1}{2}]$, $A_2 = [a, b]$;
c) $A_1 = [0, \frac{1}{4}] \cup [\frac{1}{2}, \frac{3}{4}]$, $A_2 = [0, \frac{1}{4}] \cup [\frac{3}{4}, 1]$, $A_3 = [0, \frac{1}{2}]$.
Which of them are families of independent events?

III.1.3. Let $(\Omega, \mathfrak{A}, \mathsf{P})$ be the probability space defined in the previous problem. Check that the following sequence

$$A_n = \bigcup_{k=0}^{2^{n-1}-1} \left[\frac{2k}{2^n}, \frac{2k+1}{2^n}\right], \qquad n \geq 1,$$

is a sequence of independent events.

III.1.4. Let $A_1, \dots, A_n$ be independent events and $\mathsf{P}(A_k) = p_k$, $1 \leq k \leq n$. Find the probability that
a) none of the events occurs;
b) exactly m events occur, $1 \leq m \leq n$.

III.1.5. Consider a sequence of independent trials such that a success occurs with probability p in each trial and a failure occurs with probability $q = 1-p$. Find the probability that a sequence of five consecutive "successes" occurs infinitely often with probability one.

III.1.6. Let $\{A_n, n \geq 1\}$ be a sequence of independent events. Set $B_k^{(m)} = A_k \cap A_{k+1} \cap \cdots \cap A_{k+m}$.

a) Prove that for m fixed, the events $B_k^{(m)}, k \geq 1$, occur with probability 1 infinitely often if and only if

$$(*) \qquad \sum_{k=1}^{\infty} \prod_{j=0}^{m} \mathsf{P}(A_{k+j}) = \infty.$$

b) Assume that

$$\prod_{k=n}^{\infty} \mathsf{P}(A_k) = 0.$$

Then

$$\mathsf{P}\Big(\bigcap_{m=1}^{\infty} \bigcup_{k=1}^{\infty} B_k^{(m)} \Big) = \begin{cases} 1 & \text{if condition } (*) \text{ is satisfied for all } m, \\ 0 & \text{otherwise.} \end{cases}$$

III.1.7. Let $\{A_n, n \geq 1\}$ be a sequence of independent events, and let $\mathsf{P}(A_n) = p_n$, $n \geq 1$. Find the condition under which

$$\mathsf{P}\Big(\bigcup_{n=1}^{\infty} A_n \Big) = 1.$$

Assume now that this condition is satisfied. Let ν be the smallest number n such that the event A_n occurs (ν is well defined with probability 1). Find the distribution of the random variable ν.

III.1.8. In the previous problem put $p_n = p$, $n \geq 1$, $0 < p < 1$. Then each of the events A_n, $n \geq 1$, occurs infinitely often with probability 1. Denote by $\nu_1 < \nu_2 < \cdots < \nu_k < \cdots$ the integers such that only the events with indices $\nu_1, \nu_2, \ldots$ occur. Prove that

a) for every $k \geq 1$ the distribution of the random variable $\nu_{k+1} - \nu_k$ is the same as that of ν_1.

b) the events $B_k^{(m_k)} = \{\nu_{k+1} - \nu_k = m_k\}$, $k \geq 1$, are independent for every sequence of integers $\{m_k, k \geq 1\}$.

III.1.9. A sequence $\{A_n, n \geq 1\}$ is such that

$$\sum_{n=1}^{\infty} \mathsf{P}(A \cap A_k) = +\infty$$

for every $A \in \mathfrak{A}$ with $\mathsf{P}(A) > 0$. Prove that

a) If B is an event such that $\mathsf{P}(B \cap A_n) = \varnothing$ for all n except, possibly, for a finite number, then $\mathsf{P}(B) = 0$.

b) $\mathsf{P}(\limsup_{n \to \infty} A_n) = 1$.

III.1.10. Let $\{A_n, n \geq 1\}$ be a sequence of random events. If there exists an event A such that

$$\sum_{n=1}^{\infty} \mathsf{P}(A_n \cap A) = +\infty,$$

then

$$\mathsf{P}\Big(\limsup_{n \to \infty} A_n \Big) \leq 1 - \mathsf{P}(A).$$

III.1.11. Let $\{A_n, n \geq 1\}$ be a sequence of independent events. Assume that

$$\sum_{n=1}^{\infty} \mathsf{P}(A_n) = +\infty.$$

Let A be an event for which

$$\sum_{n=1}^{\infty} \mathsf{P}(A_n \cap A) < \infty.$$

Find all possible values of $\mathsf{P}(A)$.

III.1.12. Let $\{A_n, n \geq 1\}$ be a sequence of events such that

$$\sum_{n=1}^{\infty} \mathsf{P}(A_n) = +\infty$$

and

$$\liminf_{n\to\infty} \frac{\sum_{i,k=1}^{n} \mathsf{P}(A_i \cap A_k)}{\left[\sum_{k=1}^{n} \mathsf{P}(A_k)\right]^2} = 1.$$

a) Prove that the following condition is satisfied:

$$\liminf_{n\to\infty} \mathsf{P}\left\{\left|\sum_{k=1}^{n} \alpha_k - \sum_{k=1}^{n} \mathsf{P}(A_k)\right| \geq \frac{1}{2}\sum_{k=1}^{n} \mathsf{P}(A_k)\right\} = 0,$$

where α_n is the indicator of A_n.

b) Use a) to prove that

$$\mathsf{P}\Big(\limsup_{n\to\infty} A_n\Big) = 1.$$

This generalization of the Borel-Cantelli lemma was obtained by P. Erdös and A. Rényi.

III.1.13. Let $\{A_n, n \geq 1\}$ be a sequence of pairwise independent events such that

$$\sum_{n=1}^{\infty} \mathsf{P}(A_n) = +\infty.$$

Prove that

$$\mathsf{P}\Big(\limsup_{n\to\infty} A_n\Big) = 1.$$

§III.2. A sequence of independent random variables

A sequence of random variables $\{\xi_n, n \geq 1\}$ is called *a sequence of independent variables* if $\big\{\{\xi_n < x_n\}, n \geq 1\big\}$ is a sequence of independent events for any sequence $\{x_n, n \geq 1\} \subset \mathbf{R}$. This definition is equivalent to the following. A sequence of random variables is called *a sequence of independent variables* if any collection of variables in this sequence is a collection of independent random variables.

The minimal σ-algebra of events containing all the events $\{\xi < x\}$, $x \in \mathbf{R}$, where ξ is a random variable, is called *the σ-algebra generated by the random variable ξ*.

The minimal σ-algebra of subsets of $\mathbf{R}^n$ that contains all the sets of the form

$$\{(x_1, x_2, \dots, x_n) : a_k \le x_k < b_k, 1 \le k \le n\},$$

where $a_k \in \mathbf{R}$, $b_k \in \mathbf{R}$, $1 \le k \le n$, is called *the Borel σ-algebra* in $\mathbf{R}^n$. Sets in this σ-algebra are said to be *Borel sets*.

A real function of n real variables $f : \mathbf{R}^n \to \mathbf{R}$ is said to be *a Borel function* if it is measurable with respect to the Borel σ-algebra, that is, if the set

$$\{(x_1, x_2, \dots, x_n) : f(x_1, x_2, \dots, x_n) < a\}$$

is Borel for any $a \in \mathbf{R}$.

Let $\{x_n, n \ge 1\} \subset \mathbf{R}$. Define $\limsup_{n\to\infty} x_n$ as the upper limit of the sequence $\{x_n, n \ge 1\}$ if the latter is bounded from above, and put $\limsup_{n\to\infty} x_n = +\infty$ otherwise. In the same manner define $\liminf_{n\to\infty} x_n$.

For a sequence of random variables $\{\xi_n, n \ge 1\}$ we consider

$$\liminf_{n\to\infty} \xi_n, \qquad \limsup_{n\to\infty} \xi_n.$$

These are also random variables. Indeed,

$$\limsup_{n\to\infty} \xi_n(\omega) = \begin{cases} +\infty, & \omega \in \bigcap_{m=1}^{\infty} \bigcup_{k=1}^{\infty} \{\xi_k(\omega) > m\}, \\ \limsup_{n\to\infty} \xi_n(\omega), & \omega \in \bigcup_{m=1}^{\infty} \bigcap_{k=1}^{\infty} \{\xi_k(\omega) \le m\}. \end{cases}$$

This variable is measurable with respect to $\mathfrak{A}$. The set

$$\left\{\omega : \liminf_{n\to\infty} \xi_n(\omega) = \limsup_{n\to\infty} \xi_n(\omega)\right\}$$

is measurable and means the set where the sequence $\{\xi_n, n \ge 1\}$ is convergent.

If for a sequence of random variables $\{\xi_n, n \ge 1\}$ and a sequence of positive numbers $\{c_n, n \ge 1\}$, only finitely many events $\{|\xi_n| > ac_n\}$, $n \ge 1$, occur for some $a > 0$, then we say that $\xi_n = O(c_n)$ with probability one as $n \to \infty$.

Problems

III.2.1. Prove that a sequence of random events $\{A_n, n \ge 1\}$ is a sequence of independent random events if and only if $\{I_{A_n}, n \ge 1\}$ is a sequence of independent random variables. Here I_A is the indicator of an event A.

III.2.2. Let $\{\xi_n, n \ge 1\}$ be a sequence of random variables, and $\mathfrak{A}_n$ the σ-algebra generated by the random variable ξ_n, $n \ge 1$. The sequence $\{\xi_n, n \ge 1\}$ is a sequence of independent random variables if and only if $\{\mathfrak{A}_n, n \ge 1\}$ is a sequence of independent σ-algebras.

III.2.3. If $\{\xi_n, n \ge 1\}$ is a sequence of independent random variables and $0 < n_1 < n_2 < \cdots < n_m < \cdots$, then

$$g_1(\xi_1, \dots, \xi_{n_1}), g_2(\xi_{n_1+1}, \dots, \xi_{n_2}), \dots, g_m(\xi_{n_m+1}, \dots, \xi_{n_{m+1}}), \dots$$

is a sequence of independent random variables, where g_k is a Borel function of $n_k - n_{k-1}$ variables, $k \ge 2$.

III.2.4. Prove that, for a sequence $\{\xi_n, n \geq 1\}$ of independent random variables the following random variables:

$$\liminf_{n\to\infty} \xi_n, \qquad \limsup_{n\to\infty} \xi_n, \qquad \liminf_{n\to\infty} \frac{1}{n}\sum_{k=1}^{n} \xi_k, \qquad \limsup_{n\to\infty} \frac{1}{n}\sum_{k=1}^{n} \xi_k$$

are constants (possibly infinite) with probability one. The same is true for the random variables

$$\liminf_{n\to\infty} \sqrt[n]{\prod_{k=1}^{n} \xi_k}, \qquad \limsup_{n\to\infty} \sqrt[n]{\prod_{k=1}^{n} \xi_k},$$

provided $\mathsf{P}(\xi_n \geq 0) = 1$, $n \geq 1$.

III.2.5. Let $\{\xi_n, n \geq 1\}$ be a sequence of independent random variables. Prove that

a) the sequence $\{\xi_n, n \geq 1\}$ either converges with probability one or diverges with probability one;
b) the limit of the sequence $\{\xi_n, n \geq 1\}$ is a constant with probability one;
c) the series $\sum_{n=1}^{\infty} \xi_n$ either converges with probability one or diverges with probability one.

III.2.6. Let $\{\xi_n, n \geq 1\}$ be a sequence of independent random variables. Prove that the radius of convergence of the power series

$$\sum_{n=0}^{\infty} \xi_n x^n$$

is a constant with probability one.

III.2.7. Let $\{\xi_n, n \geq 1\}$ be a sequence of independent random variables. Prove that for a random Dirichlet series

$$\sum_{n=1}^{\infty} \frac{\xi_n}{n^x}$$

there exists a constant σ, $-\infty \leq \sigma \leq +\infty$, such that the series is convergent with probability one for $x > \sigma$.

III.2.8. Let $\{\xi_n, n \geq 1\}$ be a sequence of random variables and let $\{a_n, n \geq 1\}$ be a sequence of real numbers such that $a_n \leq a_{n+1}$, $n \geq 1$, and $a_n \to +\infty$ as $n \to \infty$. Prove the following statements:

a) $\mathsf{P}(\xi_n \geq a_n$ infinitely often$) = \mathsf{P}(\max_{1\leq k\leq n} \xi_k \geq a_n$ for infinitely many $n)$.
b) If $\mathsf{P}(\xi_n \geq a_n$ infinitely often$) = 0$, then

$$\lim_{n\to\infty} \mathsf{P}\Big(\max_{1\leq k\leq n} \xi_k \geq a_n\Big) = 0.$$

III.2.9. Let $\{\xi_n, n \geq 1\}$ be a sequence of independent identically distributed random variables and $\mathsf{E}\,|\xi_1| = +\infty$. Prove that, for all $c > 0$, infinitely many events of the sequence $\{|\xi_n| > cn\}$, $n \geq 1$, occur with probability one.

III.2.10. Suppose that $\{\xi_n, n \geq 1\}$ is a sequence of independent identically distributed random variables with $\mathsf{E}\,\xi_1 = 0$. Prove that

$$\mathsf{P}\left(\frac{\xi_n}{n^{1/\alpha}} \to 0 \text{ as } n \to \infty\right) = 1, \qquad \alpha > 0,$$

if and only if $\mathsf{E}\,|\xi_1|^\alpha < \infty$.

III.2.11. Under the conditions of the preceding problem, prove that

$$\mathsf{P}\left(\frac{\xi_n}{\ln n} \to 0 \text{ as } n \to \infty\right) = 1$$

if and only if $\mathsf{E}\,e^{c|\xi_1|} < \infty$ for all $c > 0$.

III.2.12. Let $\{\xi_n, n \geq 1\}$ be a sequence of independent identically distributed random variables with an exponential distribution. Prove that

$$\mathsf{P}\big(\xi_n = O(\ln n) \text{ as } n \to \infty\big) = 1.$$

III.2.13. Let $\{\xi_n, n \geq 1\}$ be a sequence of independent Gaussian random variables with $\mathsf{E}\,\xi_n = 0$, $\mathsf{E}\,\xi_n^2 = 1$, $n \geq 1$. Prove that

$$\mathsf{P}(|\xi_n| = O(\sqrt{\ln n}) \text{ as } n \to \infty) = 1.$$

III.2.14. Let $\{\xi_n, n \geq 1\}$ be a sequence of independent random variables with $\mathsf{P}(\xi_n = 0) = \mathsf{P}(\xi_n = 1) = \frac{1}{2}$, $n \geq 1$. Define the random variable η as follows:

$$\eta = \sum_{n=1}^{\infty} \frac{\xi_n}{2^n}.$$

Prove that η has the uniform distribution on the segment $[0, 1]$.

III.2.15. Any trial of a sequence of independent experiments leads to the appearance of each digit $0, 1, 2, \ldots, 9$ with probability $\frac{1}{10}$. Let ξ_n be a result of the nth trial, and η the fraction $\eta = 0\,.\,\xi_1\xi_2\ldots\xi_n\ldots$. What is the probability of the event that η is a rational number?

III.2.16. A point is taken at random on the segment $[0, 1]$. Let x be its coordinate. Consider the decimal representation of x, that is, $x = 0\,.\,\xi_1\xi_2\ldots\xi_n\ldots$ if $x < 1$ (we exclude recurring decimals that have digit 9 as their repetend). Prove that $\{\xi_n, n \geq 1\}$ is a sequence of independent identically distributed random variables, and moreover,

$$\mathsf{P}(\xi_n = k) = \frac{1}{10}, \qquad 0 \leq k \leq 9,\ n \geq 1.$$

III.2.17. Consider a sequence of independent identically distributed random variables $\{\xi_n, n \geq 1\}$ with the Cauchy distribution. Prove that for every $x > 0$,

$$\lim_{n\to\infty} \mathsf{P}\left(\frac{1}{n} \max_{1\leq k\leq n} \xi_k < x\right) = \exp\{-1/(\pi x)\}.$$

III.2.18. Let $\{\xi_n, n \geq 1\}$ be a sequence of independent identically distributed random variables with $\mathsf{E}\,|\xi_1| < +\infty$. Prove that

$$\lim_{n\to\infty} \frac{1}{n} \mathsf{E}\left(\max_{1\leq k\leq n} |\xi_k|\right) = 0.$$

III.2.19. Let $\{\xi_n, n \geq 1\}$ be a sequence of independent random variables with

$$\sum_{n=1}^{\infty} \mathsf{E}\,\xi_n^2 < \infty.$$

Prove that

$$\mathsf{E}\Big(\sup_{n\geq 1} |\xi_n|\Big) < \infty.$$

III.2.20. Let $\nu, \xi_1, \xi_2, \ldots, \xi_n, \ldots$ be a sequence of independent random variables. Suppose that ν is integer-valued, $\mathsf{E}\,\nu^2 < \infty$, and ξ_n, $n \geq 1$, are identically distributed with $\mathsf{E}\,\xi_1^2 < \infty$. Compute the expectation and the variance of the random variable

$$S_\nu = \xi_1 + \xi_2 + \cdots + \xi_\nu.$$

III.2.21. Let $\{\xi_n, n \geq 1\}$ be a sequence of independent identically distributed random variables. Set $\nu = \min\{n \geq 2 \colon \xi_n > \xi_1\}$. Determine the distribution of ν. Does $\mathsf{E}\,\nu$ exist?

III.2.22. For the sequence defined in the preceding problem put

$$\nu = \min\Big\{n > m \colon \xi_n > \max_{1\leq k\leq m} \xi_k\Big\}.$$

Prove that $\mathsf{P}(\nu > n) = m/n$.

III.2.23. Let $\{\xi_n, n \geq 1\}$ be a sequence of independent identically distributed random variables having the exponential distribution with parameter $\lambda > 0$ and

$$\nu = \min\{n \geq 1 \colon \xi_1 + \xi_2 + \cdots + \xi_n > 1\}.$$

Find the distribution of ν.

III.2.24. Under the conditions of the preceding problem, find the distribution of ξ_ν.

III.2.25. Let $\{\xi_n, n \geq 1\}$ be a sequence of independent identically distributed random variables with $\mathsf{E}\,\xi_1 > 0$, $\mathsf{E}\,\xi_1^2 < \infty$, and

$$\nu = \min\{n \geq 1 \colon \xi_1 + \xi_2 + \cdots + \xi_n \geq 1\}.$$

Prove that the random variables ξ_ν and $\xi_{\nu+k}$ are independent for any $k \geq 1$. Determine the distribution of $\xi_{\nu+k}$.

III.2.26. Let $\{\xi_n, n \geq 1\}$ be a sequence of independent random variables uniformly distributed on the segment $[0, 1]$. Let

$$\nu = \min\{n \geq 1 \colon \xi_1 + \xi_2 + \cdots + \xi_n > 1\}.$$

Prove that $\mathsf{E}\,\nu = e$.

III.2.27. Let $\{\xi_n, n \geq 1\}$ be a sequence of independent identically distributed random variables with $\mathsf{E}\,\xi_1 = 0$ and $\mathsf{E}\,\xi_1^{2r} < \infty$ for some integer $r > 0$. Prove that there exists a number $c(r)$ such that

$$\mathsf{E}\Big(\sum_{k=1}^{n} \xi_k\Big)^{2r} \leq c(r) n^r, \qquad n \geq 1.$$

III.2.28. Let $\{\xi_n, n \geq 1\}$ be a sequence of independent identically distributed random variables with a distribution function F, and let numbers $\{a_n, n \geq 1\} \subset \mathbf{R}$ satisfy the following conditions:

$$0 = a_0 < a_n < a_{n+1}, \qquad n \geq 1,$$
$$a_n + a_m \leq a_{n+m}, \qquad n, m \geq 1.$$

If $\sum_{n=1}^{\infty} \int_{|x|>a_n} dF(x) = +\infty$ and $S_n = \xi_1 + \xi_2 + \cdots + \xi_n$, then

$$\mathsf{P}\left(\limsup_{n\to\infty}\{|S_n| > a_n\}\right) = \mathsf{P}(|S_n| > a_n \text{ for infinitely many } n) = 1.$$

Prove this statement.

III.2.29. Let $\{\xi_n, n \geq 1\}$ be a sequence of independent identically distributed random variables with $\mathsf{E}\,|\xi_1| = +\infty$. Prove that at least one of the following equalities is satisfied:

$$\mathsf{P}\left(\limsup_{n\to\infty} \frac{1}{n}\sum_{k=1}^{n} \xi_k = +\infty\right) = 1,$$
$$\mathsf{P}\left(\liminf_{n\to\infty} \frac{1}{n}\sum_{k=1}^{n} \xi_k = -\infty\right) = 1.$$

III.2.30. Random variables $\{\xi_n, n \geq 1\}$ are independent and have the Cauchy distribution. Prove that

$$\mathsf{P}\left(\limsup_{n\to\infty} \frac{1}{n}\sum_{k=1}^{n} \xi_k = +\infty\right) = 1.$$

III.2.31. Prove that for the random variables of the preceding problem the following relation is true for any $\alpha > 1$:

$$\mathsf{P}\left(\lim_{n\to\infty} \frac{1}{n^{1+\alpha}}\sum_{k=1}^{n} \xi_k = 0\right) = 1.$$

III.2.32. Let $\{\xi_n, n \geq 1\}$ be a sequence of independent random variables uniformly distributed on the segment $[0,1]$. Prove that, with probability one, the set of numbers $\{\xi_1, \xi_2, \ldots, \xi_n, \ldots\}$ is everywhere dense on $[0,1]$.

§III.3. Various modes of convergence of sequences of random variables

Let $(\Omega, \mathfrak{A}, \mathsf{P})$ be a probability space. A real function ξ defined on Ω and measurable with respect to the σ-algebra $\mathfrak{A}$ is called *a random variable*. In what follows we also consider functions on Ω with possibly infinite values $-\infty$ or $+\infty$ for some $\omega \in \Omega$. For such functions we assume that the set of ω where $|\xi(\omega)| = \infty$ is measurable and has probability zero.

A sequence of random variables $\{\xi_n, n \geq 1\}$ is said *to converge with probability one to a random variable* ξ if the set of points $\omega \in \Omega$ where either the limit $\lim_{n\to\infty} \xi_n(\omega)$ does not exist or $\lim_{n\to\infty} \xi_n(\omega) \neq \xi(\omega)$, is of probability zero, that is,

$$\mathsf{P}\left(\left\{\omega\colon \lim_{n\to\infty} \xi_n(\omega) = \xi(\omega)\right\}\right) = 1.$$

We write

$$\mathsf{P}\{\xi_n \to \xi \text{ as } n \to \infty\} = 1 \quad \text{or} \quad \mathsf{P}\left(\lim_{n\to\infty} \xi_n(\omega) = \xi(\omega)\right) = 1$$

if a sequence $\{\xi_n, n \geq 1\}$ converges with probability one to a random variable ξ.

The convergence with probability one of a sequence of random variables $\{\xi_n, n > 1\}$ is, in fact, the almost everywhere convergence of the sequence of measurable functions $\{\xi_n(\omega), \omega \in \Omega, n \geq 1\}$ with respect to the measure P.

A sequence of random variables $\{\xi_n, n \geq 1\}$ is said *to converge in probability to a random variable* ξ if for any $\varepsilon > 0$,

$$\mathsf{P}(|\xi_n - \xi| > \varepsilon) \to 0 \qquad \text{as } n \to \infty.$$

The convergence in probability of a sequence of random variables $\{\xi_n, n \geq 1\}$ is, in fact, the convergence in measure P of the sequence of measurable functions $\{\xi_n(\omega), \omega \in \Omega, n \geq 1\}$.

A sequence of random variables $\{\xi_n, n \geq 1\}$ is said *to converge in mean of order r to a random variable* ξ if $\mathsf{E}\,|\xi_n|^r < \infty$, $n \geq 1$, $\mathsf{E}\,|\xi|^r < \infty$, and

$$\mathsf{E}\,|\xi_n - \xi|^r \to 0 \quad \text{as } n \to \infty.$$

In the case $r = 2$ this convergence is said to be *mean square convergence*.

A sequence of distribution functions $\{F_n, n \geq 1\}$ is said to *converge weakly to a function* F if $F_n(x) \to F(x)$ as $n \to \infty$ for all x where F is continuous.

We say that a sequence of random variables $\{\xi_n, n \geq 1\}$ *converges in distribution to a random variable* ξ if the sequence of their distribution functions $\{F_n, n \geq 1\}$ weakly converges to the distribution function F of the variable ξ.

An ordered set of m real random variables $\xi_1, \xi_2, \ldots, \xi_m$ is said to be *a vector-valued random variable $\vec{\xi}$ with values in $\mathbf{R}^m$*.

The joint distribution function of random variables $\xi_1, \xi_2, \ldots, \xi_m$ is said to be *the distribution function of $\vec{\xi}$*, that is,

$$F(\vec{x}) = F(x_1, x_2, \ldots, x_m) = \mathsf{P}(\xi_k < x_k, 1 \leq k \leq m).$$

The weak convergence of distribution functions in $\mathbf{R}^m$ is defined in the same way as in the one-dimensional case.

A necessary and sufficient condition for the weak convergence of a sequence of distribution functions to a distribution function is given by the following theorem.

THEOREM 1. *In order that a sequence of distribution functions $\{F_n, n \geq 1\}$ weakly converge in $\mathbf{R}^m$ to a distribution function F, it is necessary and sufficient that for any bounded and continuous real function g in $\mathbf{R}^m$,*

$$\lim_{n\to\infty} \int_{\mathbf{R}^m} g(x_1, x_2, \ldots, x_m)\, dF_n = \int_{\mathbf{R}^m} g(x_1, x_2, \ldots, x_m)\, dF.$$

The definition of convergence in distribution for the case of m-dimensional random vectors is analogous to that for the one-dimensional case.

A sequence of m-dimensional random vectors $\{\vec{\xi}_n, n \geq 1\}$ is said *to converge in probability to an m-dimensional random variable $\vec{\xi}$* if the sequence of real random variables $\{\|\vec{\xi}_n - \vec{\xi}\|, n \geq 1\}$ converges to zero in probability as $n \to \infty$. Here $\|\vec{a}\|^2 = \sum_{k=1}^m a_k^2$ for $\vec{a} \in \mathbf{R}^m$.

Relationships between different notions of convergence of random variables is shown below, where the arrows mean implication.

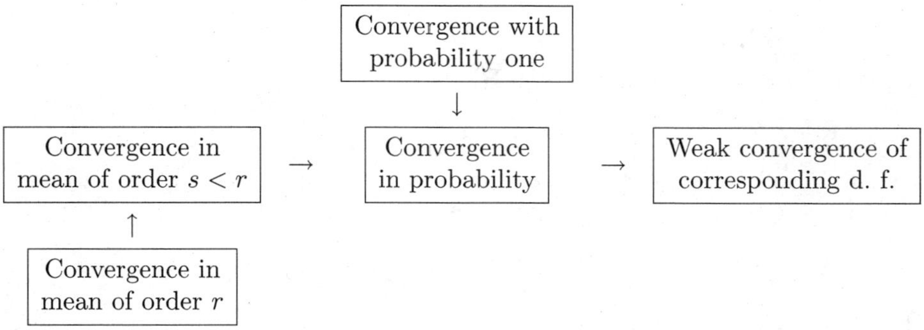

In general, if no special assumptions on the probability space or on the sequence of random variables are made, only the above relationships are valid.

In studying properties of sequences of random variables, the following theorem is often helpful.

THEOREM 2. *Let a sequence of random variables $\{\xi_n, n \geq 1\}$ converge in probability to a random variable ξ. Then there exists a subsequence $\{\xi_{n(k)}, k \geq 1\}$ converging with probability one to ξ.*

In addition to this theorem, Lebesgue's theorem on bounded convergence, Fatou's lemma, and Minkowski's inequality are useful while solving problems.

Problems

III.3.1. Let ξ, ξ_n, $n \geq 1$, be random variables. Prove that the set of points $\omega \in \Omega$ where $\xi_n(\omega) \to \xi(\omega)$ as $n \to \infty$, is measurable, that is, it belongs to $\mathfrak{A}$.

III.3.2. Prove that for a sequence of random variables $\{\xi_n, n \geq 1\}$, the set of points $\omega \in \Omega$ where the sequence $\{\xi_n(\omega), n \geq 1\}$ converges, is measurable, that is, it belongs to $\mathfrak{A}$.

III.3.3. A sequence of random variables $\{\xi_n, n \geq 1\}$ converges with probability one to a random variable ξ if and only if for any $\varepsilon > 0$,

$$\mathsf{P}\left(\limsup_{n\to\infty}\{\omega\colon |\xi_n(\omega) - \xi(\omega)| \geq \varepsilon\}\right) = 0.$$

III.3.4. A sequence of random variables $\{\xi_n, n \geq 1\}$ converges with probability one to a random variable ξ if and only if for any $\varepsilon > 0$,

$$\lim_{n\to\infty} \mathsf{P}\left\{\bigcup_{m=n}^{\infty} \{|\xi_m - \xi| \geq \varepsilon\}\right) = 0.$$

III.3.5. Suppose that for two sequences of random variables $\{\xi_n, n \geq 1\}$ and $\{\eta_n, n \geq 1\}$,

$$\mathsf{P}(\xi_n \to \xi \text{ as } n \to \infty) = 1,$$
$$\mathsf{P}(\eta_n \to \eta \text{ as } n \to \infty) = 1,$$
$$\mathsf{P}(\xi_n \neq \eta_n) = 0, \qquad n \geq 1.$$

Prove that $\mathsf{P}(\xi = \eta) = 1$.

III.3.6. Suppose that

$$\mathsf{P}(\xi_n^{(k)} \to \xi^{(k)} \text{ as } n \to \infty) = 1, \qquad 1 \le k \le m,$$

and g is a real continuous function on $\mathbf{R}^m$. Prove that

$$\mathsf{P}(g(\xi_n^{(1)}, \xi_n^{(2)}, \ldots, \xi_n^{(m)}) \to g(\xi^{(1)}, \xi^{(2)}, \ldots, \xi^{(m)}) \text{ as } n \to \infty) = 1.$$

III.3.7. Let $([0,1], \mathfrak{A}, \mathsf{P})$ be the same probability space as in Problem III.1.2. Put

$$\xi_n = I_{A_n}, \qquad n \ge 1,$$

$$A_n = \left[\frac{k}{2^m}, \frac{k+1}{2^m}\right], \qquad n = k + 2^m, \quad k = 0, 1, \ldots, 2^m - 1, \quad m = 0, 1, 2, \ldots.$$

Prove that

a) the sequence $\{\xi_n, n \ge 1\}$ converges to zero in probability;

b) $\{\xi_n, n \ge 1\}$ is not a sequence convergent with probability one.

Choose a subsequence of the sequence $\{\xi_n, n \ge 1\}$ that converges with probability one.

III.3.8. A sequence of random variables $\{\xi_n, n \ge 1\}$ converges with probability one to a random variable if and only if for every $\varepsilon > 0$,

$$\lim_{N \to \infty} \mathsf{P}\left(\bigcup_{m=1}^{\infty} \{\omega \colon |\xi_{N+m}(\omega) - \xi_N(\omega)| \ge \varepsilon\} \right) = 0.$$

III.3.9. If for a sequence of random variables $\{\xi_n, n \ge 1\}$ and for a random variable ξ the condition

$$\sum_{n=1}^{\infty} \mathsf{P}(|\xi_n - \xi| > \varepsilon) < \infty$$

is satisfied for all $\varepsilon > 0$, then $\mathsf{P}(\xi_n \to \xi \text{ as } n \to \infty) = 1$.

III.3.10. If for a sequence of random variables $\{\xi_n, n \ge 1\}$ there exists a sequence of nonnegative numbers $\{\varepsilon_n, n \ge 1\}$ such that $\sum_{n=1}^{\infty} \varepsilon_n < \infty$ and

$$\sum_{n=1}^{\infty} \mathsf{P}(|\xi_{n+1} - \xi_n| > \varepsilon_n) < \infty,$$

then $\{\xi_n, n \ge 1\}$ converges with probability one to a random variable.

III.3.11. Let $\{\xi_n, n \ge 1\}$ be a sequence of random variables such that

$$\sum_{n=1}^{\infty} \mathsf{E}\,|\xi_n|^r < \infty$$

for some $r > 0$. Prove that $\mathsf{P}(\xi_n \to 0 \text{ as } n \to \infty) = 1$.

III.3.12. Prove that a sequence of random variables $\{\xi_n, n \ge 1\}$ converges with probability one to a random variable ξ if and only if

$$\mathsf{P}\Big(\sup_{m \ge n} |\xi_m - \xi| \ge \varepsilon \Big) \to 0 \quad \text{as } n \to \infty$$

for all $\varepsilon > 0$, that is, if $\sup_{m \ge n} |\xi_m - \xi| \to 0$ as $n \to \infty$ in probability.

III.3.13. Prove that if a sequence of random variables $\{\xi_n, n \geq 1\}$ converges with probability one to a random variable ξ, then it converges to ξ in probability.

III.3.14. The convergence in probability does not imply the convergence with probability one. Construct an appropriate example.

III.3.15. A set $A \in \mathfrak{A}$ with $\mathsf{P}(A) > 0$ is said to be *an atom of the probability space* $(\Omega, \mathfrak{A}, \mathsf{P})$ if any set $B \subset A$, $B \in \mathfrak{A}$, is of probability either 0 or $\mathsf{P}(A)$. Prove that

a) if A and B are atoms, then either $\mathsf{P}(A) = \mathsf{P}(B)$ or $\mathsf{P}(A \cap B) = 0$; so we may assume that atoms do not intersect except for sets of probability 0;
b) there exists a family of disjoint atoms that is finite or denumerable;
c) $\Omega = A_0 \cup \left(\bigcup_{n=1}^{\infty} A_n\right)$, where $A_n \cap A_m = \varnothing$, $n \neq m$, A_n is an atom or empty set for $n \geq 1$, and the set A_0 is such that for any number p, $0 \leq p \leq \mathsf{P}(A_0)$, there exists $B \subset A_0$, $B \in \mathfrak{A}$, with $\mathsf{P}(B) = p$. Moreover, the above expansion of the space Ω is unique up to sets of probability 0.

III.3.16. Under the conditions of the preceding problem, prove that

a) any random variable is constant with probability one at every atom;
b) if Ω is a union of atoms, then the convergence in probability of a sequence of random variables implies the convergence with probability one;
c) if $\mathsf{P}(A_0) > 0$ in the above expansion $\Omega = A_0 \cup \left(\bigcup_{n=1}^{\infty} A_n\right)$, then the convergence in probability does not imply the convergence with probability one.

Therefore the convergence with probability one and the convergence in probability on a space $(\Omega, \mathfrak{A}, \mathsf{P})$ are equivalent if and only if Ω consists of atoms.

III.3.17. Let $\{\xi_n, n \geq 1\}$ be a sequence of random variables such that

$$\mathsf{P}(0 \leq \xi_{n+1} \leq \xi_n) = 1, \qquad n \geq 1.$$

Prove that the convergence in probability $\xi_n \to 0$ as $n \to \infty$ implies

$$\mathsf{P}(\xi_n \to 0 \text{ as } n \to \infty) = 1.$$

III.3.18. Prove that

a) if $\mathsf{P}(\xi_n \to \xi' \text{ as } n \to \infty) = 1$, $\mathsf{P}(\xi_n \to \xi'' \text{ as } n \to \infty) = 1$, then $\mathsf{P}(\xi' \neq \xi'') = 0$;
b) if $\xi_n \to \xi'$ in probability as $n \to \infty$, and $\xi_n \to \xi''$ in probability as $n \to \infty$, then $\mathsf{P}(\xi' = \xi'') = 1$.

III.3.19. Prove that a sequence of random variables $\{\xi_n, n \geq 1\}$ converges in probability to a random variable ξ if and only if any subsequence $\{\xi_{n(k)}, k \geq 1\}$ contains another subsequence that converges to ξ with probability one.

III.3.20. Consider k sequences of random variables $\{\xi_n^{(j)}, n \geq 1\}$, $1 \leq j \leq k$, and suppose that $\xi_n^{(j)} \to \xi^{(j)}$ in probability as $n \to \infty$ for all j, $1 \leq j \leq k$, where $\xi^{(j)}$ is a random variable, $1 \leq j \leq k$. Let g be a real continuous function on a Borel set $A \subset \mathbf{R}^k$. Suppose that for all $n \geq 1$

$$\mathsf{P}((\xi_n^{(1)}, \dots, \xi_n^{(k)}) \in A) = 1, \qquad \mathsf{P}((\xi^{(1)}, \dots, \xi^{(k)}) \in A) = 1.$$

Then the sequence of random variables

$$\left\{g(\xi_n^{(1)}, \dots, \xi_n^{(k)}), n \geq 1\right\}$$

converges in probability to the random variable $g(\xi_n^{(1)},\dots,\xi_n^{(k)})$.

III.3.21. Let $\xi_n \to \xi$ and $\eta_n \to \eta$ in probability as $n \to \infty$. Prove that

$$\xi_n + \eta_n \to \xi + \eta, \quad \xi_n\eta_n \to \xi\eta, \quad \max(\xi_n, \eta_n) \to \max(\xi, \eta)$$

in probability as $n \to \infty$.

III.3.22. Consider k sequences of random variables $\{\xi_n^{(j)}, n \geq 1\}$, $1 \leq j \leq k$, and suppose that $\xi_n^{(j)} \to \xi^{(j)}$ in probability as $n \to \infty$ for all j, $1 \leq j \leq k$, where $\xi^{(j)}$ is a random variable, $1 \leq j \leq k$. Let $\{x_1, x_2, \dots, x_k\} \in \mathbf{R}$ be such that $\mathsf{P}(\xi^{(j)} = x_j) = 0$ for all $1 \leq j \leq k$. By using Problem III.3.20 prove that

$$\lim_{n\to\infty} \mathsf{P}(\xi_n^{(1)} < x_1, \xi_n^{(2)} < x_2, \dots, \xi_n^{(k)} < x_k)$$
$$= \mathsf{P}(\xi^{(1)} < x_1, \xi^{(2)} < x_2, \dots, \xi^{(k)} < x_k).$$

III.3.23. A sequence of random variables $\{\xi_n, n \geq 1\}$ converges in probability to a random variable ξ. Let $F_n(x) = \mathsf{P}(\xi_n < x)$, $n \geq 1$, and $F(x) = \mathsf{P}(\xi < x)$, $x \in \mathbf{R}$. Prove that the sequence $\{F_n, n \geq 1\}$ converges weakly to the function F.

III.3.24. Prove that a sequence of random variables $\{\xi_n, n \geq 1\}$ converges in probability to a constant c if and only if the sequence of the corresponding distribution functions $F_n(x) = \mathsf{P}(\xi_n < x)$, $x \in \mathbf{R}$, $n \geq 1$, weakly converges to the function

$$F(x) = \begin{cases} 0, & x \leq c, \\ 1, & x > c. \end{cases}$$

III.3.25. Let $\{\xi_n, n \geq 1\}$ be a sequence of random variables and $\{F_n, n \geq 1\}$ be the sequence of the corresponding distribution functions. Let F be a distribution function taking more than two values. Prove that the weak convergence $F_n \to F$ as $n \to \infty$ does not imply the convergence in probability of the sequence $\{\xi_n, n \geq 1\}$.

III.3.26. Sequences $\{\xi_n^{(j)}, n \geq 1\}$, $1 \leq j \leq k$, converge in probability to random variables $\xi^{(j)}$, $1 \leq j \leq k$, respectively. Assume that the variables $\{\xi_n^{(j)}, 1 \leq j \leq k\}$ are independent for all n. Prove that the variables $\xi^{(1)}, \xi^{(2)}, \dots, \xi^{(k)}$ are independent.

III.3.27. A sequence of random variables is said to be *bounded in probability* if

$$\lim_{a\to\infty} \sup_{n\geq 1} \mathsf{P}(|\xi_n| > a) = 0.$$

Let a sequence $\{\xi_n, n \geq 1\}$ be bounded in probability and $\eta_n \to 0$ in probability as $n \to \infty$. Prove that $\xi_n\eta_n \to 0$ in probability as $n \to \infty$.

III.3.28. Let a sequence of random variables $\{\xi_n, n \geq 1\}$ possess the following property: for any sequence of real numbers $\{a_n, n \geq 1\}$ such that $a_n \to 0$ as $n \to \infty$, we have $a_n\xi_n \to 0$ in probability as $n \to \infty$. Prove that the sequence $\{\xi_n, n \geq 1\}$ is bounded in probability.

III.3.29. A sequence of random variables is said to be a *Cauchy sequence in probability* if for any $\varepsilon > 0$ and $\delta > 0$ there exists a number N such that

$$\mathsf{P}(|\xi_m - \xi_n| > \varepsilon) < \delta, \qquad m, n > N.$$

Prove that any sequence converging in probability is a Cauchy sequence in probability.

III.3.30. Consider a sequence of random variables. Assume that it is a Cauchy sequence in probability. Using Problem III.3.10, prove that this sequence converges in probability to a random variable.

III.3.31. Suppose that

$$\xi_n \to \xi, \quad \eta_n \to \eta \quad \text{as } n \to \infty$$

in probability and $\mathsf{P}(\xi_n \le \eta_n) = 1$, $n \ge 1$. Prove that $\mathsf{P}(\xi \le \eta) = 1$.

III.3.32. Suppose that

$$\xi_n \to \xi, \quad \zeta_n \to \xi \quad \text{as } n \to \infty$$

in probability and $\mathsf{P}(\xi_n \le \eta_n \le \zeta_n) \to 1$. Prove that $\eta_n \to \xi$ in probability as $n \to \infty$.

III.3.33. Suppose that

$$\xi_n \to \xi, \quad \zeta_n \to \xi \quad \text{as } n \to \infty$$

in distribution and $\mathsf{P}(\xi_n \le \eta_n \le \zeta_n) = 1$, $n \ge 1$. Prove that $\eta_n \to \xi$ in distribution as $n \to \infty$.

III.3.34. Construct an example of a sequence of probability distribution functions defined on $\mathbf{R}$ that weakly converges to a function that is not a distribution function.

III.3.35. A sequence of distribution functions $\{F_n, n \ge 1\}$ defined on $\mathbf{R}$ weakly converges to a function F. Prove that F is a distribution function if and only if

$$\sup_{n\ge1}[1 - F_n(a) + F_n(-a)] \to 0, \qquad a \to \infty.$$

III.3.36. Let $\{F_n, n \ge 1\}$ be a sequence of distribution functions of respective random variables $\{\xi_n, n \ge 1\}$ and, for some $\alpha > 0$ and a number c,

$$\int_{-\infty}^{+\infty} |x|^\alpha \, dF_n(x) = \mathsf{E}\,|\xi_n|^\alpha \le c, \qquad n \ge 1.$$

Prove that

$$\sup_{n\ge1}[1 - F_n(a) + F_n(-a)] \to 0, \qquad a \to \infty,$$

and that there exists a subsequence $\{F_{n(k)}, k \ge 1\}$ that converges to a distribution function.

III.3.37. A sequence of random variables $\{\xi_n, n \ge 1\}$ converges in mean of order r to a random variable ξ. Prove that $\xi_n \to \xi$ in probability as $n \to \infty$.

III.3.38. The limit in mean of order r is unique with probability one. Let a sequence of random variables be a Cauchy sequence in mean of order r. Prove that it converges in mean of order r to a random variable.

III.3.39. Prove that the convergence in mean of order r does not imply the convergence with probability one.

III.3.40. Prove that the convergence with probability one does not imply, in general, the convergence in mean of order r.

III.3.41. Let g be a real continuous bounded function defined on $\mathbf{R}$, and assume that a sequence of random variables $\{\xi_n, n \geq 1\}$ converges in probability to a random variable ξ. Prove that

a) $\lim_{n\to\infty} \mathsf{E}\, g(\xi_n) = \mathsf{E}\, g(\xi)$;

b) $\lim_{n\to\infty} \mathsf{E}\,|g(\xi_n) - g(\xi)|^r = 0$, $r > 0$.

III.3.42. Prove that the convergence in mean of order r implies the convergence in mean of order s for any $0 < s < r$.

III.3.43. Suppose a sequence of random variables $\{\xi_n, n \geq 1\}$ converges in mean of order $r > 0$ to a random variable ξ. Prove that

$$\lim_{n\to\infty} \mathsf{E}\,|\xi_n|^s = \mathsf{E}\,|\xi|^s, \qquad 0 < s \leq r.$$

III.3.44. Let a sequence of random variables $\{\xi_n, n \geq 1\}$ satisfy the following conditions:

a) $\mathsf{P}(\xi_n \to \xi \text{ as } n \to \infty) = 1$;

b) $\mathsf{E}\,\xi_n \to \mathsf{E}\,\xi$, $n \to \infty$.

Then it is possible that the sequence $\{\xi_n, n \geq 1\}$ does not converge in mean of order 1. Construct an appropriate example.

III.3.45. Suppose a sequence $\{\xi_n, n \geq 1\}$ converges in probability to a random variable ξ, and moreover, $\mathsf{P}(|\xi_n| > c) = 0$, $n \geq 1$, for some c. Prove that $\{\xi_n, n \geq 1\}$ converges in mean of an arbitrary order $r > 0$ to the random variable ξ.

III.3.46. Let random variables ξ and ξ_n, $n \geq 1$, satisfy the following conditions:

a) $\mathsf{P}(|\xi_n| > c) = 0$ for some c and all $n \geq 1$;

b) for any integer $k \geq 0$,

$$\mathsf{E}\,\xi_n^k \to \mathsf{E}\,\xi^k, \qquad n \to \infty.$$

Prove that the sequence $\{\xi_n, n \geq 1\}$ converges in distribution to the random variable ξ.

III.3.47. A sequence $\{\xi_n, n \geq 1\}$ converges in probability to a random variable ξ and $\mathsf{P}(|\xi_n| \leq \eta) = 1$, $n \geq 1$, for a random variable η such that $\mathsf{E}\,\eta^r < \infty$, $r > 0$. Prove that $\{\xi_n, n \geq 1\}$ converges to ξ in mean of order r.

III.3.48. A sequence $\{\xi_n, n \geq 1\}$ converges in probability to a random variable ξ and $\mathsf{E}\,|\xi_n|^r \leq c$, $n \geq 1$, for some constants c and $r > 0$. Prove that for all s, $0 < s < r$, the sequence $\{\xi_n, n \geq 1\}$ converges to ξ in mean of order s.

III.3.49. Under the conditions of the preceding problem, prove that

$$\lim_{n\to\infty} \mathsf{E}\,|\xi_n|^s = \mathsf{E}\,|\xi|^s.$$

III.3.50. Let $(\Omega, \mathfrak{A}, \mathsf{P})$ be a probability space. Two real random variables ξ and η are said to be *equivalent* if $\mathsf{P}(\xi \neq \eta) = 0$. Denote by X the set of equivalence classes, and for $\xi, \eta \in X$ determine a function d as follows:

$$d(\xi, \eta) = \mathsf{E}\left[\frac{|\xi - \eta|}{1 + |\xi - \eta|}\right].$$

(We denote an equivalence class and a representative of this class by the same symbol.) Prove that

a) d is a distance on X;
b) the convergence with respect to d is equivalent to the convergence in probability;
c) (X, d) is a complete metric space.

III.3.51. Let X be the set defined in the preceding problem and

$$d(\xi, \eta) = \mathsf{E}[\min(|\xi - \eta|, 1)].$$

Prove that (X, d) is a complete metric space and that the convergence in this space is equivalent to the convergence in probability.

III.3.52. Consider the set of all random variables ξ with $\mathsf{E}\,|\xi|^r < \infty$, where $r > 0$ is a fixed number. Let L_r be the set of all equivalence classes for these random variables. For $\xi, \eta \in L_r$ put

$$d(\xi, \eta) = \begin{cases} \mathsf{E}\,|\xi - \eta|^r, & 0 < r < 1, \\ [\mathsf{E}\,|\xi - \eta|^r]^{1/r}, & r \geq 1. \end{cases}$$

Prove the following assertions:

a) L_r is a linear space;
b) d is a metric in L_r;
c) the convergence with respect to d coincides with the convergence in mean of order r;
d) (L_r, d) is a complete metric space, and moreover, L_r equipped with the norm $\|\xi\| = d(0, \xi)$ is a Banach space for $r \geq 1$;
e) $L_r \subset L_s$ for $1 \leq s \leq r$.

III.3.53. A sequence of distribution functions $\{F_n, n \geq 1\}$ weakly converges to a continuous distribution function F. Prove that $F_n(x) \to F(x)$ as $n \to \infty$ uniformly in $x \in \mathbf{R}$.

III.3.54. Let a sequence of random variables $\{\xi_n, n \geq 1\}$ satisfy the following condition: $\mathsf{P}(|\xi_n| \geq a) = 0$, $n \geq 1$, for some constant a, and $\xi_n \to c$ in probability as $n \to \infty$, where c is a constant. Let a sequence $\{\eta_n, n \geq 1\}$ be such that $\mathsf{E}\,\eta_n = d$, $n \geq 1$, for some $d \in \mathbf{R}$ and

$$\sup_{n \geq 1} \int_{\{|\eta_n| > A\}} |\eta_n| \, d\mathsf{P} \to 0 \quad \text{as } A \to \infty.$$

Prove that $\lim_{n\to\infty} \mathsf{E}\,\xi_n \eta_n = c\,\mathsf{E}\,\eta_1$.

III.3.55. Under the conditions of the preceding problem, construct an example showing that the above statement does not hold if we do not make any additional assumption on the sequence $\{\eta_n, n \geq 1\}$ except $\mathsf{E}\,\eta_n = d$, $n \geq 1$.

III.3.56. A sequence of random variables $\{\xi_n, n \geq 1\}$ is said to be *uniformly integrable* if

$$\lim_{a\to\infty} \sup_{n \geq 1} \int_{\{|\xi_n| > a\}} |\xi_n| \, d\mathsf{P} = 0.$$

Prove that any sequence of random variables converging in mean of order $r > 0$ is uniformly integrable.

III.3.57. A sequence of random variables $\{\xi_n, n \geq 1\}$ and a random variable η are such that $\mathsf{P}(|\xi_n| \leq \eta) = 1$, $n \geq 1$, and $\mathsf{E}\,\eta < \infty$. Prove that the sequence $\{\xi_n, n \geq 1\}$ is uniformly integrable. Construct an example showing that for some uniformly integrable sequences, such a random variable η may fail to exist.

III.3.58. Let a sequence of random variables $\{\xi_n, n \geq 1\}$ be uniformly integrable. Prove that

a) $\lim_{m\to\infty} \sup_{n\geq 1} \int_{A_m} |\xi_n| \, d\mathsf{P} = 0$ for all sequences $\{A_n, n \geq 1\}$ such that

$$A_m \supset A_{m+1}, \qquad \bigcap_{m=1}^{\infty} A_m = \varnothing;$$

b) for every $\varepsilon > 0$ there exists $\delta > 0$ such that $\sup_{n\geq 1} \int_A |\xi_n| \, d\mathsf{P} < \varepsilon$ for all A with $\mathsf{P}(A) < \delta$;

c) $\sup_{n\geq 1} \mathsf{E}\,|\xi_n| < \infty$.

III.3.59. Let a sequence of random variables $\{\xi_n, n \geq 1\}$ be uniformly integrable and converge in probability to a random variable ξ. Prove that $\{\xi_n, n \geq 1\}$ converges in mean of order 1.

III.3.60. Let $\{\xi_n, n \geq 1\}$ be a sequence of random variables with $\mathsf{E}\,|\xi_n| < \infty$ for all $n \geq 1$. If there exists a nonnegative function a, increasing for $t \geq 0$, such that

$$\lim_{t\to\infty} \frac{a(t)}{t} = +\infty, \qquad \sup_{n\geq 1} \mathsf{E}\, a(|\xi_n|) < \infty,$$

then the sequence $\{\xi_n, n \geq 1\}$ is uniformly integrable. Prove this statement.

III.3.61. A sequence of random variables $\{\xi_n, n \geq 1\}$ satisfies, for some $r > 0$, the following condition:

$$\sup_{n\geq 1} \mathsf{E}\,|\xi_n|^r < \infty.$$

Prove that for all s, $0 \leq s < r$, the sequence $\{|\xi_n|^s, n \geq 1\}$ is uniformly integrable. Construct an example showing that the above statement is not satisfied for $s = r$.

III.3.62. A sequence of random variables $\{\xi_n, n \geq 1\}$ is such that

$$\mathsf{E}\,\xi_n = 0, \qquad \mathsf{E}\,\xi_n^2 = \sigma^2, \quad n \geq 1,$$
$$\mathsf{E}\,\xi_m\xi_n = 0, \qquad m \neq n.$$

Prove that the sequence

$$\left\{ \frac{S_n}{\sqrt{n}}, n \geq 1 \right\}$$

is uniformly integrable. Here $S_n = \xi_1 + \xi_2 + \cdots + \xi_n$, $n \geq 1$.

III.3.63. A sequence of random variables $\{\xi_n, n \geq 1\}$ converges in square mean to a number c, and the sequence $\{\eta_n, n \geq 1\}$ is such that

$$\mathsf{E}\,\eta_n = d \in \mathbf{R}, \qquad n \geq 1,$$
$$\sup_{n\geq 1} \mathsf{E}\,\eta_n^2 < \infty.$$

Prove that $\lim_{n\to\infty} \mathsf{E}\,\xi_n\eta_n = c\,\mathsf{E}\,\eta_1$.

III.3.64. Sequences $\{\xi_n, n \geq 1\}$ and $\{\eta_n, n \geq 1\}$ converge in square mean to random variables ξ and η, respectively. Prove that

$$\lim_{n\to\infty} \mathsf{E}\,\xi_n\eta_n = \mathsf{E}\,\xi\eta.$$

In particular, if a random variable ζ is such that $\mathsf{E}\,\zeta^2 < \infty$, then

$$\lim_{n\to\infty} \mathsf{E}\,\xi_n\zeta = \mathsf{E}\,\xi\zeta.$$

III.3.65. Let $\{\xi_n, n \geq 1\}$ be a sequence of random variables such that for some number $c \in \mathbf{R}$

$$\mathsf{P}(|\xi_n| > c) = 0, \qquad n \geq 1.$$

Under this assumption, in order that $\xi_n \to 0$ in probability as $n \to \infty$, it is necessary and sufficient that

$$\lim_{n\to\infty} \mathsf{E}\,|\xi_n| = 0.$$

III.3.66. If $\exp\{\xi_n\} \to 1$ in probability as $n \to \infty$ for a sequence $\{\xi_n, n \geq 1\}$ of random variables, then $\xi_n \to 0$ in probability as $n \to \infty$.

III.3.67. Let random variables ξ_n, $n \geq 1$, take only positive integer values, and moreover,

$$\mathsf{P}(\xi_n = k) = \frac{c_n}{k^{2+1/n}}, \qquad k \geq 1,$$

where

$$c_n = \left(\sum_{k=1}^{\infty} \frac{1}{k^{2+1/n}}\right)^{-1}.$$

Prove that $\{\xi_n, n \geq 1\}$ converges in distribution to a random variable ξ with the distribution

$$\mathsf{P}(\xi = k) = \frac{c}{k^2}, \qquad c = \left(\sum_{k=1}^{\infty} \frac{1}{k^2}\right)^{-1}.$$

Prove also that $\mathsf{E}\,\xi_n < \infty$, $n \geq 1$, but $\mathsf{E}\,\xi = +\infty$.

Construct an example of a sequence $\{\xi_n, n \geq 1\}$ that converges in distribution to a random variable ξ and $\mathsf{E}\,|\xi_n| = +\infty$, $n \geq 1$, but $\mathsf{E}\,|\xi| < \infty$.

III.3.68. Prove that any sequence of random variables $\{\xi_n, n \geq 1\}$ converging in distribution is bounded in probability, that is,

$$\lim_{a\to\infty} \sup_{n\geq 1} \mathsf{P}(|\xi_n| > a) = 0.$$

III.3.69. Assume that a sequence $\{\xi_n, n \geq 1\}$ is bounded in probability, a sequence $\{\eta_n, n \geq 1\}$ converges in probability to 0, and g is a real continuous function. Prove that

$$g(\xi_n + \eta_n) - g(\xi_n) \to 0, \qquad n \to \infty,$$

in probability. In particular,

$$(\xi_n + \eta_n)^2 - \xi_n^2 \to 0, \qquad n \to \infty,$$

in probability.

III.3.70. Let all the conditions of the preceding problem be satisfied except the condition on the sequence $\{\eta_n, n \geq 1\}$. Suppose that $\eta_n \to c$ in probability as $n \to \infty$, for a number $c \in \mathbf{R}$. Prove that both the sequence

$$\{g(\xi_n + \eta_n) - g(\xi_n + c), n \geq 1\},$$

and the sequence

$$\{g(\xi_n/\eta_n) - g(\xi_n/c), n \geq 1\}$$

for $c \neq 0$, converge in probability to 0.

III.3.71. Prove that a sequence of random variables $\{\xi_n, n \geq 1\}$ converges in distribution to a random variable ξ if and only if for every real-valued bounded and continuous function g,

$$\lim_{n\to\infty} \mathsf{E}\, g(\xi_n) = \mathsf{E}\, g(\xi).$$

III.3.72. A sequence of random variables $\{\xi_n, n \geq 1\}$ converges in distribution to a random variable ξ, and a sequence $\{\eta_n, n \geq 1\}$ converges in probability to a constant c. Prove that the sequences

$$\{\xi_n + \eta_n, n \geq 1\}, \quad \{\xi_n \eta_n, n \geq 1\},$$

as well as the sequence

$$\{\xi_n/\eta_n, n \geq 1\},$$

for $c \neq 0$, converge in distribution to the random variables $\xi + c$, $c\xi$, and ξ/c, respectively.

III.3.73. Let ζ and ξ_n, η_n, $n \geq 1$, be random variables, and $\{\xi_n, n \geq 1\}$ converge in distribution to the random variable ζ. Suppose that, with probability one,

$$(*) \qquad |\xi_n - \eta_n| \leq \varepsilon_n |\xi_n|, \qquad n \geq 1,$$

for a sequence of random variables $\{\varepsilon_n, n \geq 1\}$ converging in probability to 0. Prove that $\{\eta_n, n \geq 1\}$ converges in distribution to ζ.

III.3.74. Let all the conditions of the preceding problem be satisfied except the inequality $(*)$. Suppose that

$$\mathsf{P}(|\xi_n - \eta_n| \leq \varepsilon_n |\eta_n|) = 1, \qquad n \geq 1.$$

Prove that the sequence $\{\eta_n, n \geq 1\}$ converges in distribution to ζ.

III.3.75. A sequence of random variables $\{\xi_n, n \geq 1\}$ converges in distribution to a random variable ξ. Prove that the sequence $\{h(\xi_n), n \geq 1\}$ converges in distribution to the random variable $h(\xi)$ for any continuous function h.

III.3.76. Let $\{\nu(n), n \geq 1\}$ be a sequence of random variables such that

$$\mathsf{P}\big(\nu(n) \leq \nu(n+1)\big) = 1, \qquad n \geq 1,$$
$$\nu(n) \to +\infty, \qquad n \to \infty,$$

in probability, that is, for all $c > 0$,

$$\mathsf{P}(\nu(n) > c) \to 1, \qquad n \to \infty.$$

Prove that

$$\mathsf{P}\left(\lim_{n\to\infty} \nu(n) = +\infty\right) = 1.$$

III.3.77. A sequence of random variables $\{\nu(n), n \geq 1\}$ converges in probability to $+\infty$ as $n \to \infty$. Let

$$\varkappa(n) = \begin{cases} 1/\nu(n) & \text{if } \nu(n) \neq 0, \\ 0 & \text{if } \nu(n) = 0. \end{cases}$$

Prove that $\varkappa(n) \to 0$ in probability as $n \to \infty$.

III.3.78. A sequence of random variables $\{\xi_n, n \geq 1\}$ converges with probability 1 to a random variable ξ as $n \to \infty$. Suppose that every random variable $\nu(n)$, $n \geq 1$, takes only positive integer values and

$$\mathsf{P}\left(\lim_{n\to\infty} \nu(n) = +\infty\right) = 1.$$

Prove that

$$\mathsf{P}\left(\lim_{n\to\infty} \xi_{\nu(n)} = \xi\right) = 1.$$

III.3.79. Let $\{\xi_n, n \geq 1\}$ be a sequence of random variables such that

$$\mathsf{P}(\xi_n = 0) = 1 - \frac{1}{n}, \qquad \mathsf{P}(\xi_n = 1) = \frac{1}{n}, \qquad n \geq 1.$$

Define a sequence of random variables $\{\nu(n), n \geq 1\}$ as follows:

$$\begin{aligned} \nu(1) &= \min\{n \geq 1 \colon \xi_n = 1\}, \\ \nu(n+1) &= \min\{m > \nu(n) \colon \xi_m = 1\}, \qquad n \geq 1. \end{aligned}$$

Prove that

a) $\xi_n \to 0$ in probability as $n \to \infty$;

b) the random variables $\nu(n)$, $n \geq 1$, are well defined with probability one, and moreover,

$$n \leq \nu(n) < \nu(n+1), \qquad n \geq 1,$$
$$\mathsf{P}\left(\lim_{n\to\infty} \nu(n) = +\infty\right) = 1;$$

c) $\xi_{\nu(n)} \equiv 1$, $n \geq 1$.

III.3.80. Let a sequence of random variables $\{\xi_n, n \geq 1\}$ converge in probability to a random variable ξ, and let $\{\nu(n), n \geq 1\}$ be a sequence of positive integer-valued random variables such that $\nu(n) \to +\infty$ in probability as $n \to \infty$. Assume that the two collections $\{\xi_n, n \geq 1\}$ and $\{\nu(n), n \geq 1\}$ are independent. Prove that $\xi_{\nu(n)} \to \xi$ in probability as $n \to \infty$.

III.3.81. A sequence $\{\xi_n, n \geq 1\}$ converges in distribution to a random variable ξ, and $\{\nu(n), n \geq 1\}$ is a sequence of positive integer-valued random variables such that $\nu(n) \to +\infty$ in probability as $n \to \infty$. Assume that the two collections $\{\xi_n, n \geq 1\}$ and $\{\nu(n), n \geq 1\}$ are independent. Prove that $\xi_{\nu(n)} \to \xi$ in distribution as $n \to \infty$.

III.3.82. A sequence $\{\xi_n, n \geq 1\}$ converges in mean of order $r > 0$ to a random variable ξ, and $\{\nu(n), n \geq 1\}$ is a sequence of positive integer-valued random variables such that $\nu(n) \to +\infty$ in probability as $n \to \infty$. Assume that the two collections $\{\xi, \xi_n, n \geq 1\}$ and $\{\nu(n), n \geq 1\}$ are independent. Prove that $\xi_{\nu(n)} \to \xi$ in mean of order r as $n \to \infty$.

III.3.83. Let $\{\xi_n, n \geq 1\}$ be a sequence of independent identically distributed random variables, and f_n, $n \geq 1$, a real Borel symmetric function of n variables, that is,

$$f_n(x_1, x_2, \ldots, x_n) = f_n(x_{i_1}, x_{i_2}, \ldots, x_{i_n})$$

for any permutation $i_1, i_2, \ldots, i_n$ of numbers $1, 2, \ldots, n$. If

$$\lim_{n\to\infty} f_n(\xi_1, \xi_2, \ldots, \xi_n) = \xi$$

in probability, then $\mathsf{P}(\xi = c) = 1$, where $c \in \mathbf{R}$.

III.3.84. Let $\{F_n, n \geq 1\}$ be a sequence of distribution functions that weakly converges to a distribution function F. Prove that there exists a probability space $(\Omega, \mathfrak{A}, \mathsf{P})$ and random variables ξ and $\{\xi_n, n \geq 1\}$ defined on this space with the distribution functions F and $\{F_n, n \geq 1\}$, respectively, such that

$$\mathsf{P}\left(\lim_{n\to\infty} \xi_n = \xi\right) = 1.$$

III.3.85. Let $\{\xi_n, n \geq 1\}$ be a sequence of random variables converging in distribution to a random variable ξ. Prove that

$$\mathsf{E}\,|\xi| \leq \varliminf_{n\to\infty} \mathsf{E}\,|\xi_n|.$$

III.3.86. A sequence of distribution functions $\{F_n, n \geq 1\}$ weakly converges to a distribution function F, and the moments of order $r > 0$ of the functions F_n, $n \geq 1$, are uniformly bounded, that is,

$$\sup_{n\geq 1} \int_{-\infty}^{+\infty} |x|^r \, dF_n(t) < \infty.$$

Prove that for all s, $0 \leq s < r$,

$$\lim_{n\to\infty} \int_{-\infty}^{+\infty} |x|^s \, dF_n(t) = \int_{-\infty}^{+\infty} |x|^s \, dF(t).$$

In particular, for a positive integer m, $m < r$, we have

$$\lim_{n\to\infty} \int_{-\infty}^{+\infty} x^m \, dF_n(t) = \int_{-\infty}^{+\infty} x^m \, dF(t).$$

III.3.87. Let D be the set of all distribution functions. For $F, G \in D$, put

$$d(F,G) = \inf\{\varepsilon > 0 \colon F(x-\varepsilon) - \varepsilon \leq G(x) \leq F(x+\varepsilon) + \varepsilon, x \in \mathbf{R}\}.$$

a) Determine $d(F, G_n)$ if F is the uniform distribution on the segment $[0,1]$, and G_n, $n \geq 1$, is the distribution function of a random variable ξ_n such that

$$\mathsf{P}\left(\xi_n = \frac{k}{n}\right) = \frac{1}{n+1}, \qquad k = 0, 1, 2, \ldots, n; \quad n \geq 1.$$

Prove that

b) d is a distance on D;
c) the convergence with respect to d is equivalent to the weak convergence;
d) (D, d) is a complete metric space.

III.3.88. Prove that a sequence of m-dimensional random vectors

$$\left\{\vec{\xi}^{(n)} = \left(\xi_1^{(n)}, \xi_2^{(n)}, \dots, \xi_m^{(n)}\right), n \ge 1\right\}$$

converges in probability to an m-dimensional random variable $\vec{\xi} = (\xi_1, \xi_2, \dots, \xi_m)$ if and only if for all k, $1 \le k \le m$,

$$\xi_k^{(n)} \to \xi_k, \qquad n \to \infty,$$

in probability.

III.3.89. A sequence $\{\vec{\xi}^{(n)}, n \ge 1\}$ of random vectors in $\mathbf{R}^m$ converges in probability to a random vector $\vec{\xi} \in \mathbf{R}^m$, and g is a bounded continuous real function on $\mathbf{R}^m$. Prove that

$$\lim_{n\to\infty} \mathsf{E}\, g(\vec{\xi}^{(n)}) = \mathsf{E}\, g(\vec{\xi}).$$

III.3.90. If a sequence of random vectors converges in probability, then the sequence of the corresponding distribution functions weakly converges to a distribution function.

III.3.91. A sequence $\{\vec{\xi}^{(n)}, n \ge 1\}$ of random vectors in $\mathbf{R}^m$ converges in probability if and only if there exists a random variable $\vec{\xi} \in \mathbf{R}^m$ such that any subsequence $\{\vec{\xi}^{(n(k))}, k \ge 1\}$ contains another subsequence converging with probability one to $\vec{\xi}$ in the norm of $\mathbf{R}^m$. Prove this statement.

III.3.92. If a sequence of distribution functions $\{F_n, n \ge 1\}$ defined on $\mathbf{R}^m$ weakly converges to a continuous distribution function F on $\mathbf{R}^m$, then $F_n \to F$ uniformly on $\mathbf{R}^m$ as $n \to \infty$. Prove this statement.

III.3.93. A sequence $\{\vec{\xi}^{(n)}, n \ge 1\}$ of m-dimensional random vectors converges in distribution to an m-dimensional random vector $\vec{\xi}$, and h is a real continuous function on $\mathbf{R}^m$. Prove that the sequence $\{h(\vec{\xi}_n), n \ge 1\}$ converges in distribution to the random variable $h(\vec{\xi})$.

III.3.94. A sequence of random vectors $\{\vec{\xi}^{(n)}, n \ge 1\}$ in $\mathbf{R}^m$ converges in distribution to a random vector $\vec{\xi}$, and a sequence $\{\vec{\eta}_n, n \ge 1\}$ in $\mathbf{R}^k$ converges in probability to a constant vector $\vec{c} \in \mathbf{R}^k$. Prove that the sequence

$$\{h(\vec{\xi}^{(n)}, \vec{\eta}^{(n)}) - h(\vec{\xi}^{(n)}, c), n \ge 1\}$$

converges in probability to zero for every continuous real function h defined on $\mathbf{R}^{m+k}$.

III.3.95. Let $\{A_n, n \ge 1\}$ be a sequence of $m \times m$ matrices with random entries, $A_n = \left(\alpha_{jk}^{(n)}\right)_{j,k=1}^m$. Suppose that the sequence $\left\{\alpha_{jk}^{(n)}, n \ge 1\right\}$ converges in probability to a number c_{jk} as $n \to \infty$, for all j, k, $1 \le j, k \le m$. Put $A = (c_{jk})_{j,k=1}^m$.

Suppose a sequence $\{\vec{\xi}^{(n)}, n \geq 1\}$ of random vectors in $\mathbf{R}^m$ converges in distribution to $\vec{\xi}$ as $n \to \infty$. Prove that the sequence of random vectors

$$\left\{A_n \vec{\xi}^{(n)}, n \geq 1\right\}$$

converges in distribution to $A\vec{\xi}$.

§III.4. The law of large numbers

A sequence of random variables $\{\xi_n, n \geq 1\}$ with $\mathsf{E}\,|\xi_n| < \infty$ is said *to satisfy the law of large numbers* if

$$\frac{1}{n}\sum_{k=1}^{n} \xi_k - \frac{1}{n}\sum_{k=1}^{n} \mathsf{E}\,\xi_k \to 0, \qquad n \to \infty,$$

in probability, i.e. if for any $\varepsilon > 0$

$$\lim_{n\to\infty} \mathsf{P}\left(\left|\frac{1}{n}\sum_{k=1}^{n} \xi_k - \frac{1}{n}\sum_{k=1}^{n} \mathsf{E}\,\xi_k\right| > \varepsilon\right) = 0.$$

THEOREM 1. *A sequence $\{\xi_n, n \geq 1\}$ of independent random variables such that $\mathsf{E}\,\xi_n^2 < \infty$, $n \geq 1$, and $(1/n)\operatorname{Var}\xi_n \to 0$ as $n \to \infty$, satisfies the law of large numbers.*

THEOREM 2 (Khintchine's law of large numbers). *Any sequence $\{\xi_n, n \geq 1\}$ of independent identically distributed random variables such that $\mathsf{E}\,|\xi_1| < \infty$ satisfies the law of large numbers, namely,*

$$\frac{1}{n}\sum_{k=1}^{n} \xi_k \to \mu, \qquad n \to \infty,$$

in probability, where $\mu = \mathsf{E}\,\xi_1$.

Problems

III.4.1. A sequence of random variables $\{\xi_n, n \geq 1\}$ satisfies the condition

$$\lim_{n\to\infty} \frac{1}{n^2} \operatorname{Var}\left(\sum_{k=1}^{n} \xi_k\right) = 0.$$

Prove that this sequence satisfies the law of large numbers.

III.4.2. A sequence of random variables $\{\xi_n, n \geq 1\}$ satisfies the conditions

$$\mathsf{E}\,\xi_n = 0,$$
$$\mathsf{P}\left(\left|\frac{\xi_1 + \xi_2 + \cdots + \xi_n}{n}\right| > c\right) = 0, \qquad n \geq 1,$$

for some constant c, and the sequence

$$\frac{1}{n^2}\mathsf{E}\left(\sum_{k=1}^{n} \xi_k\right)^2, \qquad n \geq 1,$$

does not tend to zero as $n \to \infty$. Prove that the law of large numbers is not satisfied for the sequence $\{\xi_n, n \geq 1\}$.

III.4.3. Let $\{\xi_n, n \geq 1\}$ be a sequence of independent random variables uniformly distributed on the interval $[0,1]$, and f a real periodic function with period 1, continuous on $\mathbf{R}$. Prove that for any $x \in \mathbf{R}$,

$$\frac{1}{n}\sum_{k=1}^{n} f(x+\xi_k) \to \int_0^1 f(u)\,du, \qquad n \to \infty,$$

in probability.

III.4.4. For a real-valued and continuous function f on $[0,1]$ calculate the limit

$$\lim_{n\to\infty} \int_0^1 \overset{n}{\cdots} \int_0^1 f\left(\frac{x_1+x_2+\cdots+x_n}{n}\right) dx_1\,dx_2\cdots dx_n.$$

III.4.5. Let $m \in \mathbf{N}$. Calculate the limit

$$\lim_{n\to\infty} \int_0^1 \overset{n}{\cdots} \int_0^1 \cos^{2m} \frac{\pi}{2n}(x_1+\cdots+x_n)\,dx_1\,dx_2\cdots dx_n.$$

III.4.6. For a real-valued and continuous function f on $[0,1]$ prove that

$$\lim_{n\to\infty} \int_0^1 \overset{n}{\cdots} \int_0^1 f\left(\sqrt[n]{x_1x_2\cdots x_n}\right) dx_1\,dx_2\cdots dx_n = f\left(\frac{1}{e}\right).$$

III.4.7. Real-valued and continuous functions f and g on $[0,1]$ are such that $0 \leq f(x) \leq cg(x)$ for all $x \in (0,1)$ and some $c > 0$. Prove the equality

$$\lim_{n\to\infty} \int_0^1 \overset{n}{\cdots} \int_0^1 \frac{f(x_1)+f(x_2)+\cdots+f(x_n)}{g(x_1)+g(x_2)+\cdots+g(x_n)}\,dx_1\,dx_2\cdots dx_n = \frac{\int_0^1 f(x)\,dx}{\int_0^1 g(x)\,dx}.$$

III.4.8. Prove that

$$\lim_{n\to\infty} \int_0^1 \overset{n}{\cdots} \int_0^1 \frac{x_1^2+x_2^2+\cdots+x_n^2}{x_1+x_2+\cdots+x_n}\,dx_1\,dx_2\cdots dx_n = \frac{2}{3}.$$

III.4.9. A real-valued function f is twice continuously differentiable on the interval $[0,1]$. Let f'' be its second derivative. Prove that

$$\lim_{n\to\infty} n \int_0^1 \overset{n}{\cdots} \int_0^1 \left[f\left(\frac{x_1+\cdots+x_n}{n}\right) - f\left(\frac{1}{2}\right)\right] dx_1\,dx_2\cdots dx_n = \frac{f''(1/2)}{24}.$$

III.4.10. Compute the following limits:

a) $$\lim_{n\to\infty} \underset{\substack{0\leq x_i\leq 1,\,1\leq i\leq n,\\ x_1^2+x_2^2+\cdots+x_n^2\leq\sqrt{n}}}{\int \overset{n}{\cdots} \int} dx_1\,dx_2\cdots dx_n,$$

b) $$\lim_{n\to\infty} \underset{\substack{0\leq x_i\leq 1,\,1\leq i\leq n,\\ x_1^2+x_2^2+\cdots+x_n^2\leq\frac{n}{4}}}{\int \overset{n}{\cdots} \int} dx_1\,dx_2\cdots dx_n,$$

c) $$\lim_{n\to\infty} \underset{\substack{0\leq x_i\leq 1,\,1\leq i\leq n,\\ x_1^2+x_2^2+\cdots+x_n^2\leq\frac{n}{2}}}{\int \overset{n}{\cdots} \int} dx_1\,dx_2\cdots dx_n.$$

III.4.11. Let f be a real-valued nonnegative function defined on $\mathbf{R}$ and such that

$$\int_{-\infty}^{+\infty} f(x)\,dx = 1, \qquad \sigma^2 = \int_{-\infty}^{+\infty} x^2 f(x)\,dx < \infty.$$

Compute the limits

a) $$\lim_{n\to\infty} \underset{x_1^2+x_2^2+\cdots+x_n^2\le\sqrt{n}}{\int \overset{n}{\cdots} \int} f(x_1)f(x_2)\cdots f(x_n)\,dx_1\,dx_2\cdots dx_n,$$

b) $$\lim_{n\to\infty} \underset{x_1^2+x_2^2+\cdots+x_n^2\le an}{\int \overset{n}{\cdots} \int} f(x_1)f(x_2)\cdots f(x_n)\,dx_1\,dx_2\cdots dx_n, \qquad a < \sigma^2,$$

c) $$\lim_{n\to\infty} \underset{x_1^2+x_2^2+\cdots+x_n^2\le an}{\int \overset{n}{\cdots} \int} f(x_1)f(x_2)\cdots f(x_n)\,dx_1\,dx_2\cdots dx_n, \qquad a > \sigma^2.$$

III.4.12. A sequence of independent random variables $\{\xi_n, n \ge 1\}$ satisfies the conditions

$$\frac{1}{n}\sum_{k=1}^{n} \mathsf{E}\,\xi_k \to x, \qquad \frac{\operatorname{Var}\xi_n}{n} \to 0, \qquad n \to \infty.$$

Prove that for all bounded and continuous real functions f on $\mathbf{R}$,

$$\lim_{n\to\infty} \mathsf{E}\, f\left(\frac{1}{n}\sum_{k=1}^{n} \xi_k\right) = f(x).$$

III.4.13. Let $\{\xi_n, n \ge 1\}$ be a sequence of independent random variables uniformly distributed on $[0,1]$. Prove that

$$\left\{e^n \prod_{k=1}^{n} \xi_k\right\}^{1/n} \to 1, \qquad n \to \infty,$$

in probability.

III.4.14. Let $\{\xi_n, n \ge 1\}$ be a sequence of independent exponential random variables with parameter $\lambda = 1$. Prove that

$$\left\{\prod_{k=1}^{n} \xi_k\right\}^{1/n} \to e^{-c}, \qquad n \to \infty,$$

in probability. Here c is the Euler constant,

$$-c = \Gamma'(1) = \int_0^{\infty} \ln x e^{-x}\,dx.$$

III.4.15. Random variables $\xi_1, \xi_2, \ldots$ are defined by

$$\xi_n = \alpha_0\varepsilon_n + \alpha_1\varepsilon_{n-1} + \cdots + \alpha_r\varepsilon_{n-r}, \qquad n \ge 1,$$

where $\alpha_0, \alpha_1, \ldots, \alpha_r$ are real numbers and $\{\varepsilon_n, n \ge -r+1\}$ is a sequence of independent identically distributed random variables with $\mathsf{E}\,|\varepsilon_1| < \infty$. Prove that the sequence $\{\xi_n, n \ge 1\}$ satisfies the law of large numbers.

III.4.16. Random variables $\xi_1, \xi_2, \dots$ satisfy the relation

$$\xi_n + \alpha\xi_{n-1} = \varepsilon_n, \qquad n \geq 1,$$

where α, $|\alpha| < 1$, is a fixed number, ξ_0 is a random variable, and $\{\varepsilon_n, n \geq 1\}$ is a sequence of independent identically distributed random variables with $\mathsf{E}\,|\varepsilon_1| < \infty$. Prove that the sequence $\{\xi_n, n \geq 1\}$ satisfies the law of large numbers.

III.4.17. In addition to the assumptions of the preceding problem, let

$$\mathsf{E}\,\varepsilon_1 = 0, \qquad \mathsf{E}\,\varepsilon_1^4 < \infty.$$

Prove that

$$\frac{1}{n}\sum_{k=1}^{n} \xi_k\xi_{k-1}\left(\frac{1}{n}\sum_{k=1}^{n}\xi_{k-1}^2\right)^{-1} \to -\alpha, \qquad n \to \infty,$$

in probability.

III.4.18. Let $\{\xi_n, n \geq 1\}$ be a sequence of independent identically distributed random variables, and g a real-valued Borel function on $\mathbf{R}^m$ such that

$$\mathsf{E}\,|g(\xi_1, \xi_2, \dots, \xi_m)|^2 < \infty.$$

Prove that

$$\frac{1}{n}\sum_{k=1}^{n} g(\xi_k, \xi_{k+1}, \dots, \xi_{k+m-1}) \to \mathsf{E}\,g(\xi_1, \xi_2, \dots, \xi_m), \qquad n \to \infty,$$

in probability.

III.4.19. Let $\{\xi_n, n \geq 1\}$ be a sequence of independent identically distributed random variables with $\mathsf{E}\,\xi_1 = \mu$, $\operatorname{Var}\xi_1 = \sigma^2 < \infty$. Prove that

$$\bar{x} = \frac{1}{n}\sum_{k=1}^{n}\xi_k \to \mu, \qquad n \to \infty,$$

$$S^2 = \frac{1}{n}\sum_{k=1}^{n}(\xi_k - \bar{x})^2 \to \sigma^2, \qquad n \to \infty,$$

in probability.

III.4.20. Let $\{\xi_n, n \geq 1\}$ be a sequence of independent identically distributed random variables. For an integer $n \geq 1$ and a real $x \in \mathbf{R}$ define the function $F_n(x)$ by

$$F_n(x) = \frac{\#(k\colon \xi_k < x)}{n},$$

where $\#(k\colon \xi_k < x)$ stands for the total number of those ξ_k's that are less than x. For all $n \geq 1$, $F_n(x)$ is a distribution function. Find $\mathsf{E}\,F_n(x)$, $\operatorname{Var} F_n(x)$ for $x \in \mathbf{R}$, and prove that

$$F_n(x) \to F(x) = \mathsf{P}(\xi_1 < x), \qquad n \to \infty,$$

in probability for all $x \in \mathbf{R}$.

III.4.21. Let $0 < r \le 2$, and let $\{\xi_n, n \ge 1\}$ be a sequence of random variables with $\mathsf{E}\,|\xi_n|^r < \infty$, $n \ge 1$. Assume that

$$\frac{1}{n^{r/2}} \sum_{k=1}^{n} \sqrt{\mathsf{E}\,|\xi_k|^r} \to 0, \qquad n \to \infty.$$

Prove that

$$\frac{1}{n} \sum_{k=1}^{n} \xi_k \to 0, \qquad n \to \infty,$$

in probability.

III.4.22. Random variables $\{\xi_n, n \ge 1\}$ are independent and

$$\mathsf{P}(\xi_n = n^s) = \mathsf{P}(\xi_n = -n^s) = \frac{1}{2}$$

for all $n \ge 1$. Prove that the sequence $\{\xi_n, n \ge 1\}$ satisfies the law of large numbers if $s < \frac{1}{2}$.

III.4.23. Let $\{\xi_n, n \ge 1\}$ be a sequence of independent identically distributed random variables with $\mathsf{E}\,\xi_1 = \mu$, $\operatorname{Var} \xi_1 = \sigma^2 < \infty$, and $\mathsf{P}(\xi_1 = 0) = 0$. Prove that

$$\frac{\xi_1 + \xi_2 + \cdots + \xi_n}{\xi_1^2 + \xi_2^2 + \cdots + \xi_n^2} \to \frac{\mu}{\mu^2 + \sigma^2}, \qquad n \to \infty,$$

in probability.

III.4.24. A sequence of random variables $\{\eta_n(x), n \ge 1\}$, $x \in A \subset \mathbf{R}$, satisfies the conditions

$$\mathsf{E}\,\eta_n(x) = x, \quad \operatorname{Var} \eta_n(x) = \sigma^2(x), \qquad n \ge 1,\ x \in A,$$
$$\sup_{x \in A} \sigma_n^2(x) \to 0, \qquad n \to \infty.$$

Let f be a bounded function on $\mathbf{R}$ uniformly continuous on A. Prove that

$$\mathsf{E}\, f(\eta_n(x)) \to f(x), \qquad n \to \infty,$$

uniformly in $x \in A$.

III.4.25. Let f be a real-valued continuous function on $[0, 1]$. Prove that Bernstein's polynomials

$$B_n(f; x) = \sum_{k=0}^{n} \binom{n}{k} f\left(\frac{k}{n}\right) x^k (1-x)^{n-k}, \qquad n \ge 1,$$

converge to f uniformly in $x \in [0, 1]$.

III.4.26. Let f be a real-valued continuous function on $[0, +\infty)$. Prove that

$$e^{-nx} \sum_{k=0}^{\infty} f\left(\frac{k}{n}\right) \frac{(nx)^k}{k!} \to f(x), \qquad n \to \infty,$$

uniformly in x in every finite interval.

III.4.27. Prove that

$$e^{-\lambda x} \sum_{k:k\le\lambda\theta} \frac{(\lambda x)^k}{k!} \underset{\lambda\to\infty}{\longrightarrow} \begin{cases} 0, & \theta < x, \\ 1, & \theta > x. \end{cases}$$

III.4.28. Let F be a distribution function on $[0,\infty)$, and φ its Laplace transform,

$$\varphi(\lambda) = \int_0^\infty e^{-\lambda x}\, dF(x), \qquad \lambda \ge 0.$$

Using the result of the preceding problem, prove the following inverse formula for the Laplace transformation:

$$\sum_{k:k<\lambda x} \frac{(-1)^k}{k!} \lambda^k \varphi^{(k)}(\lambda) \to F(x), \qquad \lambda \to \infty,$$

for any point x where F is continuous.

III.4.29. Let $\{\xi_n, n \ge 1\}$ be a sequence of independent identically distributed random variables with $\mathsf{E}\,\xi_1 = 0$. Prove that

a) the sequence

$$\left\{ \frac{1}{n} \sum_{k=1}^{n} \xi_k, n \ge 1 \right\}$$

is uniformly integrable;

b) $\frac{1}{n}\,\mathsf{E}\,|\xi_1 + \xi_2 + \dots + \xi_n| \to 0$ as $n \to \infty$.

III.4.30. Let ν, ξ_n, $n \ge 1$, be independent random variables and $\mathsf{E}\,\xi_n = 0$, $n \ge 1$. Assume that the random variable ν has the Poisson distribution with parameter λ and

$$\frac{1}{n}\,\mathsf{E}\,\xi_n^2 \to 0, \qquad n \to \infty.$$

Prove that

$$\frac{1}{\nu+1} \sum_{k=1}^{\nu} \xi_k \to 0, \qquad \lambda \to \infty,$$

in probability.

III.4.31. A sequence of random variables $\{\xi_n, n \ge 1\}$ is such that

$$\frac{1}{n} \sum_{k=1}^{n} \xi_k \to 0, \qquad n \to \infty,$$

in probability, and $\{\nu(n), n \ge 1\}$ is a sequence of random variables such that any $\nu(n)$ assumes only positive integer values and $\nu(n) \to \infty$ in probability as $n \to \infty$. Assume that the two collections of random variables $\{\xi_n, n \ge 1\}$ and $\{\nu(n), n \ge 1\}$ are independent. Prove that

$$\frac{1}{\nu(n)} \sum_{k=1}^{\nu(n)} \xi_k \to 0, \qquad n \to \infty,$$

in probability.

§III.5. Kolmogorov's inequality and some related inequalities

The following inequalities play a principal role while studying the convergence of series of independent random variables. In what follows we use the notation $S_n = \xi_1 + \xi_2 + \cdots + \xi_n$, $n \geq 1$, for a sequence $\{\xi_n, n \geq 1\}$.

THEOREM 1 (Kolmogorov's inequality). *Let $\xi_1, \xi_2, \ldots, \xi_n$ be independent random variables and*

$$\mathsf{E}\,\xi_k = 0, \qquad \operatorname{Var}\xi_k < \infty, \qquad 1 \leq k \leq n.$$

The following inequality

$$\mathsf{P}\Big(\max_{1\leq k\leq n} |S_k| \geq a\Big) \leq \frac{\operatorname{Var} S_n}{a^2}$$

holds for all $a > 0$.

THEOREM 2. *Let $\{\xi_k, 1 \leq k \leq n\}$ be independent random variables such that $\mathsf{E}\,\xi_k = 0$ and $\mathsf{P}(|\xi_k| > c) = 0$ for all $1 \leq k \leq n$ and some $c > 0$. Then*

$$\mathsf{P}\Big(\max_{1\leq k\leq n} |S_k| \geq a\Big) \geq 1 - \frac{(a+c)^2}{\operatorname{Var} S_n}$$

for all $a > 0$.

A random variable ξ is said to be *symmetric* if the distributions of the random variables ξ and $-\xi$ coincide.

THEOREM 3 (Levy's inequality). *Let $\{\xi_k, 1 \leq k \leq n\}$ be independent symmetric random variables. The inequality*

$$\mathsf{P}\Big(\max_{1\leq k\leq n} |S_k| \geq a\Big) \leq 2\mathsf{P}(|S_n| \geq a)$$

holds for all $a > 0$.

Problems

III.5.1. Prove that for a sequence of independent random variables $\{\xi_n, n \geq 1\}$ with $\mathsf{E}\,\xi_n = 0$, $n \geq 1$, and

$$\sum_{n=1}^{\infty} \operatorname{Var}\xi_n < \infty$$

the inequality

$$\mathsf{P}\left(\sup_{n\geq 1}\left|\sum_{k=1}^{n} \xi_k\right| \geq a\right) \leq \frac{1}{a^2}\sum_{n=1}^{\infty} \operatorname{Var}\xi_n$$

holds for all $a > 0$.

III.5.2. Random variables $\{\xi_k, 1 \leq k \leq n\}$ are independent and $\mathsf{E}\,\xi_k = a_k$, $\operatorname{Var}\xi_k < \infty$, $1 \leq k \leq n$. Prove that

$$\mathsf{P}\left(\max_{1\leq k\leq n}\left|\sum_{i=1}^{k} \xi_i\right| \geq a\right) \leq \frac{\sum_{i=1}^{n} \operatorname{Var}\xi_i}{\left(a - \max_{1\leq k\leq n}\left|\sum_{i=1}^{k} a_i\right|\right)^2}$$

for all numbers a satisfying $a > \max_{1\leq k\leq n}\left|\sum_{i=1}^{k} a_i\right|$.

III.5.3. Random variables $\{\xi_k, 1 \le k \le n\}$ are independent and

$$\mathsf{P}(|\xi_k| > c) = 0, \qquad 1 \le k \le n,$$

$$\mathsf{P}\Big(\max_{m \le i \le n} \Big|\sum_{k=m}^{i} \xi_k\Big| \ge a\Big) \le \alpha$$

for some numbers c, α, a. Prove that for all m, $1 \le m \le n$, the following inequalities hold:

a) $\mathsf{P}\Big(\max_{m \le l \le n} \Big|\sum_{k=m}^{l} \xi_k\Big| \ge 2a + c\Big) \le \alpha^2;$

b) $\mathsf{P}\Big(\max_{m \le l \le n} \Big|\sum_{k=m}^{l} \xi_k\Big| \ge sa + (s-1)c\Big) \le \alpha^s, \qquad s \in \mathbf{N};$

c) $\mathsf{E}\Big(\max_{m \le l \le n} \Big|\sum_{k=m}^{l} \xi_k\Big|\Big)^r \le r\sum_{s=1}^{\infty} \alpha^{s-1}[s(a+c)]^r, \qquad r > 0.$

III.5.4. Assume that random variables $\{\xi_k, 1 \le k \le n\}$ are independent and $\mathsf{E}\,\xi_k = 0$, $1 \le k \le n$. Let g be a continuous nonnegative convex and nondecreasing function defined on $[0,\infty)$. Prove that

$$\mathsf{P}\Big(\max_{1 \le k \le n} |S_k| \ge a\Big) \le \frac{\mathsf{E}\,g(S_n)}{g(a)}$$

for all $a > 0$.

III.5.5. Independent and symmetric random variables ξ_k, $1 \le k \le n$, are such that $\xi_k^2 = 1$, $1 \le k \le n$. Prove the equality

$$\mathsf{P}\Big(\max_{1 \le k \le n} S_k \ge l\Big) = 2\mathsf{P}(S_n > l) + \mathsf{P}(S_n = l).$$

III.5.6. Random variables ξ_k, $1 \le k \le n$, satisfy the conditions of the preceding problem. We say that *the sums S_n cross an interval* $[-a,a]$, $a > 0$, *at least r times* if there exist $k_1, k_2, \dots, k_{r+1}$ such that either

$$S_{k_1} \ge a, \ S_{k_2} \le -a, \ S_{k_3} \ge a, \dots$$

or

$$S_{k_1} \le -a, \ S_{k_2} \ge a, \ S_{k_3} \le -a, \dots .$$

Let q_r be the probability that the sums $S_1, S_2, \dots, S_n$ cross an interval $[-a,a]$ at least r times. Prove that

$$q_r + q_{r+1} = \mathsf{P}\Big(\max_{1 \le k \le n} S_k \ge (2r+1)a\Big).$$

III.5.7. Let $\xi_0 = 0$ and random variables $\{\xi_k, 1 \le k \le n\}$ satisfy the conditions

$$\mathsf{E}\,\xi_k^2 < \infty, \qquad \mathsf{E}\,\{\xi_k/\xi_1, \xi_2, \dots, \xi_{k-1}\} = 0, \qquad 1 \le k \le n.$$

For every $a > 0$, prove the inequality

$$\mathsf{P}\Big(\max_{1 \le k \le n} |S_k| \ge a\Big) \le \frac{\mathsf{E}\,S_n^2}{a^2} = \frac{1}{a^2}\sum_{k=1}^{n} \mathsf{E}\,\xi_k^2.$$

III.5.8. Random variables ξ_k, $1 \le k \le n$ are independent and $\mathsf{E}\,\xi_k = 0$, $\mathsf{E}\,\xi_k^2 < \infty$, $1 \le k \le n$. For any collection of numbers $0 < a_1 \le a_1 \le \cdots \le a_n$, prove the inequality

$$\mathsf{P}(|S_k| \le a_k, 1 \le k \le n) \ge 1 - \sum_{k=1}^{n} \frac{\mathsf{E}\,\xi_k^2}{a_k^2}.$$

III.5.9. Random variables ξ_k, $1 \le k \le n$, satisfy the following conditions:

a) $\mathsf{E}\,|\xi_k| < \infty$, $1 \le k \le n$;

b) $\mathsf{E}\left\{\xi_k/\xi_1, \xi_2, \ldots, \xi_{k-1}\right\} \ge \xi_{k-1}$, $2 \le k \le n$, with probability one.

Such a sequence of random variables is said to be *a submartingale*. For every $\varepsilon > 0$, prove the inequality

$$\mathsf{P}\Big(\max_{1\le k\le n} \xi_k \ge \varepsilon\Big) \le \frac{1}{\varepsilon}\,\mathsf{E}\max(0, \xi_n).$$

III.5.10. Let ξ_k, $1 \le k \le n$, be independent random variables with $\mathsf{E}\,\xi_k = 0$, $\mathsf{E}\,\xi_k^2 < \infty$, $1 \le k \le n$. Prove the inequality

$$\mathsf{E}\Big\{\max_{1\le k\le n} |S_k|\Big\} \le 2\sqrt{\mathsf{E}\,S_n^2}.$$

§III.6. Series of independent random variables

Let $(\Omega, \mathfrak{A}, \mathsf{P})$ be a probability space, ξ and ξ_n, $n \ge 1$, random variables, and $S_n = \xi_1 + \xi_2 + \cdots + \xi_n$, $n \ge 2$. A series $\sum_{n=1}^{\infty} \xi_n$ of random variables is said *to converge in probability* (*with probability one, in mean of order r, in distribution*) *to a random variable* ξ if the sequence of random variables $\{S_n, n \ge 1\}$ converges to the random variable ξ in probability (with probability one, in mean of order r, in distribution), respectively, as $n \to \infty$. In the case of the convergence in mean of order r we assume that $\mathsf{E}\,|\xi_n|^r < \infty$, $n \ge 1$, and $\mathsf{E}\,|\xi|^r < \infty$.

The convergence results on series of independent random variables are collected in the following theorems.

THEOREM 1. *Let $\{\xi_n, n \ge 1\}$ be a sequence of independent random variables with $\mathsf{E}\,\xi_n^2 < \infty$, $n \ge 1$. If the series*

$$\sum_{n=1}^{\infty} \mathsf{E}\,\xi_n, \qquad \sum_{n=1}^{\infty} \operatorname{Var}\xi_n \tag{1}$$

are convergent, then the series

$$\sum_{n=1}^{\infty} \xi_n \tag{2}$$

converges with probability one.

On the other hand, if $\mathsf{P}(|\xi_k| > c) = 0$, $n \ge 1$, for a positive c, then the convergence with probability one of the series (2) *implies the convergence of the series* (1).

Recall that $\operatorname{Var}\xi = \mathsf{E}\,\xi^2 - (\mathsf{E}\,\xi)^2$ and note that under the conditions of the first part of Theorem 1, the series (2) is mean square convergent.

THEOREM 2 (Kolmogorov's three series theorem). *Let $\{\xi_n, n \geq 1\}$ be a sequence of independent random variables. For a number $c > 0$ and a random variable ξ, put*

$$\xi^c = \begin{cases} \xi & \text{if } |\xi| < c, \\ 0 & \text{if } |\xi| \geq c. \end{cases}$$

In order that series (1) converge with probability one, it is necessary that for all $c > 0$, and it is sufficient that for some $c > 0$ the following series converge:

$$1.\ \sum_{n=1}^{\infty} \mathsf{P}(|\xi_n| > c), \qquad 2.\ \sum_{n=1}^{\infty} \mathsf{E}\,\xi_n^c, \qquad 3.\ \sum_{n=1}^{\infty} \operatorname{Var} \xi_n^c.$$

THEOREM 3. *For a series of independent random variables, the convergence in probability and the convergence with probability one are equivalent.*

Some particular cases of Theorem 3 are considered, under additional assumptions, in problems below.

Problems

III.6.1. Let ξ and ξ_n, $n \geq 1$, be random variables. Construct the sets

a) of elementary events $\omega \in \Omega$ such that $\sum_{n=1}^{\infty} \xi_n(\omega) = \xi(\omega)$;

b) of elementary events $\omega \in \Omega$ such that the series $\sum_{n=1}^{\infty} \xi_n(\omega)$ converges.

Check whether these sets belong to $\mathfrak{A}$.

III.6.2. In order that the series with random terms

$$(*) \qquad \sum_{n=1}^{\infty} \xi_n$$

converge in probability,

a) it is necessary that $\eta_n = \sup_{m \geq n} |\xi_m| \xrightarrow{\mathsf{P}} 0$ as $n \to \infty$;

b) it is necessary and sufficient that

$$\zeta_n = \sup_{m \geq n} \left| \sum_{k=n}^{m} \xi_k \right| \xrightarrow{\mathsf{P}} 0, \qquad n \to \infty.$$

If the random variables ξ_n, $n \geq 1$, are independent, then in order that the series $(*)$ converge with probability one,

c) it is necessary that $\mathsf{P}(\lim_{n\to\infty} \eta_n = 0) = 1$;

d) it is necessary and sufficient that $\mathsf{P}(\lim_{n\to\infty} \zeta_n = 0) = 1$.

III.6.3. Let $\{\xi_n, n \geq 1\}$ be a sequence of random variables, and let a sequence of positive numbers $\{\varepsilon_n, n \geq 1\}$ be such that

$$\sum_{n=1}^{\infty} \varepsilon_n < \infty, \qquad \sum_{n=1}^{\infty} \mathsf{P}(|\xi_n| \geq \varepsilon_n) < \infty.$$

Prove that the series $\sum_{n=1}^{\infty} \xi_n$ converges with probability one.

III.6.4. Let $\{\xi_n, n \geq 1\}$ be a sequence of positive random variables such that

$$\sum_{n=1}^{\infty} \mathsf{E}\,\xi_n < \infty.$$

Prove that the series $\sum_{n=1}^{\infty} \xi_n$ converges with probability one.

III.6.5. Let $\{\xi_n, n \geq 1\}$ be a sequence of random variables such that

$$\sum_{n=1}^{\infty} \mathsf{E}\,|\xi_n| < \infty.$$

Prove that the series $\sum_{n=1}^{\infty} \xi_n$ is absolutely convergent with probability one.

III.6.6. Let $\{\xi_n, n \geq 1\}$ be a sequence of independent identically distributed random variables such that

$$\mathsf{P}(\xi_1 = -1) = \mathsf{P}(\xi_1 = 1) = \frac{1}{2}.$$

Prove that the series

$$\sum_{n=1}^{\infty} \frac{\xi_n}{n}$$

is conditionally convergent with probability one, that is, it converges, but it is not a series absolutely convergent with probability one.

III.6.7. Let $\{\xi_n, n \geq 1\}$ be a sequence of nonnegative random variables. For a random variable ξ, put $\xi^{(1)} = \min(\xi, 1)$. Prove that the series $\sum_{n=1}^{\infty} \xi_n$ converges with probability one if and only if the series $\sum_{n=1}^{\infty} \xi_n^{(1)}$ converges with probability one.

III.6.8. Let random variables ξ_n, $n \geq 1$, be orthogonal, that is, $\mathsf{E}\,\xi_m \xi_n = 0$, $m \neq n$. Prove that the series $\sum_{n=1}^{\infty} \xi_n$ converges in mean square if and only if the series $\sum_{n=1}^{\infty} \mathsf{E}\,\xi_n^2$ converges.

III.6.9. Random variables ξ and η are said to be *uncorrelated* if $\mathsf{E}\,\xi^2 < \infty$, $\mathsf{E}\,\eta^2 < \infty$, and $\mathsf{E}\,\xi\eta = \mathsf{E}\,\xi\,\mathsf{E}\,\eta$. In order that a series $\sum \xi_n$ of pairwise uncorrelated random variables ξ_n, $n \geq 1$, be mean square convergent, it necessary and sufficient that the series

$$\sum_{n=1}^{\infty} \mathsf{E}\,\xi_n, \qquad \sum_{n=1}^{\infty} \operatorname{Var} \xi_n$$

converge.

III.6.10. Construct an example of a mean square convergent series $\sum_{n=1}^{\infty} \xi_n$ such that

$$\sum_{n=1}^{\infty} |\,\mathsf{E}\,\xi_n| = +\infty, \qquad \sum_{n=1}^{\infty} \mathsf{E}\,|\xi_n| = +\infty,$$

$$\sum_{n=1}^{\infty} (\mathsf{E}\,\xi_n)^2 = +\infty, \qquad \sum_{n=1}^{\infty} \mathsf{E}\,\xi_n^2 = +\infty.$$

III.6.11. Let two series of random variables

$$\xi = \sum_{n=1}^{\infty} \xi_n, \qquad \eta = \sum_{n=1}^{\infty} \eta_n$$

converge in mean square. Prove that

$$\mathsf{E}(\xi\eta) = \lim_{n\to\infty} \sum_{k,m=1}^{n} \mathsf{E}(\xi_k \eta_m).$$

III.6.12. Let a series $\xi = \sum_{n=1}^{\infty} \xi_n$ be mean square convergent, and let η be a random variable with $\mathsf{E}\,\eta^2 < \infty$. Prove that

$$\mathsf{E}(\xi\eta) = \sum_{n=1}^{\infty} \mathsf{E}(\xi_n \eta).$$

III.6.13. Assume that a series $\xi = \sum_{n=1}^{\infty} \xi_n$ of pairwise uncorrelated random variables ξ_n, $n \geq 1$, converges in mean square (see Problem III.6.9). Prove that

$$\mathsf{E}\,\xi^2 = \sum_{n=1}^{\infty} \operatorname{Var} \xi_n + \left(\sum_{n=1}^{\infty} \mathsf{E}\,\xi_n \right)^2,$$

that is,

$$\operatorname{Var} \xi = \sum_{n=1}^{\infty} \operatorname{Var} \xi_n.$$

III.6.14. Let $\{\xi_n, n \geq 1\}$ be a sequence of independent random variables such that

$$\mathsf{E}\,\xi_n = 0, \qquad \mathsf{E}\,\xi_n^2 = 1, \qquad \mathsf{E}\,\xi_n^4 \leq c$$

for all $n \geq 1$ and some $c \in \mathbf{R}$. Let $\{a_{mn}; m, n = 1, 2, \ldots\} \subset \mathbf{R}$ be such that $a_{mn} = a_{nm}$, $n, m \geq 1$. Prove that the sequence of random variables $\{S_n, n \geq 1\}$,

$$S_n = \sum_{j,k=1}^{n} a_{jk} \xi_j \xi_k, \qquad n \geq 1,$$

converges in mean square if and only if the series $\sum_{j=1}^{\infty} a_{jj}$, $\sum_{j,k=1}^{\infty} a_{jk}^2$ converge.

III.6.15. A sequence of independent random variables $\{\xi_n, n \geq 1\}$ is such that $\mathsf{E}\,\xi_n = 0$, $n \geq 1$, and

$$\sup_{n\geq 1} \mathsf{E} \left| \sum_{k=1}^{n} \xi_k \right|^r < \infty$$

for some $r \geq 2$. Prove that the series $\sum_{n=1}^{\infty} \xi_n$ converges with probability one.

III.6.16. Let $\{\xi_n, n \geq 1\}$ be a sequence of nonnegative random variables such that $\sum_{n=1}^{\infty} \mathsf{E}\,\xi_n < \infty$. Prove that the product $\prod_{n=1}^{\infty} (1 + \xi_n)$ converges with probability one.

III.6.17. Let $\{\xi_n, n \geq 1\}$ be a sequence of independent Gaussian random variables and $\mathsf{E}\,\xi_n = 0$, $n \geq 1$. Prove that the series $\sum_{n=1}^{\infty} \xi_n^2$ converges with probability one if and only if the series $\sum_{n=1}^{\infty} \mathsf{E}\,\xi_n^2$ converges.

III.6.18. Let $\{\xi_n, n \geq 1\}$ be a sequence of independent Gaussian random variables and $\mathsf{E}\,\xi_n = 0$, $\operatorname{Var}\xi_n^2 = 1$, $n \geq 1$. Prove that the series $\sum_{n=1}^{\infty} a_n\xi_n^2$ with nonnegative numbers $\{a_n, n \geq 1\}$ is convergent with probability one if and only if $\sum_{n=1}^{\infty} a_n < \infty$.

III.6.19. Let $\{\xi_n, n \geq 1\}$ be a sequence of real independent random variables such that $\mathsf{E}\,\xi_n = 0$, $n \geq 1$, and $\sum_{n=1}^{\infty} \mathsf{E}\,\xi_n^2 < \infty$. Prove the equality

$$\mathsf{E}\exp\left\{i\sum_{n=1}^{\infty}\xi_n\right\} = \prod_{n=1}^{\infty}\mathsf{E}\exp\{i\xi_n\}.$$

III.6.20. Let $\{\xi_n, n \geq 1\}$ be a sequence of independent random variables and $\mathsf{E}\,\xi_n = 0$, $n \geq 1$. Assume that there exists a random variable ξ with $\mathsf{E}\,\xi^2 < \infty$ such that the random variables $\xi_1, \xi_2, \dots, \xi_n, \xi - \xi_1 - \xi_2 - \cdots - \xi_n$ are independent for all $n \geq 1$. Prove that $\mathsf{E}\,\xi_n^2 < \infty$ for all $n \geq 1$ and the series $\sum_{n=1}^{\infty}\xi_n$ converges with probability one.

III.6.21. Let $\{\xi_n, n \geq 1\}$ be a sequence of independent random variables and $\mathsf{E}\,\xi_n^2 < \infty$, $n \geq 1$. Let the series

$$(*) \qquad \sum_{n=1}^{\infty}\xi_n$$

be mean square convergent. Prove that the series $(*)$ is convergent with probability one.

III.6.22. Let $\{\xi_n, n \geq 1\}$ be a sequence of independent random variables such that $\mathsf{P}(|\xi_n| > c) = 0$, $n \geq 1$, for some c. Prove that the series $\sum_{n=1}^{\infty}(\xi_n - \mathsf{E}\,\xi_n)$ converges with probability one if and only if the series $\sum_{n=1}^{\infty}\operatorname{Var}\xi_n$ converges.

III.6.23. Let $\{\xi_n, n \geq 1\}$ be a sequence of independent nonnegative random variables such that $\mathsf{P}(\xi_n > c) = 0$, $n \geq 1$, for some number c. Prove that the series $\sum_{n=1}^{\infty}\xi_n$ converges with probability one if and only if the series $\sum_{n=1}^{\infty}\mathsf{E}\,\xi_n$ converges.

III.6.24. In order that a series $\sum_{n=1}^{\infty}\xi_n$ of independent random variables ξ_n, $n \geq 1$, be convergent with probability one, it is necessary and sufficient that

a) the series $\sum_{n=1}^{\infty}\mathsf{E}\,\xi_n$, $\sum_{n=1}^{\infty}\operatorname{Var}\xi_n$ be convergent in the case of Gaussian random variables ξ_n, $n \geq 1$;

b) the series $\sum_{n=1}^{\infty}\mathsf{E}\,\xi_n$ be convergent in the case of Poissonian random variables ξ_n, $n \geq 1$.

III.6.25. Let $\{\xi_n, n \geq 1\}$ be a sequence of independent identically distributed Gaussian random variables with $\mathsf{E}\,\xi_n = 0$, $\mathsf{E}\,\xi_n^2 = 1$, $n \geq 1$. Let $\{a_n, n \geq 1\}$ and $\{b_n, n \geq 1\}$ be two sequences of real numbers. If the series $\sum_{n=1}^{\infty}(a_n\xi_n + b_n)$ converges in probability, then

a) the series $\sum_{n=1}^{\infty}a_n^2$ and $\sum_{n=1}^{\infty}b_n$ converge;

b) the series $\sum_{n=1}^{\infty}(a_n\xi_n + b_n)$ converges with probability one.

III.6.26. Let $\{\xi_n, n \geq 1\}$ be a sequence of independent Gaussian random variables. Prove that the series $\sum_{n=1}^{\infty}\xi_n$ converges with probability one if and only if it converges in probability.

III.6.27. Let $\{\xi_n, n \geq 1\}$ be a sequence of independent random variables such that

$$\mathsf{P}(\xi_n = -1) = \mathsf{P}(\xi_n = 1) = \frac{1}{2}, \qquad n \geq 1.$$

Prove that the random variable

$$\eta = \sum_{n=1}^{\infty} \frac{\xi_n}{2^n}$$

is uniformly distributed on the interval $[-1, 1]$, and by using Problem III.6.19, prove the identity

$$\frac{\sin x}{x} = \prod_{n=1}^{\infty} \cos \frac{x}{2^n}.$$

III.6.28. Let $\{\xi_n, n \geq 1\}$ be a sequence of independent identically distributed random variables such that $\mathsf{E}\,\xi_n = 0$, $\mathsf{E}\,\xi_n^2 = \sigma^2$, $n \geq 1$, and let $\{a_n, n \geq 1\}$ be a sequence of real numbers with $\sum_{n=1}^{\infty} |a_n| < \infty$. Prove that the series $\xi = \sum_{n=1}^{\infty} a_n \xi_n$ converges with probability one and in mean square and that

$$\mathsf{E}\,\xi = 0, \qquad \mathsf{E}\,\xi^2 = \sigma^2 \sum_{n=1}^{\infty} a_n^2.$$

III.6.29. Sequences $\{\xi_n, n \geq 1\}$ and $\{\eta_n, n \geq 1\}$ are said to be *equivalent in Khintchine's sense* if

$$\sum_{n=1}^{\infty} \mathsf{P}(\xi_n \neq \eta_n) < \infty.$$

Prove that for random sequences $\{\xi_n, n \geq 1\}$ and $\{\eta_n, n \geq 1\}$ equivalent in Khintchine's sense, the series $\sum_{n=1}^{\infty} \xi_n$ converges with probability one if and only if the series $\sum_{n=1}^{\infty} \eta_n$ converges with probability one.

III.6.30. Let $\{\xi_n, n \geq 1\}$ be a sequence of independent random variables. In order that the series $\sum_{n=1}^{\infty} \xi_n$ converge with probability one, it is necessary and sufficient that there exist a sequence of independent random variables $\{\eta_n, n \geq 1\}$ with $\mathsf{E}\,\eta_n^2 < \infty$, $n \geq 1$, Khintchine equivalent to $\{\xi_n, n \geq 1\}$ and such that the series

$$\sum_{n=1}^{\infty} \mathsf{E}\,\eta_n, \qquad \sum_{n=1}^{\infty} \operatorname{Var} \eta_n$$

converge.

III.6.31. Let $\{\xi_n, n \geq 1\}$ be a sequence of independent random variables. Prove that the series $\sum_{n=1}^{\infty} \xi_n$ converges with probability one if and only if for some number $c > 0$ the series

$$\sum_{n=1}^{\infty} \mathsf{E}\,\xi_n^c, \qquad \sum_{n=1}^{\infty} \operatorname{Var} \xi_n^c$$

converge, where

$$\xi^c = \begin{cases} -c, & \xi \leq -c, \\ \xi, & -c < \xi < c, \\ c, & \xi \geq c. \end{cases}$$

III.6.32. Independent random variables ξ_n, $n \ge 1$, are such that $\mathsf{E}\,\xi_n = 0$, $\mathsf{E}\,\xi_n^2 = 1$, and $\mathsf{P}(|\xi_n| > c) = 0$ for some $c > 0$ and all $n \ge 1$. Let $\{a_n, n \ge 1\}$ be a sequence of real numbers. Prove that the series $\sum_{n=1}^{\infty} a_n \xi_n$ converges with probability one if and only if $\sum_{n=1}^{\infty} a_n^2 < \infty$.

III.6.33. Under the conditions of the preceding problem prove that the series $\sum_{n=1}^{\infty} a_n \xi_n$ converges with probability one if and only if the series $\sum_{n=1}^{\infty} (a_n \xi_n)^2$ converges with probability one.

III.6.34. Let $\{\xi_n, n \ge 1\}$ be a sequence of independent random variables such that $\mathsf{E}\,\xi_n = 0$ and $\mathsf{P}(|\xi_n| > c) = 0$, $n \ge 1$, for some number $c > 0$. Furthermore, let

$$\sum_{n=1}^{\infty} \mathsf{E}\,\xi_n^2 = +\infty.$$

Prove the equality

$$\mathsf{P}\left(\limsup_{n\to\infty} \left| \sum_{k=1}^{n} \xi_k \right| = +\infty \right) = 1.$$

III.6.35. Random variables ξ_n, $n \ge 1$, are independent and satisfy the conditions $\mathsf{E}\,\xi_n = 0$ and $\mathsf{P}(|\xi_n| > c) = 0$, $n \ge 1$ for some number c. If the series $\sum_{n=1}^{\infty} \xi_n$ converges with a positive probability, then $\sum_{n=1}^{\infty} \operatorname{Var} \xi_n < \infty$. Prove this statement.

III.6.36. Let $\{\xi_n, n \ge 1\}$ be a sequence of independent identically distributed random variables with an exponential distribution. Let $\{a_n, n \ge 1\}$ be a sequence of nonnegative numbers. Find a necessary and sufficient condition on $\{a_n, n \ge 1\}$ for the series $\sum_{n=1}^{\infty} a_n \xi_n$ to be convergent with probability one.

III.6.37. Let $\{\xi_n, n \ge 1\}$ be a sequence of independent identically distributed random variables with the Cauchy distribution. Find a necessary a sufficient condition on a sequence $\{a_n, n \ge 1\}$ for the series $\sum_{n=1}^{\infty} |a_n \xi_n|$ to be convergent with probability one.

III.6.38. Let $\{\xi_n, n \ge 1\}$ be a sequence of independent nonnegative random variables. Prove the following statement.

STATEMENT. *In order that a series $\sum_{n=1}^{\infty} \xi_n$ be convergent with probability one, it is sufficient that for some number $c > 0$, and necessary that for all numbers $c > 0$ the series*

$$\sum_{n=1}^{\infty} \mathsf{P}(\xi_n > c), \qquad \sum_{n=1}^{\infty} \mathsf{E}\,\xi_n^{(c)}$$

be convergent, where

$$\xi^{(c)} = \begin{cases} \xi, & |\xi| \le c, \\ 0, & |\xi| > c. \end{cases}$$

III.6.39. Let $\{\xi_n, n \ge 1\}$ be a sequence of independent nonnegative random variables. Prove that the series $\sum_{n=1}^{\infty} \xi_n$ converges with probability one if and only if the series $\sum_{n=1}^{\infty} \mathsf{E}\{\min(\xi_n, 1)\}$ converges.

III.6.40. Let $\{\xi_n, n \geq 1\}$ be a sequence of independent random variables with the distribution

$$\mathsf{P}\left(\xi_n = -\frac{1}{\sqrt{n}}\right) = \mathsf{P}\left(\xi_n = \frac{1}{\sqrt{n}}\right) = \frac{1}{2}, \qquad n \geq 1.$$

Put

$$\xi_0 = 0, \qquad \eta_n = \xi_n - \xi_{n-1}, \quad \zeta_n = \xi_n + \xi_{n-1}, \qquad n \geq 1.$$

Prove that

a) the series $\sum_{n=1}^{\infty} \eta_n$ converges with probability one;
b) the series $\sum_{n=1}^{\infty} \zeta_n$ diverges with probability one;
c) the series

$$\sum_{n=1}^{\infty} \mathsf{E}\left(\eta_n / \eta_{n-1}, \ldots, \eta_1\right), \qquad \sum_{n=1}^{\infty} \mathsf{E}\left(\zeta_n / \zeta_{n-1}, \ldots, \zeta_1\right),$$

$$\sum_{n=1}^{\infty} \operatorname{Var}\left(\eta_n / \eta_{n-1}, \ldots, \eta_1\right), \qquad \sum_{n=1}^{\infty} \operatorname{Var}\left(\zeta_n / \zeta_{n-1}, \ldots, \zeta_1\right)$$

diverge with probability one.

III.6.41. Random variables ξ_n, $n \geq 1$, are independent and symmetric. Assume that the sequence $\{S_n, n \geq 1\}$ of their partial sums is bounded in probability (see Problem III.3.68). Prove that the series $\sum_{n=1}^{\infty} \xi_n$ converges with probability one.

III.6.42. Let $\xi_1, \eta_1, \xi_2, \eta_2, \ldots, \xi_n, \eta_n, \ldots$ be a sequence of independent random variables. If the series $\sum_{n=1}^{\infty} (\xi_n + \eta_n)$ converges with probability one, then there exist numbers a_n, $n \geq 1$, such that the series $\sum_{n=1}^{\infty} (\xi_n + a_n)$ converges with probability one.

III.6.43. Prove that the convergence in probability of a series of independent terms implies the convergence of this series with probability one.

III.6.44. Let $\{\xi_n, n \geq 1\}$ be a sequence of independent Gaussian random variables. Prove that the convergence of the series $\sum_{n=1}^{\infty} \xi_n$ in distribution implies the convergence of this series with probability one.

III.6.45. Let $\{\xi_n, n \geq 1\}$ be a sequence of independent Poissonian random variables. Prove that the convergence in distribution of the series $\sum_{n=1}^{\infty} \xi_n$ implies the convergence of this series with probability one.

III.6.46. A sequence of independent symmetric random variables $\{\xi_n, n \geq 1\}$ with $\mathsf{E}\,\xi_n = 0$, $n \geq 1$, satisfies the condition

$$\sup_{n \geq 1} \mathsf{E}\left|\sum_{k=1}^{n} \xi_k\right|^r < \infty$$

for some $r > 0$. Prove that the series $\sum_{n=1}^{\infty} \xi_n$ converges with probability one.

III.6.47. Let $\{\xi_n, n \geq 1\}$ be a sequence of independent symmetric random variables. If the series $\sum_{n=1}^{\infty} \xi_n$ converges with probability 1, then so does any series obtained from $\sum_{n=1}^{\infty} \xi_n$ by rearrangement of its terms. Prove this statement.

III.6.48. Let $\{\xi_n, n \geq 1\}$ be a sequence of independent random variables, and $\{\varphi_n(t), n \geq 1\}$ the sequence of the corresponding characteristic functions:

$$\varphi_n(t) = \mathsf{E}\left(\exp\{it\xi_n\}\right), \qquad t \in \mathbf{R},\ n \geq 1.$$

Prove that the convergence with probability one of the series $\sum_{n=1}^{\infty} \xi_n$ implies that the product $\prod_{n=1}^{\infty} \varphi_n(t)$ converges uniformly for t in any finite interval.

III.6.49. Let $\{\xi_n, n \geq 1\}$ be a sequence of independent random variables such that $\mathsf{P}(|\xi_n| > c) = 0$, $n \geq 1$, for some number $c > 0$, and let

$$\sum_{n=1}^{\infty} \operatorname{Var} \xi_n = +\infty.$$

Prove that

$$\limsup_{n\to\infty} \left| \sum_{k=1}^{n} \xi_k \right| = +\infty$$

with probability one.

§III.7. The strong law of large numbers

A sequence of random variables $\{\xi_n, n \geq 1\}$ with $\mathsf{E}\,|\xi_n| < \infty$, $n \geq 1$, is said *to satisfy the strong law of large numbers* if

$$\frac{1}{n}\sum_{k=1}^{n} \xi_k - \frac{1}{n}\sum_{k=1}^{n} \mathsf{E}\,\xi_k \to 0, \qquad n \to \infty,$$

with probability one, that is,

$$\mathsf{P}\left(\sup_{n \geq N} \left| \frac{1}{n}\sum_{k=1}^{n} \xi_k - \frac{1}{n}\sum_{k=1}^{n} \mathsf{E}\,\xi_k \right| \geq \varepsilon \right) \to 0, \qquad N \to \infty,$$

for any $\varepsilon > 0$.

THEOREM 1. *Let a sequence of independent random variables $\{\xi_n, n \geq 1\}$ be such that $\mathsf{E}\,\xi_n^2 < \infty$, $n \geq 1$, and*

$$\sum_{n=1}^{\infty} \frac{\operatorname{Var} \xi_n}{n^2} < \infty.$$

Then the sequence $\{\xi_n, n \geq 1\}$ satisfies the strong law of large numbers.

THEOREM 2. *A sequence of independent identically distributed random variables $\{\xi_n, n \geq 1\}$ with $\mathsf{E}\,|\xi_1| < \infty$ satisfies the strong law of large numbers:*

$$\mathsf{P}\left(\lim_{n\to\infty} \frac{1}{n} \sum_{k=1}^{n} \xi_k = \mathsf{E}\,\xi_1 \right) = 1.$$

Problems

III.7.1. A sequence of numbers $\{a_n, n \geq 1\} \subset [0,1]$ is said to be *uniformly distributed in Weyl's sense on the interval* $[0,1]$ if for any Riemann integrable functions f on $[0,1]$ the following relation is satisfied:

$$\lim_{n\to\infty} \frac{f(a_1)+f(a_2)+\cdots+f(a_n)}{n} = \int_0^1 f(x)\,dx.$$

Let $\{\xi_n, n \geq 1\}$ be a sequence of independent random variables uniformly distributed on the interval $[0,1]$. Prove that with probability one the sequence $\{\xi_n, n \geq 1\}$ is uniformly distributed in Weil's sense on $[0,1]$.

III.7.2. Let sequences $\{\xi_n, n \geq 1\}$ and $\{\eta_n, n \geq 1\}$ be equivalent in Khintchine's sense (see Problem III.6.29). Prove that the strong law of large numbers holds for the sequence $\{\xi_n, n \geq 1\}$ if and only if the strong law of large numbers holds for the sequence $\{\eta_n, n \geq 1\}$.

III.7.3. Let $\{\xi_n, n \geq 1\}$ be a sequence of independent identically distributed random variables with either $\mathsf{E}\,\xi_1 > 0$ or $\mathsf{E}\,\xi_1 = +\infty$. Prove that

$$\mathsf{P}\Big(\limsup_{n\to\infty}(\xi_1+\xi_2+\cdots+\xi_n) = +\infty\Big) = 1.$$

III.7.4. For a sequence $\{\xi_n, n \geq 1\}$ of random variables, the series $\sum_{n=1}^{\infty} \xi_n/n$ converges with probability one. Prove that

$$\mathsf{P}\bigg(\frac{1}{n}\sum_{k=1}^{n} \xi_k \to 0 \text{ as } n\to\infty\bigg) = 1.$$

III.7.5. A sequence of independent random variables $\{\xi_n, n \geq 1\}$ is such that

$$1)\quad \sum_{n=1}^{\infty} \frac{1}{n^2}\,\mathsf{E}\,\xi_n^2 < \infty; \qquad 2)\quad \mathsf{E}\,\xi_n \to 0, \quad n\to\infty.$$

Prove that

$$\mathsf{P}\bigg(\frac{1}{n}\sum_{k=1}^{n} \xi_k \to 0 \text{ as } n\to\infty\bigg) = 1.$$

III.7.6. Prove that the condition

$$\mathsf{P}\bigg(\lim_{n\to\infty} \frac{\xi_n - \mathsf{E}\,\xi_n}{n} = 0\bigg) = 1$$

is necessary and sufficient for the strong law of large numbers to hold for a sequence $\{\xi_n, n \geq 1\}$ of independent random variables with $\mathsf{E}\,|\xi_n| < \infty$, $n \geq 1$.

III.7.7. Assume that random variables ξ_n, $n \geq 1$, with $\mathsf{E}\,\xi_n = 0$, $n \geq 1$, satisfy

$$\mathsf{P}\bigg(\frac{1}{n}\sum_{k=1}^{n} \xi_k \to 0 \text{ as } n\to\infty\bigg) = 1.$$

Prove that the series

$$\sum_{n=1}^{\infty} \frac{\xi_n}{n^{1+\alpha}}$$

converges with probability one for every $\alpha > 0$.

III.7.8. Random variables ξ_n, $n \geq 2$, are independent and such that

$$\mathsf{P}(\xi_n = -n) = \mathsf{P}(\xi_n = n) = \frac{1}{n},$$
$$\mathsf{P}(\xi_n = 0) = 1 - \frac{2}{n}, \qquad n \geq 2.$$

Prove that for the sequence $\{\xi_n, n \geq 2\}$,

a) the series $\sum_{n=2}^{\infty} \frac{1}{n^2} \operatorname{Var} \xi_n$ diverges;

b) the strong law of large numbers is not satisfied.

III.7.9. A sequence of independent random variables $\{\xi_n, n \geq 1\}$ is such that $\mathsf{E}\,\xi_n = 0$, $n \geq 1$, and

$$\sum_{n=1}^{\infty} \frac{1}{n^2} \operatorname{Var} \xi_n < \infty.$$

According to Theorem 1,

$$\mathsf{P}\left(\frac{1}{n} \sum_{k=1}^{n} \xi_k \to 0 \text{ as } n \to \infty\right) = 1.$$

Prove that there exists a sequence of independent random variables $\{\eta_n, n \geq 1\}$, Khintchine equivalent to $\{\xi_n, n \geq 1\}$ and such that

$$\sum_{n=1}^{\infty} \frac{1}{n^2} \operatorname{Var} \eta_n = +\infty.$$

III.7.10. Let $\{\xi_n, n \geq 1\}$ be a sequence of independent random variables such that $\mathsf{E}\,\xi_n = 0$, $\mathsf{P}(|\xi_n| \leq nc) = 1$, $n \geq 1$, for some number c, and

$$\mathsf{P}\left(\frac{1}{n} \sum_{k=1}^{n} \xi_k \to 0 \text{ as } n \to \infty\right) = 1.$$

Prove that the series

$$\sum_{n=1}^{\infty} \frac{1}{n^{2+\alpha}} \operatorname{Var} \xi_n$$

converges for every $\alpha > 0$.

III.7.11. For a sequence of independent identically distributed random variables $\{\xi_n, n \geq 1\}$ the following condition is satisfied:

$$\mathsf{P}\left(\frac{1}{n} \sum_{k=1}^{n} \xi_k \to 0 \text{ as } n \to \infty\right) = 1.$$

Prove that $\mathsf{E}\,|\xi_1| < \infty$.

III.7.12. Let $\{\xi_n, n \geq 1\}$ be a sequence of random variables with $\mathsf{E}\,|\xi_n| < \infty$, $n \geq 1$, and

$$\sum_{n=1}^{\infty} \frac{1}{n} \mathsf{E}\,|\xi_n - \mathsf{E}\,\xi_n| < \infty.$$

Prove that

$$\mathsf{P}\left(\lim_{n\to\infty} \frac{1}{n} \sum_{k=1}^{n} (\xi_k - \mathsf{E}\,\xi_k) = 0\right) = 1.$$

III.7.13. Let $r \geq 2$. A sequence of independent random variables $\{\xi_n, n \geq 1\}$ with $\mathsf{E}\,\xi_n = 0$, $n \geq 1$, satisfies the condition

$$\sup_{n\geq 1} \mathsf{E}\left|\sum_{k=1}^{n} \frac{\xi_k}{k}\right|^r < \infty.$$

Prove that

$$\mathsf{P}\left(\frac{1}{n}\sum_{k=1}^{n} \xi_k \to 0 \text{ as } n \to \infty\right) = 1.$$

III.7.14. Let a sequence of independent random variables $\{\xi_n, n \geq 1\}$ be such that $\mathsf{E}\,\xi_n = 0$, $\mathsf{E}\,\xi_n^4 < \infty$, $n \geq 1$. Assume also that

a) for some number c,

$$\mathsf{E}\,\xi_n^4 \leq c\left(\mathsf{E}\,\xi_n^2\right)^2, \qquad n \geq 1;$$

b) the following series converges:

$$\sum_{n=1}^{\infty}\left(\frac{1}{n^2}\sum_{k=1}^{n} \xi_k^2\right)^2 < \infty.$$

Prove that

$$\mathsf{P}\left(\frac{1}{n}\sum_{k=1}^{n} \xi_k \to 0 \text{ as } n \to \infty\right) = 1.$$

III.7.15. Let $r > 0$. A sequence of independent symmetric random variables $\{\xi_n, n \geq 1\}$ with $\mathsf{E}\,\xi_n = 0$, $n \geq 1$, satisfies the condition

$$\sup_{n\geq 1} \mathsf{E}\left|\sum_{k=1}^{n} \frac{1}{k}\xi_k\right|^r < \infty.$$

Prove that

$$\mathsf{P}\left(\frac{1}{n}\sum_{k=1}^{n} \xi_k \to 0 \text{ as } n \to \infty\right) = 1.$$

III.7.16. Prove that almost all (with respect to Lebesgue measure) numbers of the interval $[0,1]$ have the same number of digits $0,1,2,\dots,9$ in their decimal representations. More precisely, let $\nu_n(x,i)$ be the number of occurrences of a digit i among the first n digits in the decimal representation of a number x. Prove that for almost all $x \in [0,1]$ and all i, $0 \leq i \leq 9$,

$$\frac{\nu_n(x,i)}{n} \to \frac{1}{10}, \qquad n \to \infty.$$

III.7.17. Prove that, under the conditions of Problems III.4.19 and III.4.20,

$$\mathsf{P}\left(\bar{x} \to \mu \text{ as } n \to \infty\right) = 1, \qquad \mathsf{P}\left(S^2 \to \sigma^2 \text{ as } n \to \infty\right) = 1,$$
$$\mathsf{P}\left(F_n(x) \to F(x) \text{ as } n \to \infty\right) = 1.$$

III.7.18. If a sequence of independent random variables $\{\xi_n, n \geq 1\}$ satisfies the condition

$$\operatorname{Var} \xi_n \leq c \frac{n}{\ln^2 n}, \qquad n \geq 2,$$

for some number c, then it satisfies the strong law of large numbers. Prove this statement.

III.7.19. Let $\{\xi_n, n \geq 1\}$ be a sequence of independent symmetric random variables such that for all $\varepsilon > 0$,

$$\sum_{n=1}^{\infty} \mathsf{P}(|S_{2^n}| > \varepsilon 2^n) < \infty.$$

Prove that

$$\mathsf{P}\left(\frac{S_n}{n} \to 0 \text{ as } n \to \infty \right) = 1.$$

III.7.20. Let $\{\xi_n, n \geq 1\}$ be a sequence of independent Gaussian random variables. If for every $\varepsilon > 0$ the series

$$\sum_{n=1}^{\infty} \exp\left\{ -\varepsilon 2^{2n} \left(\sum_{k=1}^{2^n} \operatorname{Var} \xi_k \right)^{-1} \right\}$$

converges, then the sequence $\{\xi_n, n \geq 1\}$ satisfies the strong law of large numbers. Prove this statement.

III.7.21. Suppose a sequence of independent Gaussian random variables $\{\xi_n, n \geq 1\}$ satisfies the condition

$$\liminf_{n \to \infty} \frac{\ln n}{n} \operatorname{Var} \xi_n > 0.$$

Prove that the strong law of large numbers does not hold.

III.7.22. Let $\{\xi_n, n \geq 1\}$ be a sequence of independent Gaussian random variables with $\mathsf{E}\,\xi_n = 0$, $\mathsf{E}\,\xi_n^2 = 1$, $n \geq 1$. Prove that with probability one the limit

$$\lim_{n_2 - n_1 \to \infty} \frac{1}{n_2 - n_1} (\xi_{n_1+1} + \cdots + \xi_{n_2})$$

does not exist.

III.7.23. Let $\{\xi_n, n \geq 1\}$ be a sequence of independent random variables, $\{a_n, n \geq 1\}$ a sequence of positive numbers such that $a_n < a_{n+1}$, $n \geq 1$, and $a_n \to +\infty$ as $n \to \infty$. Under the condition that the series $\sum_{n=1}^{\infty} \xi_n / a_n$ converges with probability one, prove that

$$\mathsf{P}\left(\frac{1}{a_n} \sum_{k=1}^{n} \xi_k \to 0 \text{ as } n \to \infty \right) = 1.$$

III.7.24. Let $\{\xi_n, n \ge 1\}$ be a sequence of independent random variables such that $\mathsf{E}\,\xi_n = 0$ and $\mathsf{E}\,\xi_n^2 < \infty$, $n \ge 1$, and let $\{a_n, n \ge 1\}$ be a sequence of positive numbers such that $0 < a_n < a_{n+1}$, $n \ge 1$, and $a_n \to +\infty$ as $n \to \infty$. Assume that

$$\sum_{n=1}^{\infty} \frac{\operatorname{Var} \xi_n}{a_n^2} < \infty.$$

Prove that

$$\mathsf{P}\left(\frac{1}{a_n} \sum_{k=1}^{n} \xi_k \to 0 \text{ as } n \to \infty \right) = 1.$$

III.7.25. Let $\{\xi_n, n \ge 1\}$ be a sequence of independent random variables, and $\{a_n, n \ge 1\}$ a sequence of positive numbers such that $a_n \le c a_{n+1}$, $n \ge 1$, for some c. If

$$\mathsf{P}\left(\frac{1}{a_n} \sum_{k=1}^{n} \xi_k \to 0 \text{ as } n \to \infty \right) = 1,$$

then

$$\sum_{n=1}^{\infty} \mathsf{P}(|\xi_n| \ge a_n) < \infty.$$

Prove this statement.

III.7.26. Prove that

$$\mathsf{P}\left(\frac{1}{n^{\alpha}} \sum_{k=1}^{n} \xi_k \to 0 \text{ as } n \to \infty \right) = 1$$

for all $\alpha > \frac{1}{2}$, where $\{\xi_n, n \ge 1\}$ is a sequence of independent identically distributed random variables with $\mathsf{E}\,\xi_1 = 0$ and $\mathsf{E}\,\xi_1^2 = \sigma^2 < \infty$.

III.7.27. According to the strong law of large numbers,

$$\mathsf{P}\left(\frac{1}{n} \sum_{k=1}^{n} \xi_k \to \mu \text{ as } n \to \infty \right) = 1$$

for a sequence of independent identically distributed random variables $\{\xi_n, n \ge 1\}$ with $\mathsf{E}\,\xi_1 = \mu$ and $\mathsf{E}\,\xi_1^2 = \sigma^2 < \infty$. Prove that for all $\beta < \frac{1}{2}$,

$$\mathsf{P}\left(n^{\beta} \left(\frac{1}{n} \sum_{k=1}^{n} \xi_k - \mu \right) \to 0 \text{ as } n \to \infty \right) = 1.$$

III.7.28. (Hausdorff's estimate.) For a sequence of random variables satisfying the conditions of Problem III.7.26, prove that for all $\varepsilon > 0$,

$$\xi_1 + \xi_2 + \cdots + \xi_n = O\left(n^{1/2+\varepsilon}\right), \qquad n \to \infty.$$

III.7.29. Let $\{\xi_n, n \ge 1\}$ be a sequence of independent identically distributed random variables with the Cauchy distribution. Prove that for any $\alpha > 0$,

$$\mathsf{P}\left(\lim_{n\to\infty} \frac{1}{n^{1+\alpha}} \sum_{k=1}^{n} \xi_k = 0 \right) = 1.$$

III.7.30. Let $\{\xi_n, n \geq 1\}$ be a sequence of independent random variables and $\mathsf{E}\,\xi_n^2 < \infty$, $n \geq 1$. Put

$$\eta_n = \frac{1}{n}\sum_{k=1}^{n} \xi_k.$$

Prove that

$$\eta_n - \mathsf{E}\,\eta_n \to 0, \qquad n \to \infty,$$

in mean square, that is, $\mathsf{E}\,(\eta_n - \mathsf{E}\,\eta_n)^2 \to 0$ as $n \to \infty$, if and only if

$$\eta_{2^n} - \mathsf{E}\,\eta_{2^n} \to 0, \qquad n \to \infty,$$

in mean square.

III.7.31. Let $\{\xi_n, n \geq 1\}$ be a sequence of independent identically distributed random variables. Put

$$\eta_n = \frac{1}{n}\sum_{k=1}^{n} \xi_k.$$

Prove that $\eta_n \to 0$ in probability as $n \to \infty$ if and only if $\eta_{2^n} \to 0$ in probability as $n \to \infty$.

III.7.32. Let $\{\xi_n, n \geq 1\}$ be a sequence of independent identically distributed random variables. Put $\eta_n = S_n/n$, $n \geq 1$. Prove that

$$\mathsf{P}(\eta_n \to 0 \text{ as } n \to \infty) = 1$$

if and only if

$$\mathsf{P}(\eta_{2^n} \to 0 \text{ as } n \to \infty) = 1.$$

III.7.33. Let $\{\xi_n, n \geq 1\}$ be a sequence of random variables such that

$$\mathsf{P}\left(\frac{1}{n}\sum_{k=1}^{n} \xi_k \to 0 \text{ as } n \to \infty\right) = 1.$$

Prove that

$$\mathsf{P}\left(\frac{1}{2^n}\left(S_{2^{n+1}} - S_{2^n}\right) \to 0 \text{ as } n \to \infty\right) = 1.$$

III.7.34. Let $\{\xi_n, n \geq 1\}$ be a sequence of independent random variables. Using Problem III.7.32 for nonidentically distributed random variables, prove that

$$\mathsf{P}\left(\frac{1}{n}\sum_{k=1}^{n} \xi_k \to 0 \text{ as } n \to \infty\right) = 1$$

if and only if

$$\mathsf{P}\left(\frac{1}{2^n}\sum_{k=2^n+1}^{2^{n+1}} \xi_k \to 0 \text{ as } n \to \infty\right) = 1.$$

III.7.35. A sequence of independent random variables $\{\xi_n, n \geq 1\}$ is such that $\mathsf{E}\,\xi_n = 0$, $n \geq 1$, and

$$\sum_{n=1}^{\infty} \frac{\mathsf{E}\,|\xi_n|^{2r}}{n^{1+r}} < \infty$$

for some $r \geq 1$.

a) Prove that $\sum_{n=1}^{\infty} \mathsf{P}(\xi_n \neq \eta_n) < \infty$, where

$$\eta_n = \begin{cases} \xi_n, & |\xi_n| \le n^{(1+r)/2r}, \\ 0, & |\xi_n| > n^{(1+r)/2r}. \end{cases}$$

b) Using Problem III.7.32 for nonidentically distributed random variables, prove that

$$\mathsf{P}\left(\frac{1}{n}\sum_{k=1}^{n} \xi_k \to 0 \text{ as } n \to \infty\right) = 1.$$

III.7.36. A sequence $\{\xi_n, n \ge 1\}$ of independent random variables is such that $\mathsf{P}(|\xi_n| > c) = 0$, $n \ge 1$, for some number c. If

$$\mathsf{P}\left(\frac{1}{n^2}\sum_{k=1}^{n^2} \xi_k \to 0 \text{ as } n \to \infty\right) = 1,$$

then

$$\mathsf{P}\left(\frac{1}{n}\sum_{k=1}^{n} \xi_k \to 0 \text{ as } n \to \infty\right) = 1.$$

Prove this statement.

III.7.37. Suppose random variables ξ_n, $n \ge 1$, satisfy the following conditions:

a) $\mathsf{E}\,\xi_n = 0$, $\mathsf{E}\,\xi_n^4 \le c$, $n \ge 1$, for some number c;

b) for arbitrarily chosen natural numbers k_1, k_2, k_3, k_4 and nonnegative integers i_1, i_2, i_3, i_4 with $i_1 + i_2 + i_3 + i_4 = 4$,

$$\mathsf{E}\left(\xi_{k_1}^{i_1}\xi_{k_2}^{i_2}\xi_{k_3}^{i_3}\xi_{k_4}^{i_4}\right) = \prod_{\nu=1}^{4} \mathsf{E}\,\xi_{k_\nu}^{i_\nu}.$$

Prove that there exists a number a such that

$$\mathsf{E}\left(\xi_1 + \xi_2 + \cdots + \xi_n\right)^4 \le a n^2, \qquad n \ge 1.$$

III.7.38. Under the conditions of the preceding problem, prove that

$$\mathsf{P}\left(\frac{1}{n}\sum_{k=1}^{n} \xi_k \to 0 \text{ as } n \to \infty\right) = 1.$$

§III.8. Martingales

A sequence of random variables $\{\zeta_n, n \ge 1\}$ with finite expectations is said to be *a martingale* if

$$\mathsf{E}\left(\zeta_n / \zeta_1, \dots, \zeta_{n-1}\right) = \mathsf{E}\left(\zeta_n / F_{n-1}\right) = \zeta_{n-1}$$

with probability one for all $n > 1$, where F_n is the σ-algebra generated by the random variables $\zeta_1, \zeta_2, \dots, \zeta_n$, $n \ge 1$. A sequence of random variables $\{\zeta_n, n \ge 1\}$ is said to be *a supermartingale* if $\mathsf{E}\,|\zeta_n| < \infty$, $n \ge 1$, and

$$\mathsf{E}\left(\zeta_n / F_{n-1}\right) \le \zeta_{n-1}$$

with probability one for all $n > 1$. A sequence $\{\zeta_n, n \ge 1\}$ is said to be *a submartingale* if the sequence $\{-\zeta_n, n \ge 1\}$ is a supermartingale.

THEOREM. *If* $\{\zeta_n, n \geq 1\}$ *is a nonnegative martingale, then the limit*

$$\lim_{n \to \infty} \zeta_n = \zeta$$

exists with probability one.

Problems

III.8.1. Let $\{\xi_n, n \geq 1\}$ be independent random variables, and $\{\eta_n, n \geq 1\}$ random variables such that the collections $\{\eta_1, \eta_2, \ldots, \eta_n\}$ and $\{\xi_n, \xi_{n+1}, \ldots\}$ are independent for all $n \geq 1$. Suppose that $\mathsf{E}\,\xi_n = 0$ and $\mathsf{E}\,|\xi_n \eta_n| < \infty$, $n \geq 1$. Prove that the sequence

$$\left\{\zeta_n = \sum_{k=1}^{n} \eta_k \xi_k, n \geq 1\right\}$$

is a martingale.

III.8.2. Let $\{\xi_n, n \geq 1\}$ and $\{\eta_n, n \geq 1\}$ be two sequences of random variables such that for every $n \geq 1$ there exist the probability densities f_n and g_n of the random vectors $(\xi_1, \xi_2, \ldots, \xi_n)$ and $(\eta_1, \eta_2, \ldots, \eta_n)$, respectively. Assume that $f_n(x_1, x_2, \ldots, x_n) > 0$ for all $n \geq 1$ and $(x_1, x_2, \ldots, x_n) \in \mathbf{R}^n$. Prove that the sequence

$$\left\{\zeta_n = \frac{g_n(\xi_1, \ldots, \xi_n)}{f_n(\xi_1, \ldots, \xi_n)}, n \geq 1\right\}$$

is a martingale.

III.8.3. Let $\{\zeta_n, 1 \leq n \leq N\}$ be a martingale, and ν a random variable with values $1, 2, \ldots, N$ and such that $\{\omega \colon \nu > n\} \in F_n$. Prove that

$$\mathsf{E}\,\zeta_\nu = \mathsf{E}\,\zeta_1.$$

III.8.4. Let $\{\zeta_n, 1 \leq n \leq N\}$ be a martingale, and let $\nu = \max\{k \leq N \colon \zeta_1 \leq a, \ldots, \zeta_k \leq a\}$, $a \in \mathbf{R}$. Prove that the sequence $\{\eta_k, 1 \leq k \leq N\}$,

$$\eta_k = \begin{cases} \zeta_k, & k \leq \nu, \\ 2\zeta_\nu - \zeta_k, & \text{otherwise}, \end{cases}$$

is a martingale.

III.8.5. Using the inequality

$$\mathsf{P}\Big(\max_{k \leq n} \zeta_k \geq a\Big) \leq \mathsf{E}\left(\chi_a \zeta_n^+\right), \qquad \zeta_n^+ = \max(0, \zeta_n),$$

where

$$\chi_a = \begin{cases} 1, & \max_{k \leq n} \zeta_k > a, \\ 0, & \text{otherwise}, \end{cases}$$

prove that for $\alpha > 1$,

$$\mathsf{E}\left(\max_{k \leq n} \zeta_k^+\right)^\alpha \leq \left(\frac{\alpha}{\alpha - 1}\right)^\alpha \mathsf{E}\left(\zeta_n^+\right)^\alpha.$$

III.8.6. A sequence of random variables $\{\eta_n, n \geq 1\}$ is said to be *nondecreasing* if $\mathsf{P}(\eta_n \leq \eta_{n+1}) = 1$, $n \geq 1$. Let $\{\zeta_n, n \geq 1\}$ be a submartingale. Prove that there exists a nondecreasing sequence of random variables $\{\eta_n, n \geq 1\}$ such that $\{\zeta_n - \eta_n, n \geq 1\}$ is a martingale.

III.8.7. Let $\{\zeta_n, n \geq 1\}$ be a supermartingale. Using the representation

$$\zeta_n = \theta_n - \eta_n, \qquad n \geq 1,$$

where $\{\theta_n, n \geq 1\}$ is a martingale and $\{\eta_n, n \geq 1\}$ is a nondecreasing sequence of random variables (see Problem III.8.6), prove that

$$\mathsf{P}\Big(\max_{k \leq n} \zeta_k \geq a\Big) \leq \frac{1}{a}\left(\mathsf{E}\,\zeta_n^+ + \mathsf{E}\,\zeta_1 - \mathsf{E}\,\zeta_n\right), \qquad a > 0.$$

III.8.8. Let $\{F_n, n \geq 1\}$ be a sequence of σ-algebras such that $F_n \subset F_{n+1}$, $n \geq 1$, and let F be the minimal σ-algebra containing all the σ-algebras F_n, $n \geq 1$. Let ζ be an F-measurable random variable such that $\mathsf{P}(\zeta \geq 0) = 1$ and $\mathsf{E}\,\zeta < \infty$. Prove that

$$\mathsf{P}\left(\lim_{n\to\infty} \mathsf{E}\left(\zeta / F_n\right) = \zeta\right) = 1.$$

III.8.9. Let F and $\{F_n, n \geq 1\}$ be the σ-algebras defined in the preceding problem, and ξ a random variable with $\mathsf{E}\,|\xi| < \infty$. Prove that the sequence $\{\mathsf{E}\left(\xi / F_n\right), n \geq 1\}$ is a martingale and

$$\mathsf{P}\left(\lim_{n\to\infty} \mathsf{E}\left(\xi / F_n\right) = \mathsf{E}\left(\xi / F\right)\right) = 1.$$

III.8.10. Let $\{\xi_n, n \geq 1\}$ be a sequence of independent random variables such that $\mathsf{E}\,\xi_n = m_n \neq 0$, $n \geq 1$. Prove that the sequence

$$\left\{\zeta_n = \prod_{k=1}^{n} \frac{\xi_k}{m_k}, n \geq 1\right\}$$

is a martingale.

III.8.11. Let $\{\xi_n, n \geq 1\}$ be independent random variables such that for some p, $0 < p < 1$,

$$\mathsf{P}(\xi_n = 1) = p, \qquad \mathsf{P}(\xi_n = -1) = 1 - p = q.$$

For $n \geq 1$ put

$$\eta_n = \sum_{k=1}^{n} \xi_k, \qquad \zeta_n = \left(\frac{q}{p}\right)^{\eta_n}.$$

Prove that $\{\zeta_n, n \geq 1\}$ is a martingale.

III.8.12. Let $\{\zeta_n, n \geq 1\}$ be a submartingale. Using Jensen's inequality for conditional expectation (see Problem II.6.34), prove that $\{f(\zeta_n), n \geq 1\}$ is a submartingale for a continuous convex function f that does not decrease on the axes.

III.8.13. Let $\{\xi_n, n \geq 1\}$ be independent identically distributed random variables with $\mathsf{E}\,\xi_1 = 0$, $\mathsf{E}\,\xi_1^2 = \sigma^2$. Prove that the sequence

$$\left\{\left(\sum_{k=1}^{n} \xi_k\right)^2 - n\sigma^2, n \geq 1\right\}$$

is a martingale.

CHAPTER IV

Simplest Markov Processes

§IV.1. Moment generating functions

Let $P_0, P_1, \dots, P_n, \dots$ be a sequence of real numbers. If the series

$$P(z) = \sum_{n=0}^{\infty} P_n z^n$$

is convergent in an interval $|z| < z_0$, then the function $P(z)$ is called *the moment generating function of the sequence* $\{P_n, n \geq 1\}$.

If $P_n = \mathsf{P}(\xi = n)$, $n \geq 0$, for a random variable ξ taking nonnegative integer values, then the corresponding function

$$P_\xi(z) = \sum_{n=0}^{\infty} z^n \mathsf{P}(\xi = n) = \mathsf{E}\, z^\xi$$

is called *the moment generating function of the random variable* ξ. In this case the series $P_\xi(z)$ converges at least for $|z| \leq 1$.

Let $\xi_1, \dots, \xi_m$ be random variables taking nonnegative integer values. The function

$$P(z_1, \dots, z_m) = \sum_{n_1, \dots, n_m = 0}^{\infty} z_1^{n_1} \cdots z_m^{n_m} \mathsf{P}(\xi_1 = n_1, \dots, \xi_m = n_m) = \mathsf{E}\, z_1^{\xi_1} \cdots z_m^{\xi_m}$$

is called *the joint moment generating function of the random variables* $\xi_1, \dots, \xi_m$. The series $P(z_1, \dots, z_m)$ converges at least for $|z_i| \leq 1$, $i = 1, \dots, m$.

Problems

IV.1.1. Find the moment generating function of

a) the geometric random variable:

$$\mathsf{P}(\xi = n) = p(1-p)^n, \qquad n = 0, 1, \dots, \quad p \in [0, 1];$$

b) the binomial random variable:

$$\mathsf{P}(\xi = n) = \binom{m}{n} p^n (1-p)^{m-n}, \qquad n = 0, 1, \dots, m, \quad p \in [0, 1];$$

c) the Poisson random variable:

$$\mathsf{P}(\xi = n) = e^{-\lambda} \frac{\lambda^n}{n!}, \qquad n = 0, 1, \dots, \quad \lambda > 0;$$

d) a random variable ξ taking values $0, 1, \dots, N$ with equal probabilities $\frac{1}{N+1}$.

IV.1.2. Is it possible that the same moment generating function corresponds to two random variables with different distribution functions? Let $P(z)$ be the moment generating function of a bounded sequence $p_0, p_1, \dots, p_n, \dots$. Prove that for $-1 < z < 1$ the derivatives $P^{(n)}(z)$ exist for all $n \geq 1$ and $p_n = P^{(n)}(0)/n!$.

IV.1.3. Find sequences to which the following moment generating functions correspond:

$$\text{a) } \left(1 - z + \frac{z^2}{4}\right)^{-1}; \qquad \text{b) } \frac{a}{1 - pz^l}; \qquad \text{c) } ze^{z^l\lambda - \lambda}.$$

Which of these sequences are probability distributions of random variables with an appropriate choice of parameters?

IV.1.4. Let $\xi_1, \dots, \xi_n$ be independent random variables assuming values $0, 1, \dots, N-1$ with equal probabilities $1/N$, and let $S_n = \xi_1 + \dots + \xi_n$. Find the moment generating function of S_n and prove that

$$\mathsf{P}(S_n = j) = \frac{1}{N^n} \sum_{i=0}^{n} (-1)^{i+j+Ni} \binom{n}{i} \binom{-n}{j - Ni},$$

where for all x and any integer k,

$$\binom{x}{k} = \begin{cases} 0, & \text{for } k < 0, \\ 1, & \text{for } k = 0, \\ \dfrac{x(x-1)\cdots(x-k+1)}{k!}, & \text{for } k > 0. \end{cases}$$

IV.1.5. Find the kth term of the sequence p_n, $n \geq 0$, with the moment generating function

$$P(z) = \frac{\left(1 - \sqrt{1 - 4pqz^2}\right)}{2qz}$$

with $0 < q = 1 - p \leq 1$. Under what conditions

a) is p_n the probability distribution of a random variable ξ finite with probability one?

b) $\mathsf{E}\xi < \infty$?

Compare the answer with Problem IV.4.10.

IV.1.6. Independent random variables ξ, η, ζ assume integer values $1, 2, \dots, n$ with equal probabilities $1/n$. Find the following probabilities:

a) $\mathsf{P}(\xi + \eta = \zeta)$;

b) $\mathsf{P}(\xi + \eta = 2\zeta)$;

c) $\mathsf{P}(\xi + \eta + \zeta = n + 1)$.

IV.1.7. Let ν be a geometric random variable. Find the moment generating functions of the random variables $\nu_N^+ = \max(N, \nu)$, $\nu_N^- = \min(N, \nu)$.

IV.1.8. Let ξ be a random variable taking integer nonnegative values with the moment generating function $\varphi(z)$. Find the moment generating function of the random variable $\zeta = a\xi + b$, where a and b are nonnegative integers.

IV.1.9. Let $\varphi_\xi(z)$ be the moment generating function of a random variable ξ. Prove that the moment generating function $\psi(z)$ of the sequence $\mathsf{P}(\xi \ge n)$ is given by

$$\psi(z) = \frac{1 - z\varphi_\xi(z)}{1-z}.$$

IV.1.10. Let $\varphi_\xi(z)$ be the moment generating function of a nonnegative integer-valued random variable ξ. Find the moment generating functions of the following sequences:

$$\mathsf{P}(\xi < n), \qquad \mathsf{P}(\xi \le n), \qquad \mathsf{P}(\xi > n).$$

IV.1.11. The joint distribution of two integer-valued random variables ξ and η is given by

$$\mathsf{P}(\xi = n, \eta = k) = \begin{cases} \dfrac{e^{-\lambda}\lambda^n}{n!}\dbinom{n}{k} p^k (1-p)^{n-k}, & \text{for } n \ge k \ge 0, \\ 0, & \text{for } k > n, \end{cases}$$

where $\lambda > 0$, $p \in [0,1]$. Find the joint moment generating function of the random variables ξ and η.

IV.1.12. Random variables ξ and η are nonnegative and integer-valued. Assume that ξ is a geometric random variable and the conditional distribution of η given ξ is

$$\mathsf{P}\left(\eta = k / \xi = n\right) = \begin{cases} u(1-u)^k & \text{if } n \le N, \\ v(1-v)^k & \text{if } n > N, \end{cases} \qquad k = 0, 1, \dots,$$

where $u, v \in [0,1]$. Find the joint moment generating function of the random variables ξ and η.

IV.1.13. Let ξ and η be independent integer-valued random variables with the moment generating functions $\varphi_1(z)$ and $\varphi_2(z)$, respectively. Find the joint moment generating function of the random variables ξ and $\gamma = \xi + \eta$.

IV.1.14. Let $\varphi(z_1, z_2, \dots, z_m)$ be the moment generating function of random variables $\xi_1, \xi_2, \dots, \xi_m$, and

$$\psi(z_1, z_2, \dots, z_m) = \sum_{n_1, \dots, n_m = 0}^{\infty} \mathsf{P}(\xi_1 \le n_1, \dots, \xi_m \le n_m) z_1^{n_1} \cdots z_m^{n_m}.$$

Prove that

$$\psi(z_1, z_2, \dots, z_m) = \frac{1}{1-z_1} \cdots \frac{1}{1-z_m} \varphi(z_1, z_2, \dots, z_m).$$

IV.1.15. Let n balls be randomly put into r boxes. Denote by $\nu_i(n,r)$ the number of balls in the ith box. Find the joint moment generating function of the random variables $\nu_i(n,r)$, $i = 1, \dots, r$.

IV.1.16. Under the conditions of the preceding problem, determine the probability that the first k boxes are empty.

IV.1.17. Let ξ and η be independent random variables taking the values $0, 1, \dots, N$ with equal probabilities $(N+1)^{-1}$. Find the joint moment generating function of the random variables ξ and $\gamma = |\xi - \eta|$.

IV.1.18. Let ξ be a nonnegative integer-valued random variable and $\mathsf{E}\,|\xi| < \infty$. Prove that the derivative $\varphi'_\xi(z)$ of the moment generating function $\varphi_\xi(z)$ is continuous in $[-1,1]$ and $\varphi'_\xi(1) = \mathsf{E}\,\xi$.

IV.1.19. Let ξ be a nonnegative integer-valued random variable, and $\varphi_\xi(z)$ its moment generating function. If $\mathsf{E}\,\xi^k < \infty$ for some $k \geq 1$, then the derivative $\varphi_\xi^{(k)}(z)$ is continuous in $[-1,1]$. Moreover,

$$\varphi_\xi^{(k)}(1) = \sum_{l=0}^{k} a_{lk}\,\mathsf{E}\,\xi^l,$$

where a_{lk}, $l = 0,\dots,k$, are the coefficients of the corresponding powers of x of the polynomial $x(x-1)\cdots(x-k+1) = \sum_{l=0}^{n} a_{lk}x^l$. Write explicit expressions for $\mathsf{E}\,\xi^2$ and $\mathsf{E}\,\xi^3$ via corresponding derivatives.

IV.1.20. Using moment generating functions, find the expectations and variances for the random variables given in Problems IV.1.1, IV.1.5, IV.1.7.

IV.1.21. Let $\varphi(z_1,\dots,z_m)$ be the joint moment generating function of nonnegative integer-valued random variables $\xi_1,\dots,\xi_m$. Show that the derivative

$$\frac{\partial^2\varphi(z_1,\dots,z_m)}{\partial z_i \partial z_j}$$

is continuous in the m-dimensional cube $[-1,1]\times\cdots\times[-1,1]$, provided that $\mathsf{E}\,\xi_i\xi_j < \infty$, and moreover,

$$\mathsf{E}\,\xi_i\xi_j = \left.\frac{\partial^2\varphi(z_1,\dots,z_m)}{\partial z_i \partial z_j}\right|_{z_1=\cdots=z_m=1}.$$

Formulate a similar assertion for moments of higher orders.

IV.1.22. For the random variables $\nu_i(n,r)$, $1 \leq i \leq r$, defined in Problem IV.1.15, find $\mathsf{E}\,\nu_i(n,r)$, $\mathsf{E}\,\nu_i(n,r)\nu_j(n,r)$.

IV.1.23. Let $\xi_1,\dots,\xi_n$ be independent integer-valued nonnegative random variables and $S_n = \xi_1 + \cdots + \xi_n$. Prove that the corresponding moment generating functions are related by the equality

$$\varphi_{S_n}(z) = \varphi_{\xi_1}(z)\cdots\varphi_{\xi_n}(z).$$

IV.1.24. Prove that the sum of two independent Poisson random variables is also a Poisson random variable. Do binomial or geometric random variables possess this property?

IV.1.25. A sequence P_n, $n \geq 0$, is constructed according to the rule

$$P_{n+2} = aP_{n+1} + bP_n, \qquad n = 0,1,\dots.$$

a) Find the moment generating function $P(z)$ of the sequence P_n, $n \geq 0$.
b) Let z_1 and z_2 be the roots of the equation $bz^2 + az - 1 = 0$. Prove that for $|z_1| > |z_2|$,

$$P_n \sim \frac{1}{z_2^n}\frac{aP_0 - z_2^{-1}P_0 - P_1}{\left(z_2^{-1} - z_1^{-1}\right)bz_1z_2} \qquad \text{as } n \to \infty.$$

IV.1.26. Let the moment generating function $U(z)$ of a sequence u_n, $n \geq 0$, be represented by $U(z) = A(z)B^{-1}(z)$, where $A(z)$ and $B(z)$ are polynomials of degrees $l < m$ and m, respectively, that have no common roots. Let the roots of the polynomial $B(z)$ be distinguished. Prove that the moment generating function $U(z)$ is uniquely represented in the form

$$U(z) = \frac{a_1}{z_1 - z} + \cdots + \frac{a_m}{z_m - z}.$$

Derive from this an explicit expression for u_n and show that

$$u_n \sim \frac{a_j}{z_j^{n+1}} \quad \text{as } n \to \infty,$$

where z_j is the smallest by the absolute value root of the polynomial $B(z)$.

IV.1.27. Let f_n, $n \geq 0$, and a_n, $n \geq 0$, be two nonnegative bounded sequences with the moment generating functions $F(z)$ and $A(z)$, respectively. The sequence u_n, $n \geq 0$, is constructed as follows:

$$u_n = a_n + \sum_{k=0}^{n} f_k u_{n-k}, \qquad n \geq 0.$$

a) Find the moment generating function of the sequence u_n, $n \geq 0$.
b) Let $\sum_{n=0}^{\infty} a_n < \infty$. In order that $u = \sum_{n=0}^{\infty} u_n < \infty$, it is necessary and sufficient that

$$f = \sum_{n=0}^{\infty} f_n < 1.$$

Prove this statement.

IV.1.28. Find the moment generating function of the sequence u_n, $n \geq 0$, constructed according to the recurrence rule

$$u_{n+3} = \frac{u_{n+2} + u_{n+1} + u_n}{3}, \qquad n \geq 0.$$

Does the limit of u_n exist as $n \to \infty$? Determine it.

IV.1.29. Find the moment generating function of the sequence u_n, $n \geq 0$, constructed according to the recurrence rule

$$u_{n+k} = \frac{u_{n+k-1} + \cdots + u_n}{k}, \qquad n \geq 0.$$

IV.1.30. On a circle, $2n$ points are given such that all the distances between consecutive points are equal. These points are randomly grouped together into n pairs, and the points of each pair are connected by a chord. Find the probability that these n chords do not intersect.

IV.1.31. A sequence of independent random variables $\eta_1, \eta_2, \ldots$ taking values $a_1, \ldots, a_l$ with probabilities $p_1, \ldots, p_l$ is said to be *a Bernoulli scheme with l results* $a_1, \ldots, a_l$. Here η_n is regarded as a result of the nth experiment.

Let $\nu_i(n)$ be the number of occurrences of a_i in n trials. Find the moment generating function of the random variable $\nu_i(n)$. Find the joint moment generating function of the random variables $\nu_1(n), \ldots, \nu_l(n)$.

IV.1.32. Let a Bernoulli scheme be given (see the preceding problem). Let τ_i be the number of the trial where the first occurrence of the ith result is observed.

a) Find the moment generating function of the random variable τ_i.
b) Find the joint moment generating function of the random variables τ_1 and τ_2 in a Bernoulli scheme with two results.
c) Solve the previous problem for a Bernoulli scheme with $l > 2$ results and calculate the moment $\mathsf{E}\,\tau_1\tau_2$.

IV.1.33. Let a Bernoulli scheme be given (see Problem IV.1.31). Let τ_{ir} be the number of the trial where an ith result occurs for the rth time. Find the moment generating function of the random variable τ_{ir}, its expectation and variance.

IV.1.34. Let a Bernoulli scheme be given (see Problem IV.1.31). Let μ_{ij} be the number of occurrences of a jth result until an ith result occurs for the first time.

a) Find the moment generating function, expectation, and variance of the random variable μ_{ij}.
b) Find the joint moment generating function of random variables μ_{lj}, $j = 1, \dots, l-1$, and the moment $\mathsf{E}\,\mu_{lj}\mu_{lk}$.

IV.1.35. Let a Bernoulli scheme be given (see Problem IV.1.31). Let u_n be the probability that the number of occurrences of an ith result within n trials is even. Establish a recurrence formula for u_n and use it to find the moment generating function for the sequence u_n, $n \geq 0$.

IV.1.36. Let a Bernoulli scheme be given (see Problem IV.1.31). Let u_n be the probability that the number of occurrences of the ith result in n trials is divisible by 3, and v_n the probability that the the number of occurrences of the ith result in n trials has remainder 1 upon division by 3. Establish a system of recurrence relations for u_n and v_n and use it to find the moment generating functions for the sequences u_n, $n \geq 0$, and v_n, $n \geq 0$.

IV.1.37. Let u_n be the probability that the pair a_ia_j occurs for the first time in two consecutive trials, $(n-1)$st and nth. Find the moment generating function of the sequence $\{u_n, n \geq 0\}$.

IV.1.38. Let u_n be the probability that the sequence $a_ia_ja_k$ occurs for the first time in three consecutive trials, $(n-2)$nd, $(n-1)$st, and nth, respectively. Find the corresponding moment generating function.

IV.1.39. Let a Bernoulli scheme be given (see Problem IV.1.31). Let ν_{ir} be the number of trials until a result a_i occurs r times. Prove that the moment generating function of the random variable ν_{ir} is

$$\varphi_{ir}(z) = p_i^r z^r \frac{1 - p_i z}{1 - z + p_i^r(1-p_i)z^{r+1}}$$

and

$$\mathsf{E}\,\nu_{ir} = \frac{1 - p_i^r}{p_i^r(1-p_i)}.$$

IV.1.40. Let a Bernoulli scheme be given (see Problem IV.1.31). Let γ_{in} be the longest run consisting in a result a_i within n trials, and let

$$P_{in}(r) = \mathsf{P}(\gamma_{in} \leq r).$$

Find the moment generating function of the sequence $P_{in}(r)$, $n \geq 1$, and prove that

$$\mathsf{E}\,\gamma_{in} = \frac{\ln n}{-\ln p_i} + o(1), \qquad \operatorname{Var}\gamma_{in} = o(1).$$

IV.1.41. Let ν and $\{\xi_n, n \geq 1\}$ be nonnegative jointly independent integer-valued random variables, and let $\{\xi_n, n \geq 1\}$ be identically distributed. Let $\varphi(z)$ and $\psi(z)$ be the moment generating functions of the random variables ν and ξ_1, respectively. Consider the random variables $S_n = \xi_1 + \cdots + \xi_n$. Show that the moment generating function of the random variable S_ν is $P(z) = \varphi(\psi(z))$.

IV.1.42. Let $\{\eta_n, n \geq 1\}$ be a Bernoulli scheme with l results $a_1, \ldots, a_l$ (the probability for a_i to occur is p_i). Let ν be a random variable, independent of $\{\eta_n, n \geq 1\}$, with the moment generating function $\varphi(z)$. As in Problem IV.1.31, let $\nu_i(n)$ be the number of occurrences of the ith result in n trials, and $\alpha_i = \nu_i(\nu)$ the number of occurrences of the ith result within ν trials.

a) Find the moment generating function of ν_i and the joint moment generating function of the random variables $\alpha_1, \ldots, \alpha_n$.
b) Let ν be a binomial random variable with parameters p and m. Prove that the random variable $\alpha_i = \nu_i(\nu)$ is also binomial with parameters $p_i p$ and m.
c) Let ν be a Poisson random variable with parameter λ. Prove that the random variables $\alpha_i = \nu_i(\nu)$, $i = 1, \ldots, l$, are independent and each of them is also a Poisson random variable.
d) Let $\mathsf{E}\,\nu^k < \infty$. Prove that $\mathsf{E}\,\alpha_i^k < \infty$, $i = 1, \ldots, l$. Using moment generating functions, find $\mathsf{E}\,\alpha_i$ and $\mathsf{E}\,\alpha_i\alpha_j$.

IV.1.43. Let ν be a Poisson random variable with parameter λ, and let ν balls be randomly allocated into m cells. Prove that the probability that exactly n cells are empty is given by

$$\binom{m}{n} e^{-\lambda n/m}\left(1 - e^{\lambda(n-m)/m}\right).$$

IV.1.44. Articles are randomly tested independently of each other. Any of them may with probability p have a defect, and in this case, there is a probability α that this defect is not recognized.

a) Find the moment generating function and the distribution function of the number of defective articles that are missed by the random testing.
b) Let μ be the number of articles that are handled by the random testing until the first defective product is missed. Find the moment generating function and the distribution function of the random variable μ.

IV.1.45. Let $\{\eta_n, n \geq 1\}$ be a sequence of independent random variables taking only two values 0 and 1:

$$\eta_n = \begin{cases} 1, & \text{with probability } p_n, \\ 0, & \text{with probability } q_n = 1 - p_n. \end{cases}$$

The variable η is treated as a result of the nth trial in a Bernoulli scheme with varying probabilities (the result "1" is treated as a "success"). The number of successes within n trials may be written as the sum $S_n = \eta_1 + \cdots + \eta_n$. Find the moment generating function and expectation of the random variable S_n.

IV.1.46. Consider a Bernoulli scheme with varying probabilities. Assume that the nth trial has a success probability p' if n is even, and p'' if n is odd.

a) Let τ be the number of the trial where the first success occurs. Find the moment generating function of τ.

b) Let β be the number of successes until the first success occurs in a trial with an even number. Find the moment generating function of the random variable β.

IV.1.47. Consider a Bernoulli scheme with varying probabilities described in Problem IV.1.46. Let u_n be the probability that the number of successes within n trials is even. Find a recurrence formula for the probabilities u_n and use it to find the corresponding moment generating function.

IV.1.48. Let u_n be the probability that the pair "SF" occurs for the first time in two consecutive trials with numbers $2n-1$ and $2n$, where S stands for "success" and F stands for "failure". Find the moment generating function for the sequence $\{u_n, n \geq 1\}$.

§IV.2. Renewal scheme

Let $\{\xi_n, n \geq 1\}$ be a sequence of independent identically distributed nonnegative random variables. The sequence $\tau_0 = 0, \tau_1 = \xi_1, \tau_2 = \xi_1 + \xi_2, \ldots$ is said to be *a renewal scheme*. In this case the random variables τ_i are treated as renewal moments and ξ_n as interrenewal moments.

In view of applications in practice, the following functionals related to a renewal scheme are of interest: the number of renewals over time t,

$$\nu(t) = \max\left(n \colon \sum_{k=1}^{n} \xi_k \leq t\right),$$

the time after the last renewal before t,

$$\gamma_t^- = t - \tau_{\nu(t)},$$

the time of the first renewal after t,

$$\gamma_t^+ = \tau_{\nu(t)+1} - t,$$

and $\gamma_t = \gamma_t^- + \gamma_t^+$. An important characteristic of a renewal scheme is $N(t) = \mathsf{E}\,\nu(t)$, $t \geq 0$, that is usually called *the renewal function*.

We use the notation $F(t) = \mathsf{P}(\xi_1 < t)$, $t \geq 0$, in the problems below and assume that the distribution function $F(x)$ is not concentrated at zero, that is, $F(0+0) < 1$. The measure on the Borel σ-algebra of $[0, \infty)$ generated by the distribution function $F(t)$ is denoted by $F(A)$. This measure is uniquely determined by its values on intervals: $F([a, b)) = F(a) - F(b)$. In the renewal theory, the so-called *renewal equation* plays a crucial role:

$$z(t) = q(t) + \int_0^{t+0} z(t-s)\, F(ds).$$

Here $q(t)$ is a known measurable function bounded on every finite interval. Denote this class of functions by $\mathfrak{L}$. A solution $z(t)$ of the renewal equation is sought in the class $\mathfrak{L}$.

The so-called *renewal theorem* plays an important role in asymptotic problems of the renewal theory. This theorem describes the behavior of a solution of the renewal equation as $t \to \infty$.

THEOREM (renewal theorem). *Let a function $q(t)$ satisfy the conditions*

a) *the set R_q of its discontinuity points is of zero Lebesgue measure;*

b) *$|q(t)| \le \widetilde{q}(t)$ for all $t \ge 0$, where $\widetilde{q}(t)$ is a nonnegative nonincreasing integrable (in the Riemann sense) function on $[0, \infty)$.*

Then

1) *if the distribution function F is not arithmetic, then*

$$z(t) \to \frac{1}{\mu} \int_0^\infty q(s)\, ds \quad \text{as } t \to \infty, \tag{1}$$

where $z(t)$ is a solution of the renewal equation;

2) *if the distribution function is arithmetic with the step h, then for all $s \in [0, h]$,*

$$z(uh + x) \to \frac{h}{\mu} \sum_{k=0}^{\infty} q(kh + x) \quad \text{as } t \to \infty. \tag{2}$$

Here $\mu = \int_0^\infty x\, dF(x)$ is the expectation of F, and the limits on the right-hand sides of (1) and (2) vanish if $\mu = +\infty$.

REMARK. A distribution function F concentrated on $[0, \infty)$ is said to be *arithmetic* if there exists $h > 0$ such that

$$\sum_{n=0}^{\infty} [F(kh + 0) - F(kh)] = 1. \tag{3}$$

The maximum $h > 0$ satisfying (3) is said to be *the step of F*.

Problems

IV.2.1. Prove the following relationship between two random events:

$$\{\omega : \nu(t) \ge n\} = \{\omega : \xi_1 + \cdots + \xi_n \le t\}.$$

Use it to prove that the condition $F(0+) < 1$ is necessary and sufficient for

$$\mathsf{P}(\nu(t) < +\infty) = 1$$

to hold for all $t \ge 0$.

IV.2.2. Let $\varphi(t, z) = \mathsf{E}\, z^{\nu(t)}$ be the moment generating function of the random variable $\nu(t)$ and

$$L(s, z) = \int_0^\infty e^{-st} \varphi(t, z)\, dt,$$

$$\psi(s) = \int_0^\infty e^{-st}\, dF(t).$$

Prove that

$$L(s, z) = \frac{1 - \psi(s)}{s(1 - z\psi(s))}.$$

IV.2.3. Let the random variables $\{\xi_n, n \geq 1\}$ in a renewal scheme be exponential with the distribution function $F(t) = 1 - e^{-\lambda t}$, $t \geq 0$. Find the distribution function of the random variable $\nu(t)$ and calculate the renewal function $N(t)$.

IV.2.4. Let the random variables $\{\xi_n, n \geq 1\}$ in a renewal scheme be uniformly distributed on the interval $[0, 1]$. Find the distribution function of the random variable $\nu(t)$.

IV.2.5. Let the random variables $\{\xi_n, n \geq 1\}$ in a renewal scheme have the Erlang distribution of order r with the density

$$p(x) = \frac{1}{r!}\lambda(\lambda x)^{r-1}e^{-\lambda x}, \qquad x \geq 0.$$

Find the distribution function of the random variable $\nu(t)$.

IV.2.6. Let $\mu = \mathsf{E}\,\xi_1 = \int_0^\infty x\,dF(x) < \infty$. Prove that

$$\mathsf{P}\left(\lim_{t\to\infty}\frac{\nu(t)}{t} = \frac{1}{\mu}\right) = 1.$$

IV.2.7. Prove that

$$\mathsf{P}\left(\lim_{t\to\infty}\frac{\nu(t)}{t} = 0\right) = 1$$

if $\mu = +\infty$.

IV.2.8. Let $\{\xi_n, n \geq 1\}$ be a sequence of nonnegative random variables and $\nu_t = \max\left(n\colon \sum_{k=1}^n \xi_k \leq t\right)$. Prove that the relations

$$\frac{1}{n}\sum_{k=1}^n \xi_k \xrightarrow{\mathsf{P}} d \quad \text{as } n \to \infty$$

and

$$\frac{\nu_t}{t} \xrightarrow{\mathsf{P}} \frac{1}{d} \quad \text{as } n \to \infty$$

are equivalent. Show that an analogous assertion is valid for the convergence with probability one.

IV.2.9. Let $\mu = \mathsf{E}\,\xi_1$ and $\operatorname{Var}\xi_1 = b^2 < \infty$. Prove that

$$\mathsf{P}\left(\frac{\nu(t) - \frac{t}{\mu}}{b\mu^{-3/2}\sqrt{t}} < x\right) \to \frac{1}{\sqrt{2\pi}}\int_{-\infty}^x e^{-z^2/2}\,dz \quad \text{as } t \to \infty.$$

IV.2.10. Assume that the distribution function of a random variable ξ_1 belongs to the domain of normal attraction of a stable law with parameter $\alpha < 1$, which is equivalent to

$$\mathsf{P}\left(\frac{1}{n^{1/\alpha}}\sum_{k=1}^n \xi_k < x\right) \to \mathsf{P}(\varkappa_\alpha < x) \quad \text{as } n \to \infty,\ x \geq 0,$$

where $\varkappa_\alpha$ is a random variable with the Laplace transform $\mathsf{E}\,e^{-s\varkappa_\alpha} = e^{-cs^\alpha}$. Prove that

$$\mathsf{P}\left(\frac{\nu(t)}{t^\alpha} < x\right) \to F_\alpha(x) \quad \text{as } t \to \infty,\ x \geq 0,$$

where the distribution function $F_\alpha(x)$ is defined by

$$F_\alpha(x) = \mathsf{P}(\varkappa_\alpha > x^{-1/\alpha}), \qquad x \ge 0.$$

IV.2.11. Let $F^{*n}(t)$ be the n-fold convolution of a distribution function $F(x)$, that is,

$$F^{*n}(t) = \int_0^t F^{*(n-1)}(t-s)\,F(ds),$$

where

$$F^{*0}(t) = \begin{cases} 1, & \text{for } t > 0,\\ 0, & \text{for } t = 0.\end{cases}$$

Prove that

$$N(t) = \sum_{n=1}^{\infty} F^{*n}(t+0).$$

IV.2.12. Let

$$L(s) = \int_0^\infty e^{-st} N(t)\,dt, \qquad \psi(s) = \int_0^\infty e^{-st}\,dF(s).$$

Prove that

$$L(s) = \frac{\psi(s)}{s(1-\psi(s))}.$$

IV.2.13. Let $D^*(t) = \mathsf{E}\,\nu(t)(\nu(t)-1)$. Prove that

$$D^*(t) = \sum_{n=0}^{\infty} n(n-1)\left(F^{*n}(t) - F^{*(n+1)}(t)\right).$$

IV.2.14. Let $L^*(s) = \int_0^\infty e^{-st} D^*(t)\,dt$, where the function $D^*(t)$ is defined in Problem IV.2.13. Prove that

$$L^*(s) = \frac{2\psi(s)}{s(1-\psi(s))^2}.$$

IV.2.15. Let $F(t) = \mathsf{P}(\xi_1 < t)$. Prove that the natural correspondence between the distribution functions $F(t)$ and the renewal functions $N(t) = \mathsf{E}\,\nu(t)$ is one-to-one.

IV.2.16. Prove that $t + a(1 - e^{-t})$ is a renewal function if and only if $a = 0$.

IV.2.17. Let $\{\eta_n, n \ge 1\}$ be a sequence of independent random variables, independent of $\{\xi_n, n \ge 1\}$, such that

$$\eta_k = \begin{cases} 1, & \text{with probability } p,\\ 0, & \text{with probability } p = 1-q.\end{cases}$$

Put $\xi'_k = \eta_k \xi_k$, $k \ge 1$, and define a corresponding renewal scheme. Let $N(t)$ and $N'(t)$ be the renewal functions with respect to the sequences $\{\xi_n, n \ge 1\}$ and $\{\xi'_n, n \ge 1\}$, respectively. Prove that

$$N'(t) = \frac{1}{p}(N(t) + 1 - p), \qquad t \ge 0.$$

IV.2.18. Let random variables $\{\xi_n, n \geq 1\}$ be geometric, that is,

$$\mathsf{P}(\xi_1 = n) = p(1-p)^{n-1}, \qquad n \geq 0.$$

Find the distribution of the random variable $\nu(t)$ and the corresponding renewal function $N(t)$, $t \geq 0$.

IV.2.19. Let random variables $\{\xi_n, n \geq 1\}$ have a mixed exponential distribution with the density $\lambda e^{-\lambda(x-T)}$ for $x \geq T$. Prove that the random variable $\nu(t)$ has the distribution

$$\mathsf{P}(\nu(t) < r) = \sum_{k=0}^{r-1} e^{-\lambda(t-r)} \frac{1}{k!} (\lambda(t - rT))^k.$$

IV.2.20. Calculate the function $D(t) = \operatorname{Var} \nu(t)$, $t \geq 0$, for the renewal schemes described in Problems IV.2.3, IV.2.5, IV.2.18, IV.2.19.

IV.2.21. Let $\{\zeta_n, n \geq 1\}$ be a sequence of independent random variables uniformly distributed on the interval $[0,1]$. Let $\nu(t)$ be the number of indices n satisfying the inequality

$$\frac{1}{t} \leq \prod_{k=1}^{n} \zeta_k \leq 1.$$

Find the distribution of the random variable $\nu(t)$.

IV.2.22. Let random variables $\{\xi_n, n \geq 1\}$ have an exponential distribution with parameter λ. Prove that, for all $t, s \geq 0$, random variables $\nu(t)$ and $\nu(t+s) - \nu(t)$ are independent, and $\nu(t+s) - \nu(t)$ has the same distribution as $\nu(s)$.

IV.2.23. Prove that random variables

$$\nu(t_1), \nu(t_2) - \nu(t_1), \ldots, \nu(t_n) - \nu(t_{n-1})$$

are independent for all $0 \leq t_1 \leq t_2 \leq \cdots \leq t_n$.

IV.2.24. Calculate the function $R(t,s) = \mathsf{E}\,\nu(t)\nu(s) - \mathsf{E}\,\nu(t)\,\mathsf{E}\,\nu(s)$ for the scheme described in Problem IV.2.22.

IV.2.25. Let random variables ξ_k be defined in a renewal scheme as follows:

$$\xi_k = \begin{cases} \tau, & \text{with probability } p, \\ 0, & \text{with probability } q = 1 - p, \end{cases}$$

where $p \in [0,1]$. Find the moment generating function of the random variable $\nu(t)$ and the corresponding renewal function $N(t)$.

IV.2.26. Let $\{\xi_n, n \geq 1\}$ and $\{\xi'_n, n \geq 1\}$ be two sequences of nonnegative random variables. Assume that $\xi_k \geq \xi'_k$ with probability one for all $k \geq 1$. Prove that the random variables $\nu(t)$ and $\nu'(t)$ in the corresponding renewal schemes are related as follows:

$$\mathsf{P}(\nu(t) \leq \nu'(t)) = 1, \qquad t \geq 0.$$

IV.2.27. Prove that all the moments of the random variable $\nu(t)$ are finite:

$$\mathsf{E}\,\nu^r(t) < \infty, \qquad r \geq 1.$$

IV.2.28. Assume that $F(\tau) = 0$ for some $\tau > 0$. Prove that for all $r \geq 1$,

$$\frac{\mathsf{E}\,\nu^r(t)}{t^r} \to \frac{1}{(\mathsf{E}\,\xi_1)^r} \qquad \text{as } t \to \infty.$$

IV.2.29. Prove that $t^{-r}\,\mathsf{E}\,\nu^r(t)$ is bounded as a function of t for any $r \geq 1$.

IV.2.30. Prove that for any renewal scheme,

$$\lim_{t\to\infty} \frac{\mathsf{E}\,\nu^r(t)}{t^r} = \frac{1}{\mu^r}.$$

Here $\mu = \mathsf{E}\,\xi_1$ and the limit on the right-hand side is zero if $\mu = +\infty$.

IV.2.31. Let random variables ξ_k have a stable distribution with parameter α, $\alpha < 1$, that is, $\mathsf{E}\,e^{-s\xi_k} = e^{-cs^\alpha}$, where $c > 0$ is a constant. Prove that

$$\lim_{t\to\infty} \frac{N(t)}{t^\alpha} = \frac{1}{c}.$$

IV.2.32. Let random variables ξ_k have an integrable characteristic function $g(s)$ and $g(s) - 1 \sim cs \ln s$. Prove that

$$N(t) \sim \frac{1}{c \ln t}.$$

IV.2.33. For $n \geq 0$, let $\xi_k^{(n)}$, $k \geq 1$, be a sequence of independent identically distributed nonnegative random variables, and let $\nu^{(n)}(t)$ be the number of renewals over time t in a renewal scheme constructed with respect the random variables $\{\xi_k^{(n)}, k \geq 1\}$. Assume that $\xi_1^{(n)} \Rightarrow \xi_1^{(0)}$ as $n \to \infty$ (the symbol "$\Rightarrow$" stands for the weak convergence of distribution functions). Prove that $\nu^{(n)}(t) \Rightarrow \nu^{(0)}(t)$ as $n \to \infty$ for all t except, possibly, for a denumerable set T.

IV.2.34. Prove that any renewal function is right continuous and has at most denumerable set of discontinuity points. Prove that, under the conditions of Problem IV.2.33,

$$N^{(n)}(t) = \mathsf{E}\,\nu^{(n)}(t) \to N^{(0)}(t) = \mathsf{E}\,\nu^{(0)}(t) \quad \text{as } n \to \infty,$$

for all t, where the renewal function $N^{(0)}(t)$ is continuous.

IV.2.35. Let $U(t) = \sum_{n=0}^\infty F^{*n}(t)$. Show that $U(t) = N(t-0) + F^{*0}(t)$. Here $N(0-0) = 0$. Prove that the renewal equation has a unique solution given by

$$z(t) = \int_0^{t+0} q(t-s)\,U(ds),$$

in the class $\mathfrak{L}$. Here $U(A)$ is the measure on the Borel σ-algebra of $[0,\infty)$ uniquely determined by its values on intervals: $U([a,b)) = U(b) - U(a)$.

IV.2.36. Prove that $N(t)$ satisfies the renewal equation with $q(t) = F(t+0)$.

IV.2.37. Derive from the renewal theorem that

$$\frac{U(t+h+0) - U(t+0)}{h} = \frac{N(t+h) - N(t)}{h} \to \frac{1}{\mu} \quad \text{as } t \to \infty$$

if the distribution function $F(\cdot)$ is not arithmetic, and if it is arithmetic, then the above relation is valid for h a multiple of the step of F.

IV.2.38. Prove that the function

$$Z(t) = \sum_{n=0}^{\infty} F^{*n}(t+0) - \frac{t}{\mu}$$

satisfies the renewal equation

$$Z(t) = q(t) + \int_0^{t+0} Z(t-s)\, F(ds),$$

where

$$q(t) = \frac{1}{\mu} \int_t^{\infty} (1 - F(y+0))\, dy.$$

IV.2.39. If $\mu = \mathsf{E}\,\xi_1 < \infty$ and $\operatorname{Var}\xi_1 = \sigma^2 < \infty$, then

$$U(t) - \frac{t}{\mu} \to \frac{\sigma^2 + \mu^2}{2\mu^2} \qquad \text{as } t \to \infty.$$

IV.2.40. Let random variables ξ_k, $k \geq 1$, assume only integer values. Put $P_k(n) = \mathsf{P}(\xi_1 + \cdots + \xi_n = k)$.

a) Prove that $N(t) = \sum_{n=1}^{\infty} \sum_{k \leq t} P_k(n)$.

b) Let $q_k = \sum_{n=1}^{\infty} P_k(n)$. Prove that

$$q_k = \frac{1}{\pi i} \int_{-\pi}^{+\pi} \frac{1}{1 - \psi(s)} \sin ks\, ds,$$

where $\psi(s)$ is the characteristic function of the random variable ξ_1.

c) If $\mu = \mathsf{E}\,\xi_1 < \infty$, then

$$\lim_{k \to \infty} q_k = \frac{1}{\mu}.$$

d) If $\mu = \mathsf{E}\,\xi_1$ and $b^2 = \operatorname{Var}\xi_1 < \infty$, then

$$q_k = \frac{1}{\mu} + o\left(\frac{1}{k}\right).$$

IV.2.41. a) If the distribution function of the random variables ξ_k in a renewal scheme has a bounded density $f(t)$, then the renewal function $N(t)$ also has a density $v(t)$, which is a unique solution of the renewal equation with $q(x) = f(x)$, $x \geq 0$.

b) Find the distribution density of the renewal functions for the distribution functions $F(x)$ given in Problems IV.2.3, IV.2.4, and IV.2.5.

c) If $f(x) \leq \hat{f}(x)$, $x \in [0, \infty)$, where $\hat{f}(x)$ is a nonincreasing integrable function on $[0, \infty)$, then the density $v(t)$ of the renewal function is such that

$$v(t) \to \frac{1}{\mu} \qquad \text{as } t \to \infty,$$

where $\mu = \mathsf{E}\,\xi_1 \leq \infty$.

d) In the general case where only the boundedness of $f(t)$ is assumed,

$$v(t) - f(t) \to \frac{1}{\mu} \qquad \text{as } t \to \infty.$$

IV.2.42. Let $H_t(u) = \mathsf{P}(\gamma_t^+ > u)$. Prove that

$$\mathsf{P}(\gamma_t^- \geq v, \gamma_t^+ > u) = H_{t-v}(u+v), \qquad t - v \geq 0.$$

IV.2.43. Find the joint distribution functions of the random variables (γ_t, γ_t^+) and $(\gamma_t, \gamma_t^+, \gamma_t^-)$ defined at the beginning of this section.

IV.2.44. Find the joint distribution of the random variables (γ_t^+, γ_t^-) for the case of exponential random variables ξ_k.

IV.2.45. Find the joint distribution of the random variables (γ_t^+, γ_t^-) for the case of geometric random variables ξ_k.

IV.2.46. Find the distributions of γ_t^+, γ_t^-, γ_t and the joint distribution of $(\gamma_t^+, \gamma_t^-, \gamma_t)$ for the case of Poisson random variables ξ_k with parameter λ:

$$\mathsf{P}(\xi_k = n) = \frac{\lambda^n}{n!} e^{-\lambda}, \qquad n \geq 0.$$

IV.2.47. For exponential random variables ξ_k, $k \geq 1$, find $\mathsf{E}\,\gamma_t$ and show that

$$\lim_{t\to\infty} \mathsf{E}\,\gamma_t = 2\,\mathsf{E}\,\xi_1.$$

IV.2.48. Prove that $\nu(t)$ and γ_t^+ are independent in the case of exponential random variables ξ_k.

IV.2.49. Let ξ_k be binomial random variables,

$$\mathsf{P}(\xi_k = l) = \binom{m}{l} p^l (1-p)^{m-l}, \qquad l = 0, \dots, m,$$

where $0 < p < 1$ and m is an integer. Find the distributions of $(\gamma_t^+, \gamma_t^-, \gamma_t)$ for $t > m$.

IV.2.50. Prove that $H_t(u)$, as a function of t, satisfies the renewal equation with $q(t) = 1 - F(t + u + 0)$. Using the renewal theorem, show that

$$H_t(u) \to \frac{1}{\mu} \int_0^\infty (1 - F(s + u + 0))\, ds,$$

$$\mathsf{P}(\gamma_t^- > v, \gamma_t^+ > u) \to \frac{1}{\mu} \int_{u+v}^\infty (1 - F(s + 0))\, ds,$$

provided $\mu = \mathsf{E}\,\xi_1 < \infty$ and the distribution function F is not arithmetic.

IV.2.51. Prove that $\mathsf{E}\,\gamma_t^+ < \infty$ if $\mathsf{E}\,\xi_1 < \infty$, and that $\mathsf{E}\,\gamma_t^+$, as a function of t, satisfies the renewal equation with $q(t) = \int_0^\infty (1 - F(t + u + 0))\, du$. Using the renewal theorem, prove that

$$\lim_{t\to\infty} \mathsf{E}\,\gamma_t^+ = \frac{\mathsf{E}\,\xi_1^2}{2\,\mathsf{E}\,\xi_1}$$

provided $\mathsf{E}\,\xi_1^2 < \infty$ and the distribution function F is not arithmetic.

IV.2.52. Let F be a nonarithmetic distribution function and $\mu = \mathsf{E}\,\xi_1 < \infty$, $\mathsf{E}\,\xi_1^2 = +\infty$. Prove that

$$\lim_{t\to\infty} \mathsf{E}\,\gamma_t^+ = +\infty.$$

IV.2.53. Prove the relation

$$\mathsf{P}\big(\nu(t+h)-\nu(t)=r\big)=\delta(r,0)\mathsf{P}(\gamma_t^+>h)+\int_0^{h+0}\mathsf{P}\big(\nu(h-s)+1=r\big)\,\mathsf{P}(\gamma_h^+\in ds).$$

Show that under the conditions of Problem IV.2.52 there exists the limit

$$\lim_{t\to\infty}\mathsf{P}\big(\nu(t+h)-\nu(t)=r\big)$$

and express this limit in terms of F and the distributions of random variables $\nu(h)$, $0\le u\le h$.

IV.2.54. Prove that the limit distribution of γ_t^+ as $t\to\infty$ has the density provided $F(t)$ is a continuous function.

IV.2.55. Let the distribution function of random variables ξ_k have the density. Show that the limit joint distribution of random variables γ_t^- and γ_t^+ has the density and

$$\frac{\partial^2}{\partial u\,\partial v}\lim_{t\to\infty}\mathsf{P}(\gamma_t^+<u,\gamma_t^-<v)=\frac{p(u+v)}{\mathsf{E}\,\xi_1}$$

and the density of the limit distribution of γ_t as $\to\infty$ is

$$\frac{up(u)}{\mathsf{E}\,\xi_1}.$$

IV.2.56. Let $\tau=\inf\{t\colon\gamma_t^->v\}$ be the first moment of time when the gap between renewals is greater than v.

a) Prove that the probability $v(s)=\mathsf{P}(\tau\le s)$ satisfies the equation

$$v(t)=z(t)+\int_0^{t+0}v(t-s)\,dG(s),$$

where

$$z(t)=\begin{cases}0, & \text{for } t\le v,\\ 1-F(v), & \text{for } t>v,\end{cases}\qquad G(x)=\begin{cases}F(x), & \text{for } x\le v,\\ F(v), & \text{for } x>v.\end{cases}$$

b) Find the distribution of the random variable τ if

$$F(t)=1-e^{-\lambda t},\qquad t\ge0.$$

IV.2.57. Prove that, for all $k\ge1$, infinitely many events $\{\omega\colon\gamma_t^+=k\}$, $t=1,2,\dots$, occur with probability one, provided the random variables ξ_k are unbounded, that is, $\mathsf{P}(\xi_k\ge n)>0$ for all $n\ge1$ and their distribution is arithmetic with step one.

§IV.3. Compound Poisson process

Let $\{\tau_n,n\ge1\}$ and $\{\xi_n,n\ge1\}$ be two sequences of random variables satisfying the following conditions:

a) the sequences $\{\tau_n,n\ge1\}$ and $\{\xi_n,n\ge1\}$ are independent;

b) the random variables $\{\tau_n,n\ge1\}$ are independent and have the exponential distribution with parameter λ,

$$\mathsf{P}(\tau_1>x)=e^{-\lambda x},\qquad x\ge0;$$

c) the random variables $\{\xi_n, n \ge 1\}$ are independent and identically distributed with a distribution function $F(x)$.

Using the sequences $\{\tau_n, n \ge 1\}$ and $\{\xi_n, n \ge 1\}$ we define stochastic processes

$$\nu(t) = \max\left\{n\colon \sum_{k=1}^{n} \tau_k \le t\right\}, \qquad t \ge 0,$$

$$\xi(t) = \sum_{k=1}^{\nu(t)} \xi_k, \qquad t \ge 0,$$

that are usually called *a Poisson process* and *a compound Poisson process*, respectively.

Problems

IV.3.1. Prove that $\xi(t)$ is a process with independent increments, that is, for all $0 \le t_1 \le t_2 \le \cdots \le t_n$, the random variables $\xi(t_1), \xi(t_2) - \xi(t_1), \ldots, \xi(t_n) - \xi(t_{n-1})$ are independent.

IV.3.2. Prove that $\xi(t)$ is a homogeneous process with independent increments, that is, for all $t \ge 0$ the distribution of the random variables $\xi(s+t) - \xi(s)$ is the same for all $s \ge 0$.

IV.3.3. Prove that the characteristic function of the random variable $\xi(t)$ is of the form

$$\mathsf{E}\, e^{iz\xi(t)} = \exp\left\{ -\lambda t \int_{-\infty}^{+\infty} \left(1 - e^{izx}\right) dF(x)\right\}.$$

IV.3.4. Find the distribution of $\xi(t)$ if ξ_k are

a) exponential;

b) geometric;

c) binomial, that is, ξ_k take values 0 and 1 with probabilities p and $q = 1 - p$, respectively.

IV.3.5. Let

$$\nu(t) = \max\left\{n\colon \sum_{k=1}^{n} \tau_k \le t\right\}, \qquad t \ge 0,$$

be a Poisson process with parameter λ. The moments $\chi_n = \tau_1 + \cdots + \tau_n$, $n \ge 1$, may be treated as moments of time when requests come to a service system. Assume that the requests are processed, each independently of others, with a probability p and are refused with the probability $q = 1 - p$. Let τ'_n, $n \ge 1$, be the times between requests that are handled, and

$$\nu(t) = \max\left\{n\colon \sum_{k=1}^{n} \tau'_k \le t\right\}, \qquad t \ge 0,$$

be the renewal process generated by the random variables τ'_k, $k \ge 1$. Prove that $\nu'(t)$, $t \ge 0$, is also Poisson, with parameter λp.

IV.3.6. Let $\mathsf{E}\,|\xi_1|^k < \infty$. Prove that $\mathsf{E}\,|\xi(t)|^k < \infty$. Find explicit expressions for $\mathsf{E}\,\xi(t)$ and $\operatorname{Var}\xi(t)$ in terms of $a = \mathsf{E}\,\xi_1$ and $\sigma^2 = \operatorname{Var}\xi_1$.

IV.3.7. If $\mathsf{E}\,\xi_1 = a$, then

$$\mathsf{P}\left(\lim_{t\to\infty}\frac{\xi(t)}{t} = \lambda a\right) = 1.$$

IV.3.8. If $\mathsf{E}\,\xi_1 = a$ and $\mathsf{E}\,\xi_1^2 = \sigma^2 < \infty$, then

$$\mathsf{P}\left(\frac{\xi(t)-\lambda t}{\sqrt{\sigma^2\lambda t}} < x\right) \to \frac{1}{\sqrt{2\pi}}\int_{-\infty}^{x} e^{-z^2/2}\,dz \quad \text{as } t\to\infty.$$

IV.3.9. Prove that

$$\mathsf{P}(\xi(t) < x) = \sum_{n=0}^{\infty}\frac{1}{n!}e^{-\lambda t}(\lambda t)^n F^{*n}(x),$$

where $F^{*n}(x)$ is the n-fold convolution of the distribution function $F(x)$.

IV.3.10. Prove that the function $H(t,x) = \mathsf{P}(\xi(t) < x)$ satisfies the integral equation

$$H(t,x) = \chi_{(0,\infty)}(x)e^{-\lambda t} + \int_0^t\int_{-\infty}^{\infty}\lambda H(t-s,x-z)e^{-\lambda s}\,ds\,dF(z).$$

IV.3.11. Assume that corresponding to any interval $[a,b)$ is a random variable $\nu[a,b)$ such that

1) $\nu[a,b)$ takes only positive integer values;
2) if $a_1 < \cdots < a_n$, then the random variables $\nu[a_1,a_2),\dots,\nu[a_{n-1},a_n)$ are independent;
3) if $a < c < b$, then $\nu[a,b) = \nu[a,c) + \nu[c,b)$;
4) there exists a continuous nondecreasing function $\varphi(t)$ such that

$$\mathsf{P}(\nu[a,b) > 0) = \varphi(b) - \varphi(a) + o(\varphi(b)-\varphi(a)),$$
$$\mathsf{P}(\nu[a,b) > 1) = o(\varphi(b)-\varphi(a)),$$

uniformly as $b - a \to 0$. Prove that

$$\mathsf{P}(\nu[a,b) = r) = \frac{1}{r!}(\varphi(b)-\varphi(a))^r e^{-(\varphi(b)-\varphi(a))}.$$

IV.3.12. Let $\nu(t)$, $t \geq 0$, be a Poisson process with parameter λ. Let $\varphi(t) \geq 0$ be a nondecreasing and differentiable function. Prove that $\nu(\varphi(t))$, $t \geq 0$, is a process with independent increments, that is, for all $t_1 < t_2 < \cdots < t_n$, the random variables

$$\nu(\varphi(t_1)), \nu(\varphi(t_2)) - \nu(\varphi(t_1)), \dots, \nu(\varphi(t_n)) - \nu(\varphi(t_{n-1}))$$

are independent. Prove also that the random variables $\nu[a,b) = \nu(\varphi(b)) - \nu(\varphi(a))$ satisfy conditions 1)–4) of the preceding problem.

IV.3.13. Let $\xi(t)$ be a compound Poisson process. Assume that its jumps have a distribution $F(x)$, and λ is the parameter of the corresponding Poisson process. For a continuous function $g(x)$ put

$$\zeta_{t,x} = \int_0^t g(x+\xi(s))\,ds.$$

Then the characteristic function $U_z(t,x) = \mathsf{E}\, e^{iz\zeta_{t,x}}$ satisfies the following integral equation:

$$U_z(t,x) = \int_0^t \int_{-\infty}^{\infty} \lambda e^{-\lambda s + izg(x)s} U_z(t-s, x+y)\, dF(y)\, ds + e^{izg(x)t - \lambda t}.$$

IV.3.14. Let $\xi_k^{(1)}$, $k \geq 1$, and $\xi_k^{(2)}$, $k \geq 1$, be two independent sequences of identically distributed random variables with distribution functions $F_1(x)$ and $F_2(x)$, respectively. Also let τ_k, $k \geq 1$, be a sequence, independent of $\xi_k^{(1)}$, $k \geq 1$, and $\xi_k^{(2)}$, $k \geq 1$, of independent identically distributed exponential random variables with a parameter λ. Consider two generalized renewal processes

$$\xi_j(t) = \sum_{k=1}^{\nu(t)} \xi_k^{(j)}, \qquad j = 1, 2,$$

where $\nu(t) = \max\,(n\colon \sum_{k=1}^n \tau_k \leq t)$.

a) Is a linear combination $\xi(t) = a_1\xi_1(t) + a_2\xi_2(t)$ of $\xi_1(t)$ and $\xi_2(t)$ a generalized renewal process?
b) Find the joint characteristic function of the processes $\xi_1(t)$ and $\xi_2(t)$.

IV.3.15. Let $\tau_k^{(1)}$, $k \geq 1$, and $\tau_k^{(2)}$, $k \geq 1$, be two independent sequences of independent identically distributed random variables. The random variables $\tau_k^{(1)}$, $k \geq 1$, are exponential with a parameter λ_1, and $\tau_k^{(2)}$, $k \geq 1$, are exponential with a parameter λ_2. Also let ξ_k, $k \geq 1$, be a sequence, independent of $\tau_k^{(1)}$, $k \geq 1$, and $\tau_k^{(2)}$, $k \geq 1$, of independent identically distributed random variables with a distribution function $F(x)$. Consider two generalized renewal processes

$$\xi_j(t) = \sum_{k=1}^{\nu_j(t)} \xi_k, \qquad j = 1, 2,$$

where $\nu_j(t) = \max\big(n\colon \sum_{k=1}^n \tau_k^{(j)} \leq t\big)$ is the total number of renewals over time t in the simple renewal process generated by the random variables $\{\tau_k^{(j)}, k \geq 1\}$, $j = 1, 2$.

a) Is a linear combination $\xi(t) = a_1\xi_1(t) + a_2\xi_2(t)$, $t \geq 0$, a generalized renewal process?
b) Find the joint characteristic function of the processes $\xi_1(t)$ and $\xi_2(t)$.

IV.3.16. Let $\xi_1(t)$ and $\xi_2(t)$ be two independent generalized renewal processes with characteristic functions

$$\mathsf{E}\, e^{iz\xi_j(t)} = \exp\left\{-\lambda_j t \int_0^{\infty} \left(1 - e^{izx}\right) dF_j(x)\right\}, \qquad j = 1, 2,$$

respectively.

a) Is a linear combination $\xi(t) = a_1\xi_1(t) + a_2\xi_2(t)$, $t \geq 0$, a generalized renewal process?
b) Find the characteristic function of the process $\xi(t)$.

IV.3.17. Let $A_1, \dots, A_l$ be disjoint Borel subsets of $\mathbf{R}_1 = (-\infty, +\infty)$ and $\xi(t, A_i) = \sum_{k=1}^{\nu(t)} \chi_{A_i}(\xi_k)$, where $\chi_A(x)$ is the indicator of a set A.

a) Prove that the random variables $\xi(t, A_i)$, $i = 1, \dots, l$, are jointly independent and find the characteristic function of $\xi(t, A_i)$.

b) Let A_1, A_2 be Borel subsets of $\mathbf{R}_1$ and $A = A_1 \cap A_2$. Find the joint characteristic function of the random variables $\xi(t, A_1)$ and $\xi(t, A_2)$.

IV.3.18. Let

$$\xi_k = \begin{cases} +1, & \text{with probability } p, \\ -1, & \text{with probability } q = 1 - p, \end{cases}$$

where $p, q > 0$.

a) Find the distribution of $\xi(t)$;

b) find the Laplace transform of the random variable

$$\tau(x) = \inf(t\colon t > 0, \xi(t) = x), \qquad x = 1, 2, \dots .$$

IV.3.19. Let $\xi(t)$ be a compound Poisson process with nonnegative jumps with the distribution function $F(x)$. For $\gamma < 0$ put

$$\zeta(t) = \gamma t + \xi(t), \qquad t \ge 0.$$

Find a condition under which

$$\mathsf{P}\Big(\sup_{t\ge 0} \zeta(t) < +\infty\Big) = 1.$$

IV.3.20. Let $\nu(t)$, $t \ge 0$, be a Poisson process with parameter λ, and let $\gamma < 0$.

a) Prove that the process $\zeta(t) = \gamma t + \xi(t)$, $t \ge 0$, is bounded from above, that is,

$$\mathsf{P}\Big(\sup_{t\ge 0} \zeta(t) < +\infty\Big) = 1,$$

if and only if $|\gamma| > \lambda$.

b) Show that

$$\mathsf{P}\Big(\sup_{t\ge 0} \zeta(t) \le 0\Big) = \exp\left\{ \sum_{k=1}^{\infty} \frac{1}{k} e^{-kc} \left(1 + kc + \frac{(kc)^2}{2!} + \dots + \frac{(kc)^n}{n!} + \cdots \right) \right\},$$

where $c = \lambda/|\gamma|$.

c) For the process $\zeta(t)$ show that the function

$$Q(t, x) = \mathsf{P}\Big(\sup_{0 \le s < t} \zeta(s) < x\Big), \qquad x > 0,$$

satisfies the integral equation

$$Q(t, x) = e^{-\lambda t} + \int_{(1-x)/|\gamma|}^{t} \lambda e^{-\lambda s} Q(t - s, x - \gamma s + 1)\, ds.$$

IV.3.21. Let $\xi(t)$ be a compound Poisson process with nonnegative jumps, that is, $F(0) = 0$, and let $\gamma = \text{const} > 0$. Consider random variables

$$\Theta(x) = \inf\{n: \gamma n - \xi(n) > x\}.$$

a) Prove that the random variable $\Theta(x + y)$ has the same distribution as $\Theta(x) + \overline{\Theta}(y)$, where $\overline{\Theta}(y)$ is a random variable, independent of $\Theta(x)$, with the same distribution as $\Theta(y)$.

b) Prove that $\Theta(x) \xrightarrow{\mathsf{P}} 0$ as $x \to 0$.

c) Prove that $\mathsf{E}\, e^{-s\Theta(x)} = e^{-x\omega(s)}$.

d) Put $H(x, y) = \mathsf{P}(\Theta(x) < y)$. Prove that the probabilities $H(x, y)$, $x \geq 0$, satisfy the integral equation

$$H(x, y) = \chi_A(y) e^{-\lambda x/\gamma} + \int_0^{x/\gamma} \int_0^\infty \lambda H(x + z - u\gamma, y - u) e^{-\lambda u}\, dF(z)\, du,$$

where $A = [x/\gamma, \infty)$.

e) Let

$$\mathsf{E}\, e^{-s\xi(t)} = e^{-t\Phi(s)}, \qquad \Phi(s) = \lambda \int_0^\infty \left(1 - e^{-sx}\right) dF(x).$$

Prove that the function $\omega(s)$ satisfies the equation

$$\omega(s) = \frac{s + \Phi(\omega(s))}{\gamma}.$$

f) Prove that for every positive s the equation

$$z = \frac{s + \Phi(z)}{\gamma}$$

has only one nonnegative solution z_s, and therefore $\omega(s) = z_s$.

g) Prove that $\omega(s) \to \omega$ as $s \to 0$, where ω is the greatest nonnegative solution to the equation $z = \gamma^{-1}\Phi(z)$. Let $a = \int_0^\infty x\, dF(x)$. Prove that $\omega = 0$ if $\lambda a/\gamma \leq 1$, and $\omega > 0$ if $\lambda a/\gamma > 1$.

h) Prove that

$$\mathsf{P}\Big(\sup_{0 \leq u < t} (\gamma u - \xi(u)) \leq x\Big) = 1 - \mathsf{P}(\Theta(x) \leq t).$$

Using this relation prove that

$$\mathsf{P}\Big(\sup_{u \geq 0} (\gamma u - \xi(u)) \leq x\Big) = 1 - e^{-x\omega}.$$

IV.3.22. Let τ_k, $k \geq 1$, and ξ_k, $k \geq 1$, be two independent sequences of independent identically distributed random variables. Assume that τ_k are nonnegative and $\mathsf{P}(\tau_1 > 0) > 0$. Consider the stochastic process

$$\xi(t) = \sum_{k=1}^{\nu(t)} \xi_k, \qquad t \geq 0,$$

where

$$\nu(t) = \max\left\{n: \sum_{k=1}^{n} \tau_k \leq t\right\}.$$

a) Under which conditions is $\xi(t)$ a compound Poisson process?
b) Prove that

$$\int_0^\infty e^{-st}\,\mathsf{E}\,e^{iz\zeta(t)}\,dt = \frac{1-\varphi(s)}{s\big(1-\varphi(s)\psi(z)\big)},$$

where

$$\varphi(s) = \int_0^\infty e^{-st}\,d\mathsf{P}(\tau_1 < t), \qquad \psi(z) = \int_{-\infty}^\infty e^{izx}\,d\mathsf{P}(\xi_1 < x).$$

c) Let $\mathsf{E}\,\xi_1 = a$. Prove that $\mathsf{E}\,\xi(t) = a\,\mathsf{E}\,\nu(t)$.
d) Let $\operatorname{Var}\xi_1 = \sigma^2$. Prove that $\mathsf{E}\,\xi^2(t) = \sigma^2\,\mathsf{E}\,\nu(t) + a^2\left(\mathsf{E}\,\nu(t)\right)^2$.
e) Prove that

$$\mathsf{P}\left(\lim_{t\to\infty}\frac{\xi(t)}{\nu(t)} = a\right) = 1,$$

provided that $\mathsf{E}\,\xi_1 = a$ exists.

IV.3.23. a) Under what condition is a process $\xi(t)$, $t \ge 0$, defined in Problem IV.3.22 bounded from above, that is,

$$\mathsf{P}\Big(\sup_{t\ge 0}\xi(t) < +\infty\Big) = 1?$$

b) Let $\tau_x = \inf\{t\colon \xi(t) > x\}$ and $Q(t,x) = \mathsf{P}(\tau_x > t)$. Derive the following equation for $Q(t,x)$:

$$Q(t,x) = \mathsf{P}(\tau_1 > t) + \int_0^t dF(s)\int_{-\infty}^x dG(y)Q(t-s,x-y), \qquad x \ge 0,$$

where $F(s) = \mathsf{P}(\tau_1 < s)$, $\qquad G(y) = \mathsf{P}(\xi_1 < y)$.

IV.3.24. Let $\{(\tau_k,\xi_k)\}$, $k \ge 1$, be a sequence of independent identically distributed random vectors. Assume that the first components of the vectors are nonnegative and $\mathsf{P}(\tau_1 > 0) > 0$. Consider the stochastic process

$$\xi(t) = \sum_{k=1}^{\nu(t)}\xi_k, \qquad t \ge 0,$$

where $\nu(t) = \max\{n\colon \sum_{k=1}^n \tau_k \le t\}$.
a) Prove that

$$\Phi(s,z) = \int_0^\infty e^{-st}\,\mathsf{E}\,e^{iz\xi(t)}\,dt = \frac{1-\varphi(s,0)}{s(1-\varphi(s,z))},$$

where $\varphi(s,z) = \mathsf{E}\,e^{-s\tau_1+iz\xi_1}$.
b) Let $\mathsf{E}\,\xi_1 = a$. Prove that

$$\mathsf{E}\sum_{k=1}^{\nu(t)+1}\xi_k = a[\mathsf{E}\,\nu(t)+1].$$

c) Let $\mathsf{E}\,\xi_1 = 0$, $\mathsf{E}\,\xi_1^2 = \sigma^2 < \infty$. Prove that

$$\mathsf{E}\left[\sum_{k=1}^{\nu(t)+1}\xi_k\right]^2 = \sigma^2(\mathsf{E}\,\nu(t)+1).$$

IV.3.25. Let $\xi(t)$, $t \ge 0$, be the process defined in Problem IV.3.24.

a) Prove that

$$\mathsf{P}\left(\sup_{t\to\infty} \frac{\xi(t)}{t} = \frac{c}{a}\right) = 1$$

if there exist $\mathsf{E}\,\tau_1 = a$ and $\mathsf{E}\,\xi_1 = c$.

b) Prove that

$$\frac{\sum_{k=1}^{\nu(t)} \xi_k - \sum_{k=1}^{[ta^{-1}]} \xi_k}{\sqrt{t}} \xrightarrow{\mathsf{P}} 0 \quad \text{as } t \to \infty$$

if there exist $\mathsf{E}\,\tau_1 = a < \infty$ and $\mathsf{E}\,\xi_1 = 0$, $\mathsf{E}\,\xi_1^2 = \sigma^2 < \infty$.

c) Prove that

$$\mathsf{P}\left(\frac{\xi(t)}{\sqrt{\sigma^2 t a^{-1}}} < x\right) \to \frac{1}{\sqrt{2\pi}} \int_{-\infty}^{x} e^{-z^2/2}\,dx \quad \text{as } t \to \infty$$

if there exist $\mathsf{E}\,\tau_1 = a$, $\mathsf{E}\,\xi_1 = 0$, and $\mathsf{E}\,\xi_1^2 = \sigma^2 < \infty$.

d) Prove that

$$\mathsf{P}\left(\frac{\xi(t) - cta^{-1}}{\sqrt{\sigma^2 t a^{-1}}} < x\right) \to \frac{1}{\sqrt{2\pi}} \int_{-\infty}^{x} e^{-z^2/2}\,dx \quad \text{as } t \to \infty$$

if there exist $\mathsf{E}\,\tau_1 = a$, $\mathsf{E}\,\xi_1 = c$, and $\mathsf{E}(\xi_1 - c\tau_1 a^{-1})^2 = \sigma^2 < \infty$.

§IV.4. Random walks

Let $S_0, \xi_1, \xi_2, \dots, \xi_n, \dots$ be a sequence of independent identically distributed random variables. The sequence of partial sums

$$S_0,\ S_1 = S_0 + \xi_1,\ S_2 = S_0 + \xi_1 + \xi_2, \dots$$

is called *a random walk*. The random variables are treated as jumps of the walk, and S_n as the state of the walk after nth jump.

A walk is called *integer-valued* if its jumps ξ_n and the initial state S_0 are integer-valued random variables. In what follows we assume that $S_0 \equiv 0$.

A sequence $\{S_n, n \ge 1\}$ is called *a Bernoulli walk* if

$$\xi_n = \begin{cases} +1, & \text{with probability } p, \\ -1, & \text{with probability } q = 1 - p, \end{cases}$$

where $0 < p < 1$. If $p = \frac{1}{2}$, then $\{S_n, n \ge 1\}$ is said to be *a simple Bernoulli walk*.

Problems

IV.4.1. Let $\{S_n, n \ge 1\}$ be a simple Bernoulli walk. Prove that for $|y| \le n$,

$$P_{n,y} = \mathsf{P}(S_n = y) = \begin{cases} 0 & \text{if } n - y \text{ is odd}, \\ 2^{-n}\dbinom{n}{n-y} & \text{if } n - y \text{ is even}. \end{cases}$$

IV.4.2. Let $\{S_n, n \ge 1\}$ be a simple Bernoulli walk. Prove that for $x, y > 0$,

$$\mathsf{P}\Big(S_n = y,\ \min_{1\le k\le n} S_k \le 0 / S_0 = x\Big) = \mathsf{P}\left(S_n = y / S_0 = -x\right) = P_{n,y+x},$$

where $P_{n,y}$ is defined in the preceding problem.

IV.4.3. Let $\{S_n, n \geq 1\}$ be a simple Bernoulli walk. Prove that for $0 < y \leq n$,

$$\mathsf{P}(S_1 > 0, \dots, S_{n-1} > 0, S_n = y) = \frac{1}{2}(P_{n-1,y-1} - P_{n-1,y+1})$$
$$= \begin{cases} 0 & \text{if } n - y \text{ is odd,} \\ 2^{-n}\dfrac{y}{n}\dbinom{n}{n-y} & \text{if } n - y \text{ is even.} \end{cases}$$

IV.4.4. Let $\{S_n, n \geq 1\}$ be a simple Bernoulli walk. Show that for $0 < y \leq n$,

$$\mathsf{P}(S_1 < S_n, S_2 < S_n, \dots, S_{n-1} < S_n = y) = \mathsf{P}(S_1 > 0, \dots, S_{n-1} > 0, S_n = y).$$

IV.4.5. Let $\{S_n, n \geq 1\}$ be a simple Bernoulli walk. Show that

$$\mathsf{P}(S_1 > 0, S_2 > 0, \dots, S_{2n-1} > 0, S_{2n} = 0) = \frac{1}{n}\frac{1}{2^{2n}}\binom{2n-2}{n-1},$$
$$\mathsf{P}(S_1 \geq 0, S_2 \geq 0, \dots, S_{2n-1} \geq 0, S_{2n} = 0) = \frac{1}{(n+1)}\frac{1}{2^{2n}}\binom{2n}{n}.$$

IV.4.6. Let $\{S_n, n \geq 1\}$ be a simple Bernoulli walk. Show that

$$U_{2n} = \mathsf{P}(S_{2n} = 0) = \mathsf{P}(S_1 \neq 0, S_2 \neq 0, \dots, S_{2n} \neq 0)$$
$$= \mathsf{P}(S_1 \geq 0, S_2 \geq 0, \dots, S_{2n} \geq 0) = \binom{2n}{n}2^{-2n}.$$

IV.4.7. Let $\{S_n, n \geq 1\}$ be a simple Bernoulli walk. Show that

$$f_{2n} = \mathsf{P}(S_1 \neq 0, S_2 \neq 0, \dots, S_{2n-1} \neq 0, S_{2n} = 0)$$
$$= \mathsf{P}(S_1 \geq 0, S_2 \geq 0, \dots, S_{2n-2} \geq 0, S_{2n-1} < 0) = U_{2n-2} - U_{2n}$$
$$= \frac{1}{2n}U_{2n-2} = \frac{1}{2n}\binom{2n-2}{n-1}2^{-2n},$$

where U_n is defined in the preceding problem.

IV.4.8. Let $\{S_n, n \geq 1\}$ be a simple Bernoulli walk. Use Problems IV.4.4 and IV.4.3 to show that the probability of the event that a walk with the initial state $S_0 = 0$ reaches a point $0 \leq y \leq n$ for the first time at the moment $2n - y$ is

$$f_{2n}^{(y)} = \frac{y}{2n-y}\binom{2n-y}{n}2^{-2n+y}.$$

IV.4.9. Let $\{S_n, n \geq 1\}$ be a Bernoulli walk. Prove that for $|y| \leq n$,

$$\mathsf{P}\left(S_n = y / S_0 = 0\right) = \begin{cases} 0 & \text{if } n - y \text{ is odd,} \\ \dbinom{n}{\frac{n+y}{2}} p^{(n+y)/2} q^{(n-y)/2} & \text{if } n - y \text{ is even.} \end{cases}$$

REMARK. In Problems IV.4.9–IV.4.18, S_n denotes an integer-valued Bernoulli walk, i.e., a walk with steps

$$\xi_n = \begin{cases} +1, & \text{with probability } p, \\ -1, & \text{with probability } q, \end{cases}$$

where $p, q > 0$, $p + q = 1$.

IV.4.10. Consider S_n, $n \geq 0$, and $x > 0$. Let $\tau_x = \min\{n \geq 1 \colon S_n = x\}$ be the moment when the random walk S_n reaches the point x. Prove that the distribution of the random variable τ_x, $x > 0$, coincides with that of $\tau^{(1)} + \cdots + \tau^{(x)}$, where $\tau^{(i)}$ are independent random variables having the same distribution as τ_1. Find the moment generating function for the random variables τ_1 and τ_x.

IV.4.11. Find the moment generating function of the random variable τ_0. Prove that $\tau_0 < \infty$ with probability one if and only if $p = q = \frac{1}{2}$.

IV.4.12. Let P_x be the probability of the event that a random walk S_n ever reaches a point x. Prove that

$$P_x = \begin{cases} 1 & \text{if } p \geq q, \\ (p/q)^x & \text{if } p < q. \end{cases}$$

IV.4.13. Let $S_0 = 0$, and let η^+ be the maximal value of a walk S_n, $n \geq 0$. Prove that $\mathsf{P}(\eta^+ = \infty) = 1$ for $p \geq q$, and if $p < q$, then

$$\mathsf{P}(\eta^+ = x) = \left(\frac{p}{q}\right)^x \left(1 - \frac{p}{q}\right), \qquad x = 0, 1, \ldots .$$

IV.4.14. Denote by

$$P_x(y, z) = \mathsf{P}(\tau_y < \tau_z / S_0 = x)$$

the probability of the event that a random walk S_n starting at x reaches y before it reaches z. Prove that

$$P_x(y, z) = P_{x-y}(0, z - y)$$

and

$$P_x(0, z) = \frac{(q/p)^x - (q/p)^z}{1 - (q/p)^z}, \qquad x = 1, 2, \ldots, z - 1.$$

IV.4.15. Let $q \geq p$, and let μ_z be the number of visits to a point z over time τ_0 by a random walk S_n, that is,

$$\mu_z = \sum_{k=1}^{\tau_0} \delta_{S_k}^z,$$

where δ_i^j is the Kronecker symbol. Also let $r = q/p$ and

$$P_{xz}^{(n)} = \mathsf{P}(\mu_z = n / S_0 = x), \qquad n \geq 0, \ 0 < x < z.$$

Prove that

$$P_{xz}^{(n)} = \begin{cases} \dfrac{r^x - r^z}{1 - r^z}, & \text{for } n = 0, \\ \dfrac{1 - r^x}{1 - r^z} \left[1 - q\left(1 - \dfrac{1 - r^{z-1}}{1 - r^z}\right)\right]^{n-1} q\left(1 - \dfrac{1 - r^{z-1}}{1 - r^z}\right), & \text{for } n \geq 1. \end{cases}$$

IV.4.16. Suppose $S_0 = 1$ with probability one. For the random variables μ_z, $z \geq 1$, defined in the preceding problem, prove the following relation:

$$\mathsf{P}(\mu_y = n, \mu_z = m) = \mathsf{P}\left(\mu'_y = n, \sum_{n=1}^{\mu_y} \mu_{z-y}^{(k)} = m\right),$$

where $n, m \geq 0$ and

a) the random variables $\mu_{z-y}^{(k)}$, $k \geq 1$, are jointly independent;

b) the random variable μ'_y has the same distribution as the random variable μ_y;

c) the random variables $\mu_{z-y}^{(k)}$, $k \geq 1$, have the same distribution as μ_{z-y}.

i) Using the above relation, prove that the moment generating function of the random variables μ_z, $z \geq 1$, satisfies the equality

$$\mathsf{E}\, u^{\mu_z} = \varphi_z(u) = \varphi_y(\varphi_{z-y}(u)),$$

and that the joint moment generating function of the random variables μ_y and μ_z, $y < z$, satisfies the equality

$$\mathsf{E}\, u^{\mu_y} v^{\mu_z} = \varphi_{y,z}(u, v) = \varphi_y(u\varphi_{z-y}(v)).$$

ii) Using Problem IV.4.15, prove that the moment generating function $\varphi_z(u)$ is given by

$$\varphi_z(u) = \frac{q/p - (q/p)^z}{1 - (q/p)^z} + \frac{1 - q/p}{1 - (q/p)^z} q \left[1 - \frac{1 - (q/p)^{z-1}}{1 - (q/p)^z}\right] \frac{u}{\lambda_z},$$

where

$$\lambda_z = 1 - u\left[1 - q\frac{(q/p)^{z-1} - (q/p)^z}{1 - (q/p)^z}\right].$$

iii) Find the moment functions $\mathsf{E}\,\mu_z$ and $\mathsf{E}\,\mu_y\mu_z$ for $y \leq z$.

IV.4.17. Denote by

$$P_x^{(n)}(0, z) = \mathsf{P}\left(\tau_0 = n, \tau_z \geq n / S_0 = x\right), \qquad x = 1, 2, \ldots, z-1,\ n \geq 1,$$

the probability that a random walk S_k starting at x does not visit z over n steps and reaches zero at the nth step. Also let

$$\varphi_x(s, z) = \sum_{n=0}^{\infty} P_x^{(n)}(0, z) s^n$$

be the moment generating function of the sequence $P_x^{(n)}(0, z)$, $n \geq 0$. We put $P_x^{(0)}(0, z) = 0$ for $x = 1, \ldots, z-1$.

a) Prove that the functions $\varphi_x(s, z)$ satisfy the system of linear equations

$$\text{(a)}\qquad \begin{gathered}\varphi_x(s, z) = sq\varphi_{x-1}(s, z) + sp\varphi_{x+1}(s, z), \qquad x = 1, \ldots, z-1,\\ \varphi_0(s, z) = 1, \qquad \varphi_z(s, z) = 0.\end{gathered}$$

Prove that this system has a unique solution for every $|s| \leq 1$.

b) In order that the functions $\varphi_x(s) = \lambda^x(s)$ satisfy the recurrence equation

$$\text{(b)}\qquad \varphi_x(s) = qs\varphi_{x-1}(s) + ps\varphi_{x+1}(s),$$

it is necessary that the function $\lambda(s)$ satisfy the quadratic equation

$$\lambda(s) = ps\lambda^2(s) + qs. \tag{c}$$

c) Let $\lambda_1(s)$ and $\lambda_2(s)$ be solutions of (c). Prove that the functions $\varphi_x(s) = u_1(s)\lambda_1^x(s) + u_2(s)\lambda_2^x(s)$ satisfy relation (b) for arbitrary functions $u_1(s)$ and $u_2(s)$.

Using boundary conditions $\varphi_0(s,z) = 1$ and $\varphi_z(1,z) = 0$, find $u_1(s)$ and $u_2(s)$ such that $\varphi_x(s,z) \equiv \varphi_x(s)$, $x = 0,1,\dots,z$, and find an explicit expression for $\varphi_x(s,z)$.

IV.4.18. For $0 < y < z$ denote by

$$Q_x^{(n)}(y,0,z) = \mathsf{P}(S_n = y, n > \max(\tau_0,\tau_z)/S_0 = x), \qquad x = 1,2,\dots,z-1,\ n \ge 1,$$

the probability that a random walk S_n starting at x does not visit 0 and z over n steps and reaches y at the nth step. Also let

$$\psi_x(s,y,z) = \sum_{n=0}^{\infty} Q_x^{(n)}(y,0,z)s^n$$

be the moment generating function of the sequence $Q_x^{(n)}(y,0,z)$, $n \ge 0$, where $Q_x^{(0)}(y,0,z) = 1$ for $x = y$ and $Q_x^{(0)}(y,0,z) = 0$ for $x \ne y$. Determine $\psi_x(s,y,z)$.

IV.4.19. Consider a random walk with steps ξ_k, $k \ge 1$, having the distribution

$$\xi_k = \begin{cases} +1, & \text{with probability } p, \\ 0, & \text{with probability } q, \\ -1, & \text{with probability } r, \end{cases}$$

where $p,q > 0$, $r \ge 0$, $p+q+r = 1$.

For this random walk, prove statements given in Problems IV.4.9–IV.4.18 for a simple Bernoulli random walk.

IV.4.20. Let $\mathfrak{M}_n$ be the minimal σ-algebra generated by the random variables $S_0, S_1, \dots, S_n$. A random variable τ taking values $0,1,2,\dots,+\infty$ is *a Markov time for the random walk* S_n, $n \ge 0$, if $\{\omega : \tau > n\} \in \mathfrak{M}_n$ for all $n \ge 1$. For any Markov time τ, prove the relations

$$\begin{aligned}\mathsf{P}&\left(S_{\tau+k} \in A_k, k = 1,\dots,r/\tau < \infty, S_\tau = x, S_1 \in B_1,\dots,S_{\tau-1} \in B_{\tau-1}\right)\\ &= \mathsf{P}\left(S_{\tau+k} \in A_k, k = 1,\dots,r/\tau < \infty, S_\tau = x\right)\\ &= \mathsf{P}(S_k + x \in A_k, k = 1,\dots,r).\end{aligned}$$

REMARK. In Problems IV.4.20–IV.4.45, S_n denotes an arbitrary random walk with steps having a distribution function $F(x)$ such that $F(0) > 0$ and $1 - F(0+) > 0$, so that the steps of the walk may assume both negative and positive values.

IV.4.21. a) Let $\bar\tau^+ = \min\{n \ge 1 : S_n \ge 0\}$ be the first time when a random walk S_n enters the set $[0,\infty)$. The random variables $\bar\tau^+$ and $\bar\gamma^+ = S_{\bar\tau^+}$ are said to be *the first weak ladder epoch* and *the first weak ladder height*, respectively. Let $u = \mathsf{P}(\bar\tau^+ < \infty, \bar\gamma^+ = 0)$. Prove that $u \in [0,1)$ and

$$u = \sum_{n=1}^{\infty} \mathsf{P}(S_1 < 0,\dots,S_{n-1} < 0, S_n = 0).$$

b) Let $\tau^+ = \min\{n \geq 1: S_n > 0\}$ be the first time when a random walk S_n enters the set $(0, \infty)$. The random variables τ^+ and $\gamma^+ = S_{\tau^+}$ are said to be *the first strong ladder epoch* and *the first strong ladder height*, respectively. Prove that

$$\mathsf{P}(\tau^+ = +\infty) = \mathsf{P}(\bar{\tau}^+ < +\infty) + u\mathsf{P}(\bar{\tau}^+ = +\infty) + \cdots = \frac{1}{1-u}\mathsf{P}(\bar{\tau}^+ = +\infty).$$

IV.4.22. Prove that the random variables $\bar{\tau}^+$ and τ^+ coincide with probability one if the distribution of steps of the random walk is continuous.

IV.4.23. Put

$$\bar{\tau}_0^+ = 0, \qquad \bar{\tau}_k^+ = \min\{n > \bar{\tau}_{k-1}^+ : S_n \geq S_{\bar{\tau}_{k-1}^+}\}, \quad k \geq 1,$$
$$\tau_0^+ = 0, \qquad \tau_k^+ = \min\{n > \tau_{k-1}^+ : S_n > S_{\tau_{k-1}^+}\}, \quad k \geq 1.$$

The random variables $\bar{\tau}_k^+$ and τ_k^+ are said to be *the kth weak ladder epoch* and *the kth strong ladder epoch*, respectively. Prove that

$$\mathsf{P}(\tau_k^+ < +\infty) = (1 - \mathsf{P}(\tau^+ = +\infty))^k, \qquad k \geq 1,$$
$$\mathsf{P}(\bar{\tau}_k^+ < +\infty) = (1 - \mathsf{P}(\bar{\tau}^+ = +\infty))^k, \qquad k \geq 1.$$

IV.4.24. Prove that the random variables $\bar{\tau}_k^+$, $k \geq 1$, and τ_k^+, $k \geq 1$, are Markov times for the random walk S_n, $n \geq 0$, and check that both $\bar{\varkappa}_k^+ = \bar{\tau}_k^+ - \bar{\tau}_{k-1}^+$, $k \geq 1$, and $\varkappa_k^+ = \tau_k^+ - \tau_{k-1}^+$, $k \geq 1$, form sequences of independent identically distributed random variables. Here $\bar{\tau}_0^+$ and τ_0^+ are zero and the random variables $\bar{\varkappa}_k^+(\omega)$ and $\varkappa_k^+(\omega)$ are defined only for those elementary events ω that satisfy $\bar{\tau}_k^+(\omega) < \infty$ and $\tau_k^+(\omega) < \infty$, respectively.

IV.4.25. Let $\bar{\nu}^+ = \max\{n: \bar{\tau}_n^+ < \infty\}$ and $\nu^+ = \max\{n: \tau_n^+ < \infty\}$ are, respectively, the total number of weak and strong ladder epochs of a random walk S_n, $n \geq 0$. Prove that both random variables $\bar{\nu}^+$ and ν^+ are either finite or improper with probability one, and

$$\mathsf{P}(\bar{\nu}^+ \geq n) = \mathsf{P}(\bar{\tau}_n^+ < +\infty) = \big(1 - \mathsf{P}(\bar{\tau}^+ = +\infty)\big)^n, \qquad n \geq 1,$$
$$\mathsf{P}(\nu^+ \geq n) = \mathsf{P}(\tau_n^+ < +\infty) = \big(1 - \mathsf{P}(\tau^+ = +\infty)\big)^n, \qquad n \geq 1.$$

Recall that a random variable ν is said to be *improper* if $\mathsf{P}(|\nu| = +\infty) > 0$.

IV.4.26. Let $\bar{\gamma}_k^+ = S_{\bar{\tau}_k^+}$, $k \geq 0$, and $\gamma_k^+ = S_{\tau_k^+}$, $k \geq 0$, be the kth weak and strong ladder heights of a random walk, respectively. In the general case, the random variables $\bar{\gamma}_k^+$ and γ_k^+ may be improper and are defined for the same elementary events as $\bar{\tau}_k^+$ and τ_k^+, respectively. Prove that both

$$\bar{\alpha}_k^+ = \bar{\gamma}_k^+ - \bar{\gamma}_{k-1}^+, \qquad k \geq 1,$$

and

$$\alpha_k^+ = \gamma_k^+ - \gamma_{k-1}^+, \qquad k \geq 1,$$

form sequences of independent identically distributed random variables.

IV.4.27. Let $\eta^+ = \max_{n \geq 0} S_n$ be the maximal value of a random walk S_n, $n \geq 0$. If $\eta^+ < +\infty$ with probability one, then the random walk is said to be *bounded from above*. Prove that a random walk is bounded from above if and only if the random variables $\bar{\tau}^+$ and τ^+ are improper.

IV.4.28. Prove that if a random walk is not bounded from above, then $\mathsf{P}(\eta^+ = +\infty) = 1$.

IV.4.29. Let random variables $S_k^* = S_n - S_{n-k}$, $k = 0, 1, \dots, n$, be constructed for a random walk $S_0, \dots, S_n$. Prove that the joint distributions of the random variables $S_0, \dots, S_n$ and $S_0^*, \dots, S_n^*$ coincide.

IV.4.30. Prove that

$$\mathsf{P}(S_n \geq S_0, \dots, S_n \geq S_{n-1}) = \mathsf{P}(S_1^* \geq 0, \dots, S_n^* \geq 0).$$

IV.4.31. Define the so-called "negative" ladder epochs as follows:

$$\bar{\tau}^- = \min\{n \geq 1 \colon S_n \leq 0\}, \qquad \tau^- = \min\{n \geq 1 \colon S_n < 0\}.$$

Prove that $\mathsf{E}\,\bar{\tau}^- < \infty$ if and only if the random variable $\bar{\tau}^+$ is improper, and that in this case

$$\mathsf{E}\,\bar{\tau}^- = \frac{1 - \mathsf{P}(\bar{\tau}^+ = +\infty)}{\mathsf{P}(\bar{\tau}^+ = +\infty)}.$$

IV.4.32. Prove that $\mathsf{E}\,\tau^- < \infty$ if there exists $\mathsf{E}\,\xi_1 = a$ and $a < 0$.

IV.4.33. Let $\eta^- = \min_{n \geq 1} S_n$ be the minimal value of a random walk. Prove that $\eta^- > -\infty$ with probability one (*the random walk is bounded from below*) and $\mathsf{E}\,\bar{\tau}^+ < \infty$, $\mathsf{E}\,\tau^+ < \infty$ if there exists the expectation $\mathsf{E}\,\xi_1 = a$ and $a > 0$.

IV.4.34. Let τ be a random variable assuming nonnegative integer values. Suppose that the random event $\{\omega \colon \tau > n\}$ does not depend on the variables $\xi_{n+1}, \xi_{n+2}, \dots$ for any n. Further let $\mathsf{E}\,|\xi_1| < \infty$ and $\mathsf{E}\,\tau < \infty$. Prove that $\mathsf{E}\,|S_\tau| < \infty$ and $\mathsf{E}\,S_\tau = \mathsf{E}\,\xi_1\,\mathsf{E}\,\tau$.

IV.4.35. Let $\bar{\tau}^+$ and $\bar{\gamma}^+$ be the first weak ladder epoch and the first weak ladder height, respectively. Assume that there exists $\mathsf{E}\,\xi_1 = a > 0$.

a) Prove that there exists $\mathsf{E}\,\bar{\gamma}^+$ and $\mathsf{E}\,\bar{\gamma}^+ = \mathsf{E}\,\bar{\tau}^+\,\mathsf{E}\,\xi_1$.

b) Derive analogous equalities for the random variables τ^+ and γ^+.

IV.4.36. Assume that the expectation $\mathsf{E}\,|\xi_1|$ is finite. Prove that $\mathsf{E}\,\xi_1 = 0$ if and only if $\mathsf{P}(\bar{\tau}^+ = +\infty) = \mathsf{P}(\bar{\tau}^- = +\infty) = 0$. On the other hand, if $\mathsf{E}\,\xi_1 = 0$, then $\mathsf{E}\,\bar{\tau}^+ = \mathsf{E}\,\bar{\tau}^- = +\infty$.

IV.4.37. Suppose that $\mathsf{E}\,|\xi_1| < \infty$. Prove that a random walk is not bounded from above and from below if and only if $\mathsf{E}\,\xi_1 = 0$.

IV.4.38. Show that a random walk is bounded both from above and from below if and only if $\mathsf{P}(\xi_1 = 0) = 1$.

IV.4.39. An integer-valued random walk is said to be *recurrent* if $P_x = 0$ for all $x = 0, \pm 1, \dots$, where $P_x = \mathsf{P}(S_n \neq x, n = 1, 2, \dots)$. Prove that a recurrent random walk is unbounded both from above and from below if $\mathsf{P}(\xi_1 = 0) < 1$.

IV.4.40. For $n \geq 1$ and $A \subset [0, \infty)$, prove that

$$\mathsf{P}(S_1 > 0, \dots, S_{n-1} > 0, S_n \in A) = \mathsf{P}(S_n > S_1, \dots, S_n > S_{n-1}, S_n \in A).$$

IV.4.41. Let γ be the first positive value of a random walk. Prove that

$$\mathsf{P}(\gamma < x) = F(x) + \int_0^\infty \psi(dy)F(x-y), \qquad x < 0,$$

where

$$\psi(A) = \sum_{n=0}^{\infty} \mathsf{P}(S_n > S_1, \dots, S_n > S_{n-1}, S_n \in A), \qquad F(x) = \mathsf{P}(\xi_1 < x).$$

IV.4.42. Let the left tail of a distribution $F(\cdot)$ be exponential, that is, $F(x) = qe^{\beta x}$ for $x \le 0$. Let γ_- be the first negative value of a random walk. Prove that

a) $\mathsf{P}(\gamma_- < x) = Ce^{\beta x}$, $x \le 0$.

b) $C = 1$ if $\mathsf{E}\,\xi_1 \le 0$.

IV.4.43. Let the right tail of a distribution $F(\cdot)$ be exponential, that is, $F(x) = 1 - pe^{-\alpha x}$ for $x > 0$, and let $\mathsf{E}\,\xi_1 = a \ge 0$. Prove that $\psi((0, x]) = \alpha x$, $x > 0$, where $\psi(A)$ is defined in Problem IV.4.41.

IV.4.44. Let the right tail of a distribution $F(\cdot)$ be exponential, that is, $F(x) = 1 - pe^{-\alpha x}$ for $x > 0$, and let $\mathsf{E}\,\xi_1 = a \ge 0$. Prove that the distribution of the random variable γ (defined in Problem IV.4.41) is given by

$$\mathsf{P}(\gamma < x) = F(x) + \alpha \int_0^\infty F(x-y)\,dy, \qquad x \le 0.$$

Prove that in this case

$$\mathsf{P}(\gamma < 0) = 1 - \alpha a.$$

IV.4.45. Determine the distribution of the random variable γ (defined in Problem IV.4.41) for a random walk with steps having the distribution

$$\mathsf{P}(\xi_1 = k) = q\beta^k, \qquad k = 0, 1, \dots .$$

IV.4.46. a) Let $\tau(x)$ be the first time when a random walk reaches the point x, $\tau(x) = \min\{n \ge 1 : S_n = x\}$. Let $x_1, x_2, \dots$ be a sequence of integer points. Define the random variables

$$\tau^{(k)} = \begin{cases} 0, & \text{for } k = 0, \\ \tau(x_1), & \text{for } k = 1, \\ \min\{n > \tau^{(k-1)} : S_n = x_k\}, & \text{for } k > 1. \end{cases}$$

Prove that any random variable $\tau^{(k)}$ is a Markov time.

b) Put $\chi_k = \tau^{(k)} - \tau^{(k-1)}$, $k \ge 1$. Prove that the random variables χ_n, $n \ge 1$, are independent, that is,

$$\begin{aligned} &\mathsf{P}(\chi_1 = n_1, \dots, \chi_k = n_k) \\ &\quad = \mathsf{P}(\chi_1 = n_1)\mathsf{P}\left(\chi_2 = n_2 / \tau^{(1)} < \infty\right) \cdots \mathsf{P}\left(\chi_k = n_k / \tau^{(k-1)} < \infty\right). \end{aligned}$$

Note that in general the random variable χ_k is improper and is defined only for those elementary events ω where $\tau^{(n)} < \infty$.

REMARK. In Problems IV.4.47–IV.4.67 integer-valued random walks are considered.

IV.4.47. Prove that

$$\mathsf{P}\left(\chi_k = n/\tau^{(k-1)} < \infty\right) = \mathsf{P}(\tau(x_k - x_{k-1}) = n), \qquad k \geq 1.$$

IV.4.48. Prove that the joint distributions of two collections of random variables S_k, $k = 0, 1, \dots, n$, and $S_n^* = S_n - \xi_1 - \dots - \xi_{n-k}$, $k = 0, 1, \dots, n$, coincide. (Here S_0 is allowed to be a random variable.) Use this property to show that

$$\begin{aligned}\mathsf{P}\big(S_0 = 0, S_1 \notin \{0,x\}, \dots, S_{n-1} \notin \{0,x\}, S_n = 0\big) \\ = \mathsf{P}\big(S_0 = x, S_1 \notin \{0,x\}, S_{n-1} \notin \{0,x\}, S_n = x\big).\end{aligned}$$

IV.4.49. Put

$${}_yf_{zx} = \mathsf{P}\left(\tau(x) < \tau(y)/S_0 = z\right),$$

where the random variables $\tau(\cdot)$ are defined in Problem IV.4.46. Prove that ${}_xf_{00} = {}_0f_{xx}$.

IV.4.50. Let ν_{xy} be the total number of visits of a point y over time $\tau(x)$. Prove that

$$\mathsf{P}\left(\nu_{xy} = n/S_0 = z\right) = \begin{cases} 0, & \text{with probability } 1 - {}_xf_{zy}, \\ n, & \text{with probability } {}_xf_{zy} \cdot {}_xf_{yy}^{n-1}(1 - {}_xf_{yy}),\ n \geq 1. \end{cases}$$

Show also that ${}_xf_{yy} < 1$ for $x, y = 0, \pm 1, \dots$ if $\mathsf{P}\left(\tau(0) < \infty/S_0 = 0\right) = 1$, and therefore all the moments of the random variable ν_{xy} are finite:

$$\mathsf{E}\left(\nu_{xy}^r/S_0 = z\right) < \infty, \qquad r \geq 1.$$

IV.4.51. Suppose that $\mathsf{P}\left(\tau(0) < \infty/S_0 = 0\right) = 1$. Under this condition, prove that the expectation of the total number of visits of the point y between two consecutive visits of the point x is

$$\mathsf{E}\left(\nu_{xy}/S_0 = x\right) = 1.$$

IV.4.52. Let τ be the time of the first return of a random walk S_n, $n \geq 0$, to the origin (τ is usually called *the return time*), i.e., $\tau = \min\{n \geq 1 \colon S_n = 0\}$. Prove that τ is a Markov time for the random walk S_n and derive the equality

$$\mathsf{P}(\tau = n) = \sum_{k=1}^{n} \mathsf{P}(\tau = k)\mathsf{P}(S_{n-k} = 0), \qquad n \geq 1.$$

IV.4.53. A random walk S_n, $n \geq 0$, is said to be *recurrent* if $\mathsf{P}(\tau < \infty) = 1$. Prove that a random walk is recurrent if and only if

$$\sum_{n=1}^{\infty} \mathsf{P}(S_n = 0) = +\infty.$$

IV.4.54. Let the steps ξ_k of a random walk S_n assume only two values:

$$\xi_k = \begin{cases} +1, & \text{with probability } p, \\ -1, & \text{with probability } q. \end{cases}$$

Prove that

$$\mathsf{P}(S_{2n} = 0) = \binom{2n}{n} p^n q^n.$$

Using the Stirling formula, derive from this equality that a random walk is recurrent if and only if $p = q = \frac{1}{2}$.

IV.4.55. Let

$$p_k = \mathsf{P}(\xi_1 = k), \qquad k = 0, \pm 1, \dots,$$
$$p_n^{(k)} = \mathsf{P}(S_n = k), \qquad k = 0, \pm 1, \dots,$$
$$\varphi(z) = \mathsf{E}\, e^{iz\xi_1} = \sum_{k=-\infty}^{+\infty} p_k e^{izk}.$$

Prove that

$$\sum_{n=0}^{\infty} \lambda^n p_n^{(k)} = \frac{1}{2\pi} \int_{-\pi}^{\pi} \frac{e^{-izk}}{1 - \lambda\varphi(z)}\, dz.$$

IV.4.56. Prove that a random walk is recurrent if and only if

$$\lim_{\lambda \uparrow 1} \int_{-\pi}^{\pi} \operatorname{Re} \frac{1}{1 - \lambda\varphi(z)}\, dz = +\infty.$$

IV.4.57. Prove that

$$\int_{-\pi}^{\pi} \operatorname{Re} \frac{1}{1 - \varphi(z)}\, dz < \infty$$

provided the random walk is not recurrent.

IV.4.58. Prove that a random walk is recurrent if $\mathsf{E}\,|\xi_1| < \infty$ and $\mathsf{E}\,\xi_1 = 0$.

IV.4.59. Prove the following theorem. *In order that a random walk be recurrent, it is necessary and sufficient that*

$$\int_{-\pi}^{\pi} \operatorname{Re} \frac{1}{1 - \varphi(z)}\, dz = +\infty.$$

Using this theorem prove that a random walk is not recurrent if $\mathsf{E}\,|\xi_1| < \infty$ and $\mathsf{E}\,\xi_1 \neq 0$.

IV.4.60. Let random variables ξ_k have a symmetric distribution. Prove that the random walk is recurrent if for some $\delta > 0$

$$\int_{|z|<\delta} \frac{1}{1 - \varphi(z)}\, dz = +\infty.$$

IV.4.61. Let the steps ξ_k of a random walk S_n have a symmetric distribution and $\mathsf{P}(\xi_k = n) \sim cn^{-\alpha}$ as $n \to \infty$. Prove that the random walk is not recurrent for $1 < \alpha < 2$ and is recurrent for $\alpha > 2$.

IV.4.62. For what real α is the function

$$\varphi_\alpha(s) = 1 - \left|\sin \frac{s}{2}\right|^\alpha$$

a characteristic function of a step of an integer-valued random walk? Prove that there exists α such that a corresponding random walk with the characteristic function $\varphi_\alpha(s)$ has the following property: for any integer a the random walk $S_n + an$, $n \geq 0$, is recurrent.

IV.4.63. Consider a random walk S_n with steps $\{\xi_k\}$ having the distribution

$$\xi_1 = \begin{cases} -1, & \text{with probability } q, \\ k, & \text{with probability } p_k,\ k = 1, 2, \dots, \end{cases}$$

where $q + p_1 + p_2 + \cdots = 1$. Let u_r be the probability that the first positive value of the sequence S_n, $n \geq 1$, equals r. Let

$$U(s) = \sum_{r=1}^{\infty} s^r u_r, \qquad P(s) = \sum_{r=1}^{\infty} s^r p_r.$$

Prove that

$$U(s) = 1 - \frac{P(s) + qs^{-1} - 1}{u_1 q + qs^{-1} - 1}, \qquad 0 < s < 1.$$

IV.4.64. Consider an integer-valued random walk S_n such that

$$\xi_1 = \begin{cases} +1, & \text{with probability } p, \\ -k, & \text{with probability } q_k,\ k = 1, 2, \dots, \end{cases}$$

where $p + q_1 + q_2 + \cdots = 1$. Let τ_k be the first time when the random walk reaches a level k, i.e., $\tau_k = \min\{n \geq 1 \colon S_n = k\}$.

a) Prove that the random variable τ_k has the same distribution as $\tau^{(1)} + \cdots + \tau^{(k)}$, where $\tau^{(i)}$ are independent random variables having the same distribution as τ_1.

b) Find the distribution of the random variable $\eta^+ = \max_{n \geq 0} S_n$.

c) Let

$$\varphi(s) = \mathsf{E}\, s^{\tau_1} = \sum_{k=1}^{\infty} s^k \mathsf{P}(\tau_1 = k), \qquad Q(s) = \sum_{k=0}^{\infty} q_k s^k.$$

Prove that $\varphi(s)$ satisfies the equation

$$\varphi(s) = s\left[p + \varphi(s) Q(\varphi(s))\right].$$

d) Find the moment generating function $\varphi(s)$ in the case $q_2 = q_3 = \cdots = 0$.

e) Prove that $\mathsf{P}(\tau_1 < +\infty) = 0$ if

$$\mu = p - \sum_{n=0}^{\infty} k q_k \geq 0.$$

Otherwise, $\varphi_0 = \mathsf{P}(\tau_1 < +\infty)$ is the unique solution of the equation

$$\varphi = p + \varphi Q(\varphi)$$

such that $0 < \varphi < 1$.

f) Find the probability φ_0 for the case $q_k = 0$, $k \geq 2$.

IV.4.65. Consider an integer-valued random walk S_n with steps having the distribution $p_k = \mathsf{P}(\xi_1 = k)$, $k = 0, \pm 1, \dots$, such that $p_k = 0$ for $|k| > 2$.

Let u_1 and u_2 be the probabilities that the first positive value of the sequence S_n, $n \geq 1$, equals either 1 or 2, respectively (u_i, $i = 1, 2$, is *the distribution of the strong ladder height*). Using the formula of total probability show that

$$u_1 = p_1 + p_0 u_1 + p_{-1} u_2 + p_{-1} u_1 u_1 + p_{-2} u_2 u_1 + p_{-2} u_1 u_2 + p_{-2} u_1 u_1 u_1,$$
$$u_2 = p_2 + p_0 u_2 + p_{-1} u_1 u_2 + p_{-2} u_2 u_2 + p_{-2} u_1 u_1 u_2.$$

§IV.5. Markov chains

A sequence of random variables $\{\eta_n, n \geq 0\}$ taking values in a finite or denumerable set $H = \{a_1, a_2, \dots\}$ is said to be *a discrete Markov chain* if

$$\mathsf{P}\left(\eta_{n+1} = a_j / \eta_n = a_i, \eta_k = a_{i_k}, 1 \leq k \leq n-1\right) = \mathsf{P}\left(\eta_{n+1} = a_j / \eta_n = a_i\right) = p_{ij}(n)$$

for all $i, j, i_1, \dots, i_{n-1} \in H$ and $n \geq 0$. Unless otherwise stated, in the sequel we assume that $a_i = i$.

The probabilities $p_{ij}(n)$ are said to be *transition probabilities of the Markov chain* η_n, $n \geq 0$, at the nth step. If transition probabilities $p_{ij}(n)$ do not depend on n, then we say that the Markov chain η_n, $n \geq 0$, is *homogeneous*. The corresponding matrix $P = \|p_{ij}\|$, $i, j \in H$, is said to be *the matrix of transition probabilities*.

A distribution $p_i = \mathsf{P}(\eta_0 = i)$, $i \in H$, is said to be *the initial distribution* of a Markov chain η_n, $n \geq 0$.

A Markov chain is said to be *finite* or *denumerable* if its set of states H is finite or denumerable, respectively. Unless otherwise stated, we consider a homogeneous Markov chain with the set of states $H = \{1, 2, \dots\}$ and the matrix of transition probabilities $P = \|p_{ij}\|$, $i, j \in H$.

Problems

IV.5.1. Denote by $p_{ij}^{(n)}$ the transition probabilities of a Markov chain η_n, $n \geq 0$, over n steps, $p_{ij}^{(n)} = \mathsf{P}\left(\eta_n = j / \eta_0 = i\right)$. Using induction prove that the matrix $P^{(n)} = \|p_{ij}^{(n)}\|$, $i, j \in H$, is the nth power of the matrix of the transition probabilities P.

IV.5.2. Prove that for positive integers k and r and for all states $i, j \in H$,

$$p_{ij}^{(k+r)} = \sum_{l \in H} p_{il}^{(k)} p_{lj}^{(r)}.$$

IV.5.3. A matrix P is said to be *stochastic* if all its entries are nonnegative and the sum of all entries in any row equals 1. Prove that every matrix of transition probabilities is stochastic. Also prove that if P is a matrix of transition probabilities, then for all n, the matrix $P^{(n)}$ is also stochastic.

IV.5.4. Consider a Markov chain η_n, $n \geq 0$, with two states and the matrix of transition probabilities $\left\| \begin{smallmatrix} 1-\alpha & \alpha \\ \beta & 1-\beta \end{smallmatrix} \right\|$, where $0 < \alpha, \beta < 1$. Find the moment generating functions $p_{11}^{(n)}$, $n \geq 0$, and $p_{21}^{(n)}$, $n \geq 0$, and explicit expressions for the probabilities $p_{ii}^{(n)}$. Calculate $\lim_{n\to\infty} p_{i1}^{(n)} = q_1$.

IV.5.5. Express the conditional probability $\mathsf{P}\left(\eta_0 = i / \eta_n = j\right)$ in terms of corresponding transition probabilities over k steps.

IV.5.6. Express the conditional probability $\mathsf{P}\left(\eta_r = l / \eta_0 = i, \eta_n = j\right)$ in terms of corresponding transition probabilities over k steps.

IV.5.7. Let r be a positive integer. Prove that the sequence $\zeta_n = \eta_{nr}$, $n \geq 0$, is a homogeneous Markov chain. Determine the matrix of transition probabilities for this chain.

IV.5.8. Consider a Markov chain related to the Bernoulli scheme with two possible results S and F (the probability of a success S is p and that of a failure F is $q = 1 - p$). The relationship between the chain and the Bernoulli scheme can be described as follows. The Markov chain stays in a state 1 if the $(n-1)$st and the nth trials give a pattern SS. Analogously, states 2, 3, and 4 correspond to patterns SF, FS, and FF, respectively. Find the matrix of transition probabilities of the Markov chain η_n, $n \geq 0$.

IV.5.9. Let $\xi_0, \xi_1, \dots$ be a sequence of independent random variables taking values ± 1 with equal probabilities $\frac{1}{2}$. Define the random variables

$$\eta_n = \frac{\xi_n + \xi_{n+1}}{2}.$$

Find the transition probabilities $p_{jk}(m,n) = \mathsf{P}\left(\eta_n = k / \eta_m = j\right)$ for $m < n$, $j, k = -1, 0, 1$. Prove that the sequence η_n, $n \geq 0$, is not a Markov chain.

IV.5.10. Let $\xi_0, \xi_1, \dots$ be independent identically distributed random variables taking values -1 and $+1$ with probabilities $1 - p$ and p, respectively. Put $\eta_n = \xi_n \xi_{n+1}$. Is the sequence η_n, $n \geq 0$, a Markov chain? Does the limit of probabilities $\mathsf{P}(\eta_n = -1)$ exist as $n \to \infty$?

IV.5.11. Random variables $\xi_0, \xi_1, \dots$ are defined in the same way as in the preceding problem. Let

$$\eta_n = \max_{0 \leq k \leq n} \sum_{i=1}^{k} \xi_i - \min_{0 \leq k \leq n} \sum_{i=1}^{k} \xi_i.$$

Is the sequence η_n, $n \geq 0$, a Markov chain?

IV.5.12. From the definition of a Markov chain derive the following relations:

$$\begin{aligned} &\mathsf{P}\left(\eta_{n+k} = j_k, 1 \leq k \leq r, / \eta_n = i, \eta_k = i_k, 1 \leq k \leq n-1\right) \\ &\quad = \mathsf{P}\left(\eta_{n+k} = j_k, 1 \leq k \leq r / \eta_n = i\right) = p_{ij_1} p_{j_1 j_2} \cdots p_{j_{r-1} j_r}. \end{aligned}$$

IV.5.13. Let $\mathfrak{M}_{u,v}$ be the minimal σ-algebra generated by random variables η_n, $n = u, u+1, \dots, v$, where $u \leq v \leq \infty$. Let $A \in \mathfrak{M}_{0,n}$ and $B \in \mathfrak{M}_{n+1,\infty}$. Prove that

$$\mathsf{P}\left(B / \eta_n = i, A\right) = \mathsf{P}\left(B / \eta_n = i\right).$$

IV.5.14. Prove that

$$\mathsf{P}(\eta_k \in A_k, 0 \leq k \leq n) = \sum_{i_0 \in A_0, \dots, i_n \in A_n} p_{i_0} p_{i_1 i_2} \cdots p_{i_{n-1} i_n}.$$

IV.5.15. Let η_n, $n \geq 0$, be a Markov chain with the set of states $H = \{1, 2, 3\}$ and let

$$\begin{Vmatrix} 0 & 1-\alpha & \alpha \\ 1-\beta & 0 & \beta \\ \frac{1}{3} & \frac{1}{3} & \frac{1}{3} \end{Vmatrix}$$

be the matrix of transition probabilities. Define the sequence ζ_n, $n \geq 0$, as follows:

$$\zeta_n = \begin{cases} 1 & \text{if either } \eta_n = 1 \text{ or } \eta_n = 2, \\ 2 & \text{if } \eta_n = 3. \end{cases}$$

Under what condition is the sequence ζ_n, $n \geq 0$, a Markov chain?

IV.5.16. Let η_n, $n \geq 0$, be a homogeneous Markov chain with the set of states $H = \{1, 2, \dots\}$ and with the matrix of transition probabilities $\|p_{ij}\|$, $i, j \in H$. Also let $A_1, \dots, A_m$ be disjoint subsets of H such that $A_1 \cup A_2 \cup \dots \cup A_m = H$. Define the sequence ζ_n, $n \geq 0$, as follows: if $\eta_n \in A_i$ for some $1 \leq i \leq m$, then $\zeta_n = i$. Prove that the sequence ζ_n, $n \geq 0$, is a Markov chain with the set of states $H' = \{1, 2, \dots, m\}$ and the matrix of transition probabilities $\|Q_{rl}\|_{r,l=1}^{m}$ provided the transition probabilities of the Markov chain η_n, $n \geq 0$, satisfy the conditions $p_{ij} = Q_{rl}$ for all $1 \leq r, l \leq m$ if $i \in A_r$ and $j \in A_l$.

IV.5.17. A random variable τ assuming values $0, 1, \dots, +\infty$ is said to be *a Markov time for a Markov chain* η_n, $n \geq 0$, if $\{\omega : \tau > n\} \in \mathfrak{M}_{0,n}$ for all $n \geq 0$. Prove that

$$\begin{aligned} &\mathsf{P}\left(\eta_{r+k} = j_k, 1 \leq k \leq r / \eta_r = i, \tau < \infty, \eta_k = i_k, 1 \leq k \leq \tau - 1\right) \\ &\qquad = \mathsf{P}\left(\eta_k = j_k, 1 \leq k \leq r / \eta_0 = i\right) = p_{ii_1} \cdots p_{i_{r-1}i_r} \end{aligned}$$

for any Markov time τ.

IV.5.18. Prove that $p_{ij}^{(n)} > 0$ if and only if there exists a sequence of states $i = i_0, i_1, \dots, i_n = j$ such that

$$p_{i_0i_1}p_{i_1i_2} \cdots p_{i_{n-1}i_n} > 0.$$

IV.5.19. A state j is said to be *attainable from a state* i (in this case we write $i \to j$) if there exists n such that $p_{ij}^{(n)} > 0$. Prove that the relations $i \to j$ and $j \to k$ imply $i \to k$.

IV.5.20. Let the set of states $H = \{1, 2, \dots, m\}$ of a Markov chain η_n, $n \geq 0$, be finite. Prove that there exists $n \leq m$ such that $p_{ij}^{(n)} > 0$ provided $i \to j$.

IV.5.21. Let η_n, $n \geq 0$, be a Markov chain. Consider the first time τ_j when the chain reaches j, $\tau_j = \min\{n \geq 1 : \eta_n = j\}$. Prove that τ_j is a Markov time.

IV.5.22. Let $f_{ij}^{(n)} = \mathsf{P}\left(\tau_j = n / \eta_0 = i\right)$, $n \geq 1$. Derive the relation

$$p_{ij}^{(n)} = \delta_i^j + \sum_{k=0}^{n} f_{ij}^{(k)} p_{jj}^{(n-k)}, \qquad n = 0, 1, \dots,$$

where $f_{ij}^{(0)} = 0$, $p_{ij}^{(0)} = \delta_i^j$. Establish the relationship between the moment generating functions for $p_{ij}^{(n)}$, $p_{jj}^{(n)}$, and $f_{ij}^{(n)}$.

IV.5.23. Let

$$f_{jj} = \mathsf{P}\left(\tau_j < \infty / \eta_0 = j\right) = \sum_{n=1}^{\infty} f_{jj}^{(n)}.$$

A state j is said to be *recurrent* if $f_{jj} = 1$. A state j is said to be *nonrecurrent* if $f_{jj} \neq 1$. Prove that a state j is recurrent if and only if

$$\sum_{n=0}^{\infty} p_{jj}^{(n)} = +\infty.$$

IV.5.24. A state i is said to be *essential* if $j \to i$ for every j such that $i \to j$. A state i is said to be *nonessential* if there exists j such that $i \to j$ but $j \not\to i$. Prove that every nonessential state is nonrecurrent.

IV.5.25. Prove that if j is a nonessential state, then $p_{ij}^{(n)} \to 0$ as $n \to \infty$ for any $i \in H$.

IV.5.26. Prove that it is not possible that all the states of a finite Markov chain are nonessential.

IV.5.27. Construct an example of a denumerable Markov chain such that all its states are nonessential.

IV.5.28. States i and j are said to be *communicating* (in this case we write $i \leftrightarrow j$) if both $i \to j$ and $j \to i$. Let A be a set of states such that $i \leftrightarrow j$ for all $i, j \in A$ and $i \not\to j$ for all $i \in A$ and $j \notin A$. A class with such a property is said to be *an essential class of states*. By definition, all the states of an essential class are essential.

Let i be an essential state. Prove that the set A_i of states that are communicating with the state i is an essential class.

IV.5.29. Let η_n, $n \geq 0$, be a finite Markov chain and let H be the set of its states. Prove that there exists a unique partition A_j, $0 \leq j \leq r$, of the set H such that the sets A_j, $1 \leq j \leq r$, are essential (at least one of them is nonempty) and A_0 is a set of nonessential states (this set may be empty).

IV.5.30. Six matrices of transition probabilities of six Markov chains are given below. Could you recognize what states of these chains are essential? Describe sets of nonessential states for each of these chains.

$$\begin{Vmatrix} \frac13 & \frac13 & 0 & \frac13 & 0 \\ 0 & 0 & \frac12 & 0 & \frac12 \\ \frac12 & 0 & 0 & 0 & \frac12 \\ 0 & 0 & 0 & \frac12 & \frac12 \\ 1 & 0 & 0 & 0 & 0 \end{Vmatrix}, \quad \begin{Vmatrix} 0 & \frac12 & 0 & \frac12 & 0 \\ \frac12 & 0 & 0 & \frac12 & 0 \\ 0 & 0 & 1 & 0 & 0 \\ 0 & 0 & \frac12 & \frac12 & 0 \\ 0 & 0 & 0 & 0 & 1 \end{Vmatrix}, \quad \begin{Vmatrix} \frac12 & 0 & \frac12 & 0 & 0 \\ \frac12 & \frac12 & 0 & 0 & 0 \\ \frac12 & 0 & \frac12 & 0 & 0 \\ 0 & 0 & \frac13 & \frac13 & \frac13 \\ 0 & 0 & 0 & 0 & 1 \end{Vmatrix},$$

$$\begin{Vmatrix} \frac12 & 0 & \frac12 & 0 & 0 \\ 0 & \frac12 & 0 & 0 & \frac12 \\ 0 & 0 & 0 & 1 & 0 \\ \frac13 & 0 & \frac13 & \frac13 & 0 \\ \frac12 & 0 & 0 & 0 & \frac12 \end{Vmatrix}, \quad \begin{Vmatrix} 0 & 0 & \frac13 & \frac13 & \frac13 \\ 0 & \frac13 & \frac13 & \frac13 & 0 \\ \frac14 & \frac14 & \frac14 & \frac14 & 0 \\ 0 & 0 & 0 & \frac12 & \frac12 \\ 0 & 0 & 0 & \frac12 & \frac12 \end{Vmatrix}, \quad \begin{Vmatrix} 0 & 0 & 0 & \frac12 & \frac12 \\ 0 & \frac12 & \frac12 & 0 & 0 \\ 0 & 1 & 0 & 0 & 0 \\ 0 & 1 & 0 & 0 & 0 \\ 0 & 0 & 0 & 1 & 0 \end{Vmatrix}.$$

IV.5.31. Prove that the matrix of transition probabilities that corresponds to transitions within some essential class of states is stochastic.

IV.5.32. Let $i \leftrightarrow j$. The greatest common divisor d_i of those n for which $p_{ij}^{(n)} > 0$, is called *the period of the state* i. Prove that all the states in an essential class have the same period.

IV.5.33. Let the set of all states H of a Markov chain η_n, $n \geq 0$, is an essential class. The period d of states $i \in H$ is called *the period of the Markov chain* η_n, $n \geq 0$. If $d = 1$, then we say that a Markov chain is *aperiodic*. The following result holds: *if d is the period of a Markov chain η_n, $n \geq 0$,, then the set H of all possible states of the chain uniquely splits into the classes $C_1, \dots, C_d$ such that $C_i \cap C_j = \varnothing$ for $i \neq j$ and*

$$\sum_{j \in C_{n+r}} p_{ij}^{(n)} = 1, \qquad n = 1, 2, \dots,$$

if $i \in C_r$ *for all* $1 \le r \le d$, *where* $C_{r+nd} = C_r$, $1 \le r \le d$, $n \ge 1$.

The sets $C_1, \dots, C_d$ defined above are called *cyclic subclasses of the domain* H.

Let $\eta_0 \in C_r$ with probability one. Prove that the sequence $\eta_n = \eta_{nd}$, $n \ge 0$, is also a homogeneous Markov chain with the set of states C_r and the matrix of transition probabilities $\|p'_{ij}\| = \|p_{ij}^{(d)}\|_{i,j \in C_r}$. Moreover, prove that the period of this chain is 1.

IV.5.34. Three matrices of transition probabilities of three Markov chains are given below. Looking at these matrices, determine the periods of the corresponding chains η_n, $n \ge 0$.

$$\left\| \begin{matrix} \frac13 & 0 & 0 & \frac13 & \frac13 \\ 0 & 0 & 1 & 0 & 0 \\ 0 & 0 & 0 & 1 & 0 \\ 0 & \frac12 & 0 & 0 & \frac12 \\ 1 & 0 & 0 & 0 & 0 \end{matrix} \right\|, \qquad \left\| \begin{matrix} 0 & 0 & \frac12 & 0 & \frac12 \\ 0 & 0 & 0 & 0 & 1 \\ \frac13 & \frac13 & 0 & \frac13 & 0 \\ 0 & 0 & 1 & 0 & 0 \\ \frac12 & \frac12 & 0 & 0 & 0 \end{matrix} \right\|, \qquad \left\| \begin{matrix} 0 & 0 & \frac12 & 0 & \frac12 \\ 0 & 0 & 1 & 0 & 0 \\ 0 & 0 & 0 & 1 & 0 \\ \frac12 & \frac12 & 0 & 0 & 0 \\ 0 & 0 & 0 & 0 & 1 \end{matrix} \right\|.$$

Find cyclic subclasses of these Markov chains. Calculate the matrices of transition probabilities of the Markov chains η_{nd}, $n \ge 0$, depending on the class that contains the initial state.

IV.5.35. Let $\eta_n = \xi_1 + \dots + \xi_n$, $n \ge 0$, where $\{\xi_n, n \ge 1\}$ is a sequence of independent identically distributed random variables. It is clear that η_n, $n \ge 0$, is a Markov chain. Prove that

a) in order that all the states of the Markov chain be communicating, it is necessary and sufficient that the characteristic function $\varphi(z) = \mathsf{E} \exp\{iz\xi_1\}$ differ from 1 for $z \in (0, 2\pi)$;

b) in order that the Markov chain be aperiodic, it is necessary and sufficient that $|\varphi(z)| < 1$ for $z \in (0, 2\pi)$;

c) in order that the Markov chain have period d, it is necessary and sufficient that $|\varphi(z)| < 1$ for $z \in (0, 2\pi/d)$ and $|\varphi(2\pi/d)| = 1$.

IV.5.36. Let η_n, $n \ge 0$, be a Markov chain, and τ a Markov time. Put

$$\widetilde{\tau} = \min(n > \tau : \eta_n \in D).$$

Prove that $\widetilde{\tau}$ is also a Markov time.

IV.5.37. Let $i_1, \dots, i_N, \dots$ be the sequence of states of a Markov chain η_n, $n \ge 0$. Define random variables

$$\tau^{(N)} = \min\left(n > \tau^{(N-1)} : \eta_n = i_N\right), \qquad N = 1, 2, \dots .$$

Prove that $\tau^{(N)}$, $N \ge 1$, are Markov times and show that

$$\mathsf{P}\left(\tau^{(N)} < \infty / \eta_0 = i_0\right) = \prod_{r=1}^{N} \mathsf{P}\left(\tau_{i_r} < \infty / \eta_0 = i_{r-1}\right).$$

IV.5.38. Let j be a state of a Markov chain η_n, $n \ge 0$. Let τ_j be the first time when the chain reaches j, that is, $\tau_j = \min(n \ge 1 : \eta_n = j)$. Denote by ${}_kf_{ij} = \mathsf{P}\left(\tau_j < \tau_k / \eta_0 = i\right)$ the conditional probability that the Markov chain starting at the state i reaches the state j before it reaches a state k. Prove that if $i \to j$, then either ${}_kf_{ij} > 0$ or ${}_if_{kj} > 0$ for every state $k \ne j$.

IV.5.39. A Markov chain η_n, $n \geq 0$, has the set of states $H = \{1, 2, \ldots, N\}$ and the matrix of transition probabilities

$$\left\|\begin{matrix} q & p & 0 & 0 & \cdot & \cdot & \cdot & \cdot \\ q & 0 & p & 0 & \cdot & \cdot & \cdot & \cdot \\ 0 & q & 0 & p & \cdot & \cdot & \cdot & \cdot \\ \cdot & \cdot & \cdot & \cdot & \cdot & \cdot & \cdot & \cdot \\ \cdot & \cdot & \cdot & \cdot & \cdot & q & 0 & p \\ \cdot & \cdot & \cdot & \cdot & \cdot & 0 & q & p \end{matrix}\right\|.$$

Find the probability ${}_1f_{iN} = \mathsf{P}\left(\tau_N < \tau_1/\eta_0 = i\right)$.

IV.5.40. Let $\tau_j^{(N)}$ be the moment when a Markov chain η_n, $n \geq 0$, reaches a state j for the Nth time. Prove that

$$\mathsf{P}\left(\tau_i^{(N)} < \tau_j/\eta_0 = k\right) = {}_jf_{ii}^{N-1} \cdot {}_jf_{ki}.$$

IV.5.41. Let μ_{jk} be the number of visits of a state $k \neq j$ by a Markov chain η_n, $n \geq 0$, over the time τ_j.

a) Prove that $\mathsf{P}\left(\mu_{jk} > n/\eta_0 = i\right) = {}_jf_{ik} \cdot {}_jf_{kk}^{n-1}$.

b) Prove that if $k \to j$, then $\mathsf{E}\left(\mu_{jk}^r/\eta_0 = i\right) < \infty$ for all $i \in H$ and $r \geq 1$.

IV.5.42. Let η_n, $n \geq 0$, be a finite Markov chain such that its set of states is an essential class. Prove that all the moments of random variables τ_j, $j \in H$, are finite:

$$\mathsf{E}\left(\tau_j^r/\eta_0 = i\right) < \infty, \qquad r \geq 1,\ i \in H.$$

IV.5.43. Let η_n, $n \geq 0$, be a Markov chain such that its set of states is an essential class. As in Problem IV.5.23, we say that a state i is *recurrent* if ${}_if_i = 1 - f_{ii} = 0$. Prove that if a state i is recurrent, then for all $j \in H$

$${}_jf_i = \mathsf{P}\left(\tau_j = +\infty/\eta_0 = i\right) = 0.$$

IV.5.44. Let η_n, $n \geq 0$, be a Markov chain such that its set of states is an essential class. Prove that if a state i is recurrent, then ${}_jf_i = 0$ for any state j, $j \neq i$.

IV.5.45. Let η_n, $n \geq 0$, be a Markov chain such that its set of states is an essential class. Prove that either all the states of the Markov chain are recurrent or none of them is recurrent (in the first case the Markov chain is said to be *recurrent*, while in the second it is said to be *nonrecurrent*).

IV.5.46. Denote by

$$\mu_j = \sum_{k=1}^{\infty} \delta_{\eta_k}^j$$

the total number of visits of a state j by a Markov chain η_n, $n \geq 0$. Prove that

$$\mathsf{P}\left(\mu_j \geq n/\eta_0 = i\right) = (1 - {}_jf_i) f_{jj}^{n-1}, \qquad n \geq 1.$$

Derive that $\mathsf{P}(\mu_j = +\infty) = 1$ for a recurrent chain and $\mathsf{P}(\mu_j = +\infty) = 0$ for a nonrecurrent chain.

IV.5.47. Let p_n, $n \geq 0$, be a probability distribution. The greatest integer d such that $p_n = 0$ for all n, except possibly for $n = 0, d, 2d, \dots$, is said to be *the step of the distribution* p_n, $n \geq 0$.

Prove that if a recurrent Markov chain η_n, $n \geq 0$, has the period d, then the step of the distribution $f_{jj}^{(n)} = \mathsf{P}\left(\tau_j = n / \eta_0 = j\right)$, $n \geq 0$, equals d.

IV.5.48. For $k \in D$ put

$$_D p_{ik}^{(n)} = \mathsf{P}\left(\eta_n = k, \tau_D > k / \eta_0 = i\right).$$

Show that the probabilities $_D p_{ik}^{(n)}$ satisfy the following system of recurrent equations:

$$_D p_{ik}^{(n)} = \sum_{j \in \overline{D}} p_{ij} \cdot {_D p_{jk}^{(n-1)}}, \qquad n \geq 1,$$

where $_D p_{ik}^{(0)} = \delta(i,k)$.

IV.5.49. Let τ_D be the first time when a Markov chain η_n, $n \geq 0$, reaches a domain D, i.e., $\tau_D = \min(n \geq 1 \colon \eta_n \in D)$. Put

$$f_i(R, D) = \mathsf{P}\left(\tau(D) < \infty, \eta_{\tau_D} \in R / \eta_0 = i\right), \qquad R \subseteq D,\ i \in H.$$

a) Show that the probabilities $f_i(R, D)$ satisfy the following system of linear equations:

$$f_i(R, D) = f_i(R) + \sum_{j \in \overline{D}} p_{ij} f_i(R, D), \qquad i \in H, \tag{a}$$

where $f_i(R) = \sum_{i \in R} p_{ij}$.

b) Show that

$$f_i(R, D) = \sum_{n=0}^{\infty} \sum_{k \in \overline{D}} {_D p_{ik}^{(n)}} f_k(R). \tag{b}$$

c) From (a) and (b) derive that the probabilities $f_i(R, D)$, $i \in H$, form the minimal solution of a system (a) in the class $\mathcal{L}$ of nonnegative functions that are uniformly bounded with respect to i. The minimality means that $x_i \geq f_i(R, D)$, $i \in H$, for every solution x_i, $i \in H$, of the system (a).

d) Prove that $f_i(R, D)$, $i \in H$, is the unique solution of the system (a) in the class $\mathcal{L}$ if and only if

$$_D f_i = \mathsf{P}\left(\tau(D) = +\infty / \eta_0 = i\right) \equiv 0, \qquad i \in H.$$

IV.5.50. Put

$$\varphi_i(z) = \mathsf{E}\left(z^{\tau(D)} / \eta_0 = i\right), \qquad i \in H,\ |z| \leq 1.$$

a) Prove that the moment generating functions $\varphi_i(z)$ satisfy the following system of linear equations:

$$\varphi_i(z) = z\left(p_i(D) + \sum_{j \in \overline{D}} p_{ij} \varphi_j(z)\right), \tag{a}$$

where $p_i(D) = \sum_{k \in D} p_{ik}$.

b) Show that

$$\varphi_i(z) = \sum_{n=0}^{\infty} \sum_{k \in \overline{D}} z^{n+1}\, {}_D p_{ik}^{(n)}\, p_i(D).$$

c) Show for $|z| < 1$ that $\varphi_i(z)$, $i \in H$, is the unique solution of the system (a) that is uniformly bounded with respect to i.

IV.5.51. Put $m_i(D) = \mathsf{E}\left(\tau(D)/\eta_0 = i\right)$.

a) Show that the expectation $m_i(D)$ satisfies the system of linear equations

(a) $$m_i(D) = 1 + \sum_{j \in \overline{D}} p_{ij} m_j(D), \qquad i \in H,$$

where we assume $p_{ij} m_j(D)$ to be zero if $p_{ij} = 0$ and $m_j(D) = +\infty$.

b) Show that

$$m_i(D) = 1 + \sum_{n=1}^{\infty} \sum_{j \in \overline{D}} {}_D p_{ij}^{(n)}.$$

c) Using a) and b) show that $m_i(D)$, $i \in H$, is the minimal solution of the system (a) in the class $\mathcal{N}$ of vectors $(x_j, j \in H)$ with nonnegative coordinates x_j, $j \in H$. The minimality means that $x_i \geq m_i(D)$, $i \in H$, for any solution $(x_j, j \in H) \in \mathcal{N}$ of the system (a).

IV.5.52. Let a Markov chain η_n, $n \geq 0$, have the set of states $H = \{0, 1, \ldots, N\}$ and the matrix of transition probabilities

$$\left\| \begin{matrix} q_0 & p_0 & 0 & 0 & 0 & \cdot & \cdot & \cdot & \cdot & \cdot \\ q_1 & 0 & p_1 & 0 & 0 & \cdot & \cdot & \cdot & \cdot & \cdot \\ 0 & q_2 & 0 & p_2 & 0 & \cdot & \cdot & \cdot & \cdot & \cdot \\ \cdot & \cdot & \cdot & \cdot & \cdot & \cdot & \cdot & \cdot & \cdot & \cdot \\ \cdot & \cdot & \cdot & \cdot & \cdot & \cdot & q_{N-2} & 0 & p_{N-2} & 0 \\ \cdot & \cdot & \cdot & \cdot & \cdot & \cdot & 0 & q_{N-1} & 0 & p_{N-1} \\ \cdot & \cdot & \cdot & \cdot & \cdot & \cdot & 0 & 0 & q_N & p_N \end{matrix} \right\|,$$

where $0 < p_i, q_i < 1$ and $p_i + q_i = 1$ for $0 \leq i \leq N$.

a) Let $D = \{0, N\}$. Determine the probabilities $f_i(N, D)$ that the Markov chain reaches the state N before it reaches the state 0.

b) Let $D = \{N\}$. Find the expectations $m_i(D)$.

c) Solve the problems a) and b) in the case $p_i = p = 1 - q$ for all $0 \leq i \leq N$. Find the moment generating functions $\varphi_i(z)$ in this case for the domain D specified in the problem b).

IV.5.53. Put

$$m_i^{(k)}(D) = \mathsf{E}\left(\tau^k(D)/\eta_0 = i\right), \qquad i \in H.$$

Show that the expectations $m_i^{(k)}(D)$, $i \in H$, satisfy the system of linear equations

$$m_i^{(k)}(D) = 1 + \sum_{j \in \overline{D}} \sum_{r=1}^{k-1} \binom{k}{r} m_j^r + \sum_{j \in \overline{D}} p_{ij} m_j^{(k)}(D), \qquad i \in H.$$

IV.5.54. Let η_n, $n \geq 0$, be a recurrent Markov chain, and $\tau_i^{(n)}$ the consecutive moments when the chain assumes a value i, that is, $\tau_i^{(0)} = 0$ and $\tau_i^{(n)} = \min\left(k > \tau_i^{(n-1)} : \eta_k = i\right)$, $n \geq 1$. Prove that the random variables $\varkappa_i^{(n)} = \tau_i^{(n+1)} - \tau_i^{(n)}$, $n \geq 0$, are independent and finite with probability one. Moreover, the random variables $\varkappa_i^{(n)}$, $n \geq 1$, are identically distributed and

$$\mathsf{P}\left(\varkappa_i^{(1)} = k\right) = f_{ii}^{(k)} = \mathsf{P}\left(\tau_i = k/\eta_0 = k\right).$$

IV.5.55. Show that if $m_{ii} = \mathsf{E}\left(\tau_i/\eta_0 = i\right) < \infty$, then

$$\frac{\tau_i^{(n)}}{n} \to m_{ii} \quad \text{as } n \to \infty$$

for an arbitrary initial distribution of the Markov chain η_n, $n \geq 0$.

IV.5.56. Let η_n, $n \geq 0$, be a Markov chain. Define random variables

$$\delta_i(n) = \begin{cases} 1 & \text{if } \eta_n = i, \\ 0 & \text{if } \eta_n \neq i. \end{cases}$$

Let $\nu_i(n) = \sum_{k=1}^n \delta_i(k)$ be the number of visits of a state i in n steps.

a) Prove that if $m_{ii} < \infty$, then with probability one

$$\frac{\nu_i(n)}{n} \to \frac{1}{m_{ii}}, \qquad n \to \infty,$$

whatever the initial distribution of the chain.

b) Prove that if $m_{ii} = +\infty$, then with probability one

$$\frac{\nu_i(n)}{n} \to 0, \qquad n \to \infty.$$

IV.5.57. Let η_n, $n \geq 0$, be a Markov chain. Assume that $\mathsf{P}(\eta_0 = i) = 1$, and let

$$\mu_{ij}^{(n)} = \sum_{k=\tau_i^{(n)}+1}^{\tau_i^{(n+1)}} \delta_i(k)$$

be the number of visits of the state j by the chain between its nth and $(n+1)$st visits of the state i. Prove that the random variables $\mu_{ij}^{(n)}$, $n \geq 0$, are independent and $\mu_{ij}^{(n)}$, $n \geq 1$, are identically distributed. Moreover,

$$\mathsf{P}\left(\mu_{ij}^{(1)} = k\right) = \mathsf{P}\left(\mu_{ij} = k/\eta_0 = i\right), \qquad k \geq 0,$$

where μ_{ij} stands for the number of visits of the state j by the chain η_n, $n \geq 0$, in time τ_j (these random variables were introduced in Problem IV.5.21).

IV.5.58. Let η_n, $n \geq 0$, be a Markov chain. Assume that $m_{ii} < \infty$. Prove that for any initial distribution,

$$\frac{1}{n} \sum_{k=1}^{\nu_i(n)} \mu_{ij}^{(k-1)} \to \frac{\mathsf{E}\,\mu_{ij}}{m_{ii}}, \qquad n \to \infty,$$

with probability one.

IV.5.59. Use the preceding problem to show that

$$\mathsf{E}\,\mu_{ij} = \frac{m_{ii}}{m_{jj}}$$

provided $m_{ii} < \infty$.

IV.5.60. Let $\eta_n, n \geq 0$, be a recurrent Markov chain. Put $m_{ij} = \mathsf{E}\left(\tau_j / \eta_0 = i\right)$.
a) Show that if $m_{ii} < \infty$, then $m_{ki} < \infty$ for all $k \neq i$.
b) Show that either $m_{ii} = +\infty$ for all $i \in H$, or $m_{ii} < \infty$ for all $i \in H$.
c) Show that if $m_{ii} < \infty$, then

$$\sum_{j \in H} \frac{1}{m_{jj}} = 1.$$

IV.5.61. A Markov chain η_n, $n \geq 0$, is said to be *ergodic* if for all $i, j \in H$ there exist positive limits

$$\lim_{n \to \infty} \frac{p_{ij}^{(1)} + \cdots + p_{ij}^{(n)}}{n} = q_j$$

such that $\sum_{j \in H} q_j = 1$. The sequence $\{q_j, j \in H\}$ is said to be *the stationary distribution of the Markov chain* $\{\eta_n, n \geq 0\}$.

Use Problems IV.5.56 and IV.5.60 to show that the condition $m_{ii} < \infty$ is necessary and sufficient for a Markov chain η_n, $n \geq 0$, to be ergodic. Note that the condition above is satisfied either for all $i \in H$ or for no $i \in H$ (Problem IV.5.60 b)). Moreover, under the condition $m_{ii} < \infty$, the stationary distribution of the chain is given by

$$q_j = \frac{1}{m_{jj}}, \qquad j \in H.$$

IV.5.62. Show that the relation for probabilities $p_{ij}^{(n)}$ in Problem IV.5.22 is a particular case of the renewal equation for $i = j$ (see the beginning of §IV.2). Use the renewal theorem for arithmetic distributions (loc. cit.) to show that

a) If a Markov chain η_n, $n \geq 0$, is recurrent and aperiodic, then

$$p_{jj}^{(n)} \to q_j, \qquad n \to \infty.$$

b) If a Markov chain η_n, $n \geq 0$, is recurrent and has period $d > 1$, then

$$p_{jj}^{(nd)} \to dq_j, \qquad n \to \infty.$$

c) If a Markov chain η_n, $n \geq 0$, is aperiodic, then for all $i \in H$ we have

$$p_{ij}^{(dn)} \to dq_j, \qquad n \to \infty, \quad j \in H,$$

while if a Markov chain η_n, $n \geq 0$, is recurrent and has period $d > 1$, then for states $i \in C_k$ and $j \in C_{k+r}$ we have

$$p_{ij}^{(nd+r)} \to \frac{d}{m_{ij}}, \qquad n \to \infty.$$

Here $C_1, \dots, C_d$ are corresponding cyclic classes, and $C_{d+1} = C_1, C_{d+2} = C_2, \dots$.

d) Show that if a state j of a Markov chain η_n, $n \geq 0$, is nonrecurrent, then for any $i \in H$,

$$p_{ij}^{(n)} \to 0, \qquad n \to \infty.$$

Therefore, a nonrecurrent Markov chain cannot be ergodic.

IV.5.63. Use Problem IV.5.59 to show that the stationary probabilities of an ergodic Markov chain η_n, $n \geq 0$, satisfy the following system of linear equations:

$$\begin{cases} q_j = \sum_{i\in H} q_i p_{ij}, & j \in H, \\ \sum_{i \in H} q_i = 1. \end{cases}$$

IV.5.64. Show that if a Markov chain η_n, $n \geq 0$, is ergodic, then the system of linear equations given in Problem IV.5.63 has a unique nonnegative solution.

REMARK. The uniqueness of a nonnegative solution of the system given in Problem IV.5.63 is also a necessary condition for a Markov chain to be ergodic.

IV.5.65. Show that a Markov chain with a finite set of states is ergodic if and only if all its states are communicating.

IV.5.66. Suppose a Markov chain has a set of states $H = \{1, 2, \dots, N\}$ and the matrix of transition probabilities

$$\begin{Vmatrix} q & p & 0 & 0 & 0 & \cdots & \cdot & \cdot & \cdot & \cdot \\ q & 0 & p & 0 & 0 & \cdots & \cdot & \cdot & \cdot & \cdot \\ 0 & q & 0 & p & 0 & \cdots & \cdot & \cdot & \cdot & \cdot \\ \cdot & \cdot & \cdot & \cdot & \cdot & \cdots & q & 0 & p & 0 \\ \cdot & \cdot & \cdot & \cdot & \cdot & \cdots & 0 & q & 0 & p \\ \cdot & \cdot & \cdot & \cdot & \cdot & \cdots & 0 & 0 & q & p \end{Vmatrix}.$$

Find the stationary distribution q_j, $1 \leq j \leq N$.

IV.5.67. Let η_n, $n \geq 0$, be a finite aperiodic Markov chain that has only one essential class of states. Assume that its matrix of transition probabilities $\mathbf{P} = \|p_{ij}\|_{i,j=1}^{m}$ satisfies the condition

$$\sum_{i=1}^{m} p_{ij} = 1, \qquad 1 \leq j \leq m.$$

Prove that the corresponding stationary probabilities q_j are given by

$$q_1 = \cdots = q_m = \frac{1}{m}.$$

IV.5.68. Let $\xi_0, \xi_1, \dots$ be a sequence of independent identically distributed random variables, each being binomial, that is,

$$\mathsf{P}(\xi_n = k) = \binom{m}{k} p^k (1-p)^{m-k}, \qquad 0 \leq k \leq m.$$

Define a sequence η_n, $n \geq 0$, as follows:

$$\eta_n = \begin{cases} \eta_{n-1} - k + \xi_{n-1} & \text{if } \eta_{n-1} \geq k, \\ \eta_{n-1} + \xi_{n-1} & \text{if } \eta_{n-1} < k, \end{cases}$$

where k is a fixed positive integer. Prove that the sequence η_n, $n \geq 0$, is a Markov chain, and moreover, it is ergodic provided $k < mp$.

IV.5.69. Each of two urns contains N white and N black balls. The number of black balls in the first urn is treated as a state of a Markov chain. At any stage we draw at random one ball from the first urn and one ball from the second urn and then put each of them to the other urn. Find the transition probabilities p_{ik} and the stationary probabilities and prove that q_k equals the probability that among N balls drawn at random from a collection of N black and N white balls, there will be exactly k black balls.

IV.5.70. A Markov chain η_n, $n \geq 0$, has a set of states $H = \{0, 1, \dots\}$ and the transition probabilities

$$p_{ij} = e^{-\lambda} \sum_{k=0}^{\min(i,j)} \binom{i}{k} (1-q)^k q^{i-k} \frac{\lambda^{j-k}}{(j-k)!}.$$

Prove that

$$p_{jj}^{(n)} \to e^{-\lambda/q} \frac{(\lambda/q)^j}{j!}, \qquad n \to \infty.$$

REMARK. Such chains appear in statistical mechanics.

IV.5.71. The matrix of transition probabilities of a Markov chain is given by

$$\begin{Vmatrix} q & p & 0 & 0 & 0 & \cdots \\ q & 0 & p & 0 & 0 & \cdots \\ 0 & q & 0 & p & 0 & \cdots \\ \cdot & \cdot & \cdot & \cdot & \cdot & \cdots \end{Vmatrix}.$$

a) Show that $m_{ij} < \infty$, $i \in H$, if $q > p$, and use the system of equations given in Problem IV.5.51 to calculate m_{ij}.

b) Prove that if $p = q = \frac{1}{2}$, then $m_{i0} = +\infty$ and the Markov chain is recurrent, while if $q < p$, then the Markov chain is nonrecurrent. Determine the probability ${}_{i-1}f_i$ for this case.

c) Find the stationary probabilities of the Markov chain if $q > p$.

IV.5.72. Let η_n, $n \geq 0$, be a Markov chain with the matrix of transition probabilities

$$\begin{Vmatrix} p_0 & p_1 & p_2 & p_3 & \cdots \\ 1 & 0 & 0 & 0 & \cdots \\ 0 & 1 & 0 & 0 & \cdots \\ 0 & 0 & 1 & 0 & \cdots \\ \cdot & \cdot & \cdot & \cdot & \cdots \end{Vmatrix}.$$

Find conditions for the chain to be recurrent and ergodic. Check if the stationary distribution exists, and determine it if it does.

IV.5.73. The matrix of transition probabilities of a Markov chain η_n, $n \geq 0$, is given by

$$\begin{Vmatrix} q_0 & p_0 & 0 & 0 & \cdots \\ q_1 & 0 & p_1 & 0 & \cdots \\ q_2 & 0 & 0 & p_2 & \cdots \\ \cdot & \cdot & \cdot & \cdot & \cdots \end{Vmatrix}.$$

Prove that this chain is not recurrent if and only if

$$\sum_{i=0}^{\infty} q_i = 1.$$

Find the conditions for the chain to be ergodic, and if it is, find the stationary distribution. For the case where the chain is not recurrent, find the probability ${}_0f_i = \mathsf{P}\left(\tau_0 = +\infty / \eta_0 = i\right)$.

IV.5.74. Consider a Markov chain with the matrix of transition probabilities

$$\left\| \begin{matrix} p_0 & p_1 & p_2 & p_3 & \cdots \\ p_0 & p_1 & p_2 & p_3 & \cdots \\ 0 & p_0 & p_1 & p_2 & \cdots \\ \cdot & \cdot & \cdot & \cdot & \cdots \end{matrix} \right\|.$$

Use moment generating functions to investigate whether or not the conditions for recurrence and ergodicity are satisfied. Find the moment generating function of the stationary distribution if it exists.

IV.5.75. Consider a Markov chain with a set of states $H = \{0, 1, \dots\}$ and the matrix of transition probabilities

$$\left\| \begin{matrix} q+r & p & 0 & 0 & 0 & \cdots \\ q & r & p & 0 & 0 & \cdots \\ 0 & q & r & p & 0 & \cdots \\ \cdot & \cdot & \cdot & \cdot & \cdot & \cdots \end{matrix} \right\|,$$

where $p > 0$, $q > 0$, $r \geq 0$, $q + r + p = 1$. Prove that

a) the chain is ergodic if $q > p$, and find its stationary distribution;

b) the chain is recurrent if $q = p$, but $m_{i,i-1} = +\infty$;

c) the chain is not recurrent if $q < p$, and find the probabilities ${}_{i-1}f_i$ and ${}_{i-k}f_i$.

IV.5.76. Let a Markov chain η_n, $n \geq 0$, have the set of states $H = \{0, 1, \dots\}$ and the matrix of transition probabilities given in Problem IV.5.75. Assume that $q \geq p$. Let $\varphi(z) = \mathsf{E}\left(z^{\tau_{i-1}} / \eta_0 = i\right)$ be the moment generating function of the chain. Prove that

$$\varphi(z) = \frac{1 - rz - \sqrt{(1 - rz)^2 - 4pqz^2}}{2zp}.$$

IV.5.77. Assume that the distribution function of a random variable $\varkappa_i^{(1)}$ belongs to the domain of normal attraction of a stable law with parameter $\alpha < 1$, that is, for $x \geq 0$,

$$\text{(a)} \qquad \mathsf{P}\left(\frac{1}{n^{1/\alpha}} \sum_{k=1}^{n} \varkappa_i^{(k)}\right) \to \mathsf{P}(\varkappa_\alpha < x), \qquad n \to \infty,$$

where $\varkappa_i^{(k)}$ are copies of $\varkappa_i^{(1)}$, and $\varkappa_\alpha$ is a nonnegative random variable with the Laplace transform

$$\mathsf{E}\, e^{-s\varkappa_\alpha} = e^{-cs^\alpha}, \qquad s \geq 0.$$

Prove that in this case

$$\mathsf{P}\left(\frac{\nu_i(n)}{n^\alpha} < x\right) \to F_\alpha(x), \qquad n \to \infty,$$

where the distribution function $F_\alpha(x)$ is defined by

$$F_\alpha(x) = \mathsf{P}\left(\varkappa_\alpha > \frac{1}{x^{1/\alpha}}\right), \qquad x \geq 0.$$

IV.5.78. Let η_n, $n \ge 0$, be the Markov chain defined in Problem IV.5.75, and $q \ge p$. Prove that its moment generating function is given by

$$\mathsf{E}\left(z^{\tau_0}/\eta_0 = 0\right) = z(r+q) + \frac{1}{2}\left(1 - rz - \sqrt{(1-rz)^2 - 4p^2q^2z^2}\right).$$

Derive from this equality that for $p = q = (1-r)/2$ the Laplace transform of the random variable $\tau_i^{(n)} = \varkappa_i^{(0)} + \cdots + \varkappa_i^{(n-1)}$ satisfies the relation

$$\mathsf{E}\exp\left\{-\frac{s}{n^2}\sum_{k=1}^{n}\varkappa_i^{(k)}\right\} = \left(\frac{1}{2} + \frac{1}{2}e^{-s/n^2} - \frac{1}{2}\sqrt{1 - e^{2s/n^2}}\right)^n \to \exp\left\{\sqrt{\frac{s}{2}}\right\}$$

as $n \to \infty$. Note that by the continuity theorem, this relation is equivalent to relation (a) in the preceding problem.

IV.5.79. Let $\mathsf{E}\left(\tau_i^2/\eta_0 = i\right) = m_{ii}^{(2)} < \infty$. Prove that in this case

$$\mathsf{P}\left(\frac{\nu_i(n) - n/m_{ii}}{\sigma_i\sqrt{n}} < x\right) \to \frac{1}{\sqrt{2\pi}}\int_{-\infty}^{x} e^{-z^2/2}\,dz, \qquad n \to \infty,$$

where $\sigma_i^2 = m_{ii}^{(2)} - m_{ii}^2$.

IV.5.80. Let a Markov chain η_n, $n \ge 0$, be ergodic, and let q_j, $j \in H$, be its stationary distribution. Assume that the initial distribution $\mathsf{P}(\eta_0 = j) = p_j$ coincides with the stationary distribution: $p_j = q_j$, $j \in H$. Prove that in this case the sequence η_n, $n \ge 0$, is stationary in the sense that for all $n \ge 0$,

$$\mathsf{P}(\eta_n = i_1, \ldots, \eta_{n+k} = i_k) = q_{i_1}p_{i_1i_2}\cdots p_{i_{k-1}i_k}, \qquad i_1, \ldots, i_k \in H, \quad k \ge 1.$$

IV.5.81. a) Let η_n, $n \ge 0$, be a homogeneous Markov chain with the set of states $H = \{1, 2, \ldots\}$ and the matrix of transition probabilities $P = \|p_{ij}\|_{i,j\in\mathbf{N}}$. Prove that the sequence $\overline{\eta}_n = (\eta_n, \eta_{n+1}, \ldots, \eta_{n+k})$, $n \ge 0$, is also a homogeneous Markov chain with the set of states $H_k = \{(i_0, \ldots, i_k)\colon i_0, \ldots, i_k \in H, p_{i_0i_1}\cdots p_{i_{k-1}i_k} > 0\}$ and the transition probabilities

$$\overline{p}_{ij} = \mathsf{P}\left(\overline{\eta}_{n+1} = \overline{j}/\overline{\eta}_n = \overline{i}\right) = \begin{cases} p_{i_kj_k} & \text{if } (i_1, \ldots, i_k) = (j_0, \ldots, j_{k-1}), \\ 0 & \text{if } (i_1, \ldots, i_k) \ne (j_0, \ldots, j_{k-1}), \end{cases}$$

where $\overline{j} = (j_0, \ldots, j_k)$, $\overline{i} = (i_0, \ldots, i_k)$.

b) If a Markov chain η_n, $n \ge 0$, is ergodic and q_j, $j \in H$, is its stationary distribution, then the Markov chain $\overline{\eta}_n$, $n \ge 0$, is also ergodic and its stationary distribution is given by

$$\overline{q}_j = q_{j_0}p_{j_0j_1}\cdots p_{j_{k-1}j_k}, \qquad (j_0, \ldots, j_k) \in H_k.$$

IV.5.82. Let a Markov chain $\eta_n^{(\varepsilon)}$, $n \ge 0$, be given for every $\varepsilon > 0$, with the set of states $H = \{1, 2, \ldots\}$ and the matrix of transition probabilities $\|p_{ij}(\varepsilon)\|$. Let

$$p_{ij}(\varepsilon) \to p_{ij}(0) \quad \text{as } \varepsilon \to 0, \qquad i, j \in H.$$

a) Prove that if the limit Markov chain $\eta_n^{(0)}$, $n \ge 0$, is recurrent, then for any domain D the probability distributions of random variables

$$\tau_D^{(\varepsilon)} = \min\left(n \ge 1\colon \eta_n^{(\varepsilon)} \in D\right)$$

weakly converge as $\varepsilon \to 0$.

b) $f_i^\varepsilon(D,R) = \mathsf{P}\{\eta^\varepsilon_{\tau_D}(\varepsilon) \in R, \tau_D^\varepsilon < \infty / \eta_0^\varepsilon = 0\} \to f_i^{(0)}(D,R)$.

c) Construct an example showing that there are ergodic Markov chains $\eta_n^{(\varepsilon)}$, $n \geq 0$, such that

$$q_j^{(\varepsilon)} \not\to q_j^{(0)} \quad \text{as } \varepsilon \to 0.$$

CHAPTER V

Limit Theorems of Probability Theory

§V.1. Characteristic functions

The characteristic function of a random variable. Let ξ be a random variable with the distribution function $F(x)$. The function

$$\varphi(z) = \mathsf{E}\, e^{iz\xi} = \int_{-\infty}^{\infty} e^{izx}\, dF(x)$$

defined for all real z is called *the characteristic function of the random variable* ξ.

If a random variable ξ has the distribution density $p(x)$, then

$$\varphi(z) = \int_{-\infty}^{\infty} e^{izx} p(x)\, dx.$$

All characteristic functions $\varphi(z)$ possess the following properties:

a) $\varphi(0) = 1$;

b) $\varphi(z)$ is uniformly continuous on $(-\infty, \infty)$;

c) $\varphi(z)$ is positive definite, that is, for all complex numbers $c_1, c_2, \dots, c_r$ and all real numbers $z_1, z_2, \dots, z_r$, $r \geq 1$, the following condition is satisfied:

$$\sum_{k,j=1}^{r} c_k \bar{c}_j \varphi(z_k - z_j) \geq 0.$$

The characteristic function of the sum of independent random variables equals the product of their characteristic functions.

BOCHNER–KHINTCHINE THEOREM. *In order that a function $\varphi(z)$ be the characteristic function of a probability distribution, it is necessary and sufficient that it be continuous, positive definite, and satisfy the condition $\varphi(0) = 1$.*

Continuity theorems for characteristic functions (Lévy's theorems).

THEOREM 1. *If a sequence of distribution functions $\{F_n(x), n \geq 1\}$ weakly converges to a distribution function $F(x)$, then the sequence of corresponding characteristic functions*

$$\varphi_n(z) = \int_{-\infty}^{\infty} e^{izx}\, dF_n(x), \qquad n \geq 1,$$

converges at any point z to the characteristic function

$$\varphi(z) = \int_{-\infty}^{\infty} e^{izx}\, dF(x).$$

Moreover, the convergence is uniform with respect to z in any finite interval.

THEOREM 2. *Let a sequence of characteristic functions $\{\varphi_n(z), n \geq 1\}$ converges at any point z to a continuous function $\varphi(z)$. Then $\varphi(z)$ is a characteristic function, that is, $\varphi(z) = \int_{-\infty}^{\infty} e^{izx}\, dF(x)$, where $F(x)$ is a distribution function. Moreover, the corresponding sequence of distribution functions $\{F_n(x), n \geq 1\}$ weakly converges to the distribution function $F(x)$.*

Laplace transform. For nonnegative random variables ξ $(F(0) = 0)$, it is often convenient to consider their *Laplace transforms*

$$\psi(s) = \mathsf{E}\, e^{-s\xi} = \int_{-\infty}^{\infty} e^{-sx}\, dF(x), \qquad s \geq 0,$$

instead of their characteristic functions.

Characteristic functions of random vectors. Let $\bar{\xi} = (\xi_1, \xi_2, \dots, \xi_m)$ be a random vector. Then the function

$$\varphi(\bar{z}) = \mathsf{E}\, e^{(\bar{z},\bar{\xi})} = \int \cdots \int_{\mathbf{R}^m} e^{i(\bar{z},\bar{x})}\, dF(x_1, x_2, \dots, x_m)$$

is called *the characteristic function of the random vector* $\bar{\xi}$, where $(\bar{z}, \bar{\xi}) = \sum_{k=1}^{m} z_k \xi_k$ and $F(x_1, x_2, \dots, x_m)$ is the distribution function of $\bar{\xi}$.

Problems

V.1.1. Let a random variable ξ assume only values 1 and -1 with equal probabilities $\frac{1}{2}$, that is, ξ is a Bernoulli random variable. Calculate the characteristic function of ξ.

V.1.2. Prove that $\varphi(z) = \cos^2 z$ is a characteristic function and determine the corresponding probability distribution.

V.1.3. Let a random variable ξ assume only values -1, 0, and 1 with equal probabilities $\frac{1}{3}$. Calculate the characteristic function of ξ.

V.1.4. Prove that both $\varphi_1(z) = \sum_{k=1}^{\infty} a_k \cos kz$ and $\varphi_2(z) = \sum_{k=0}^{\infty} a_k e^{i\lambda_k z}$, where $a_k \geq 0$ and $\sum_{k=0}^{\infty} a_k = 1$, are characteristic functions and determine the corresponding probability distributions.

V.1.5. a) Let $\xi_1, \xi_2, \dots, \xi_n$ be independent random variables assuming values -1 and 1 with probabilities $\frac{1}{2}$. Calculate the characteristic function of their sum $S_n = \xi_1 + \xi_2 + \cdots + \xi_n$.

b) Prove that for all integers n, $\varphi(z) = \cos^n z$ is a characteristic function.

V.1.6. Calculate the characteristic function for

a) the Poisson distribution with parameter λ;

b) the binomial distribution with parameters p and m;

c) the geometric distribution with parameter p.

V.1.7. Let ξ be a random variable uniformly distributed on $[-a, a]$. Calculate the characteristic function of ξ.

V.1.8. Let ξ be a random variable uniformly distributed on $[a,b]$. Prove that its characteristic function equals

$$\varphi_\xi(z) = \frac{e^{ibz} - e^{iaz}}{i(b-a)z}.$$

V.1.9. Let $\varphi(z)$ be the characteristic function of a Gaussian random variable with parameters 0 and 1.

a) Using differentiation and integration by parts prove that $\varphi'(z) = -z\varphi(z)$ and obtain that $\varphi(z) = e^{-z^2/2}$.

b) Prove that the characteristic function of a Gaussian distribution with parameters a and σ^2 equals $\varphi(z) = e^{iza - z^2\sigma^2/2}$.

V.1.10. A random variable ξ has the exponential distribution with parameter λ. Calculate the characteristic function.

V.1.11. a) Let ξ_1 and ξ_2 be independent identically distributed random variables with the characteristic function $\varphi(z)$. Calculate the characteristic function of the difference $\xi_1 - \xi_2$.

b) If $\varphi(z)$ is a characteristic function, then so is $|\varphi(z)|^2$. Prove this statement.

V.1.12. If $\varphi(z)$ is the characteristic function of a random variable ξ, then the random variable $\eta = a + b\xi$ has the characteristic function $e^{iaz}\varphi(zb)$. Prove this statement.

V.1.13. Let a random variable ξ be nonnegative. Prove that its Laplace transform $\mathsf{E}\, e^{-s\xi} = \psi(s)$ is defined for all complex s such that $\operatorname{Re} s \geq 0$. In this domain the function $\psi(s)$ is continuous and bounded. Moreover, it is an analytic function in the domain $\operatorname{Re} s > 0$. The characteristic function of the random variable ξ equals

$$\varphi(z) = \psi(-iz), \qquad z \in \mathbf{R}_1.$$

V.1.14. Suppose a random variable ξ takes only integer values, its moment generating function is $h(z) = \mathsf{E}\, z^\xi$, and its characteristic function is $\varphi(z)$. Prove that $\varphi(z) = h(e^{iz})$. Prove also that the probabilities $p_k = \mathsf{P}(\xi = k)$, $k = 0, \pm 1, \dots$, can be represented in terms of $h(z)$ as follows:

$$p_k = \frac{1}{2\pi i}\int_{C_1} h(z) z^{-k-1}\, dz = \frac{1}{2\pi}\int_{-\infty}^{\infty} \varphi(z) e^{izk}\, dz,$$

where C_1 is the unit circle centered at the point $z = 0$.

V.1.15. Let $F(x)$ and $G(x)$ be two distribution functions with characteristic functions $\varphi(z)$ and $\psi(z)$, respectively. Prove the so-called *Parseval equality*:

$$\int_{-\infty}^{\infty} e^{-izt}\varphi(z)\, dG(z) = \int_{-\infty}^{\infty} \psi(x-t)\, dF(x).$$

V.1.16. Prove that if the characteristic function $\varphi(z)$ of a random variable ξ is absolutely integrable, that is,

$$\int_{-\infty}^{\infty} |\varphi(z)|\, dz < \infty,$$

then the random variable ξ has the distribution density $p(x)$, and moreover,

$$p(x) = \frac{1}{2\pi}\int_{-\infty}^{\infty} e^{-izx}\varphi(z)\,dz.$$

V.1.17. Let $\varphi(z)$ be the characteristic function corresponding to a distribution function $F(x)$. Prove that

$$\frac{F(x+h)-F(x)}{h} = \frac{1}{2\pi}\int_{-\infty}^{\infty} \varphi(z)\frac{1-e^{-izh}}{izh}\,dz,$$

provided the integral on the right-hand side is convergent.

V.1.18. Let $\varphi(z)$ be the characteristic function of a distribution $F(x)$, and y and $y+h$ points where $F(x)$ is continuous. Prove that the following inversion formula holds for $F(x)$:

$$\frac{F(y+h)-F(y)}{h} = \lim_{a\to 0}\frac{1}{2\pi}\int_{-\infty}^{\infty} \varphi(z)e^{-a^2z^2/2}\frac{1-e^{-izh}}{izh}e^{-izy}\,dz.$$

V.1.19. Let $\varphi(z)$ be the characteristic function of a distribution $F(x)$. Prove that

$$F(x) = \lim_{a\to 0}\int_{-\infty}^{x}\frac{1}{2\pi}\int_{-\infty}^{\infty} e^{-z^2a^2/2}\varphi(z)e^{-izy}\,dz\,dy$$

at any point x where F is continuous.

V.1.20. Prove that the natural correspondence between distribution functions and characteristic functions is one-to-one.

V.1.21. Prove that the characteristic function of a random variable ξ with a distribution function $F(x)$ assumes only real values if and only if the distribution F is symmetric, that is,

$$F(-x+0) = 1 - F(x), \qquad x \ge 0.$$

V.1.22. a) Let a random variable ξ have the two-sided exponential distribution with the probability density

$$p(x) = \frac{1}{2}e^{-|x|}, \qquad x \in \mathbf{R}_1.$$

Show that its characteristic function is given by

$$\varphi(z) = \frac{1}{1+z^2}.$$

b) A random variable ξ has the Cauchy distribution with the probability density

$$p(x) = \frac{1}{\pi}\frac{a}{a^2+x^2}, \qquad a > 0.$$

Prove that the characteristic function of ξ is

$$\varphi(z) = e^{-a|z|}.$$

V.1.23. a) A random variable ξ has the probability density

$$p(x) = \begin{cases} 0, & \text{for } |x| > a, \\ \dfrac{1}{a}\left(1 - \dfrac{|x|}{a}\right), & \text{for } |x| \le a. \end{cases}$$

Prove that the characteristic function of ξ is

$$\varphi(z) = 2\frac{1 - \cos az}{a^2 z^2}.$$

b) A random variable ξ has the probability density

$$p(x) = \frac{1 - \cos az}{\pi a^2 z^2}.$$

Show that its characteristic function is

$$\varphi(z) = \begin{cases} 0, & \text{for } |z| > a, \\ 1 - \dfrac{|z|}{a}, & \text{for } |z| \le a. \end{cases}$$

c) Prove that a function $\varphi(z)$ which is periodic with period $2a$ and satisfies the condition $\varphi(z) = 1 - |z|/a$ for $|z| \le a$, is characteristic.

V.1.24. Using the Bochner–Khintchine theorem prove that if a characteristic function $\varphi(z)$ vanishes for $|z| > a$, a function $g(z)$ has a period $2a$, and $g(z) = \varphi(z)$ for $|z| \le a$, then $g(z)$ is a characteristic function.

V.1.25. Prove that there are different characteristic functions $\varphi_1(z)$ and $\varphi_2(z)$ such that $\varphi_1^2(z) = \varphi_2^2(z)$.

V.1.26. Prove that there exist random variables ξ_1, ξ_2, ξ_3 with different distribution functions such that the distributions of the random variables $\xi_1 + \xi_2$ and $\xi_2 + \xi_3$ coincide.

V.1.27. Using characteristic functions prove the following statement. If ξ_1 and ξ_2 are independent Gaussian random variables with parameters $N(a_1, \sigma_1^2)$ and $N(a_2, \sigma_2^2)$, respectively, then the random variable $\eta = \xi_1 + \xi_2$ is Gaussian with parameters $N(a_1 + a_2, \sigma_1^2 + \sigma_2^2)$.

V.1.28. Let ξ_1 and ξ_2 be two independent Poisson random variables with parameters λ_1 and λ_2, respectively. Prove that $\eta = \xi_1 + \xi_2$ is a Poisson random variable with the parameter $\lambda_1 + \lambda_2$.

V.1.29. a) A random variable ξ has the gamma distribution with parameters (α, β) (see Problem II.3.17). Prove that the characteristic function of ξ is

$$\varphi(z) = \left(\frac{\beta}{\beta - iz}\right)^{\alpha}.$$

b) Random variables ξ_1 and ξ_2 are independent and have the gamma distribution with parameters (α_1, β_1) and (α_2, β_2), respectively. Prove that $\eta = \xi_1 + \xi_2$ has the gamma-distribution with parameters $(\alpha_1 + \alpha_2, \beta)$.

V.1.30. Calculate the characteristic function of a random variable that has the χ^2 distribution with n degrees of freedom (see Problem II.3.121).

V.1.31. Let ξ and η be independent random variables with the distribution functions $F(x)$ and $G(x)$, respectively, and let $\varphi(z)$ and $\gamma(z)$ be their characteristic functions. Prove that the characteristic function of $\xi\eta$ equals

$$\int_{-\infty}^{\infty} \gamma(zx)\,dF(x) = \int_{-\infty}^{\infty} \varphi(zx)\,dG(x).$$

V.1.32. Let ξ be a random variable with the characteristic function $\varphi(z)$, and let η be a random variable that is independent of ξ and uniformly distributed on $[0,1]$. Calculate the characteristic function of $\xi\eta$.

V.1.33. Prove that if $\varphi(z)$ is a characteristic function, then so is

$$\varphi(z) = \frac{1}{z}\int_0^z \varphi(u)\,du.$$

V.1.34. Let ξ_1 and ξ_2 be independent Gaussian random variables with parameters $N(0,1)$. Calculate the characteristic function of the random variable $\eta = \frac{1}{2}(\xi_1^2 - \xi_2^2)$.

V.1.35. Construct an example of dependent random variables ξ and η such that the characteristic function of $\xi + \eta$ coincides with the product of the characteristic functions of ξ and η.

V.1.36. A random variable ξ is called a *lattice* if its values $\{a_r\}$ can be represented as $a_r = a + k(r)h$, where $k(r)$ is an integer. The maximal possible value of h is said to be *the maximal step of the distribution*. Prove that if $\varphi(z)$ is the characteristic function of a random variable ξ and $|\varphi(z)| = 1$ for some $z \neq 0$, then ξ is a lattice.

V.1.37. Give a probability interpretation of the equality

$$\frac{\sin z}{z} = \frac{\sin(z/2)}{z/2}\cos\frac{z}{2}.$$

V.1.38. Give a probability interpretation of the equality

$$\frac{\sin z}{z} = \prod_{k=1}^{\infty} \cos\frac{z}{2^k}.$$

V.1.39. Let $F(x)$ and $\varphi(z)$ be the distribution function and the characteristic function, respectively, of a random variable that has a finite second moment

$$M_2 = \int_{-\infty}^{\infty} x^2\,dF(x).$$

a) Prove that there exists $\varphi''(z)$ and that $\psi(z) = -\varphi''(z)/M_2$ is a characteristic function. Find its distribution function.

b) Prove that $(1-t^2)e^{-t^2/2}$ is a characteristic function. Find its distribution function.

V.1.40. Let ξ be a random variable with the characteristic function $\varphi(z)$ and

$$m_n = \mathsf{E}\,\xi^n, \qquad M_n = \mathsf{E}\,|\xi|^n.$$

Prove that if $M_n < \infty$, then there exists the nth derivative $\varphi^{(n)}(z)$ and

$$m_n = \frac{\varphi^{(n)}(0)}{i^n}.$$

V.1.41. Let ξ be a random variable. Prove that if there exists $\mathsf{E}\,|\xi|^n = M_n < \infty$, then the following Taylor expansion is valid for the characteristic function $\varphi(z)$:

$$\varphi(z+t) = \varphi(z) + t\varphi'(z) + \cdots + \frac{t^{n-1}}{(n-1)!}\varphi^{(n-1)}(z) + \theta\frac{M_n t^n}{n!},$$

where $|\theta| \le 1$.

V.1.42. Let ξ be a random variable with the distribution function $F(x)$ and the characteristic function $\varphi(z)$. If all the moments $M_n = \mathsf{E}\,|\xi|^n$ exist and

$$\limsup_{n\to\infty} \frac{1}{n}\sqrt[n]{M_n} = \lambda < \infty,$$

then the distribution $F(x)$ is uniquely determined by its moments and the characteristic function $\varphi(z)$ is analytic in a neighborhood of the point $z = 0$. Prove this statement.

V.1.43. Prove that a probability distribution $F(x)$ with support in a finite interval (i.e., $F(a) = 0$ and $F(b) = 1$ for some $-\infty < a < b < \infty$) is uniquely determined by its moments.

V.1.44. Let ξ be a random variable and $\varphi(z)$ its characteristic function. Prove that if $\varphi''(0)$ exists, then ξ has the finite second moment.

V.1.45. Let ξ be a random variable and $\varphi(z)$ its characteristic function. Prove that if $\varphi^{(2k)}(0)$ exists, then the moment $\mathsf{E}\,|\xi|^{2k}$ is finite.

V.1.46. Let ξ be a random variable and $\varphi(z)$ its characteristic function. The existence of $\varphi'(0)$ does not imply $\mathsf{E}\,|\xi| < \infty$. Construct an appropriate example.

V.1.47. Let ξ be a random variable and $\varphi(z)$ its characteristic function. Prove that

a) if $\mathsf{E}\,\xi = 0$, then

$$\mathsf{E}\,|\xi| = \frac{1}{\pi}\int_{-\infty}^{\infty} \frac{1 - \operatorname{Re}\varphi(z)}{z^2}\,dz;$$

b) if there exists $\operatorname{Var}\xi$, then

$$\mathsf{E}\,|\xi| = -\frac{2}{\pi}\int_{0}^{\infty} \frac{\operatorname{Re}(\varphi'(z))}{z}\,dz.$$

V.1.48. Which of the following functions are characteristic: a) $\cos z$; b) $\sin z$; c) $(1+\cos z)/2$; d) e^{-z^4}; e) $\sin z/z$; f) $(1+z^2)^{-1}$; g) $\cos z^2$?

V.1.49. Prove that the function

$$\varphi(z) = \begin{cases} \sqrt{1-z^2}, & \text{for } |z| \le 1,\\ 0, & \text{for } |z| > 1 \end{cases}$$

is not characteristic.

V.1.50. A sequence of random variables $\{\xi_n, n \geq 1\}$ tends to zero in probability as $n \to \infty$ if and only if the corresponding sequence of characteristic functions $\{\varphi_n(z), n \geq 1\}$ is such that $\varphi_n(z) \to 1$ as $n \to \infty$, for all $z \in \mathbf{R}$. Prove this assertion.

V.1.51. Let $\{\varphi_n(z), n \geq 0\}$ be a sequence of characteristic functions such that $\varphi_n(z) \to 1$ as $n \to \infty$, for all z with $|z| < \Delta$. Prove that $\varphi_n(z) \to 1$ as $n \to \infty$ for all $z \in \mathbf{R}$.

V.1.52. Let $\{p_n, n \geq 1\}$ be a sequence of nonnegative numbers such that $\sum_{n=1}^{\infty} p_n = 1$, and $\{\varphi_n(z), n \geq 1\}$ a sequence of characteristic functions. Prove that $\varphi(z) = \sum_{n=1}^{\infty} \varphi_n(z) p_n$ is the characteristic function of a certain random variable.

V.1.53. Prove that any real-valued continuous function $\omega(z)$ is characteristic provided it satisfies the following conditions:

a) $\omega(z)$ is even and convex on $(0, \infty)$;

b) $\omega(0) = 1$;

c) $\lim_{z\to\infty} \omega(z) = 0$.

V.1.54. Let $\varphi_0(z)$ and $\{\varphi_n, n \geq 1\}$ be integrable characteristic functions, and let $f_0(x)$ and $\{f_n(x), n \geq 1\}$ be the corresponding probability densities, respectively. Prove that if

$$\int_{-\infty}^{\infty} |\varphi_n(z) - \varphi_0(z)|\, dz \to 0 \qquad \text{as } n \to \infty,$$

then

$$f_n(x) \to f_0(x) \qquad \text{as } n \to \infty$$

uniformly in $x \in \mathbf{R}$.

V.1.55. Let a probability density $f(x)$ be an even function, and let $\varphi(z)$ be the corresponding characteristic function. Assume that $\varphi(z)$ is strictly positive. Put

$$\varphi_a(z) = \frac{2\varphi(z) - \varphi(z+a) - \varphi(z-a)}{2(1-\varphi(a))}.$$

Check that $\varphi_z(a)$ is a characteristic function and prove that the corresponding probability density $f_a(x)$ is given by

$$f_a(x) = \frac{f(x)(1-\cos ax)}{1-\varphi(a)}.$$

Also show that $\varphi_a(\cdot) \to \varphi(\cdot)$ as $a \to \infty$, but $f_a(\cdot)$ does not tend to $f(\cdot)$.

V.1.56. Let $\varphi(z)$ be a characteristic function and $F(x)$ the corresponding probability distribution. Suppose that $|\varphi(z)| < 1$ for $0 < z < \lambda$, and $|\varphi(z)| = 1$ for other values of z. Prove that there exists a real number b such that

$$\sum_{k=-\infty}^{\infty} \left[F\left(b + k\frac{\lambda}{2\pi} + 0\right) - F\left(b + k\frac{\lambda}{2\pi}\right)\right] = 1.$$

V.1.57. Prove that under the conditions of the preceding problem, $\varphi(z)$ is a periodic function with the period λ.

V.1.58. Prove that if $|\varphi(z)| = 1$ for all $z \in \mathbf{R}$, then $\varphi(z) = e^{izb}$ for some $b \in \mathbf{R}$.

V.1.59. Prove that if $\varphi(z)$ is a characteristic function, then so is $e^{\varphi(z)-1}$.

V.1.60. Let $\varphi(z)$ be a real-valued characteristic function. Prove the following inequalities:

$$1-\varphi(2z)\le 4(1-\varphi(z)),\qquad 1+\varphi(2z)\ge 2\varphi^2(z).$$

V.1.61. Let $\varphi(z)$ be a real-valued characteristic function. Prove the following inequalities:

$$\begin{gathered}|\varphi(z_1)-\varphi(z_2)|\le\sqrt{2\operatorname{Re}(1-\varphi(z_1-z_2))},\\ 1-\varphi^2(z)\le 8(1-\varphi(z)),\\ 1-\operatorname{Re}\varphi(2z)\le 4(1-\operatorname{Re}\varphi(z)).\end{gathered}$$

V.1.62. Let $\xi_{p,n}$ be a binomial random variable with parameters p and n. Prove that if $p\to 0$ in such a way that $pn\to\lambda$, then the distributions of $\xi_{p,n}$ weakly converge to the distribution of a Poisson random variable ξ_λ with parameter λ.

V.1.63. Let ξ_λ be a Poisson random variable with parameter $\lambda<\infty$. Prove that the distributions of random variables $(\xi_\lambda-\lambda)/\sqrt{\lambda}$ weakly converge as $\lambda\to\infty$ to the distribution of a Gaussian random variable with parameters $N(0,1)$.

V.1.64. Let ν and $\{\xi_n, n\ge 1\}$ be jointly independent random variables. Assume that ν is integer-valued and $\{\xi_n, n\ge 1\}$ are identically distributed. Let $h(s)$ be the moment generating function of ν, and $\varphi(z)$ the characteristic function of ξ_1. Consider the random variable $S_\nu=\sum_{k=1}^{\nu}\xi_k$. Prove that the characteristic function of S_ν is

$$\psi(z)=h(\varphi(z)).$$

V.1.65. Let $\varphi(z)$ be a characteristic function. Prove that

$$\frac{p}{1-(1-p)\varphi(z)},\qquad p\in(0,1),$$

is also a characteristic function.

V.1.66. Let $\{\xi_n, n\ge 1\}$ be independent identically distributed random variables. Let a geometric random variable ν_p be independent of $\{\xi_n, n\ge 1\}$ and have parameter $p>0$. Consider the random variable $\zeta_p=\sum_{k=1}^{\nu_p}\xi_k$.

a) Let $\mathsf{E}\,\xi_1=a>0$. Prove that the distribution of the random variable $p\zeta_p$ weakly converges as $p\to\infty$, to the distribution of an exponential random variable ζ with parameter a^{-1}.

b) Let $\mathsf{E}\,\xi_1=0$ and $\operatorname{Var}\xi_1=\sigma^2>0$. Prove that the distribution of the random variable $\sqrt{p}\zeta_p$ weakly converges as $p\to\infty$ to the distribution of a random variable with the characteristic function

$$\frac{1}{1+\frac{1}{2}z^2\sigma^2}.$$

V.1.67. Let ξ and τ be independent random variables. Assume that ξ is Gaussian with parameters $N(0,\sigma^2)$, and τ is exponential with parameter λ. Calculate the characteristic function of the random variable $\sqrt{\tau}\xi+a\tau$.

V.1.68. Let ξ be a Poisson random variable with parameter λ, where λ is a random variable with the characteristic function $\Phi(t)$. Prove that

$$\Phi\left(\frac{e^{it}-1}{t}\right)$$

is the characteristic function of ξ.

V.1.69. Let ξ be a nonnegative random variable such that $\mathsf{E}\,\xi^k < \infty$. Prove that there exists $\varphi^{(k)}(s)$ and

$$(-1)^k \varphi^{(k)}(0) = \mathsf{E}\,\xi^k,$$

where $\varphi(s) = \mathsf{E}\,e^{-s\xi}$.

V.1.70. Let ξ be a nonnegative random variable and $\varphi(z)$ be the Laplace transform of its distribution. Prove that if $\mathsf{P}(\xi > 0) > 0$, then $\varphi(z) < 1$ for $z > 0$. Prove also that the Laplace transform of a random variable ξ taking nonnegative values and, possibly, the value $+\infty$, tends to 1 as $z \to 0$ if and only if $\mathsf{P}(\xi = +\infty) = 0$.

V.1.71. Let $\varphi(s)$ be the Laplace transform of a probability distribution $F(x)$. Prove that

$$\frac{\varphi(s+h)}{\varphi(h)}, \qquad h \geq 0,$$

is also the Laplace transform of a certain probability distribution $G(x)$.

V.1.72. Let $F(x)$ and $\varphi(s)$ be the distribution function and Laplace transform of a nonnegative random variable with the finite nth moment $\mu_n = \mathsf{E}\,\xi^n < \infty$. Prove that there exists $\varphi^{(n)}(s)$, and

$$(-1)^n \frac{\varphi^{(n)}(s)}{\mu_n}$$

is the Laplace transform of a distribution function $G(x)$. Determine $G(x)$.

V.1.73. Let ξ and η be nonnegative random variables with the Laplace transforms $\varphi(s)$ and e^{-s^α}, respectively. Prove that the Laplace transform of the random variable $\xi^{1/\alpha}\eta$ is $\varphi(s^\alpha)$.

V.1.74. Prove that the characteristic function $\varphi(z)$ of a random vector $\bar{\xi} = (\xi_1, \xi_2, \ldots, \xi_m)$ is positive definite, that is, for all n, all real vectors $\bar{z}_1, \ldots, \bar{z}_n \in \mathbf{R}_m$, and all complex numbers $\alpha_1, \ldots, \alpha_n$, we have

$$\sum_{k,j=1}^{n} \varphi\left(\bar{z}_k - \bar{z}_j\right) \alpha_k \bar{\alpha}_j \geq 0.$$

V.1.75. The characteristic function $\varphi(\bar{z})$ of a random vector is uniformly continuous and

$$|\varphi(\bar{z}_1) - \varphi(\bar{z}_2)| \leq \sqrt{2\,\mathrm{Re}(1 - \varphi(\bar{z}_1 - \bar{z}_2))}.$$

Prove this statement.

V.1.76. Prove that if a function $g(\bar{z})$ is positive definite and $\varphi(\bar{z})$ is the characteristic function of a random vector, then $g(\bar{z})\varphi(\bar{z})$ is also positive definite.

V.1.77. If a function $g(\bar{z})$ is continuous, positive definite, and absolutely integrable on $\mathbf{R}_m$, then the function

$$\rho(\bar{x}) = \int_{\mathbf{R}_m} e^{-i(\bar{z},\bar{x})} g(\bar{z})\, dz_1 \cdots dz_m$$

is nonnegative and integrable. Prove also that

$$g(\bar{z}) = \left(\frac{1}{2\pi}\right)^m \int_{\mathbf{R}_m} e^{i(\bar{z},\bar{x})} \rho(\bar{x})\, dx_1 \cdots dx_m.$$

V.1.78. Calculate the characteristic function of a two-dimensional Cauchy distribution that has the probability density

$$\rho(x_1, x_2) = \frac{1}{2\pi\sqrt{(t^2 + x_1^2 + x_2^2)^3}},$$

where $t > 0$.

V.1.79. Random variables have a joint probability density given by

$$\rho(x,y) = \begin{cases} \frac{1}{4}(1+xy)(x^2-y^2), & \text{for } (x,y) \in [-1,1] \times [-1,1], \\ 0, & \text{otherwise.} \end{cases}$$

Determine the characteristic function of the random vector (ξ, η).

V.1.80. Let $\varphi(z_1, \ldots, z_m)$ be the characteristic function of a random vector $\bar{\xi} = (\xi_1, \ldots, \xi_m)$. Let $\bar{\eta} = (\eta_1, \ldots, \eta_l)$ be the random vector obtained by a linear transformation of the vector $\bar{\xi}$, i.e.,

$$\eta_i = \xi_i a_{i1} + \cdots + \xi_m a_{im} + b_i, \qquad i = 1, \ldots, l.$$

Find the characteristic function of $\bar{\eta}$.

V.1.81. Prove that the characteristic functions $\varphi(z_1, \ldots, z_m)$ and $\varphi_{\bar{z}}(\lambda)$ of a random vector $\bar{\xi} = (\xi_1, \ldots, \xi_m)$ and a random variable $\eta_{\bar{z}} = z_1\xi_1 + \cdots + z_m\xi_m$ are related by

$$\varphi(\lambda z_1, \ldots, \lambda z_m) = \varphi_{\bar{z}}(\lambda).$$

V.1.82. The probability density of a random vector $\bar{\xi} = (\xi_1, \ldots, \xi_m)$ is given by

$$p(x_1, \ldots, x_m) = \sqrt{\frac{\det R^*}{(2\pi)^m}} \exp\left\{-\frac{1}{2} Q(x_1, \ldots, x_m)\right\}.$$

Here $\mu_i = \mathsf{E}\,\xi_i$, $i = 1, \ldots, m$, R^* is the inverse matrix of

$$R = \|r_{ij}\| = \big\| \mathsf{E}(\xi_i - \mu_i)(\xi_j - \mu_j) \big\|_{i,j=1}^m,$$

and

$$Q(x_1, \ldots, x_m) = \sum_{i,j=1}^m r_{ij}^*(x_i - \mu_i)(x_j - \mu_j).$$

Prove that the characteristic function of $\bar{\xi}$ is

$$\varphi(z_1, \ldots, z_m) = \exp\left\{ i \sum_{j=1}^m z_j \mu_j - \frac{1}{2} \sum_{i,j=1}^m r_{ij} z_i z_j \right\}.$$

V.1.83. Let ξ_1, ξ_2, ξ_3 be Gaussian independent random variables with parameters $N(0,1)$. Calculate the characteristic functions of the random vector (η_1, η_2), where

$$\eta_1 = \xi_1 + \xi_2, \qquad \eta_2 = \xi_1 + \xi_3.$$

V.1.84. Prove that if for a random vector $\bar{\xi} = (\xi_1, \dots, \xi_m)$ we have

$$\mathsf{E}\left|\xi_{i_1}^{k_1} \cdots \xi_{i_l}^{k_l}\right| < \infty,$$

then there exists the derivative

$$\frac{\partial^{k_1+\cdots+k_l} \varphi(z_1, \dots, z_m)}{\partial z_{i_1}^{k_1} \cdots \partial z_{i_l}^{k_l}}$$

of the characteristic function $\varphi(z_1, \dots, z_m)$ of the vector $\bar{\xi}$, and

$$i^{k_1+\cdots+k_l}\, \mathsf{E}\, \xi_{i_1}^{k_1} \cdots \xi_{i_l}^{k_l} = \left.\frac{\partial^{k_1+\cdots+k_l} \varphi(z_1, \dots, z_m)}{\partial z_{i_1}^{k_1} \cdots \partial z_{i_l}^{k_l}}\right|_{z_1=\cdots=z_m=0}.$$

V.1.85. Random variables ξ_1, ξ_2, ξ_3, and ξ_4 have a joint Gaussian distribution,

$$\begin{aligned} \mathsf{E}\,\xi_i &= 0, \qquad i = 1,2,3,4; \\ \mathsf{E}\,\xi_i\xi_j &= r_{ij}, \qquad i,j = 1,2,3,4. \end{aligned}$$

Find the following moments: a) $\mathsf{E}\,\xi_1\xi_2\xi_3\xi_4$; b) $\mathsf{E}\,\xi_1\xi_2\xi_3$; c) $\mathsf{E}(\xi_1\xi_2\xi_3)^2$.

§V.2. Central limit theorems

Let $\xi_1, \xi_2, \dots$ be random variables and $S_n = \xi_1 + \cdots + \xi_n$, $n \ge 2$. Under rather general conditions, the Gaussian distribution is the limit of distributions of sums

$$\frac{S_n - A_n}{B_n}$$

centered and normalized in a proper way. More precisely, for $x \in \mathbf{R}$,

$$\mathsf{P}\left(\frac{S_n - A_n}{B_n} < x\right) \to \Phi(x) = \frac{1}{\sqrt{2\pi}} \int_{-\infty}^{x} e^{-\frac{u^2}{2}}\, du \tag{1}$$

as $n \to \infty$.

Theorems asserting that the limit of distributions of sums of independent random variables is a Gaussian distribution, are called *central limit theorems*.

Basic tools in proving central limit theorems include a method based on the continuity theorem. According to this theorem, the relation (1) is equivalent to the convergence of the corresponding characteristic functions

$$\mathsf{E} \exp\left\{iz \frac{S_n - A_n}{B_n}\right\} \to \exp\left\{-\frac{z^2}{2}\right\}$$

as $n \to \infty$ for all $z \in \mathbf{R}$.

So-called *local limit theorems* are also of theoretical and practical value. Local limit theorems deal with random variables having either arithmetic distributions or probability density. In the case of random variables with arithmetic distributions, local limit theorems describe the asymptotic behavior of the probability that a sum of independent random variables takes a given value. Conditions for probability densities to converge to the Gaussian density are studied in local limit theorems for probability densities.

Problems

V.2.1. Let $\{\xi_n, n \geq 1\}$ be a sequence of independent identically distributed random variables such that $\mathsf{E}\,\xi_1 = a$ and $\operatorname{Var}\xi_1 = \sigma^2 < \infty$. Prove that the central limit theorem holds for sums S_n with $A_n = an$ and $B_n = \sqrt{n}\sigma$.

V.2.2. Let D be a domain of volume v in l-dimensional space, and $f(x_1, \ldots, x_l)$ a function defined on D and such that $|f(x)| \leq h$. To calculate the integral $\int_D f(x_1, \ldots, x_l)\,dx_1 \cdots dx_l$ by *the Monte Carlo method* one takes at random and independently n points $\bar{x}_1, \ldots, \bar{x}_n$ and one regards

$$I_n = \frac{1}{n}\sum_{k=1}^{n} f(\bar{x}_k)$$

as an approximate value of the integral. Find $\mathsf{E}\,I_n$. Estimate $\operatorname{Var} I_n$ and determine the limit distribution for $\sqrt{n}(I_n - I)$ as $n \to \infty$.

V.2.3. Consider a sequence of independent trials. Any trial may lead to either "success" or "failure". The probability of "success" in a trial is p. Let $\nu_N(p)$ be the frequency of "successes" in N trials. Prove that $\mathsf{E}\,\nu_N(p) = p$ and $\sup_p \operatorname{Var}\nu_N(p) = \frac{1}{4N}$.

Using the central limit theorem, find an estimate for the probability of the event that the absolute value of the difference between $\nu_N(p)$ and p is at most ε. Determine N such that this probability for $\varepsilon = 0.01$ does not exceed a) 0.1; b) 0.05; c) 0.01.

V.2.4. Let $\{\xi_n, n \geq 1\}$ be a sequence of independent identically distributed symmetric random variables. In order that there exist a sequence $\{B_n, n \geq 1\}$ such that the sequence $\frac{1}{n}\sum_{k=1}^{n}\xi_k$ has the Gaussian limit distribution, it is sufficient that

$$\lim_{t\to\infty}\frac{t^2\int_{|x|>t} dF(x)}{\int_{|x|\leq t} x^2\,dF(x)} = 0,$$

where $F(x) = \mathsf{P}(\xi_1 < x)$.

V.2.5. Let $\{\xi_n, n \geq 1\}$ be a sequence of independent identically distributed random variables such that $\mathsf{E}\,\xi_1 = 0$ and $\operatorname{Var}\xi_1 = \sigma^2 < \infty$. For each m let ν_m be a random variable independent of the sequence $\{\xi_n, n \geq 1\}$. Assume that ν_n, $n \geq 1$, is integer-valued and

$$\mathsf{P}\left(\frac{\nu_n}{k_n} \leq x\right) \to F(x) \quad \text{as } n \to \infty$$

at points x where the limit function $F(x)$ is continuous. Here $\{k_n\}$ is a sequence of real numbers such that $k_n \to \infty$ as $n \to \infty$. If $F(0) = 0$, then for all $x \in \mathbf{R}$,

$$\mathsf{P}\biggl(\frac{1}{\sqrt{k_n}}\sum_{k=1}^{\nu_n}\xi_k < x\biggr) \to G(x)$$

as $n \to \infty$, where $G(x)$ is the distribution function with the characteristic function

$$\psi(z) = \int_0^\infty e^{-z^2\sigma^2 x/2}\, dF(x).$$

V.2.6. Let ξ be a random variable, $F(x)$ its distribution function, and $\varphi(z)$ its characteristic function. Assume that there exist $\mathsf{E}\,\xi = a$ and $\operatorname{Var}\xi = b$.

a) Prove that

$$\varphi(z)e^{-iza} = 1 - \frac{bz^2}{2} + z^2\alpha(z),$$

where

$$\alpha(z) = \frac{1}{z^2}\int_{-\infty}^{\infty}\left(e^{iz(x-a)} - 1 - iz(x-a) + \frac{z^2(x-a)^2}{2}\right) dF(x).$$

b) Prove that for all $\varepsilon > 0$ and $B > 0$,

$$\alpha(z) \le \int_{|x|\le\varepsilon B}(x-a)^2\, dF(x) + \frac{\varepsilon B|z|}{6} b.$$

c) Let $\xi_1, \xi_2, \dots$ be independent random variables such that $\mathsf{E}\,\xi_k = a_k$ and $\operatorname{Var}\xi_k = b_k^2$ exist. In order that the central limit theorem be valid for sums S_n with $A_n = \sum_{k=1}^n a_k$ and $B_n = \left(\sum_{k=1}^n b_k^2\right)^{1/2}$, it is sufficient that

$$\lim_{n\to\infty}\frac{1}{B_n^2}\sum_{k=1}^n\int_{|x-a_k|>\varepsilon B_n}(x-a_k)^2\, dF_k(x) = 0,$$

where $F_k(x) = \mathsf{P}(\xi_k < x)$.

d) Suppose for some $\delta > 0$ there exists

$$\mathsf{E}\,|\xi_k - a_k|^{2+\delta} = c_k < \infty, \qquad k \ge 1.$$

In order that the central limit theorem be valid for sums S_n with $A_n = \sum_{k=1}^n a_k$ and $B_n = \left(\sum_{k=1}^n b_k^2\right)^{1/2}$, it is sufficient that

$$\lim_{n\to\infty}\frac{1}{B_n^{2+\delta}}\sum_{k=1}^n c_k = 0.$$

REMARK. The statements c) and d) above are called *the Lindeberg theorem* and *the Lyapunov theorem*, respectively.

V.2.7. Let ξ_n, $n \ge 1$, be independent random variables such that

$$\mathsf{P}(\xi_k = \pm 1) = \frac{1}{2}\left(1 - \frac{1}{k^2}\right) \quad\text{and}\quad \mathsf{P}(\xi_k = \pm\sqrt{k}\,) = \frac{1}{4k^2}.$$

Is the central limit theorem valid for sums S_n?

V.2.8. Let ξ_n, $n \ge 1$, be independent random variables such that

$$\mathsf{P}(\xi_k = \pm 1) = \frac{1}{2}\left(1 - \frac{1}{k}\right) \quad\text{and}\quad \mathsf{P}(\xi_k = \pm\sqrt{k}\,) = \frac{1}{2k}.$$

Prove that the distribution of S_n/B_n does not converge to $\Phi(x)$ whatever normalizing constants B_n.

V.2.9. Let ξ_n, $n \geq 1$, be independent random variables with the distribution $\mathsf{P}(\xi_k = \pm 2^k) = \frac{1}{2}$. Is the central limit theorem valid for sums S_n?

V.2.10. Let ξ_n, $n \geq 1$, be independent random variables with the distribution $\mathsf{P}(\xi_k = \pm x_k) = \frac{1}{2}$. What restriction should be imposed on $\{x_n, n \geq 1\}$ in order that the central limit theorem for S_n/B_n with some sequence $\{B_n, n \geq 1\}$ hold? Prove that under this restriction,

$$\frac{1}{B_n} \max_{1 \leq k \leq n} |x_k| \to 0 \qquad \text{as } n \to \infty.$$

V.2.11. A number ξ is taken at random from the segment $[0, 1]$. Consider the decimal expansion of this number:

$$\xi = \sum_{n=1}^{\infty} \frac{e_n(\xi)}{10^n}.$$

Prove that there exists a sequence $\{B_n, n \geq 1\}$ such that the distribution of $(1/B_n) \sum_{k=1}^{n} e_k(\xi)$ tends to the Gaussian distribution.

V.2.12. Let $\bar{\xi}_n = (\eta_n^{(1)}, \ldots, \eta_n^{(m)})$, $n \geq 1$, be a sequence of independent random vectors and $S_n = \xi_1 + \cdots + \xi_n$. Prove that the distribution of vectors S_n/B_n tends to the Gaussian distribution with the characteristic function

$$\varphi(\bar{z}) = \exp\left\{ i\,(\bar{z}, \bar{a}) - \frac{1}{2}(B\bar{z}, \bar{z}) \right\}$$

as $n \to \infty$ if and only if for all $\bar{z} = (z_1, \ldots, z_m) \in \mathbf{R}_m$ the distribution function of random variables

$$S_n(z) = \sum_{k=1}^{n} \frac{(\bar{\xi}_k, \bar{z})}{B_n}$$

tends, as $n \to \infty$, to the Gaussian distribution on the line with the mean $(\bar{a}, \bar{z})$ and variance $(B\bar{z}, \bar{z})$. Here $\bar{a} = (a_1, \ldots, a_m)$ is the vector of expectations of the Gaussian random vector, and $B = \left\| b_{ij} \right\|_{i,j=1}^{m}$ is its covariance matrix.

V.2.13. Let $\{\xi_n, n \geq 1\}$ be a sequence of independent identically distributed random variables such that $\mathsf{E}\,\xi_1 = 0$ and $\operatorname{Var} \xi_1 = 1$. Prove that for all $0 < t_1 < \cdots < t_m$ the joint distribution function of random variables

$$\zeta_n(t_i) = \frac{1}{\sqrt{n}} \sum_{k \leq t_i n} \xi_k, \qquad i = 1, \ldots, m,$$

converges as $n \to \infty$ to the distribution of the Gaussian random vector with zero mean and the covariance matrix

$$B(\bar{t}) = \left\| \min(t_i, t_j) \right\|_{i,j=1}^{m}.$$

V.2.14. For each $n \geq 1$ let $\xi_{1n}, \ldots, \xi_{k_n n}$ be a series of independent random variables and $k_n \to \infty$ as $n \to \infty$. Put

$$S_n = \xi_{1n} + \cdots + \xi_{k_n n}$$

and

$$S_{\varepsilon n} = \xi_{1n}^{(\varepsilon)} + \cdots + \xi_{k_n n}^{(\varepsilon)},$$

where

$$\xi^{(\varepsilon)} = \begin{cases} 0 & \text{if } |\xi| > \varepsilon, \\ \xi & \text{if } |\xi| \le \varepsilon. \end{cases}$$

a) Prove that there exists a sequence of positive numbers $\varepsilon_n \downarrow$ as $n \to \infty$ such that

(a) $$\sum_{k=1}^{k_n} \mathsf{P}(|\xi_{kn}| > \varepsilon_n) \to 0 \quad \text{as } n \to \infty,$$

provided that for any fixed $\varepsilon > 0$,

$$\sum_{k=1}^{k_n} \mathsf{P}(|\xi_{kn}| > \varepsilon) \to 0 \quad \text{as } n \to \infty.$$

b) Prove that, under the condition (a),

$$S_n - S_{\varepsilon_n n} \to 0 \qquad \text{as } n \to \infty$$

in probability.

c) Prove that if (a) is satisfied for some sequence $\{\varepsilon_n, n \ge 1\}$, then for each $\varepsilon > 0$

$$\mathsf{E}\, S_{\varepsilon n} - \mathsf{E}\, S_{\varepsilon_n n} \to 0 \quad \text{as } n \to \infty,$$
$$\operatorname{Var} S_{\varepsilon n} - \operatorname{Var} S_{\varepsilon_n n} \to 0 \quad \text{as } n \to \infty.$$

V.2.15. Central limit theorem in a scheme of series. For each $n \ge 1$ let $\xi_{1n}, \dots, \xi_{k_n n}$ be a series of independent random variables and $k_n \to \infty$ as $n \to \infty$. Put

$$S_n = \xi_{1n} + \dots + \xi_{k_n n}.$$

Show that

$$\mathsf{P}(S_n < x) \to \frac{1}{\sqrt{2\pi}} \int_{-\infty}^{x} \exp\left\{-\frac{(u-a)^2}{2\sigma^2}\right\} du = \Phi_{a,\sigma^2}(x)$$

provided that

(1) $$\sum_{k=1}^{k_n} \mathsf{P}(|\xi_{kn}| > \varepsilon) \to 0 \quad \text{as } n \to \infty,$$

(2) $$\sum_{k=1}^{k_n} \mathsf{E}\, \xi_{kn}^{(\varepsilon)} \to a \quad \text{as } n \to \infty,$$

(3) $$\sum_{k=1}^{k_n} \operatorname{Var} \xi_{kn}^{(\varepsilon)} \to \sigma^2 \quad \text{as } n \to \infty,$$

where $-\infty < a < \infty$, $\sigma^2 \ge 0$.

REMARK. Conditions (1)–(3) are called *the central criteria for the normal convergence*. These conditions are not only sufficient but also necessary for the convergence of distributions of sums S_n to the Gaussian law as $n \to \infty$ under the so-called *infinitesimal condition*:

$$\max_{1\le k\le k_n} \mathsf{P}(|\xi_{kn}| > \varepsilon) \to 0 \quad \text{as } n \to \infty \text{ for all } \varepsilon > 0.$$

For $\sigma^2 = 0$, conditions (1)–(3) gives a convergence criteria to the degenerate law.

V.2.16. If condition (2) in the central criteria for the normal convergence is omitted, then

$$\mathsf{P}(S_n - a_n < x) \to \Phi_{0,\sigma^2}(x),$$

where for a fixed $\varepsilon > 0$ we put

$$a_n = \sum_{k=1}^{k_n} \mathsf{E}\,\xi_{kn}^{(\varepsilon)}.$$

Prove this statement.

V.2.17. For each $n \ge 1$ let $\xi_{1n}, \dots, \xi_{k_n n}$ be a series of independent random variables and $k_n \to \infty$ as $n \to \infty$.

a) Let $\mathsf{E}\,\xi_{kn}^2 < \infty$, $k = 1, \dots, k_n$, and

$$\sum_{k=1}^{k_n} \operatorname{Var} \xi_{kn} \to 1 \quad \text{as } n \to \infty,$$

$$\lim_{n\to\infty} \sum_{k=1}^{k_n} \int_{|x|>\varepsilon} (x - \mathsf{E}\,\xi_{kn})^2 \, dF_{kn}(x) = 0 \tag{a}$$

for a fixed $\varepsilon > 0$. Here $F_{kn}(x)$ is the distribution function of ξ_{kn}. Put

$$S_n = \sum_{k=1}^{k_n} \xi_{kn}, \qquad a_n = \sum_{k=1}^{k_n} \mathsf{E}\,\xi_{kn}.$$

Then

$$\mathsf{P}(S_n - a_n < x) \to \Phi_{0,1}(x).$$

b) Show that condition (a) can be replaced by

$$\lim_{n\to\infty} \sum_{k=1}^{k_n} \mathsf{E}\,|\xi_{kn} - \mathsf{E}\,\xi_{kn}|^{2+\delta} = 0 \quad \text{for some } \delta > 0.$$

V.2.18. For each $n \ge 1$ let $\xi_{1n}, \dots, \xi_{k_n n}$ be a series of independent random variables and $k_n \to \infty$ as $n \to \infty$. Assume that $\sum_{k=1}^{k_n} \xi_{kn} \to 0$ in probability as $n \to \infty$. Then for any subsequence of natural numbers $\{l_n, n \ge 1\}$, $l_n \le k_n$, there exists a sequence $\{\alpha_n, n \ge 1\}$ such that

$$\sum_{k=1}^{l_n} \xi_{kn} - \alpha_n \to 0$$

in probability as $n \to \infty$. Prove this statement.

V.2.19. For each $n \geq 1$ let $\xi_{1n}, \dots, \xi_{k_n n}$ be a series of independent nonnegative random variables and $k_n \to \infty$ as $n \to \infty$. In order that $\sum_{k=1}^{k_n} \xi_{kn} \to 0$ in probability as $n \to \infty$, it is necessary and sufficient that for all $\varepsilon > 0$

$$\text{a)} \quad \lim_{n\to\infty} \sum_{k=1}^{k_n} \mathsf{P}(|\xi_{kn}| > \varepsilon) = 0,$$

$$\text{b)} \quad \lim_{n\to\infty} \sum_{k=1}^{k_n} \int_{|x|\leq\varepsilon} x\, dF_{kn}(x) = 0,$$

where $F_{kn}(x)$ is the distribution function of ξ_{kn}.

V.2.20. Prove that conditions a) and b) in the preceding problem are necessary for the convergence $\sum_{k=1}^{k_n} \xi_{kn} \to 0$ in probability as $n \to \infty$.

V.2.21. Let a random variable ξ have a distribution density and $\mathsf{E}\,\xi = 0$, $\operatorname{Var}\xi = b^2 < \infty$. Prove that there exists $\delta > 0$ such that

$$\sup_{|z|>u} |\varphi(z)| \leq e^{-\delta u^2} \quad \text{for } 0 < u < 1$$

(here $\varphi(z)$ is the characteristic function of ξ).

V.2.22. a) Let $\{\xi_n, n \geq 1\}$ be a sequence of independent identically distributed random variables such that $\mathsf{E}\,\xi_1 = 0$, $\operatorname{Var}\xi_1 = 1$, and let $p(x)$ be their common distribution density. Assume that $\varphi(z)$ is the characteristic function of ξ_1 such that $|\varphi(z)|^\nu$ is integrable on $(-\infty, \infty)$ for some $\nu \geq 1$. Prove that

$$\text{(a)} \qquad p_n(x) \to \frac{1}{2\pi} e^{-x^2/2} \quad \text{as } n \to \infty,$$

where $p_n(x)$ is the distribution density for the random variable $n^{-1/2}(\xi_1 + \cdots + \xi_n)$.

b) Let $\{\xi_n, n \geq 1\}$ be a sequence of independent identically distributed random variables such that $\mathsf{E}\,\xi_1 = 0$, $\operatorname{Var}\xi_1 = 1$, and their common distribution density $p(x)$ is bounded. Prove (a) in this case. Prove also that (a) holds if the assumption of boundedness is replaced by a weaker one:

1) there exists n_0 such that the random variable $\sum_{k=1}^{n_0} \xi_k$ has a bounded distribution density; or
2) $\int_{-\infty}^{\infty} p^2(x)\, dx < \infty$.

V.2.23. Prove that under the conditions of Problem V.2.22 b), the density $p_n(x)$ tends to the Gaussian density uniformly on any finite interval and

$$\lim_{n\to\infty} \int_{-\infty}^{\infty} \left| p_n(x) - \frac{1}{\sqrt{2\pi}} e^{-x^2/2} \right| dx = 0.$$

V.2.24. Prove that if $\mathsf{E}\,|\xi_1|^{2+\delta} < \infty$ for some $0 < \delta < 1$ and the conditions of Problem V.2.22 b) are satisfied, then

$$\left| p_n(x) - \frac{1}{\sqrt{2\pi}} e^{-x^2/2} \right| \leq \frac{c}{n^{\delta/2}},$$

where c is a constant.

V.2.25. Let the distribution function $F(x)$ of a random variable ξ be of the form $F(x) = pF_1(x) + qF_2(x)$ for some positive p, q such that $p + q = 1$, where $F_1(x)$ and $F_2(x)$ are distribution functions. Assume that F_1 has the distribution density. Let

$$\int_{-\infty}^{\infty} x\,dF_1(x) = \int_{-\infty}^{\infty} x\,dF_2(x) = 0,$$
$$\int_{-\infty}^{\infty} x^2\,dF_1(x) = \int_{-\infty}^{\infty} x^2\,dF_2(x) = 1.$$

Denote the density of $F_1^{*n}(x)$ by $f_n^{(1)}(x)$, where $F_1^{*n}(x)$ is the n-fold convolution of F_1 with itself. Let $f_n(x)$ be the density of the absolutely continuous component of the distribution function $F^{*n}(x)$. Prove that

$$f_n(x) \geq \sum_{k=1}^{n} \binom{n}{k} p^k q^{n-k} \int_{-\infty}^{\infty} f_k^{(1)}(x-y)\,dF_2^{*(n-k)}(y).$$

V.2.26. a) Prove that under the conditions of Problem V.2.25,

$$\lim_{n\to\infty} \int_{-\infty}^{\infty} \sum_{\left|\frac{k}{n}-1\right|\geq\varepsilon} \binom{n}{k} p^k q^{n-k} \int_{-\infty}^{\infty} f_k^{(1)}(x-y)\,dF_2^{*(n-k)}(y) = 0$$

for all $\varepsilon > 0$.

b) Prove that if the conditions of Problem V.2.25 are satisfied and the function $p(x)$ is square integrable, then

$$\lim_{m,n\to\infty} \sqrt{m+n} \int_{\left|\frac{m}{n}\right|<\frac{1}{\delta}} f_n^{(1)}\left(x\sqrt{m+n} - y\right)\,dF_2^{*m}(y) = \frac{1}{\sqrt{2\pi}} e^{-x^2/2}.$$

V.2.27. a) Using Problems V.2.25 and V.2.26 prove that for the density $p_n(x)$ of an absolutely continuous component of the distribution function of

$$\frac{\xi_1 + \cdots + \xi_n}{\sqrt{n}}$$

we have

$$\lim_{n\to\infty} \int_{-\infty}^{\infty} \left| p_n(x) - \frac{1}{\sqrt{2\pi}} e^{-x^2/2} \right| dx = 0,$$

where $\{\xi_n, n \geq 1\}$ is a sequence of independent identically distributed random variables such that $\mathsf{E}\,\xi_1 = 0$, $\operatorname{Var}\xi_1 = 1$, and their distribution function $F(x)$ has a nonzero absolutely continuous component.

b) Prove the same equality in the case where there exists n_0 such that the distribution function of the random variable $\xi_1 + \cdots + \xi_{n_0}$ has an absolutely continuous component.

V.2.28. Random variables ξ_k, $k \geq 1$, are independent and

$$\mathsf{P}(\xi_k = 0) = q_k, \qquad \mathsf{P}(\xi_k = 1) = p_k = 1 - q_k.$$

Put $S_n = \xi_1 + \cdots + \xi_n$ and assume that $\sum_{n=1}^{\infty} p_n q_n = +\infty$. Prove that for any constant c we have

$$\mathsf{P}(S_n = m) \sim \frac{1}{\sqrt{2\pi b_n}} \exp\left\{ \frac{(m-a_n)^2}{2b_n} \right\}$$

for $|m - a_n| \le cb_n$, where $a_n = \sum_{k=1}^n p_k$, $b_n = \sum_{k=1}^n p_k q_k$.

V.2.29. Give an asymptotic expansion for the probability

$$P_n(m) = \mathsf{P}\biggl(\sum_{k=1}^n \xi_k = m\biggr),$$

where ξ_k, $k \ge 1$, are independent random variables such that

$$\mathsf{P}(\xi_k = 0) = q, \qquad \mathsf{P}(\xi_k = 1) = p = 1 - q.$$

V.2.30. Random variables ξ_k, $k \ge 1$, have the distribution with the density

$$p(x) = \frac{1}{2\sqrt{2\pi}}\left(e^{-(x-1)^2/2} + e^{-(x+1)^2/2}\right).$$

Give an asymptotic expansion for the density $p_n(x)$ of the random variable

$$\frac{\xi_1 + \cdots + \xi_n}{\sqrt{2n}}.$$

§V.3. Infinitely divisible and stable distributions

A distribution function $F(x)$ is called *an infinitely divisible distribution*[1] if it can be represented for all $n \ge 1$ as a n-fold convolution of a distribution function $F_n(x)$, that is,

$$F(x) = F_n^{*n}(x).$$

In terms of characteristic functions, this means that the characteristic function $\varphi(z)$ of an infinitely divisible distribution function is represented as $\varphi(z) = \varphi_n^n(z)$, where $\varphi_n(z)$ is a characteristic function. The characteristic function of an infinitely divisible distribution function is called *an infinitely divisible characteristic function*.

THEOREM 1. *If $\varphi(z)$ is an infinitely divisible characteristic function, then $\varphi(z)$ does not vanish for any $z \in \mathbf{R}$, and its logarithm can be uniquely represented in the form*

$$\text{(a)} \qquad \ln\varphi(z) = iza + \int_{-\infty}^{\infty}\left(e^{iza} - 1 - \frac{izx}{1+x^2}\right)\frac{1+x^2}{x^2}\,d\psi(x),$$

where $a \in \mathbf{R}$, $\psi(z)$ is a nondecreasing right continuous function such that $\psi(-\infty) = 0$, $\psi(+\infty) < \infty$, and the expression under the integral sign is $-z^2/2$ at $x = 0$.

Representation (a) is called *the canonical representation of the logarithm of an infinitely divisible law*.

Infinitely divisible laws play an important role in probability theory. In particular, they are the only limit distributions for sums of independent random variables in schemes of series.

For each $n \ge 1$ let $\xi_{1n}, \dots, \xi_{k_n n}$ be independent random variables such that for all $\varepsilon > 0$,

$$\max_{1\le k\le k_n} \mathsf{P}(|\xi_{kn}| > \varepsilon) \to 0$$

as $n \to \infty$ and $k_n \to \infty$ as $n \to \infty$. The condition above is called *the infinitesimal condition*.

[1] Or an infinitely divisible law.

THEOREM 2 (The central criterion for convergence). *The family of limit laws for the sums of independent random variables satisfying the infinitesimal condition coincides with the family of infinitely divisible laws.*

In order that distributions of sums S_n weakly converge to an infinitely divisible law with the logarithm of its characteristic function given by the canonical representation (a), *it is necessary and sufficient that*

a) *At any continuity point x of $\psi(x)$,*

$$\sum_{k=1}^{k_n} F_{kn}(x) \to \int_{-\infty}^{x} \frac{1+y^2}{y^2}\, d\psi(y) \quad \textit{for } x<0,$$

$$\sum_{k=1}^{k_n} (1-F_{kn}(x)) \to \int_{x}^{\infty} \frac{1+y^2}{y^2}\, d\psi(y) \quad \textit{for } x>0.$$

Here $F_{kn}(x) = \mathsf{P}(\xi_{kn} < x)$.

b) *For any fixed $\tau > 0$ such that $\pm\tau$ are continuity points of $\psi(x)$,*

$$\sum_{k=1}^{k_n} \int_{|x|<\tau} dF_{nk}(x) \to a + \int_{|y|<\tau} y\, d\psi(y) - \int_{|y|>\tau} \frac{1}{y}\, d\psi(y)$$

as $n \to \infty$.

c) *For $\xi_{kn}^{(\varepsilon)} = \xi_{kn} I(|\xi_{kn}| < \varepsilon)$, where $I(A)$ stands for the indicator function of an event A, we have*

$$\lim_{\varepsilon\to 0} \limsup_{n\to\infty} \left[\sum_{k=1}^{k_n} \operatorname{Var} \xi_{kn}^{(\varepsilon)} - (\psi(0+)+\psi(0-))\right] = 0.$$

Infinitely divisible laws are the distributions of a wide class of processes with independent increments. Stable laws play an important role among infinitely divisible distributions. A probability law with the characteristic function $\varphi(z)$ is said to be *stable* if for all $b', b'' > 0$ there exist $a, b > 0$ such that

$$\varphi(b'z)\,\varphi(b''z) = e^{-iza}\varphi(bz).$$

The characteristic function of a stable law also is called stable.

THEOREM 3. *In order that a characteristic function $\varphi(z)$ be stable, it is necessary and sufficient that its logarithm be represented in the form*

$$\text{(b)} \qquad \ln \varphi(z) = iaz - c|z|^{\alpha}\left(1 - i\beta \frac{z}{|z|}\omega(z,\alpha)\right),$$

where $a \in \mathbf{R}$, $c \geq 0$, $-1 \leq \beta \leq 1$, $0 < \alpha \leq 2$, and

$$\omega(z,\alpha) = \begin{cases} \tan \frac{\pi}{2}\alpha, & \textit{for } \alpha \neq 1, \\ \frac{2}{\pi} \ln|z|, & \textit{for } \alpha = 1. \end{cases}$$

The parameter α is said to be *the characteristic exponent of a stable law.*

Stable distributions are the only limit laws for normalized sums of independent identically distributed random variables. Let $\{\xi_n, n \geq 1\}$ be a sequence of independent identically distributed random variables and

$$\zeta_n = \frac{\xi_1 + \cdots + \xi_n}{B_n} - A_n,$$

where $B_n > 0$ and A_n are normalizing and centering sequences of real numbers.

THEOREM 4. *The family of limit laws for distributions of sums of independent and identically distributed random variables ζ_n coincides with the family of stable laws.*

In order that the distributions of ζ_n converge to the Gaussian law with the characteristic function $(2\pi)^{-1/2}\exp\{-z^2/2\}$, it is necessary and sufficient that

$$\lim_{x\to+\infty}\frac{x^2\int_{|y|>x}dF(y)}{\int_{|y|\le x}y^2dF(y)}=0,$$

where $F(x)=\mathsf{P}(\xi_1<x)$. We can take $A_n=n\,\mathsf{E}\,\xi_1$ and choose B_n in such a way that $B_n\to\infty$ as $n\to\infty$ and

$$\lim_{n\to\infty}\left[\frac{n}{B_n^2}-\int_{|x-\mathsf{E}\,\xi_1|\le B_n}(x-\mathsf{E}\,\xi_1)^2\,dF(x)\right]=1.$$

In order that the distributions of ζ_n converge to a stable law with characteristic exponent $\alpha<2$, it is necessary and sufficient that

1) $$\frac{1-F(x)}{1-F(x)+F(-x)}\to p \quad \text{as } x\to+\infty,$$
$$\frac{F(-x)}{1-F(x)+F(-x)}\to q \quad \text{as } x\to+\infty, \text{ and}$$

2) $$1-F(x)+F(-x)\sim\frac{2-\alpha}{\alpha}x^{-\alpha}L(x) \quad \text{as } x\to+\infty,$$

where $L(x)$ is a slowly varying function, that is, a function such that for any fixed $y>0$,

$$\frac{L(yx)}{L(x)}\to 1 \quad \text{as } x\to+\infty.$$

The sequences $\{A_n, n\ge1\}$ and $\{B_n, n\ge1\}$ can be chosen to satisfy the conditions

$$\lim_{n\to\infty}n\big(F(-B_n)+1-F(B_n)\big)=1,$$

$$A_n=\begin{cases}0, & \text{for } 0<\alpha<1,\\ n\int_{|x|<B_n}x\,dF(x), & \text{for } \alpha=1,\\ n\,\mathsf{E}\,\xi_1, & \text{for } 1<\alpha<2.\end{cases}$$

Problems

V.3.1. Prove that any infinitely divisible characteristic function $\varphi(z)$ has the following property: $\varphi(z)\ne0$ for $z\in\mathbf{R}$.

V.3.2. Prove that the following probability distributions are infinitely divisible:

a) the Gaussian distribution,
b) the Poisson distribution,
c) the degenerate distribution,
d) the exponential distribution.

Determine a and $\psi(z)$ in these cases.

V.3.3. Prove that

$$\varphi(z) = \frac{1}{(1 - iz)^\alpha}$$

is an infinitely divisible characteristic function for all $\alpha > 0$. Find the canonical representation for $\ln \varphi(z)$.

V.3.4. Prove that $\varphi(z) = e^{-|z|^\alpha}$ is an infinitely divisible characteristic function for all $0 \le \alpha < 2$. Find the canonical representation for $\ln \varphi(z)$.

V.3.5. Prove that a Cauchy random variable is infinitely divisible. The Cauchy distribution function is given by

$$F(x) = \frac{1}{\pi}\left(\frac{\pi}{2} + \arctan \frac{x - b}{a}\right).$$

V.3.6. Using the representation (a) (see the beginning of this section), obtain the following representation of an infinitely divisible characteristic function:

$$\ln \varphi(z) = ibz - \frac{\sigma^2}{2} z^2 + \int_{-\infty}^{0} \left(e^{izx} - 1 - \frac{izx}{1+x^2}\right) dM(x)$$
$$+ \int_{0}^{\infty} \left(e^{izx} - 1 - \frac{izx}{1+x^2}\right) dN(x),$$

where $b \in \mathbf{R}$, $\sigma^2 \ge 0$, and the functions $M(x)$ and $N(x)$ possess the following properties:

a) M and N are nondecreasing on $(-\infty, 0)$ and $(0, +\infty)$, respectively;
b) $M(-\infty) = N(+\infty) = 0$;
c) for all $\varepsilon > 0$,

$$\int_{-\varepsilon}^{0} x^2 \, dM(x) + \int_{0}^{\varepsilon} x^2 \, dN(x) < \infty.$$

Find a relation between (a, ψ) in (a) and (b, σ^2, M, N) in (b).

V.3.7. Let $\varphi(z)$ be an infinitely divisible characteristic function of a random variable ξ having the second moment. Transform (a) into the representation

$$\text{(c)} \qquad \ln \varphi(z) = izc + \int_{-\infty}^{\infty} \left(e^{izx} - 1 - izx\right) \frac{1}{x^2} \, dK(x),$$

where $c \in \mathbf{R}$ and $K(x)$ is a nondecreasing function such that $K(-\infty) = 0$, $K(+\infty) < \infty$.

Find a relationship between (a, ψ) in (a) and (c, K) in (c). Prove that $\operatorname{Var} \xi = K(+\infty)$.

V.3.8. Prove that if ξ is a nonnegative random variable with the infinitely divisible distribution function, then

$$\text{(d)} \qquad \mathsf{E}\, e^{iz\xi} = \exp\left\{iaz + \int_{0}^{\infty} \left(e^{izx} - 1\right) dG(x)\right\},$$

where $a \ge 0$ and $G(x)$ is a nondecreasing function such that

$$\int_{0}^{\infty} \frac{1}{1+x} \, dG(x) < \infty.$$

On the other hand, a random variable ξ having the characteristic function given by (d) with $a \geq 0$ is nonnegative with probability one.

V.3.9. Let $a \in \mathbf{R}$, $\varphi_n(z)$ infinitely divisible characteristic functions, and $\psi_n(x)$ the corresponding functions in (a) such that $\psi_n(+\infty) < \infty$. In order that $\varphi_n(z) \to \varphi_0(z)$ as $n \to \infty$, $z \in \mathbf{R}$, it is sufficient that

a) $a_n \to a_0$ as $n \to \infty$;
b) $\psi_n(x) \to \psi_0(x)$ as $n \to \infty$ at all continuity points of $\psi_0(x)$;
a) $\psi_n(+\infty) \to \psi_0(+\infty)$ as $n \to \infty$.

V.3.10. Prove that if $\varphi(z)$ is an infinitely divisible characteristic function, then so is $\varphi^\gamma(z)$ for all $\gamma > 0$.

V.3.11. Prove that the family of infinitely divisible distributions coincides with the family of limit distributions for the weak convergence of distribution functions of random variables ζ_n, $n \geq 1$,

$$\zeta_n = \sum_{k=1}^{n} \xi_k,$$

where $\{\xi_n, n \geq 1\}$ is a sequence of independent Poisson random variables.

V.3.12. If ξ has an infinitely divisible distribution function and there exists $c > 0$ such that $\mathsf{P}(|\xi| > c) = 0$, then $\mathsf{P}(\xi = \mathsf{E}\,\xi) = 1$.

V.3.13. Prove that the function

$$f(z) = \frac{1-\beta}{1-\alpha} \cdot \frac{1+\alpha e^{-iz}}{1-\beta e^{iz}}, \qquad 0 < \alpha \leq \beta < 1,$$

is not infinitely divisible, but $|f(z)|$ is.

V.3.14. A random variable ξ has an arithmetic distribution with step h. If its distribution is infinitely divisible, then there exist $q_k \geq 0$ and an integer n such that

$$\mathsf{E}\,e^{iz\xi} = \exp\left\{inhz + \sum_{k=-\infty}^{\infty} q_k \left(e^{izkh} - 1\right)\right\}.$$

V.3.15. If the distribution function of a random variable ξ is infinitely divisible and arithmetic with step h, then

$$\xi = \sum_{k=-\infty}^{\infty} kh\nu_k,$$

where ν_k are independent Poisson random variables.

V.3.16. Let $\{\xi_n, n \geq 1\}$ be a sequence of independent identically distributed random variables with the density $\frac{1}{2}e^{-|x|}$ and

$$\eta = \sum_{n=1}^{\infty} \frac{\xi_n}{n}.$$

Prove that the distribution function of η is infinitely divisible and find its canonical representation.

V.3.17. If ξ is a Gaussian random variable and $\xi = \xi_1 + \xi_2$, where ξ_1 and ξ_2 are independent random variables with an infinitely divisible distribution, then ξ_1 and ξ_2 are Gaussian random variables.

V.3.18. If ξ is a Poisson random variable and $\xi = \xi_1 + \xi_2$, where ξ_1 and ξ_2 are independent random variables with an infinitely divisible distribution, then ξ_1 and ξ_2 are Poisson random variables.

V.3.19. Let $\varphi(z)$ be an infinitely divisible characteristic function. Then there exist constants a and b such that $|\ln \varphi(z)| < a + bz^2$ for all $z \in \mathbf{R}$.

V.3.20. Let $P(s)$ be the moment generating function of a distribution $p_k \geq 0$, $\sum_{k=0}^{\infty} p_k = 1$. Assume that $p_0 > 0$ and $\ln P(s) - \ln p_0$ can be expanded into a power series with positive coefficients. If $\varphi(z)$ is the characteristic function of a distribution $F(x)$, then $P(\varphi(z))$ is an infinitely divisible characteristic function. Find the canonical representation for $\ln P(\varphi(z))$ using convolutions $F^{*n}(x)$.

V.3.21. Let ν and $\xi_1, \xi_2, \ldots$ be independent random variables. Assume that ν takes only nonnegative integer values and has an infinitely divisible distribution function. Set $S_0 = 0$, $S_n = \xi_1 + \cdots + \xi_n$. Then the random variable S_ν has an infinitely divisible distribution. Find the canonical representation for the logarithm of its characteristic function.

V.3.22. 1) Formulate the central criterion for convergence in terms of representation (b).

2) From the central criterion for convergence obtain necessary and sufficient conditions for the convergence to

a) the Gaussian distribution;
b) the degenerate distribution;
c) the Poisson distribution.

3) In order that the random variables $S_n - a_n$ have a limit distribution, it is necessary and sufficient that the conditions a) and c) of the central criterion for convergence be satisfied and the constants a_n be of the form

$$a_n = \sum_{k=1}^{k_n} \int_{|y|<\tau} y \, dF_{kn}(y) - a - \int_{|y|<\tau} y \, d\psi(y) + \int_{|y|\geq\tau} \frac{1}{y} \, d\psi(y) + o(1)$$

for all continuity points τ of the function $\psi(x)$.

V.3.23. Prove that the characteristic function $e^{-|z|^\alpha}$ is stable for $0 < \alpha \leq 2$.

V.3.24. Prove that distribution functions with densities

$$p_1(x) = \frac{1}{\pi(1+x^2)}, \qquad p_2(x) = \frac{1}{\sqrt{\pi}} x^{3/2} e^{-1/x}$$

are stable.

V.3.25. a) Prove that characteristic functions that admit the representation of Theorem 3 are stable.

b) Prove that stable laws are infinitely divisible.

V.3.26. Prove that nondegenerate stable distributions have densities.

V.3.27. Prove that the support of the density of a stable law with parameters $\gamma = 0$, $\alpha < 1$, $\beta = +1$ is concentrated on the positive semiaxis.

V.3.28. Let the distribution function of a random variable ξ be stable with a characteristic exponent α. Prove that the moments $\mathsf{E}\,|\xi|^\gamma$ exist for $\gamma < \alpha$.

V.3.29. Using Theorem 4 prove that

a) The asymptotic relation

$$\int_{-x}^{x} y^2\,dF(y) \sim x^{2-\alpha} L(x) \quad \text{as } x \to \infty$$

is equivalent to the condition a) of Theorem 2 if $\alpha = 2$, or to the condition b) if $\alpha < 2$.

b) The following representation is valid for the logarithm of the characteristic function of any stable law with characteristic exponent $\alpha < 2$:

$$\ln \varphi(z) = iaz + \int_{-\infty}^{0} \left(e^{izx} - 1 - \frac{izx}{1+x^2} \right) dM(x) + \int_{0}^{\infty} \left(e^{izx} - 1 - \frac{izx}{1+x^2} \right) dN(x),$$

where

$$M(x) = \frac{c_1}{|x|^\alpha}, \qquad N(x) = -\frac{c_2}{x^\alpha},$$

$c_1, c_2 \ge 0$, and $c_1 + c_2 > 0$.

c) The normalizing coefficients B_n satisfy the condition $B_n/B_{n+1} \to 1$ as $n \to \infty$.

V.3.30. Let $\{\xi_n, n \ge 1\}$ be a sequence of independent identically distributed random variables and

$$\zeta_n = \frac{\xi_1 + \cdots + \xi_n}{B_n} - A_n.$$

If the distributions of ζ_n converge to a stable law with characteristic exponent $\alpha \le 2$, then $\mathsf{E}\,|\xi_1|^\gamma < \infty$ for all $\gamma < \alpha$.

§V.4. Wiener processes

A process $w(t)$ depending on a continuous time parameter $t \ge 0$ is called *Brownian motion* or *a Wiener process* if

(a) $w(t)$ is a process with independent increments, that is, for all $n \ge 1$ and $0 \le t_1 \le \cdots \le t_n$ the random variables $w(t_1), w(t_2) - w(t_1), \dots, w(t_n) - w(t_{n-1})$ are jointly independent;

(b) $w(t)$ is a continuous function of t with probability one (in fact, $w(t)$ is a function of two variables, t and an elementary event ω of the initial probability space $(\Omega, \mathfrak{A}, \mathsf{P})$; thus, it is required that $w(t, \omega)$ be a continuous function of t for almost all ω with respect to measure P);

(c) the process $w(t)$ is homogeneous. This means that for $h > 0$, the distribution of $w(t+h) - w(t)$ does not depend on t.

THEOREM. *If $w(t)$ is a Wiener process, then there exist constants $a \in \mathbf{R}$ and $b \ge 0$ such that for all $t \ge 0$,*

$$\mathsf{E}\,e^{izw(t)} = e^{izat - bz^2 t/2},$$

that is, for all $t \geq 0$ the random variable $w(t)$ has the Gaussian distribution with the mean at and variance bt.

The coefficients a and b are called *the shift* and *the diffusion* of the process $w(t)$, respectively.

The variable $w(0)$ is called *the initial state of the process $w(t)$*. A Wiener process that with probability one has the initial state $w(0) = 0$, shift $a = 0$, and diffusion $b = 1$ is called *the standard Wiener process*.

REMARK. The Wiener process is a mathematical model for the motion of microparticles in a liquid or gas under the action of molecules in chaotic movement. This phenomenon was first observed by a biologist Brown in 1827. A mathematical description for it was given by N. Wiener in 1918.

Problems

V.4.1. A proof of Theorem 1 (see the beginning of this chapter) follows from the fact that under the assumptions (a) and (b), the distributions of sums of independent identically distributed random variables

$$w(t) = \sum_{k=1}^{n} \left[w\left(t\frac{k}{n}\right) - w\left(t\frac{k-1}{n}\right) \right]$$

satisfy conditions (1)–(3) of Problem V.2.15, and therefore converge to the Gaussian distribution. Using properties (a)–(c), check the first of those conditions, namely,

$$n\mathsf{P}\left(\left| w\left(\frac{t}{n}\right) - w(0) \right| > \varepsilon \right) \to 0 \quad \text{as } n \to \infty.$$

V.4.2. Let $w(t)$ be a Wiener process with shift a and diffusion b, and let $w(0) = 0$ with probability one. Prove that its expectation $m(t)$ and correlation function $R(t, s)$ are given by

$$m(t) = at, \qquad R(t,s) = b \min(t, s).$$

V.4.3. Let $w(t)$ be a Wiener process with shift a and diffusion $b \neq 0$. Prove that

$$\frac{w(t) - w(0) - at}{b}$$

is the standard Wiener process. Conversely, if $w(t)$ is the standard Wiener process, then $at + bw(t)$, $t \geq 0$, is a Wiener process with shift a and diffusion b.

V.4.4. Let $w(t)$ be a Wiener process with shift a and diffusion b. Let τ be a random variable, independent of $w(t)$, that has the exponential distribution with parameter λ. Find the characteristic function of the random variable $w(\tau)$.

V.4.5. Let $w(t)$ be a Wiener process with shift a and diffusion b. Assume that $w(0) = 0$ and denote by $p(x, t, y)$ the density of the random variable $w(t + x)$. Prove that for $x \in \mathbf{R}$, $t > 0$, the function $p(x, t, y)$ satisfies the following differential equation:

$$\frac{\partial p(x,t,y)}{\partial t} = a\frac{\partial p(x,t,y)}{\partial x} + \frac{b}{2}\frac{\partial^2 p(x,t,y)}{\partial x^2}.$$

V.4.6. Let $f(x)$ be a bounded continuous function. Prove that the function $U(t,x) = \mathsf{E}\, f(w(t+x))$ satisfies the equation

$$\frac{\partial U(t,x)}{\partial t} = a\frac{\partial U(t,x)}{\partial x} + \frac{b}{2}\frac{\partial^2 U(t,x)}{\partial x^2}.$$

V.4.7. Let $\xi(t)$, $t \geq 0$, be a real-valued stochastic process such that

1) $\xi(t)$, $t \geq 0$, is continuous with probability one;
2) $\xi(t)$, $t \geq 0$, is Gaussian, that is, for all $n \geq 1$ and $0 \leq t_1 \leq \cdots \leq t_n$, the random vector $(\xi(t_1), \ldots, \xi(t_n))$ has a multidimensional Gaussian distribution.

Prove that

a) the expectation $m(t) = \mathsf{E}\,\xi(t)$, $t \geq 0$, and the correlation function $R(t,s) = \mathsf{E}(\xi(t) - m(t))(\xi(s) - m(s))$ are continuous;
b) if $R(t,s) = R(\min(t,s))$ for all $t, s \geq 0$, where $R(t) = R(t,t)$, then $\xi(t)$, $t \geq 0$, is a process with independent increments;
c) a continuous Gaussian process $\xi(t)$ is a Wiener process if and only if its expectation and correlation functions, $m(t)$ and $R(t,s)$, are of the form

$$m(t) = at, \quad t \geq 0, \qquad R(t,s) = b\min(t,s), \quad t,s \geq 0.$$

V.4.8. Let $w(t)$ be the standard Wiener process. For $0 < t_1 < \cdots < t_n$ find the joint probability density of random variables $w(t_1), \ldots, w(t_n)$.

V.4.9. Find the conditional distribution function $\mathsf{P}\big(w(t) < x / w(s) = y\big)$ for $t < s$ and the corresponding conditional density.

V.4.10. Prove that the conditional density of the variable $w(t)$ given $w(t_1) = y_1$ and $w(t_2) = y_2$ for $t_1 < t < t_2$, is Gaussian with mean $y_1 + (t-t_1)(y_2-y_1)/(t_2-t_1)$ and variance $(t_2 - t)(t - t_1)/(t_2 - t_1)$.

V.4.11. Let $w(t)$ be the standard Wiener process, and $f(x)$ a bounded measurable function. Prove that

$$\mathsf{E}\int_0^T f(w(s))\,ds = \int_0^T \frac{1}{\sqrt{2\pi s}} \int_{-\infty}^{\infty} f(y) e^{-y^2/(2s)}\,dy\,ds.$$

V.4.12. Let $w(t)$ be a Wiener process with shift a and diffusion b. Assume that $g_1(t), \ldots, g_k(t)$ are integrable functions on $[0,T]$. Prove that the random vector $\eta = (\eta_1, \ldots, \eta_k)$ with

$$\eta_i = \int_0^T g_i(t) w(t)\,dt, \qquad i = 1, \ldots, k,$$

has the multidimensional Gaussian distribution with the vector of expectations $\bar{d} = (d_1, \ldots, d_k)$ and the correlation matrix $C = \big\|c_{ij}\big\|_{i,j=1}^k$ given by

$$d_i = a\int_0^T g_i(t) t\,dt, \qquad i = 1, \ldots, k,$$

$$c_{ij} = b\int_0^T \int_0^T g_i(t) g_j(s) \min(t,s)\,dt\,ds, \qquad i,j = 1, \ldots, k.$$

V.4.13. Let $w(t)$ be the standard Wiener process, and $g(t)$ a continuous differentiable function. Put

$$\int_0^T g(s)\,dw(s) = w(T)g(T) - \int_0^T g'(s)w(s)\,ds.$$

Prove that if $g_1(t)$ and $g_2(t)$ are continuous differentiable functions, then

$$\mathsf{E}\int_0^T g_1(s)\,dw(s)\int_0^T g_2(s)\,dw(s) = \int_0^T g_1(s)g_2(s)\,ds.$$

V.4.14. Let $w(t)$ be the standard Wiener process, and $\{\varphi_k(t), k \geq 1\}$ a sequence of continuously differentiable functions that are mutually orthogonal in $L_2[0,T]$, that is,

$$\int_0^T \varphi_k(t)\varphi_j(t)\,dt = 0, \qquad k \neq j.$$

Prove that

$$\eta_k = \int_0^T \varphi_k(t)\,dw(t), \qquad k = 1, 2, \ldots,$$

are independent Gaussian random variables such that

$$\mathsf{E}\,\eta_k = 0, \qquad \operatorname{Var}\eta_k = \int_0^T \varphi_k^2(t)\,dt.$$

V.4.15. Let $w(t)$ be the standard Wiener process, and $\{g_k(t), k \geq 1\}$ a complete orthonormal sequence of functions in $L_2[0,T]$ satisfying the condition

$$(1)\qquad \int_0^T\int_0^T g_k(t)g_j(t)\min(t,s)\,dt\,ds = \begin{cases} 0, & \text{for } k \neq j, \\ 1, & \text{for } k = j. \end{cases}$$

a) Prove that the random variables

$$\zeta_k = \int_0^T w(s)g_k(s)\,ds$$

are independent and Gaussian with zero mean and unit variance. Moreover,

$$w(t) = \sum_{n=1}^{\infty} \zeta_k g_k(t),$$

where the series on the right converges in the sense that

$$\lim_{n\to\infty}\int_0^T \mathsf{E}\left[w(t) - \sum_{k=1}^{n}\zeta_k g_k(t)\right]^2 dt = 0.$$

b) Prove that if functions $g_k(t)$ form a complete orthonormal system in $L_2[0,T]$ and satisfy condition (1), then they are eigenfunctions of the integral equation

$$(2)\qquad g_k(t) = \lambda_k \int_0^T \min(t,s)g_k(s)\,ds.$$

c) Find all eigenvalues and eigenfunctions for equation (2).

V.4.16. Let random variables ζ_k be defined as in Problem V.4.15. Then

$$\int_0^T w^2(t)\,dt = \sum_{k=1}^{\infty} \zeta_k^2.$$

Use this result to calculate the Laplace transform

$$\mathsf{E}\exp\left\{-s\int_0^T w^2(t)\,dt\right\} = \prod_{k=1}^{\infty} \frac{1}{\sqrt{1+2s/\lambda_k}},$$

where

$$\lambda_k = \frac{\pi^2}{4T^2}(4k^2-1).$$

V.4.17. Let $w(t)$ be the standard Wiener process, and let a sequence $0 = t_{0n} < t_{1n} < \cdots < t_{nn} = 1$ of partitions of the segment $[0,1]$ be such that

$$\max_{1\le i\le n} (t_{in} - t_{i-1,n}) \to 0 \quad \text{as } n\to\infty.$$

Put

$$u_n = \sum_{i=0}^{n-1} \bigl(w(t_{in}) - w(t_{i+1,n})\bigr)^2.$$

Prove that $\mathsf{E}\,u_n = 1$ for all n and $\lim_{n\to\infty} \mathsf{E}\,u_n^2 = 1$.

V.4.18. A process $\bar{w}(t) = (w_1(t),\dots,w_k(t))$ with a continuous time parameter $t\ge 0$ assuming values in $\mathbf{R}_k$ is called *a k-dimensional Wiener process* if for all $\bar{z}\in\mathbf{R}_k$, the process $(\bar{z},\bar{w}(t))$, $t\ge 0$, is a real-valued Wiener process. Prove that there exists a vector $\bar{a}\in\mathbf{R}_k$ and a positive semidefinite matrix $B = \|b_{ij}\|_{i,j=1}^k$ such that

$$\mathsf{E}\exp\{i\,(\bar{z},\bar{w}(t))\} = \exp\left\{i\,(\bar{a},\bar{w}(t))\,t - \frac{1}{2}t\,(B\bar{z},\bar{z})\right\}$$

for all $t\ge 0$. The vector $\bar{a}$ is *the shift vector*, and the matrix B is said to be *the diffusion operator* of a k-dimensional Wiener process $\bar{w}(t)$.

V.4.19. Let $w_i(t)$, $i=1,\dots,k$, be the standard Wiener processes that are jointly independent, that is, for $t_1\le t_2\le\cdots\le t_n$, the families of random vectors $(w_i(t_1),\dots,w_i(t_n))$, $i=1,\dots,k$, are independent. Define the process $w_i'(t)$:

$$w_i'(t) = a_i t + \sum_{j=1}^{k} b_{ij} w_j(t), \qquad i=1,\dots,l.$$

Prove that $\bar{w}(t)' = (w_1'(t),\dots,w_l'(t))$, $t\ge 0$, is a k-dimensional Wiener process and determine its shift vector and diffusion operator.

V.4.20. Let $\bar{w}(t) = (w_1(t),\dots,w_l(t))$, $t\ge 0$, be a k-dimensional Wiener process with the shift vector $\bar{a}$ and the diffusion operator $B = \|b_{ij}\|_{i,j=1}^k$. For all $t\ge 0$ define the process

$$w_i'(t) = d_i t + \sum_{j=1}^{k} c_{ij} w_j(t), \qquad i=1,\dots,k,$$

where d_i, $1 \le i \le k$, and c_{ij}, $1 \le i, j \le k$, are real numbers. Prove that $\bar{w}'(t) = (w_1'(t), \ldots, w_l'(t))$, $t \ge 0$, is a k-dimensional Wiener process and find its shift vector and diffusion operator.

V.4.21. Let $\bar{w}(0) = 0$. Find a condition under which the processes $w_i(t)$, $i = 1, \ldots, k$, defined in the preceding problem are independent standard Wiener processes.

§V.5. Functionals of Wiener processes

In what follows, $w(t)$ denotes the standard Wiener process.

Problems

V.5.1. Let τ_a be the time of the first passage across a level a, i.e., $\tau_a = \inf\{t\colon w(t) > a\}$. Show that for all $0 \le t_1 \le \cdots \le t_l$, $s \le \infty$,

$$\mathsf{P}\left(w(\tau_a + t_i) - w(\tau_a) < x_i, i = 1, \ldots, l / \tau_a < s\right) = \mathsf{P}\big(w(t_i) < x_i, i = 1, \ldots, l\big).$$

V.5.2. Let

$$\widetilde{w}(t) = \begin{cases} w(t), & \text{for } t < \tau_a, \\ 2a - w(t), & \text{otherwise}, \end{cases}$$

where τ_a is defined in the preceding problem. Prove that $\widetilde{w}(t)$ is a Wiener process.

V.5.3. Prove that for $a \ge x$,

$$\mathsf{P}\Big(\max_{t \le T} w(t) < a, w(T) < x\Big) = \frac{1}{\sqrt{2\pi T}}\left(\int_{-\infty}^{x} e^{-u^2/(2T)}\, du - \int_{2a-x}^{\infty} e^{-u^2/(2T)}\, du\right).$$

V.5.4. Let τ_a be defined as in Problem V.5.1. Prove that $\mathsf{P}(\tau_a < \infty) = 1$ for all $a > 0$. Find the distribution of the random variable τ_a and show that

$$\mathsf{E}\, e^{-s\tau_a} = e^{-\sqrt{2s}}.$$

V.5.5. Let τ_a be defined as in Problem V.5.1. Prove that τ_a as a function of a is a process with independent increments, that is, for all $0 \le a_1 \le \cdots \le a_l$, $l \ge 1$, the random variables $\tau_{a_1}, \tau_{a_2} - \tau_{a_1}, \ldots, \tau_{a_l} - \tau_{a_{l-1}}$ are independent. The distribution of the random variable $\tau_{a+b} - \tau_a$ for all $a \ge 0$ coincides with the distribution of τ_b.

V.5.6. Find the distribution of the random variable

$$\sup_{a \le s \le b} w(s).$$

V.5.7. Let $t \le T$. Prove that

$$\mathsf{P}\Big(\sup_{0 \le s \le T} w(s) \le a, w(t) < y\Big)$$
$$= \int_{-\infty}^{y} \mathsf{P}\Big(\sup_{0 \le s \le T-t} w(s) < a - x\Big) \frac{1}{\sqrt{2\pi t}} \left(e^{-x^2/(2t)} - e^{-(2a-x)^2/(2t)}\right) dx.$$

V.5.8. Put $\tau_a' = \{\sup t\colon w(t) < a\}$. Prove that for all $a > 0$,

$$\mathsf{P}(\tau_a' = \tau_a) = 1.$$

V.5.9. Find the probability $P_{t_1 t_2}$ that the process $w(t)$ reaches zero at least once on the interval $[t_1, t_2]$.

V.5.10. Let ξ be the greatest t such that $w(t) = 0$ and $t \le T$. Prove that

$$\mathsf{P}(\xi < t) = \frac{2}{\pi} \arcsin \sqrt{\frac{t}{T}}.$$

V.5.11. Let η be the least t such that $w(t) = 0$ and $t \ge T$. Prove that

$$\mathsf{P}(\eta < t) = \frac{2}{\pi} \arccos \sqrt{\frac{T}{t}}.$$

V.5.12. Let ξ be the random variable defined in Problem V.5.10, and η the random variable defined in Problem V.5.11. Prove that

$$\mathsf{P}(\xi < t_1, \eta < t_2) = \frac{2}{\pi} \arccos \sqrt{\frac{t_1}{t_2}}.$$

V.5.13. Let $w^+(t) = \sup_{0 \le s \le t} w(s)$. Prove that

$$\mathsf{P}\left(w^+(t) > x / w(T) = w^+(T)\right) = \exp\left\{-\frac{x^2}{2T}\right\}.$$

V.5.14. Let $w^+(t)$ be defined as in the preceding problem. Prove that $X(t) = w^+(t) - w(t)$ is a Markov process, that is, for all $0 \le t_1 \le \cdots \le t_l \le t$ and $h > 0$,

$$\mathsf{P}\left(X(t+h) < x / X(t) = y, X(t_1), \ldots, X(t_l)\right) = \mathsf{P}\left(X(t+h) < x / X(t) = y\right).$$

V.5.15. Prove that $|w(t)|$ is a Markov process (see Problem V.5.14 for the definition of a Markov process).

V.5.16. Let $X(t)$ be defined as in Problem V.5.14. Prove that the random processes $|w(t)|$ and $X(t)$ have the same finite-dimensional distributions.

V.5.17. Find the conditional probability that the process $w(t)$ does not vanish on the interval (t_1, t_2), $0 < t_1 \le t_2 \le t$.

V.5.18. Let $0 < t_1 < t_2$. Prove that the probability that the process $w(t)$ does not vanish in an interval $(0, t_2)$ given that $w(t)$ does not vanish in the interval $(0, t_1)$, equals $\sqrt{t_1 t_2^{-1}}$.

V.5.19. Prove that for $a, b > 0$,

$$\mathsf{P}\left(w(s) \ne 0, s \in (u, u+t) / w(u) = a, w(u+t) = b\right) = 1 - e^{-2abt}.$$

V.5.20. Prove that

$$\mathsf{P}\left(w(1) < x / w(u) \ge 0, 0 \le u \le 1\right) = 1 - \exp\left\{-\frac{x^2}{2}\right\}.$$

V.5.21. Let $w^+(t) = \sup_{0 \le s \le t} w(s)$ and

$$q_a(t, x) = \mathsf{P}\left(w^+(t) \le a - x, w(t) < y - x\right).$$

Prove that the function $q_a(t, x)$ satisfies the differential equation

$$\frac{\partial}{\partial t} q_a(t, x) = \frac{1}{2} \frac{\partial^2}{\partial x^2} q_a(t, x), \qquad x < a,\ t > 0,$$

and the boundary conditions

$$\lim_{x\uparrow a} q_a(t,x) = 0, \qquad \lim_{t\downarrow 0} q_a(t,x) = \begin{cases} 1, & \text{for } x < y, \\ 0, & \text{for } x > y. \end{cases}$$

V.5.22. Let $u(x)$ be the probability that the process $w(t)$ starting at the point x reaches level b before it reaches level c, $c < b$. Prove that for all $c < x < b$ the function $u(x)$ satisfies the relation

$$u(x) = \frac{1}{\sqrt{2\pi t}} \int_c^b e^{-(x-y)^2/(2t)} u(y)\, dy + o(t^2) \quad \text{as } t \to 0.$$

Use this relation to prove that $u(x)$ satisfies the differential equation $u''(x) = 0$ with the boundary conditions $u(c) = 0$, $u(b) = 1$. Find $u(x)$.

V.5.23. For $a < x < b$ and $t > 0$, let

$$q_{a,b}(t,x) = \mathsf{P}\Big(\sup_{0\le s\le t} w(s) < b - x,\ \inf_{0\le s\le t} w(s) \ge a - x\Big).$$

Prove that the function $q_{a,b}(t,x)$ satisfies the differential equation

$$\frac{\partial}{\partial t} q_{a,b}(t,x) = \frac{1}{2}\frac{\partial^2}{\partial x^2} q_{a,b}(t,x)$$

and the boundary conditions

$$\lim_{t\downarrow 0} q_{a,b}(t,x) = 1,$$

$$\lim_{x\downarrow a} q_{a,b}(t,x) = \lim_{x\uparrow b} q_{a,b}(t,x) = 0.$$

V.5.24. Let τ_a be defined as in Problem V.5.1 and $\tau_a^* = \inf\{t\colon w(t) < a\}$. Put $\zeta_{a,b} = \min(\tau_a^*, \tau_b)$. Prove that the function

$$u_\lambda(x) = \mathsf{E}\exp\{-\lambda \zeta_{a-x,b-x}\}, \qquad \lambda > 0,$$

satisfies the differential equation

$$\lambda u_\lambda(x) = \frac{1}{2} u_\lambda''(x), \qquad a < x < b,$$

and the boundary conditions

$$u_\lambda(a) = u_\lambda(b) = 1.$$

V.5.25. Using the results of the preceding problem prove the formula

$$u_\lambda(x) = \frac{\cosh\sqrt{2\lambda}\,(x - (a+b)/2)}{\cosh\sqrt{2\lambda}(a-b)/2}.$$

V.5.26. Consider the event $A_k = \{\omega\colon \sup_{a^k < t < a^{k+1}} w(t) > c_k\}$. Prove that only finitely many events A_k occur with probability one if $\varepsilon > 0$ and

$$1 < a < (1+\varepsilon)^2, \qquad c_k = \sqrt{2(1+\varepsilon)^2 a^k \ln\ln a^k}.$$

Derive that

$$\mathsf{P}\left(\limsup_{t\to\infty} \frac{w(t)}{\sqrt{2t\ln\ln t}} \le 1+\varepsilon\right) = 1, \qquad \varepsilon > 0.$$

V.5.27. Consider the events $B_k = \{\omega: w(a^{k+1}) - w(a^k) > c_k\}$. Prove that infinitely many events B_k occur with probability one if $\varepsilon > 0$ and

$$\frac{a}{a-1}\left(1-\frac{\varepsilon}{2}\right)^2 < 1, \qquad c_k = \left(1-\frac{\varepsilon}{2}\right)\sqrt{2a^{k+1}\ln\ln a^k}.$$

Derive that

$$\mathsf{P}\left(\limsup_{t\to\infty}\frac{w(a^k)}{\sqrt{2a^k\ln\ln t}} > 1-\varepsilon\right) = 1, \qquad \varepsilon > 0.$$

V.5.28. Using the results of Problems V.5.26 and V.5.27 prove that

$$\mathsf{P}\left(\limsup_{t\to\infty}\frac{w(t)}{\sqrt{2t\ln\ln t}} = 1\right) = 1.$$

V.5.29. Prove that

a) $\mathsf{P}\left(\lim\limits_{t\to\infty}\dfrac{1}{\sqrt{2t\ln\ln t}}\sup\limits_{0\le s\le t} w(s) = 1\right) = 1;$

b) $\mathsf{P}\left(\limsup\limits_{t\to\infty}\dfrac{|w(t)|}{\sqrt{2t\ln\ln t}} = 1\right) = 1.$

V.5.30. Let $f(x)$ be a continuous bounded function and

$$u_z(t,x) = \mathsf{E}\exp\left\{iz\int_0^t f(w(s))\,ds\right\}.$$

Prove that the function $u_z(t,x)$ satisfies the differential equation

$$\frac{\partial}{\partial t}u_z(t,x) = \frac{1}{2}\frac{\partial^2}{\partial x^2}u_z(t,x) + zf(x)u_z(t,x)$$

with the initial condition $u_z(0,x) = 1$.

V.5.31. Let $u_z(t,x)$ be defined as in the preceding problem. Put

$$\Phi_{z,\lambda}(x) = \int_0^\infty u_z(t,x)e^{-\lambda t}\,dt.$$

Prove that

$$\text{(a)} \qquad \lambda\Phi_{z,\lambda}(x) - 1 = \frac{1}{2}\frac{\partial^2}{\partial x^2}\Phi_{z,\lambda}(x) + zf(x)\Phi_{z,\lambda}(x).$$

V.5.32. Taking the limit, prove that (a) is valid at all continuity points x of the function $f(x)\Phi_{z,\lambda}(x)$. Note that it is allowed that some of these points belong to the set of discontinuities of the function f. Prove that in this case $\frac{\partial}{\partial x}\Phi_{z,\lambda}(x)$ is continuous.

V.5.33. Let

$$f(x) = \begin{cases} 1, & \text{for } x > 0, \\ 0, & \text{for } x \le 0. \end{cases}$$

Find $\Phi_{z,\lambda}(x)$ for this function $f(x)$. Using $\Phi_{z,\lambda}(x)$ prove the relation

$$\lim_{t\to\infty}\mathsf{P}\left(\frac{1}{t}\int_0^t f(w(s))\,ds < x\right) = \frac{2}{\pi}\arcsin\sqrt{x}.$$

CHAPTER VI

Solutions, Hints, Answers

Solutions to Chapter I

§**1.**

I.1.3. ANSWER: $A = \varnothing$, $B = (0, \frac{1}{2}) \cup (\frac{1}{2}, 1)$.

I.1.18. Consider the intersection of all algebras containing the class K and use the assertion of Problem I.1.16.

I.1.19. ANSWER: $\mathfrak{M}_0(K) = \{\Omega, \varnothing, A, \overline{A}\}$.

I.1.26. Consider the intersection of all σ-algebras containing the class K and use the assertion of Problem I.1.25.

I.1.28. ANSWER:

$$\text{a) } \{a\} = \bigcap_{n=1}^{\infty} \left[a, a + \frac{1}{n}\right); \qquad \text{c) } (a, b) = [a, b) \setminus \{a\}.$$

I.1.33. Obviously, every σ-algebra is a monotone class. Conversely, let $\mathfrak{M}$ be an algebra and a monotone class. Consider the sequence of sets A_k in $\mathfrak{M}$. Since $\mathfrak{M}$ is an algebra, the sequence $B_n = \bigcup_{k=1}^{n} A_k$ is a monotone sequence of sets in $\mathfrak{M}$ ($B_n \in \mathfrak{M}$). Since $\mathfrak{M}$ is a monotone class, we have $\lim B_n = \bigcup_{k=1}^{\infty} A_k \in \mathfrak{M}$. Therefore, $\mathfrak{M}$ is a σ-algebra.

I.1.35. There is at least one monotone class that contains K, for instance, the class of all subsets of Ω. Consider the intersection $\mathfrak{M}_0(K)$ of all monotone classes that contain the class K. According to the assertion of Problem I.1.34, $\mathfrak{M}_0(K)$ is a monotone class. For any monotone class M containing K we have $\mathfrak{M}_0(K) \subset M$. Therefore, $\mathfrak{M}_0(K)$ is the minimal monotone class that contains K.

I.1.36. Since every σ-algebra is a monotone class, we have $M_0(K) \subset F_0(K)$.

I.1.37. a) We have $\mathfrak{M} \subset \widetilde{M} \subset M$ (if $A \in \mathfrak{M}$ and $\mathfrak{M}$ is an algebra, then $\overline{A} \in \mathfrak{M}$, i.e., $A \in \mathfrak{M} \subset M$, $\overline{A} \in \mathfrak{M} \subset M$, and $A \in \widetilde{M}$, whence $A \in M$). The class M is a monotone class. Indeed, if $B_n \in \widetilde{M}$, $B_n \subseteq B_{n+1}$, then $\lim B_n = \bigcup_{k=1}^{\infty} B_k \subset M$ since $B_k \in M$, and

$$\overline{\lim B_n} = \overline{\bigcup_{k=1}^{\infty} B_k} = \bigcap_{k=1}^{\infty} \overline{B}_k \in M$$

since $B_k \in M$. But this means just that $\lim B_n \in \widetilde{M}$. Using similar arguments, it can be proved that $B_n \in \widetilde{M}$ and $B_n \supset B_{n+1}$ imply

$$\lim B_n = \bigcap_{k=1}^{\infty} B_k \in \widetilde{M}.$$

Since $\widetilde{M}$ is a monotone class that contains $\mathfrak{M}$, we have $M \subset \widetilde{M}$. Thus $M = \widetilde{M}$.

b) From the equality $\lim(A \cap B_n) = A \cap \lim B_n$, which is valid for any monotone sequence of events, it follows that M_A is a monotone class. If $A \in \mathfrak{M}$, then any set $B \in \mathfrak{M}$ belongs to M_A. Thus, if $A \in \mathfrak{M}$, then $\mathfrak{M} \subset M_A \subset M$. Since M_A is a monotone class that contains $\mathfrak{M}$, for $A \in \mathfrak{M}$ we have $M \subset M_A$, whence $M = M_A$. Since, by the definition of the class M_A, the assertions $A \in M_B$ and $B \in M_A$ are equivalent, we have $A \in M_B$ for every $A \in \mathfrak{M}$ and every $B \in M = M_A$. Thus $\mathfrak{M} \subset M_B \subset M$ for all $B \in M$, and hence $M_B = M$ for all $B \in M$.

c) From b) it follows that M is an algebra of sets. But every algebra that is a monotone class is also a σ-algebra (Problem I.1.33).

d) Since $M_0(\mathfrak{M}) = M$ is a σ-algebra of sets that contains the class K, it follows that $F_0(\mathfrak{M}) \subset M_0(\mathfrak{M})$. According to the assertion of Problem I.1.36, we have $M_0(\mathfrak{M}) \subset F_0(\mathfrak{M})$. Therefore, $F_0(\mathfrak{M}) = M_0(\mathfrak{M})$.

§**2.**

I.2.1. ANSWER: $m \times n$.

I.2.3. ANSWER: $17 \cdot 16 \cdot 15$.

I.2.7. ANSWER: $10 \cdot 9 \cdot 8 \cdot 7 \cdot 6 \cdot 5$.

I.2.9. ANSWER: $(m+1)(n+1)$.

I.2.11. ANSWER: $\frac{1}{2}n(n-3)$.

I.2.13. If there is no straight line on the plane, then the number of parts is equal to 1. Evidently, in order that the number of parts be maximal, it is necessary that no three straight lines intersect at one point and no two lines be parallel. Let us draw straight lines consecutively, keeping in mind the above conditions. Note that each additional line increases the number of parts by 1 plus the number of new points of intersection. Therefore, the nth straight line increases the number of parts by n plus the number of new points of intersection, which is equal to $n(n-1)/2$. Thus, the largest number of parts into which n straight lines can subdivide the plane is

$$1 + n + \frac{n(n-1)}{2} = 1 + \frac{n(n+1)}{2}.$$

I.2.14. Let A_n be the largest number of parts into which n planes can subdivide the space. Using the previous problem, prove that

$$A_{n+1} = A_n + \frac{n(n+1)}{2} + 1.$$

ANSWER: $A_n = \frac{1}{6}(n^3 + 5n + 6)$.

I.2.16. Let A_n be the largest number of parts into which n spheres can subdivide the space. Prove that

$$A_{n+1} = A_n + n^2 - n + 2.$$

I.2.17. ANSWER: $n^2(n-1)^2$.

I.2.21. ANSWER: $(\alpha_1 + 1)\cdots(\alpha_n + 1)$.

I.2.23. All tables that possess the required property can be obtained as follows. Write $+1$ or -1 arbitrarily everywhere, except the last row and the last column. This can be done in $2^{(m-1)(n-1)}$ ways. Let p be the product of all numbers written. Now write $+1$ or -1 at the end of each row (the nth column) so that the product of all numbers in every row be equal to 1. Denote by x the product of these $m-1$ numbers in the nth column. Now write $+1$ or -1 at the end of each column (the mth row) so that the product of numbers in every column be equal to 1. Denote by y the product of these $n-1$ numbers in the mth row. Note that x and y have the same sign. Indeed, $px = 1$, $py = 1$, and hence $p^2xy = 1$, so that $xy > 0$. As the last entry of the mth row and the nth column take 1 with the sign of x (or y, which is the same). Then the product of numbers in the nth column (and in the mth row) is also 1. Thus we have obtained the table that possesses the required property. The number of all such tables is $2^{(m-1)(n-1)}$.

I.2.26. To every point of intersection of two diagonals there correspond four vertices of the n-gon, and to every four vertices there corresponds one point of intersection, namely, the point of intersection of diagonals of the quadrangle with vertices at the given four points. Therefore, the number of all points of intersection is equal to the number of ways in which four vertices can be chosen of n vertices, i.e., to $\binom{n}{4}$.

I.2.27. Each shortest path from the point $(0,0)$ to the point (m,n) consists of $m+n$ segments, m horizontal and n vertical. Various paths differ only in the order of horizontal and vertical segments. Therefore, the total number of shortest paths from the point $(0,0)$ to the point (m,n) is equal to the number of ways in which n vertical segments can be chosen of $m+n$ segments, i.e., to $\binom{m+n}{n}$. We could consider also the number of ways in which m horizontal, rather than n vertical, segments could be chosen of $m+n$ segments to obtain the number $\binom{m+n}{m}$. Thus we have proved geometrically that $\binom{m+n}{n} = \binom{m+n}{m}$.

I.2.28. a) Consider shortest paths from the point $(0,0)$ to the point $(k, n-k)$. Divide all these paths into those that pass through the point $(k-1, n-k)$ and those that pass through the point $(k, n-k-1)$.

b) The number of shortest paths from the point $O(0,0)$ to the point $A(n,n)$ is equal to $\binom{2n}{n}$. Each of these paths passes through one and only one of the points $A_k(k, n-k)$. The number of shortest paths from the point O to the point A_k is equal to $\binom{k+(n-k)}{k} = \binom{n}{k}$, whereas the number of shortest paths from the point A_k to the point A is equal to $\binom{n-k+k}{n} = \binom{n}{k}$. Therefore, the number of shortest paths from O to B passing through A_k is equal to $\binom{n}{k}\binom{n}{k} = \binom{n}{k}^2$. Adding the numbers of paths that pass through every point A_k, $k = 0, 1, \ldots, n$, we get that the total number of shortest paths from O to A is equal to $\binom{2n}{n}$.

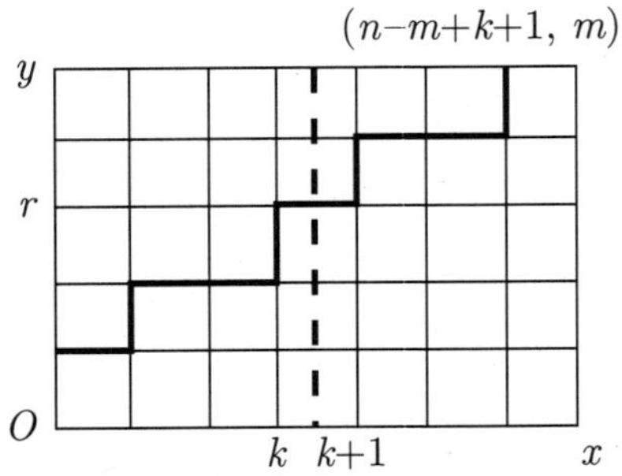

FIGURE 8

c) Consider all shortest paths from the point $(0,0)$ to the point $(n-m+k+1, m)$. Divide all these paths into classes $L_0, L_1, \ldots, L_m$ by assigning to the class L_r all the paths that cross the straight line $x = k + \frac{1}{2}$ at the point $\left(k+\frac{1}{2}, r\right)$ (Figure 8). Since every polygonal line in L_r can be divided into three parts (the polygonal line that joins $(0,0)$ to (k,r), the horizontal segment that joins (k,r) to $(k+1,r)$, and the polygonal line that joins $(k+1,r)$ to $(n-m+k+1,m)$), the total number of polygonal lines in the class L_r is equal to

$$\binom{k+r}{r}\binom{n-r}{m-r},$$

and the total number of shortest paths from $(0,0)$ to $(n-m+k+1,m)$ is equal to $\binom{n+k+1}{m}$.

d) Consider all shortest polygonal lines that join the point $(0,0)$ to the point $(r, n-r)$. The number of all such polygonal lines is equal to $\binom{n}{r}$. Assign to the class B_k all polygonal lines that cross the straight line $x = \frac{1}{2}$ at the point $\left(\frac{1}{2}, k\right)$, $k = 0, 1, \ldots, n-r$. Obviously, the class B_k consists of $\binom{n-k-1}{r-1}$ polygonal lines. Hence

$$\binom{n}{r} = \sum_{k=0}^{n-r} \binom{n-k-1}{r-1}.$$

I.2.29. Consider the symmetry about the straight line CD. ANSWER:

$$\binom{n+m}{n} - \binom{n+m}{d}.$$

I.2.30. For every $m-1$ members of the committee, there must be a lock that they cannot open. At the same time, *each* of the remaining $n-m+1$ members of the committee has a key to this lock (the appearance of any of them makes it possible to unlock the safe). Therefore, the minimum number of locks is equal to $\binom{n}{m-1}$, and the minimum number of keys is equal to $(n-m+1)\binom{n}{m-1}$.

It is quite clear that in order to satisfy the condition of the problem, the keys must be distributed in a special way. Namely, for every group of $m-1$ members of the committee one of $\binom{n}{m-1}$ locks is chosen, while $n-m+1$ keys to this lock are distributed among the other $n-m+1$ members. To different groups there should correspond different locks. Thus every member of the committee will have $n^{-1}\binom{n}{m-1}(n-m+1)$ keys.

I.2.31. If no diagonal is drawn, then we have one part. Let us draw diagonals consecutively. Note that the drawing of each diagonal increases the number of parts by 1 plus the number of points of intersection with the diagonals that have been drawn earlier. Therefore, the number of parts formed after drawing all diagonals is equal to one plus the number of points of intersection plus the number of all diagonals. If no three diagonals intersect at one point, then the number of points of intersection is equal to $\binom{n}{4}$ (see Problem I.2.26). The number of diagonals is equal to $n(n-3)/2$. Thus the number of parts is equal to

$$1+\binom{n}{4}+\frac{n(n-3)}{2}=\frac{(n-1)(n-2)(n^2-3n+12)}{24}.$$

I.2.34. Let $1 \le k \le p-1$. We have

$$\binom{p}{k}=\frac{p(p-1)\cdots(p-k+1)}{k!}.$$

Since p is a prime number, it is not divisible by $k!$. Therefore, $(p-1)\cdots(p-k+1)$ is divisible by $k!$ and $\binom{p}{k}$ is divisible by p.

I.2.35. For $a=1$ the theorem is valid. Assume that a^p-a is divisible by p and prove that then $(a+1)^p-(a+1)$ is divisible by p. Indeed,

$$(a+1)^p-(a+1)=(a^p-a)+\binom{p}{1}a^{p-1}+\binom{p}{2}a^{p-2}+\cdots+\binom{p}{p-2}a^2+\binom{p}{p-1}a$$

is divisible by p since a^p-a is divisible by p by assumption and $\binom{p}{k}$, $1\le k\le p-1$, is divisible by p (Problem I.2.34).

I.2.37. ANSWER: $\binom{n}{0}+\binom{n}{1}+\cdots+\binom{n}{n}=(1+1)^n=2^n$.

I.2.39. There will be $\binom{n}{k}$ walkers at the point $(k, n-k)$.

I.2.40. ANSWER: $\binom{n}{3}-\binom{m}{3}$.

I.2.41. The number of ways in which four books can be arranged on a bookshelf is equal to the number of ways of ordering a set of four elements, i.e., to

$$P_4=1\cdot 2\cdot 3\cdot 4=24.$$

I.2.42. Even numbers can be put to even positions (there are n such positions) in $n!$ ways. To every distribution of even numbers in even positions there correspond $n!$ distributions of odd numbers in odd positions. Therefore, according to the multiplication rule, the total number of permutations satisfying the condition of the problem is equal to $n!\cdot n!=(n!)^2$.

I.2.43. Let us find the number of permutations is which two given elements, a and b, are placed next to each other. The following cases are possible: a occupies the first place, a occupies the second place, ... , a occupies the $(n-1)$th place, and b is to the right of a. The number of such cases is $n-1$. Besides, a and b can be interchanged, and hence there are $2(n-1)$ ways of placing a and b side by side. To each of them there correspond $(n-2)!$ permutations of other elements. Thus, the number of permutations in which a and b are next to each other is equal to $2(n-1)(n-2)!=2(n-1)!$, and the required number of permutations is equal to $n!-2(n-1)!=(n-1)!\,(n-2)$.

I.2.44. If 8 castles are placed so that they cannot take one another, then there is only one castle on every vertical line and on every horizontal line. Consider one of such arrangements of the castles. Let the castle of the first horizontal line stand in the a_1th vertical line, the castle of the second horizontal line in the a_2th vertical line, etc., and the castle of the last horizontal line in the a_8th vertical line. Among the numbers $(a_1, \dots, a_8)$ no two are equal, since otherwise there would be two castles standing on the same vertical line. Therefore, $(a_1, \dots, a_8)$ is a permutation of the set $(1, \dots, 8)$. Thus, to every arrangement of the castles there corresponds a certain permutation of the numbers $1, 2, \dots, 8$. Conversely, to every permutation of the numbers $1, 2, \dots, 8$ there corresponds an arrangement of the castles where none of them can take another. Therefore, the required number of arrangements is equal to

$$P_8 = 1 \cdot 2 \cdot 3 \cdot 4 \cdot 5 \cdot 6 \cdot 7 \cdot 8 = 40\,320.$$

I.2.45. The required number is equal to the number of arrangements of 25 elements taken 4 at a time:

$$A_{25}^4 = 25 \cdot 24 \cdot 23 \cdot 22 = 303\,600.$$

I.2.46. The desired number is equal to the number of 4-element ordered subsets (the days of taking examinations) of a set of eight elements, i.e., to $A_8^4 = 8 \cdot 7 \cdot 6 \cdot 5 = 1680$. If the last examination is known to be taken on the eighth day, then the number is $4 \cdot A_7^3 = 7 \cdot 6 \cdot 5 \cdot 4 = 840$.

I.2.53. ANSWER:

$$\frac{10!}{2!\,3!\,2!} = 151200.$$

I.2.54. ANSWER:

$$\frac{n!}{k_1! \cdots k_m!}.$$

I.2.60. a) Mark every ball by the letter a (all balls are identical). Arrange n letters a in succession. Put a bar in front of the first letter and after the last one. Then place $N-1$ bars in the blanks between letters and interpret the space between any two consecutive bars as the content of the corresponding urn. For instance, the sequence of symbols

$$|\,aa\,|\,|\,a\,|\,|\,aaa\,|$$

means that 6 balls are distributed among 5 urns so that the first urn contains 2 balls, the second is empty, the third contains 1 ball, the fourth is empty, and the fifth contains 3 balls. Such sequences always begin and finish with a bar, and the other $N-1$ bars and n letters a are arranged in an arbitrary order. The distribution of the balls is completely determined by the choice of places for $N-1$ bars. It remains to observe that $N-1$ places for bars can be chosen of $N-n+n$ places in $\binom{N+n-1}{N-1}$ ways.

b) It is necessary to calculate the number of ways in which $N-1$ bars can be placed so that there will be a letter on either side of every bar. There are $n-1$ blanks between the letters, and any $N-1$ of them can be chosen as places for bars.

I.2.61. a) Let n identical balls be distributed among N urns so that the first urn contains x_1 balls, ... , the Nth urn contains x_N balls. Then

$$x_1 + x_2 + \dots + x_N = n.$$

Conversely, any solution $(x_1, \dots, x_N)$ of this equation determines a certain arrangement of n balls in N urns (x_1 balls in the first urn, ..., and x_N balls in the Nth urn). Thus there is a one-to-one correspondence between the set of all nonnegative solutions of the equation $x_1 + x_2 + \cdots + x_N = n$ and the set of all possible arrangements of n identical balls in N urns. Therefore, the number of solutions is equal to $\binom{N+n-1}{n-1} = \binom{N+n-1}{N}$.

b) To positive solutions of the equation $x_1 + x_2 + \cdots + x_N = n$ there correspond arrangements of n balls in N urns such that none of the urns is empty.

ANSWER: $\binom{N-1}{n-1}$.

I.2.62. ANSWER: $\binom{N+n-1}{n}$.

I.2.64. The following combinations can be formed of three elements, a, b, and c, taken two at a time with repetitions:

$$aa \quad ac \quad bc \quad ab \quad bb \quad cc.$$

The number of different combinations of m elements taken n at a time is

$$f_m^n = \binom{m+n-1}{m-1} = \binom{m+n-1}{n}.$$

Indeed, every combination is completely determined by specifying how many times an element of every type appears in the combination. With each combination we associate a sequence of zeros and ones formed according to the following rule: write as many ones as there are elements of the first type in the combination, then insert zero and after that write as many ones as there are elements of the second type in the combination, and so on. For example, with the above combinations of three elements taken two at a time with repetitions, the following sequences will be associated:

$$\begin{matrix} 1100 & 1001 & 0101 \\ 1010 & 0110 & 0011. \end{matrix}$$

Thus, to every combination of m elements taken n at a time with repetitions there corresponds a sequence of n ones and $m-1$ zeros, and conversely, every such sequence uniquely determines a certain combination. Therefore, the number of combinations of m elements taken n at a time with repetitions is equal to the number of sequences of n ones and $m-1$ zeros, i.e., to $\binom{m+n-1}{m-1}$.

I.2.65. ANSWER: $\binom{16}{10}$.

I.2.66. ANSWER: $f_{r+1}^2 = \binom{r+2}{r}$.

I.2.70. Consider tables of values of such functions. ANSWER: n^k.

I.2.71. c) It suffices to demonstrate that each element in $A_1 \cup A_2 \cup \cdots \cup A_n$ is counted on the right-hand side of the equality c) exactly once. Consider an arbitrary element a of $A_1 \cup A_2 \cup \cdots \cup A_n$. Assume that a enters exactly m sets A_i. Then the element a is counted on the right-hand side of c)

$$\binom{m}{1} - \binom{m}{2} + \binom{m}{3} - \cdots + (-1)^{m-1}\binom{m}{m}$$

times. But

$$\binom{m}{1}-\binom{m}{2}+\binom{m}{3}-\cdots+(-1)^{m-1}\binom{m}{m}$$
$$=1-\left[1-\binom{m}{1}+\binom{m}{2}-\cdots+(-1)^m\binom{m}{m}\right]=1.$$

Thus each element a of $A_1\cup A_2\cup\cdots\cup A_n$ is counted only once. This proves c).

The assertion of the problem can also be proved by induction.

I.2.75. Denote by A_k the set of those permutations in which k occupies the kth place. Put $D_n=N(A_1\cup A_2\cup\cdots\cup A_n)$. The set $A_{i_1}\cap A_{i_2}\cap\cdots\cap A_{i_k}$ contains those permutations in which the numbers $i_1,i_2,\dots,i_k$ occupy the i_1th, i_2th, ..., i_kth places, while the other $n-k$ places are occupied arbitrarily by the other $n-k$ numbers. For this reason

$$N(A_{i_1}\cap A_{i_2}\cap\cdots\cap A_{i_k})=(n-k)!$$

and

$$\sum_{1\le i_1<i_2<\cdots<i_k\le n} N(A_{i_1}\cap A_{i_2}\cap\cdots\cap A_{i_k})=\binom{n}{k}(n-k)!=\frac{n!}{k!}.$$

The equality c) in Problem I.2.71 implies that

$$N(A_1\cup A_2\cup\cdots\cup A_n)=n\left(\frac{1}{1!}-\frac{1}{2!}+\frac{1}{3!}-\cdots+(-1)^{n-1}\frac{1}{n!}\right).$$

I.2.80. Let A_i be the set of those permutations in which i occupies the ith position. Then we have

$$\begin{aligned}&N_{[m]}(A_1\cup A_2\cup\cdots\cup A_n)\\&\quad=\binom{m}{m}\binom{n}{m}(n-m)!-\binom{m+1}{m}\binom{m+1}{n}(n-m+1)!+\cdots\\&\qquad+(-1)^{n-m}\binom{n}{m}\binom{n}{n}\\&\quad=\frac{n!}{m!}\left[\frac{1}{2!}-\frac{1}{3!}+\cdots+(-1)^{n-m}\frac{1}{(n-m)!}\right].\end{aligned}$$

I.2.81. See the solution of Problem I.2.82.

I.2.82. To every permutation of elements of the set A_r attach a permutation of elements of the set $A\setminus A_r$. This gives $i_r!\,(n-i_r)!$ permutations of elements of the set A. Do the same for each of the sets $A_1,\dots,A_k$ to obtain

$$\sum_{r=1}^{k} i_r!\,(n-i_r)!$$

permutations of elements of A. By the condition of the problem, all these permutations are different. Therefore,

$$\sum_{r=1}^{k} i_r!\,(n-i_r)!\le n!,$$

whence

$$\sum_{r=1}^{k} \binom{n}{i_r}^{-1} \le 1.$$

The assertion just proved implies the assertion of the previous theorem (Sperner's theorem). Indeed, since $\binom{n}{i_r} \le \binom{n}{m}$ for $m = [n/2]$, we have

$$1 \ge \sum_{r=1}^{k} \binom{n}{i_r}^{-1} \ge k\binom{n}{m}^{-1},$$

whence $k \le \binom{n}{m}$.

I.2.83. The assertion of the problem follows from Sperner's theorem. Associate with every sum of the form $\sum_{k=1}^{n} \varepsilon_k x_k$, the subset S of the set $A = \{1, 2, \dots, n\}$ such that $k \in S$ if and only if $\varepsilon_k = 1$. Now it remains to take into consideration the following fact: if the sums $\sum_{k=1}^{n} \varepsilon_k x_k$ and $\sum_{k=1}^{n} \varepsilon'_k x_k$ lie inside an interval of length 2, then none of the sets S and S' is part of the other.

§3.

I.3.1. Answer:

$$\Omega = \{\text{HH, TH, HT, TT}\}, \quad A = \{\text{HH, TH, HT}\}, \quad B = \{\text{HH, TH}\}.$$

I.3.3. Answer:

$$\Omega = \{(i,j)\colon 1 \le i \le 6, 1 \le j \le 6\}, \qquad A = \{(2,6), (3,5), (4,4), (5,3), (6,2)\},$$
$$B = \{(i,6), 1 \le i \le 5; (6,6); (6,j), 1 \le j \le 5\}.$$

I.3.5. Answer:

$$\Omega = \{\text{H, TH, TTH}, \dots, \underbrace{\text{TT}\cdots\text{T}}_{n \text{ times}}\text{H}, \dots\}.$$

I.3.7. Answer: $\Omega = [0,1] \times [0,1]$.

I.3.12. Ω consists of all possible batches containing m items (the number of such batches is equal to $\binom{N}{m}$); A consists of batches of m items among which exactly l are defective. The number of such batches is equal to $\binom{n}{l}\binom{N-n}{m-l}$.

Answer:

$$\mathsf{P}(A) = \frac{\binom{n}{l}\binom{N-n}{m-l}}{\binom{n}{m}}.$$

I.3.17. Answer: a) $\frac{5}{14}$; b) $\frac{5}{14}$; c) $\frac{2}{7}$.

I.3.36. b) If $\omega_k \in A \cup B \cup C$, then the number p_k is taken into account on the right-hand side of the equality once; if $\omega_k \notin A \cup B \cup C$, then p_k is not accounted for. The assertions a) and c) are proved in a similar way.

§4.

I.4.1. Let A be the event that among the digits selected there is no digit 0, and B the event that among the digits selected there is no digit 1. Then

$$\mathsf{P}(A) = \frac{9^k}{10^k}, \qquad \mathsf{P}(B) = \frac{9^k}{10^k}, \qquad \mathsf{P}(A \cap B) = \frac{8^k}{10^k},$$
$$\mathsf{P}(A \cup B) = \mathsf{P}(A) + \mathsf{P}(B) - \mathsf{P}(A \cap B) = \frac{2 \cdot 9^k - 8^k}{10^k}.$$

I.4.4. ANSWER: $\dfrac{6!}{6^6} \approx 0.0154$.

I.4.5. ANSWER: $\dfrac{A_{10}^7}{10^7} \approx 0.06$.

I.4.6. The probability that 1 will not show up at a throw of four dice is equal to $p = 5^4/6^4$. The probability that the pair $(1,1)$ will not appear at 24 throws of two dice is equal to $Q = (35/36)^{24}$. Considering that $(1+x)^n \geq 1+nx$ for $x > -1$, we get

$$\sqrt[4]{Q} = \left(\frac{35}{36}\right)^6 = \left(1 - \frac{1}{36}\right)^6 > 1 - \frac{1}{6} = \sqrt[4]{p},$$

whence $Q > p$, i.e., $1 - Q < 1 - p$.

I.4.7. ANSWER: $\dfrac{12!}{12^{12}} \approx 0.0005$.

I.4.8. ANSWER: $\dfrac{30!}{2^6 6^6} \cdot \dbinom{12}{6} \cdot \dfrac{1}{2^{30}} \approx 0.00035$.

I.4.9. ANSWER: $\dfrac{2(n-r-1)}{n(n-1)}$.

I.4.10. a) Possible outcomes of the experiment are sets of k balls drawn from $n+m$ balls. Their number is equal to $\binom{m+n}{k}$. Let A_r be the event that there are r white balls among the k balls drawn. The outcomes of the experiment that produce A_r are those sets of k balls among which there are exactly r white balls. Their number is equal to $\binom{n}{r}\binom{m}{k-r}$. Therefore,

$$\mathsf{P}(A_r) = \frac{\binom{n}{r}\binom{m}{k-r}}{\binom{n+m}{k}}.$$

It is evident that $r \leq n$, $r \leq k$, $k - r \leq m$, that is, $\max\{0, k-m\} = r \leq \min\{n, k\}$.

b) Since one of the events A_r will definitely occur, we have

$$\mathsf{P}\Bigg(\bigcup_{r=\max\{0,k-m\}}^{\min\{n,k\}} A_r\Bigg) = 1.$$

I.4.12. ANSWER:

$$\sum_{s=1}^{r} \frac{\binom{m}{s}\binom{n-m}{r-s}}{\binom{n}{r}}.$$

I.4.14. Let p be the probability of guessing all 6 sports. It is equal to $\binom{49}{6}^{-1}$; the probability of guessing i sports is equal to

$$\frac{\binom{6}{i}\binom{43}{6-i}}{\binom{49}{6}};$$

the probability of receiving a prize in "Sportloto" is equal to

$$\sum_{i=3}^{6}\frac{\binom{6}{i}\binom{43}{6-i}}{\binom{49}{6}}.$$

I.4.15. ANSWER:

$$\text{a) } \frac{\binom{2}{1}\binom{2n-2}{n-1}}{\binom{2n}{n}} = \frac{n}{2n-1}; \qquad \text{b) } \frac{2\binom{2}{2}\binom{2n-2}{n-2}}{\binom{2n}{n}} = \frac{n-1}{2n-1}; \qquad \text{c) } \frac{\binom{4}{2}\binom{2n-4}{n-2}}{\binom{2n}{n}}.$$

I.4.16. ANSWER:

$$\text{a) } \frac{\binom{n}{k}}{\binom{N}{k}}, \quad n = \left[\frac{N}{q}\right]; \qquad \text{b) } \frac{2\binom{n}{k}}{\binom{N}{k}}, \quad n = \left[\frac{N}{q_1}\right] + \left[\frac{N}{q_2}\right] - \left[\frac{N}{q_1q_2}\right];$$

$$\text{c) } 1 - \frac{\binom{N-n}{k}}{\binom{N}{k}}, \quad n = \left[\frac{N}{q}\right].$$

I.4.17. ANSWER:

$$\text{a) } \frac{n^k}{N^k}, \quad n = \left[\frac{N}{q}\right]; \qquad \text{b) } \frac{n^k}{N^k}, \quad n = \left[\frac{N}{q_1}\right] + \left[\frac{N}{q_2}\right] - \left[\frac{N}{q_1q_2}\right];$$

$$\text{c) } 1 - \frac{(N-n)^k}{N^k}, \quad n = \left[\frac{N}{q}\right].$$

I.4.18. ANSWER:

$$\text{a) } \frac{N-k+1}{\binom{N}{k}}; \qquad \text{b) } \frac{N-k+1}{A_n^k}; \qquad \text{c) } \frac{\binom{N}{k}}{\binom{N}{k}k!} = \frac{1}{k!}.$$

I.4.19. ANSWER:

$$\text{a) } \frac{(N-k+1)k!}{N^k}; \qquad \text{b) } \frac{N-k+1}{N^k}; \qquad \text{c) } \frac{A_N^k}{N^k}; \qquad \text{d) } \frac{\binom{N}{k}}{N^k}.$$

I.4.20. ANSWER: $\dfrac{12!}{2^6 6^{12}} \approx 0.003438.$

I.4.21. ANSWER: $\dfrac{n!}{n_1!\,n_2!\cdots n_6!}\dfrac{1}{6^n}.$

I.4.23. a) Possible outcomes of the experiment can be interpreted as permutations of the set $1, 2, \ldots, n$. Their number is equal to $n!$. Let A be the event that at least one letter will be mailed correctly. This event is obtained from those permutations in which at least one number is in its place. As computed in Problem I.2.80, the number of such permutations is equal to

$$n!\left(1 - \frac{1}{2!} + \cdots + \frac{(-1)^{n-1}}{n!}\right).$$

Therefore,

$$\mathsf{P}(A) = 1 - \frac{1}{2!} + \frac{1}{3!} - \cdots + \frac{(-1)^{n-1}}{n!}.$$

Taking the Taylor expansion of e^{-x}, one sees that

$$\mathsf{P}(A) \to 1 - \frac{1}{e} \quad \text{as } n \to \infty.$$

b) Use the result of Problem I.2.80. ANSWER:

$$\frac{1}{m!}\left[\frac{1}{2!} - \frac{1}{3!} + \cdots + (-1)^{n-m}\frac{1}{(n-m)!}\right].$$

I.4.25. ANSWER:

$$\frac{1}{2!} - \frac{1}{3!} + \cdots + \frac{(-1)^n}{n!}.$$

I.4.26. a) ANSWER: $\dfrac{n!}{n_1!\,n_2!\cdots n_N!}\cdot\dfrac{1}{N^n}$.

b) Every particle can be in one of N cells. Therefore, there are

$$\underbrace{N \times \cdots \times N}_{n \text{ times}} = N^n$$

ways to distribute n particles among N cells. The number of ways in which the particles can be distributed so that there will be k particles in a given cell, can be computed in the following manner: for a given cell, k particles can be chosen in $\binom{n}{k}$ ways, while the other $n-k$ particles can be distributed among $N-1$ cells in $(N-k)^{n-k}$ ways. Hence

$$p_k = \binom{n}{k}\frac{(N-1)^{n-k}}{N^n} = \frac{1}{N^k}\binom{n}{k}\left(1 - \frac{1}{N}\right)^{n-k}.$$

To find the most probable value of k, it is necessary to study the behavior of the ratio p_{k+1}/p_k as k increases from 0 to n. Then we will see that the most probable value of k can be found from the inequalities

$$\frac{n-N+1}{N} < k < \frac{n+1}{N}.$$

d) We find the probability that at least one cell will be empty. Let $A_1, \dots, A_N$ be the distributions of n particles among N cells such that the first, the second, $\dots$, the Nth cell is empty. Use the result of Problem I.2.75 to obtain

$$N(A_1 \cup \cdots \cup A_N) = \binom{N}{1}(N-1)^n - \binom{N}{2}(N-2)^n + \cdots + (-1)^{N-1}\binom{N}{N-1}1^n.$$

Therefore, the probability that all the cells are occupied is equal to

$$\frac{1}{N^n}\left(N^n - \binom{N}{1}(N-1)^n + \binom{N}{2}(N-2)^n - \cdots + (-1)^{N-1}\binom{N}{N-1}1^n\right).$$

e) Using the result of the previous problem, we see that among N^n distributions of n particles among N cells there are

$$\binom{n}{r}\left[r^n - \binom{r}{1}(r-1)^n + \binom{r}{2}(r-2)^n - \cdots + (-1)^{r-1}\binom{r}{r-1}1^n\right]$$

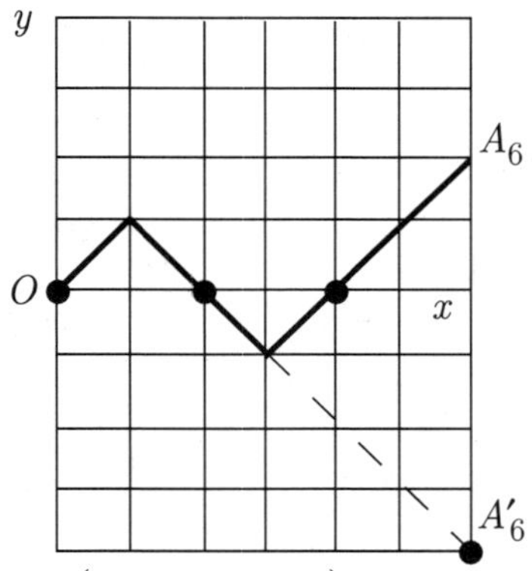

FIGURE 9

distributions such that exactly r cells are occupied. Therefore, the required probability is equal to

$$\frac{1}{N^n}\binom{n}{r}\left[r^n - \binom{r}{1}(r-1)^n + \binom{r}{2}(r-2)^n - \cdots + (-1)^{r-1}\binom{r}{r-1}1^n\right].$$

I.4.29. ANSWER:

$$\text{a) } \binom{9}{3}\frac{2^6}{3^9}; \qquad \text{b) } \binom{9}{3}\binom{6}{3}\binom{3}{3}\frac{1}{3^9}; \qquad \text{c) } \binom{9}{4}\binom{5}{3}\binom{2}{2}\frac{3!}{3^9}.$$

I.4.30. a) Use the result of Problem I.2.63. ANSWER: $\binom{N+n-1}{n}^{-1}$.
b) Use the result of Problem I.2.60. ANSWER:

$$\frac{\binom{N}{m}\binom{N-m-1}{n-1}}{\binom{N+n-1}{n}}.$$

I.4.31. ANSWER: $\binom{N}{n}^{-1}$.

I.4.32. a) Assign a number to each person in the line. Let ε_i be equal to 1 if the ith person has 50 copecks, and $\varepsilon_i = -1$ if the ith person has a rouble. Consider the sum $S_k = \varepsilon_1 + \cdots + \varepsilon_k$, which is the difference between the number of 50 copeck coins and the number of rouble bills given to the cashier by the first k people. In the coordinate system xOy, consider the points $A_k = (k, S_k)$, $k = 1, 2, \dots, m+n$, and draw the polygonal line that joins the point O to the point $A_{n+m} = (m+n, n-m)$ and passes through the points $A_1, A_2, \dots, A_{m+n-1}$ (Figure 9). This polygonal line will be called the trajectory corresponding to a given arrangement of people in the line. Every trajectory consists of $m+n$ segments, n of which go up and m go down. Any trajectory is completely determined by specifying the segments that go up. The number of different arrangements of people in the line equals $\binom{m+n}{n}$.

The trajectories that correspond to those arrangements of people in the line for which no one has to wait for change are those that do not cross the straight line $y = -1$. In order to count the number of such trajectories, for every trajectory T that crosses or touches the line $y = -1$, we construct the new trajectory T' according to the following rule: up to the first point of contact with the line $y = -1$, the trajectory T' coincides with T, and then T' is the symmetric image of the trajectory

T about the line $y = -1$. (In Figure 9 the trajectory T' is represented by the dashed line.) All trajectories T' terminate at the point $A'_{m+n} = (m+n, m-n-2)$, which is the symmetric image of the point A_{m+n} about the line $y = -1$. The correspondence described is one-to-one. Hence the required number of trajectories is equal to the number of polygonal lines joining O to A'_{m+n}. This number can be easily counted. If a polygonal line consists of y segments that go down and x segments that go up, then $x + y = m + n$; $y - x = n + 2 - m$, whence $y = n + 1$. The number of trajectories leading from O to A_{m+n} is equal to $\binom{m+n}{n+1}$. The number of trajectories that do not cross the line $y = -1$ is equal to $\binom{m+n}{n} - \binom{m+n}{n+1}$. Thus the probability that nobody will have to wait for change is equal to

$$\frac{\binom{m+n}{n} - \binom{m+n}{n+1}}{\binom{m+n}{n}} = \frac{n+1-m}{n+1}.$$

b) The problem reduces to the calculation of the number of trajectories from the point O to the point $(m+n, n-m)$ that do not cross the straight line $y = -(p+1)$. By Theorem 2, the number of the trajectories that do not cross this line is equal to the number of trajectories from the point $(O, -2(p+1))$ to the point $(n+m, n-m)$ i.e., to $\binom{m+n}{p+n+1}$. The desired number of trajectories is

$$\binom{m+n}{n} - \binom{m+n}{m-p-1},$$

and the probability that no one will have to wait for his change is

$$\frac{\binom{m+n}{m} - \binom{m+n}{m-p-1}}{\binom{m+n}{m}}.$$

REMARK. Problems I.4.33–I.4.37 can be solved using geometrical considerations similar to those used in solving Problems I.4.32 and I.4.33. These considerations are based on the count of the number of trajectories, which becomes much easier if the following assertions are kept in mind.

THEOREM 1. *Let $N_{x,y}$ be the number of trajectories from the point O to the point (x, y). Then*

$$N_{x,y} = \frac{x!}{\left(\frac{x+y}{2}\right)!\left(\frac{x-y}{2}\right)!}$$

if x and y are of the same parity, and $N_{x,y} = 0$ if not.

Indeed, if a trajectory consists of p segments that go up and q segments that go down, then $p+q = x$ and $p-q = y$, whence $p = (x+y)/2$ and $q = (x-y)/2$. Since a trajectory is completely determined if the segments that go up are specified, we have

$$N_{x,y} = \binom{x}{\frac{x+y}{2}} = \frac{x!}{\left(\frac{x+y}{2}\right)!\left(\frac{x-y}{2}\right)!}.$$

THEOREM 2 (Reflection principle). *Let $A = (a, \alpha)$ and $B = (b, \beta)$ be points with integer-valued coordinates such that $b > a \geq 0$, $\alpha > 0$, $\beta > 0$, and let $A' = (a, -\alpha)$ be the mirror image of A about the Ox axis. Then the number of trajectories from*

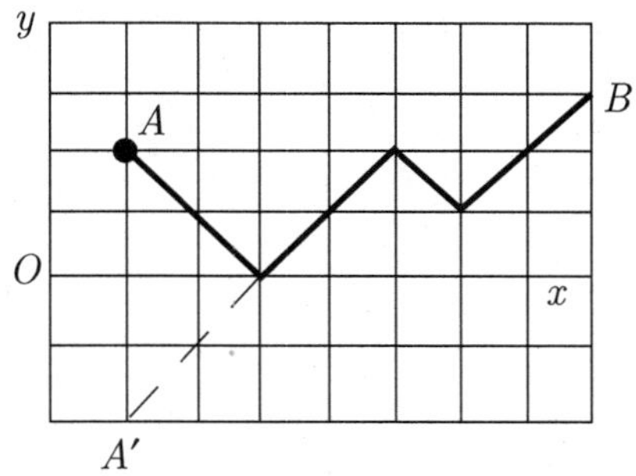

FIGURE 10

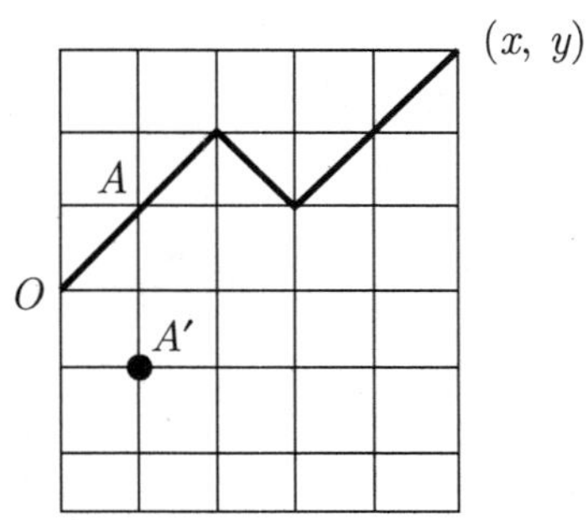

FIGURE 11

A to B that either touch or cross the Ox axis is equal to the number of trajectories from A' to B.

With every trajectory T from A to B that either touches or crosses the Ox axis we associate the trajectory from A' to B according to the following rule (Figure 10): the section of T up to its first contact with the Ox axis is reflected about the axis, and from this point the trajectories T' and T coincide. Thus to every trajectory T from A to B that crosses the Ox axis there corresponds a certain trajectory T' from A' to B. Conversely, to every trajectory from A' to B there corresponds exactly one trajectory from A to B that crosses the Ox axis (take the section of the trajectory from A' to B to its first intersection with the Ox axis and reflect it about Ox). Thus there is a one-to-one correspondence between the set of trajectories from A to B that either touch or cross the Ox axis and the set of trajectories from A' and B. The theorem is proved.

THEOREM 3. *Let $x > 0$ and $y > 0$. Then the number of trajectories from O to (x, y) that have no vertices on the Ox axis (except the point O) is equal to $yx^{-1}N_{x,y}$.*

All trajectories from O to (x, y) that do not cross the Ox axis pass through the point $A(1, 1)$ (see Figure 11). By Theorem 2, the total number of trajectories from A to B that do not cross the Ox axis is equal to the number of trajectories from

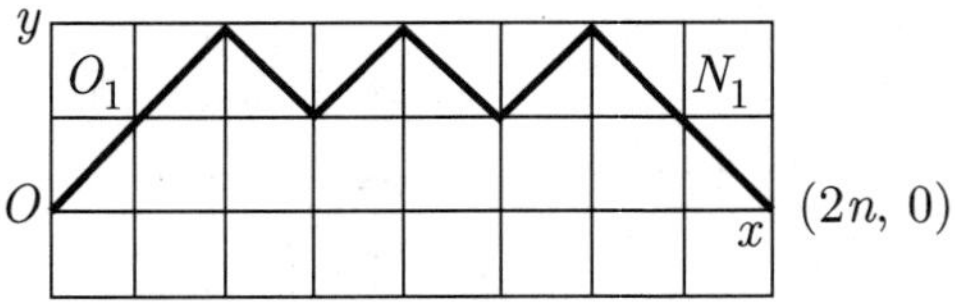

FIGURE 12

A' to B, i.e., to $N_{x-1,y+1}$. Thus the desired number of trajectories is equal to

$$N_{x-1,y-1} - N_{x-1,y+1} = \frac{(x-1)!}{\left(\frac{x+y}{2}-1\right)!\,\left(\frac{x-y}{2}\right)!} - \frac{(x-1)!}{\left(\frac{x+y}{2}\right)!\,\left(\frac{x-y}{2}-1\right)!}$$
$$= \frac{y}{x}\frac{x!}{\left(\frac{x+y}{2}\right)!\,\left(\frac{x-y}{2}\right)!} = \frac{y}{x}N_{x,y}.$$

Now we establish some properties of trajectories that join the point O to the point $(2n, 0)$ on the Ox axis.

Put

$$L_{2n} = \frac{1}{n+1}\binom{2n}{n}.$$

THEOREM 4. *Among $\binom{2n}{n}$ trajectories that join the point O to the point $(2n, 0)$, there are*

a) *exactly L_{2n-2} trajectories that lie above the Ox axis and have no common points with Ox except O and $(2n, 0)$;*
b) *exactly L_{2n} trajectories that have no vertices below the Ox axis.*

All the trajectories from O to $(2n, 0)$ that lie above the Ox axis and have no common points with the Ox axis pass through the point $(2n-1, 0)$. According to Theorem 3, the number of trajectories that join O to $(2n-1, 1)$ and do not cross the Ox axis is equal to

$$\frac{N_{2n-1,1}}{2n-1} = \frac{\binom{2n-1}{n}}{2n-1} = \frac{\binom{2n-2}{n-1}}{n-1} = L_{2n-2}.$$

Consider a trajectory that joins O to $(2, 0)$ and has no common points with the Ox axis. Delete the first and the last segment of this trajectory (Figure 12) to obtain the trajectory that joins Q_1 to N_1 and has no vertices below the Ox axis. We translate the origin to the point O_1. Then the number of trajectories that join O_1 to $(2n-2, 0)$ (now the point N_1 has the coordinates $(2n-2, 0)$) and have no vertices below the Ox axis is equal to L_{2n-2}. Therefore, the number of trajectories that join O to $(2n, 0)$ and have no vertices below the Ox axis is equal to L_{2n}.

I.4.33. *The first method.* Let ε_i be equal to $+1$ if the ith person voted for A, and -1 if the vote was for B. Put $S_k = \varepsilon_1 + \cdots + \varepsilon_k$ and consider, in the xOy coordinate system (Figure 13), the polygonal line that joins the points O, $(1, S_1)$, ... , $(a+b, S_{a+b})$. Obviously, $S_{a+b} = a - b$. To every outcome of voting there corresponds a certain polygonal line (a trajectory) that joins the point O to $(a+b, a-b)$. The trajectory consists of $a+b$ segments of which a segments go up. Therefore, the total number of trajectories is $\binom{a+b}{a}$. The candidate A will be constantly ahead of B if the corresponding trajectory passes through the point $(1, 1)$ (the first vote must be given for A) and does not cross the Ox axis. The

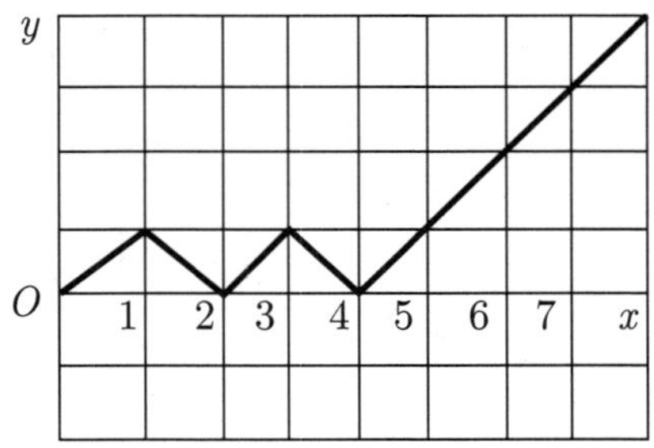

$(a = 5,\ b = 2)$

FIGURE 13

number of such trajectories was counted in the solution of the previous problem (in the case under consideration we must put $n = a - 1$ and $m = b$).

Thus, the number of outcomes of voting is

$$\binom{a+b-1}{a-1}\frac{a-1+1-b}{a-1+1} = \frac{a-b}{a+b}\binom{a+b}{b}.$$

The probability that the candidate A will constantly be ahead of B is $(a-b)/(a+b)$.

The second method. Consider an arrangement of a symbols A and b symbols B on a circle. Let us find the number of those initial positions of the letter A starting at which (and moving, for instance, in the clockwise direction) we shall always encounter letters A at least as often as letters B. In order to find these positions, cross out consecutively all neighboring pairs. (It is quite possible that this will require more than one turn along the circle.) As a result, $a - b$ symbols A will remain, which will be just the sought-for initial positions. Thus the desired probability is equal to $(a - b)/(a + b)$.

I.4.36. ANSWER: $(n + 1)^{-1}$.

I.4.37. ANSWER: $2^n \binom{2n}{n}^{-1}$.

§**5.**

I.5.1. ANSWER: $1 - (1 - k)^2$.

I.5.3. ANSWER: $(a - d)^2 a^{-2}$.

I.5.5. ANSWER:

$$\Omega = \{\,(x, y, z){:}\, 0 \le x \le a, 0 \le y \le a, 0 \le z \le a\,\};$$

$$A = \{\,(x, y, z){:}\, x + y > z, x + z > y, y + z > x\,\}; \qquad \mathsf{P}(A) = \frac{1}{2}.$$

I.5.7. ANSWER: 6/19.

I.5.11. ANSWER: 1/4.

I.5.13. ANSWER: 2/3.

I.5.14. ANSWER: $\left(1 - r^3/R^3\right)^N$.

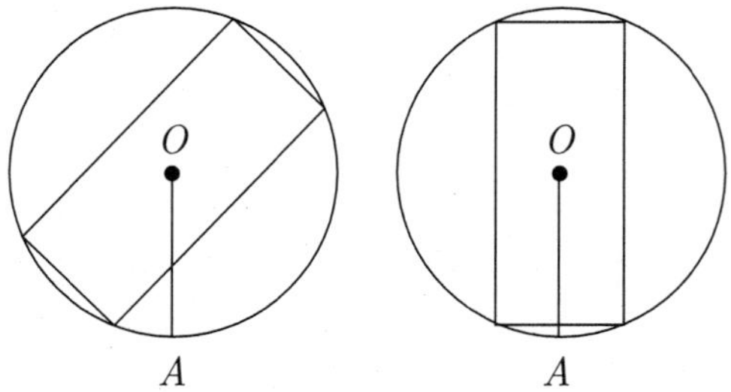

FIGURE 14

I.5.15. a) Let x be the distance from the middle of the needle to the nearest line, and let y be the angle formed by the needle with this line. Then $\Omega = \{(x, y): 0 \leq x \leq a, 0 \leq y \leq \pi\}$, $A = \{(x, y): x \leq l \sin y\}$, and hence

$$\mathsf{P}(A) = \frac{1}{a\pi} \int_0^\pi l \sin \varphi \, d\varphi = \frac{2l}{a\pi}.$$

I.5.16. Imagine a cylinder inscribed in a sphere whose center coincides with the center of mass of the coin. Let us choose at random a point on the surface of the sphere. If a radius drawn from the center to this point intersects the lateral surface of the cylinder, then the coin is considered to have fallen on the rim (Figure 14).

Since the area of a spheric zone is equal to the product of the height of the zone by the length of a great circle, we conclude that the thickness of the coin must be equal to $\frac{2}{3}$ of the sphere radius. If r is the radius of the coin and R that of the sphere, then by the Pythagorean theorem $R^2 = \frac{1}{9}R^2 + r^2$, whence $\frac{1}{3}R = \frac{\sqrt{2}}{4}r \approx 0.354r$, i.e., the thickness of the coin must be equal to $\frac{\sqrt{2}}{4} \approx 0.354$ of its diameter.

§6.

I.6.6. a) ANSWER: $1 - \mathsf{P}(A \cap B)$.

I.6.10. ANSWER: $p_0 = -\mathsf{P}(A) - \mathsf{P}(B) + \mathsf{P}(A \cap B)$; $p_1 = \mathsf{P}(A) + \mathsf{P}(B) - 2\mathsf{P}(A \cap B)$; $p_2 = \mathsf{P}(A \cap B)$.

I.6.12. ANSWER: $\mathsf{P}(A \cap B) = \mathsf{P}(A) + \mathsf{P}(B) - \mathsf{P}(A \cup B) \geq 0.8 + 0.8 - 1 = 0.6$, since $\mathsf{P}(A \cup B) \leq 1$.

I.6.14. Use the induction and the inequality

$$\mathsf{P}(A_1 \cap A_2) = \mathsf{P}(A_1) + \mathsf{P}(A_2) - \mathsf{P}(A_1 \cup A_2) \geq \mathsf{P}(A_1) + \mathsf{P}(A_2) - 1.$$

I.6.16. Let $\mathsf{P}(A) \geq \mathsf{P}(B)$. Since $\mathsf{P}(A \cap B) \leq \mathsf{P}(B)$, we have

$$\mathsf{P}(A \cap B) - \mathsf{P}(A)\mathsf{P}(B) \leq \mathsf{P}(B)[1 - \mathsf{P}(A)] \leq \mathsf{P}(A)[1 - \mathsf{P}(A)] \leq \frac{1}{4}.$$

Put $x = \mathsf{P}(A \cap B)$, $a = \mathsf{P}(A \cap \overline{B})$, and $b = \mathsf{P}(\overline{A} \cap B)$. Then

$$\begin{aligned}\mathsf{P}(A \cap B) - \mathsf{P}(A)\mathsf{P}(B) &= x - (a + x)(b + x) = x - [ab + x(a + b + x)] \\ &\geq x - ab - x = -ab \geq -\frac{1}{4},\end{aligned}$$

since

$$a \leq \mathsf{P}(A), \qquad b \leq \mathsf{P}(\overline{A}), \qquad ab \leq \mathsf{P}(A)[1 - \mathsf{P}(A)] \leq \frac{1}{4}.$$

I.6.18. Note that

$$\mathsf{P}(A\cap B)+\mathsf{P}(A\cap\overline{B})+\mathsf{P}(\overline{A}\cap B)+\mathsf{P}(\overline{A}\cap\overline{B})=1.$$

Put $\mathsf{P}(A\cap B)=x+\frac{1}{4}$, $\mathsf{P}(A\cap\overline{B})=y+\frac{1}{4}$, $\mathsf{P}(\overline{A}\cap B)=z+\frac{1}{4}$, and $\mathsf{P}(\overline{A}\cap\overline{B})=w+\frac{1}{4}$. Then $x+y+z+w=0$. Therefore,

$$\begin{aligned}&\left(x+\frac{1}{4}\right)^2+\left(y+\frac{1}{4}\right)^2+\left(z+\frac{1}{4}\right)^2+\left(w+\frac{1}{4}\right)^2\\&\qquad=x^2+y^2+z^2+w^2+\frac{x+y+z+w}{2}+\frac{1}{4}\geq\frac{1}{4},\end{aligned}$$

with the equality if and only if $x=y=z=w=0$.

I.6.21. *Necessity.* The inequalities are true since their left-hand sides are the probabilities of the random events $\overline{A}\cap\overline{B}$, $A\cap\overline{B}$, $\overline{A}\cap B$, and $A\cap B$.

Sufficiency. The inequalities, which are assumed to hold, imply that

$$0\leq p_{12}\leq p_1\leq p_1+p_2-p_{12}\leq 1,\qquad 0\leq p_{12}\leq p_2\leq p_1+p_2-p_{12}\leq 1.$$

Let $\Omega=[0,1]$, $\mathfrak{A}$ the σ-algebra of measurable sets, and $P(\cdot)$ Lebesgue measure. Consider the probability space $(\Omega,\mathfrak{A},\mathsf{P})$. Let $A=(0,p_1)$ and $B=(p_1-p_{12},p_1+p_2-p_{12})$. Then $\mathsf{P}(A)=p_1$, $\mathsf{P}(B)=p_2$, and $\mathsf{P}(A\cap B)=p_{12}$.

I.6.24. The algebra $\mathfrak{M}$ consists of 2^{2^n} events, each being a sum of some number of pairwise disjoint "fundamental" events of the form

$$\gamma=A_{i_1}\cap\cdots\cap A_{i_k}\cap\overline{A}_{j_1}\cap\cdots\cap\overline{A}_{j_{n-k}}.$$

It is evident that the number of "fundamental" events is equal to 2^n.

I.6.25. Since every event B_k in the algebra $\mathfrak{M}$ is a sum of some number of pairwise disjoint events (see Problem I.6.24), the inequality

$$\sum_{k=1}^{m}C_k\mathsf{P}(B_k)\geq 0 \tag{1}$$

is equivalent to the inequality

$$\sum_{k=1}^{m}C_k\mathsf{P}(B_k)=\sum_{\gamma}\lambda_\gamma\mathsf{P}(\gamma)\geq 0, \tag{2}$$

in which all 2^n "fundamental" events are added together and

$$\lambda_\gamma=\sum_{\gamma\in B_k}C_k. \tag{3}$$

The sum (3) contains only those events, for which the "fundamental" event γ enters the representation of B_k. Note that the coefficients λ_γ in (2) depend on the numbers C_k and on the functional dependence of the events B_k on the events A_i and do not depend on the numerical values of probabilities $\mathsf{P}(A_i)$. If each of the numbers $\mathsf{P}(A_i)$, $i=1,\dots,n$, is either 0 or 1 (and there is at least one 1), then only one of the "fundamental" events has probability 1, while the probabilities of all others are 0. Therefore, the assertion $\lambda_\gamma\geq 0$ for all "fundamental" events γ is equivalent to the following assertion: the inequality (2) holds for all sets of events $\{A_i\}$ in which $\mathsf{P}(A_i)$ is either 0 or 1.

I.6.27. According to Problem I.6.25, it is sufficient to check the equality for an arbitrary group of events $\{A_i\}$, each having probability either 0 or 1. Suppose exactly l events among $A_1, \ldots, A_n$ have probability 1. Then it is necessary to check the equality

$$\sum_{k=0}^{n-r}(-1)^k\binom{r+k}{k}\binom{l}{r+k} = \begin{cases} 1 & \text{if } l = r, \\ 0 & \text{if } l \neq r. \end{cases}$$

If $l < r$, then all terms on the right-hand side are equal to 0; if $l = r$, then the left-hand side is equal to 1; and if $l > r$, then the left-hand side can be written in the form

$$\binom{l}{r}(l-1)^{l-r} = 0, \qquad l > r.$$

I.6.29. Use the assertion of Problem I.6.25. If among the events $A_1, \ldots, A_n$ exactly l events have probability 1 and the probabilities of other events are zero, then the problem reduces to the inequality

$$\binom{l}{r+1} \Big/ \binom{n}{r+1} \leq \binom{l}{r} \Big/ \binom{n}{r}.$$

§7.

I.7.1. ANSWER:

$$\Omega = \{\text{HH, HT, TH, TT}\}, \qquad A = \{\text{HH, HT}\}, \qquad B = \{\text{HH, TH}\},$$
$$\mathsf{P}(A) = \frac{2}{4} = \frac{1}{2}, \qquad \mathsf{P}(B) = \frac{2}{4} = \frac{1}{2}, \qquad \mathsf{P}(A \cap B) = \frac{1}{4}, \qquad \mathsf{P}(A/B) = \frac{1}{2}.$$

I.7.2. ANSWER:

$$\{\text{HHH, HHT, HTH, HTT, THH, THT, TTH, TTT}\},$$
$$A = \{\text{HHT, HTH, THH}\},$$
$$B = \{\text{HHH, HHT, HTH, HTT, THH, THT, TTH}\},$$
$$A \cap B = \{\text{HHT, HTH, THH}\},$$
$$\mathsf{P}(A) = \frac{3}{8}, \qquad \mathsf{P}(B) = \frac{7}{8}, \qquad \mathsf{P}(A \cap B) = \frac{3}{8}, \qquad \mathsf{P}(A/B) = \frac{3}{7}.$$

I.7.3. The space of elementary events is of the form $\Omega = \{\text{BB, BG, GB, GG}\}$. $A = \{\text{BB}\}$ is the event that there are two boys in the family; $B = \{\text{BB, BG, GB}\}$ is the event that there is at least one boy in the family; and $C = \{\text{BB, BG}\}$ is the event that the eldest child is a boy. We have

$$\mathsf{P}(A/B) = \frac{1}{3} \quad \text{and} \quad \mathsf{P}(A/C) = \frac{1}{2}.$$

I.7.5. ANSWER: $1 - \frac{5\cdot4\cdot3}{6\cdot5\cdot4} = \frac{1}{2}$.

I.7.7. ANSWER: $1 - \frac{10\cdot5^9}{6^{10}-5^{10}} \approx 0.61$.

I.7.8. Consider the random events A: the chosen person is a man; B: the chosen person is a woman; and C: the chosen person is color-blind. Then

$$\mathsf{P}(A/C) = \frac{\mathsf{P}(A \cap C)}{\mathsf{P}(C)} = \frac{\mathsf{P}(A \cap C)}{\mathsf{P}(C \cap A) + \mathsf{P}(C \cap B)} = \frac{0.05}{0.05 + 0.0025} = \frac{20}{21}.$$

I.7.10. ANSWER: $\frac{21}{46}$.

I.7.11. We have

$$\mathsf{P}(A/B) = \frac{\mathsf{P}(A \cap B)}{\mathsf{P}(B)} \leq \frac{\mathsf{P}(A)}{\mathsf{P}(B)} = \frac{p}{1-\varepsilon}.$$

On the other hand, using the inequality

$$\mathsf{P}(A \cap B) \geq \mathsf{P}(A) + \mathsf{P}(B) - 1,$$

we get

$$\mathsf{P}(A/B) \geq \frac{p-\varepsilon}{1-\varepsilon}.$$

I.7.14. ANSWER:

$$\begin{gathered} \Omega = \{\text{BB, BG, GB, GG}\}, \\ A = \{\text{BG, GB}\}, \qquad B = \{\text{BB, BG, GB}\}, \qquad A \cap B = \{\text{BG, GB}\}, \\ \mathsf{P}(A) = \frac{1}{2}, \qquad \mathsf{P}(B) = \frac{3}{4}, \qquad \mathsf{P}(A \cap B) = \frac{1}{2}, \\ \mathsf{P}(A \cap B) \neq \mathsf{P}(A)\mathsf{P}(B). \end{gathered}$$

I.7.27. If the events A, B, and C are independent, then the events $\widetilde{A}$, $\widetilde{B}$, and $\widetilde{C}$, where $\widetilde{A}$ is either A or $\overline{A}$, $\widetilde{B}$ is either B or $\overline{B}$, and $\widetilde{C}$ is either C or $\overline{C}$, are independent. A similar assertion is true for an arbitrary collection of independent events. It is easily seen that in order to prove this assertion, it suffices to establish the independence of A and $\overline{B}$, for independent events A and B. But this follows from the equalities

$$\begin{aligned} \mathsf{P}(A \cap \overline{B}) &= \mathsf{P}(A \setminus A \cap B) = \mathsf{P}(A) - \mathsf{P}(A \cap B) = \mathsf{P}(A) - \mathsf{P}(A)\mathsf{P}(B) \\ &= \mathsf{P}(A)[1 - \mathsf{P}(B)] = \mathsf{P}(A)\mathsf{P}(\overline{B}). \end{aligned}$$

Thus, if A, B, C are independent events, then so are the events A, $\overline{B}$, $\overline{C}$. Hence

$$\begin{aligned} \mathsf{P}(A \cap (B \cup C)) &= \mathsf{P}(A \setminus A \cap \overline{(B \cup C)}) = \mathsf{P}(A) - \mathsf{P}(A \cap \overline{B} \cap \overline{C}) \\ &= \mathsf{P}(A) - \mathsf{P}(A)\mathsf{P}(\overline{B})\mathsf{P}(\overline{C}) = \mathsf{P}(A)[1 - \mathsf{P}(\overline{B} \cap \overline{C})] \\ &= \mathsf{P}(A)[1 - \mathsf{P}(\overline{B \cup C})] = \mathsf{P}(A)\mathsf{P}(B \cup C). \end{aligned}$$

Similarly,

$$\begin{aligned} \mathsf{P}(A \cap (B \setminus C)) &= \mathsf{P}(A \cap B \cap \overline{C}) = \mathsf{P}(A)\mathsf{P}(B)\mathsf{P}(\overline{C}) \\ &= \mathsf{P}(A)\mathsf{P}(B \cap \overline{C}) = \mathsf{P}(A)\mathsf{P}(B/C). \end{aligned}$$

Note that in this case it is not sufficient that the events A, B, and C be pairwise independent.

I.7.28. The assertion follows from the equalities

$$\begin{aligned} \mathsf{P}(A \cap (B \setminus C)) &= \mathsf{P}\{A \cap B \cap \overline{C}\} = \mathsf{P}\{(A \cap B) \setminus (A \cap B \cap C)\} \\ &= \mathsf{P}(A \cap B) - \mathsf{P}(A \cap B \cap C) = \mathsf{P}(A \cap B) - \mathsf{P}(A \cap C) \\ &= \mathsf{P}(A)[\mathsf{P}(B) - \mathsf{P}(C)] = \mathsf{P}(A)\mathsf{P}(B \setminus C) \end{aligned}$$

if $B \supset C$, or from the equalities

$$\begin{aligned}\mathsf{P}\{A \cap (B \cup C)\} &= \mathsf{P}(A)\mathsf{P}(B) + \mathsf{P}(A)\mathsf{P}(C) \\ &= \mathsf{P}(A)[\mathsf{P}(B) + \mathsf{P}(C)] = \mathsf{P}(A)\mathsf{P}(B \cup C)\end{aligned}$$

if $B \cap C = \varnothing$.

I.7.29. Each of the events A and B is either Ω or $\varnothing$.

I.7.30. It follows from the previous problem that the events A and $B_1 \cup B_2$ are independent. Then by induction we get that the events A and $\bigcup_{k=1}^n B_k$ are independent for every n, that is,

$$\mathsf{P}\left\{A \cap \left(\bigcup_{k=1}^n B_k\right)\right\} = \mathsf{P}(A)\mathsf{P}\left(\bigcup_{k=1}^n B_k\right).$$

The sequences of events $\{A \cap (\bigcup_{k=1}^n B_k), n \geq 1\}$ and $\{\bigcup_{k=1}^n B_k, n \geq 1\}$ are monotone, that is,

$$A \cap \left(\bigcup_{k=1}^n B_k\right) \subset A \cap \left(\bigcup_{k=1}^{n+1} B_k\right) \quad \text{and} \quad \bigcup_{k=1}^n B_k \subset \bigcup_{k=1}^{n+1} B_k, \qquad n \geq 1.$$

Then, by the probability continuity theorem, we can pass to the limit as $n \to \infty$ in the equality obtained.

I.7.31. ANSWER:

a) $\mathsf{P}(\bar{A}_1 \cap \dots \cap \bar{A}_n) = \prod_{k=1}^n \mathsf{P}(\bar{A}_k) = \prod_{k=1}^n [1 - \mathsf{P}(A_k)]$;

b) $\mathsf{P}\left(\bigcup_{k=1}^n A_k\right) = 1 - \mathsf{P}\left(\overline{\bigcup_{k=1}^n A_k}\right) = 1 - \mathsf{P}\left(\bigcap_{k=1}^n \bar{A}_k\right) = 1 - \prod_{k=1}^n (1 - p_k)$;

c) $\prod_{k=1}^n (1 - p_k) \sum_{k=1}^n p_k/(1 - p_k)$ if $p_k < 1$.

I.7.32. ANSWER: $1 - (1 - p)^n$.

I.7.33. ANSWER: $\mathsf{P}(A) = 1 - (1 - p)^{mn}$, $\mathsf{P}(B) = [1 - (1 - p)^n]^m$.

I.7.35. ANSWER: a) $1 - (1 - p_1 p)^{n-1}$; b) $1 - (1 - p_1 p)^{n-2}$.

I.7.36. ANSWER: $1 - \left(\frac{35}{36}\right)^n \geq \frac{1}{2}$ for $n \geq 25$.

I.7.39. ANSWER:

a) $A = A_1 \cup A_2$, $\mathsf{P}(A) = 1 - \mathsf{P}(\bar{A} \cap \bar{A}_2) = 1 - (1 - p_1)(1 - p_2)$;

b) $A = A_1 \cap A_2$, $\mathsf{P}(A) = p_1 p_2$;

c) $A = \bigcap_{i=1}^n A_i$, $\mathsf{P}(A) = \prod_{i=1}^n p_i$;

d) $A = \bigcup_{i=1}^n A_i$, $\mathsf{P}(A) = 1 - \prod_{i=1}^n (1 - p_i)$, $A = A_1 \cup (A_2 \cap A_3 \cap A_4) \cup A_5$.

I.7.40. ANSWER: p^n.

I.7.41. ANSWER: b) $1-(1-p)(1-pp_1)$.

I.7.42. ANSWER: a) $1-(1-p)^n$; b) $1-(1-p)(1-p_1p)^{n-1}$.

I.7.44. In case a) the reliability of the system is equal to $P_a = [1-(1-p)^2]^n$. In case b) it is equal to $P_b = 1-(1-p^n)^2$. We prove that $P_a > P_b$ for $0 < p < 1$. Since

$$P_a = [1-(1-p)^2]^n = p^n(2-p)^n \quad \text{and} \quad P_b = 1-(1-p^2)^n = p^n(2-p^n),$$

it is sufficient to prove the inequality $(2-p)^n > 2-p^n$. Put $p = 1-q$. Then the inequality to be proved takes the form

$$(1+q)^n + (1-q)^n > 2. \tag{1}$$

To prove (1), note that the function $y = x^n$ is concave. Then for any x_1 and x_2 $(x_1 \neq x_2)$,

$$\frac{x_1^n + x_2^n}{2} > \left(\frac{x_1+x_2}{2}\right)^n. \tag{2}$$

Putting $x_1 = 1+q$ and $x_2 = 1-q$ in (2), we get (1).

I.7.45. ANSWER:

a) $\displaystyle\sum_{n=0}^{\infty}(2\alpha\beta)^n\alpha^2 = \frac{\alpha^2}{(1-2\alpha\beta)}$;

b) $\displaystyle\sum_{n=0}^{\infty}(2\alpha\beta)^n\beta^2 = \frac{\beta^2}{(1-2\alpha\beta)}$;

c) $\displaystyle\alpha - \frac{\alpha^2}{(1-2\alpha\beta)} = \alpha\beta\frac{\beta-\alpha}{1-2\alpha\beta} < 0$

(if $\alpha > \beta$, then it is more advantageous for A to play the whole match);

d) 0.

I.7.47. The probability that A will win is equal to

$$\mathsf{P}(A) = \sum_{n=0}^{\infty}[(1-p_1)(1-p_2)]^n p_1 = \frac{p_1}{1-(1-p_1)(1-p_2)};$$

the probability for B to win is equal to

$$\mathsf{P}(B) = (1-p_1)\frac{p_2}{1-(1-p_1)(1-p_2)}.$$

The probability that the contest will continue infinitely long is equal to

$$1-\mathsf{P}(A)-\mathsf{P}(B) = 0.$$

I.7.50. Let A be the event that the first student draws a lucky card, and B the event that the second student draws a lucky card. Obviously, $\mathsf{P}(A) = n/N$. The probability $\mathsf{P}(B)$ can be calculated by the law of total probability. Two cases are possible, H_1: the first student has taken a lucky card ($\mathsf{P}(H_1) = n/N$), and H_2: the first student has taken an unlucky card ($\mathsf{P}(H_2) = (N-n)/N$). By the law of total probability,

$$\mathsf{P}(B) = \mathsf{P}(H_1)\mathsf{P}(B/H_1) + \mathsf{P}(H_2)\mathsf{P}(B/H_2) = \frac{n}{N}\frac{n-1}{N-1} + \frac{N-n}{N}\frac{n}{N-1} = \frac{n}{N}.$$

I.7.52. Let A be the event that a white ball is drawn, and let H_i be the event that the urn number i, $i = 1, \dots, N$, is selected. Then $H_1, \dots, H_N$ form a complete group of events, $\mathsf{P}(H_i) = 1/N$, and, by the law of total probability,

$$\mathsf{P}(A) = \sum_{i=1}^{N} \mathsf{P}(H_i)\mathsf{P}(A/H_i) = \frac{1}{N}\sum_{i=1}^{N}\frac{m_i}{n_i}.$$

I.7.54. Denote by A the event that the ball drawn from the second urn is white. Let H_1 denote the hypothesis that a white ball has been transferred, and let H_2 denote the hypothesis that the ball transferred is not white. It is obvious that

$$\mathsf{P}(H_1) = \frac{m_1}{n_1} \quad \text{and} \quad \mathsf{P}(H_2) = \frac{n_1 - m_2}{n_1} = 1 - \frac{m_1}{n_1}.$$

The events H_1 and H_2 form a complete group of events. Then, by the law of total probability,

$$\mathsf{P}(A) = \mathsf{P}(H_1)\mathsf{P}(A/H_1) + \mathsf{P}(H_2)\mathsf{P}(A/H_2) = \frac{m_1}{n_1}\cdot\frac{m_2+1}{n_2+1} + \left(1 - \frac{m_1}{n_1}\right)\frac{m_2}{n_2+1}.$$

I.7.55. ANSWER: $\frac{2}{3}$.

I.7.56. The probability of detecting the object in one scanning cycle is

$$pp_1 + (1-p)p_0;$$

the probability of detecting the object in n scanning cycles is

$$1 - [1 - pp_1 - (1-p)p_0]^n.$$

I.7.58. a) The probability that there are k boys in the family is equal to

$$\sum_{n=k}^{\infty} \alpha p^n \binom{n}{k}\frac{1}{2^n}.$$

Then use the equality

$$\sum_{n=k}^{\infty}\binom{n}{k}s^n = \frac{s^k}{(1-s)^{k+1}}.$$

b) ANSWER: $\frac{p}{1-p}$.

I.7.62. Let H_i be the random event that the urn contains i $(i = 0, 1, \dots, n)$ white balls, and let B be the random event that a white ball is drawn from the urn. Under the condition of the problem $\mathsf{P}(H_i) = 1/(n+1)$. Using the Bayes formula, we get

$$\mathsf{P}(H_i/B) = \frac{\mathsf{P}(H_i)\mathsf{P}(B/H_i)}{\sum_{k=1}^{n} \mathsf{P}(H_k)\mathsf{P}(B/H_k)} = \frac{\frac{1}{n+1}\frac{i}{n}}{\frac{1}{n+1}\sum_{k=1}^{n}\frac{k}{n}} = \frac{2i}{n(n+1)}, \qquad i = 0, 1, \dots, n.$$

The most probable is the hypothesis H_n:

$$\mathsf{P}(H_n) = \frac{2}{n+1}.$$

I.7.65. Let D be the random event that there were two hits, and let C be the random event that the third shot hits the target. Then

$$\mathsf{P}(C/D) = \frac{\mathsf{P}(C \cap D)}{\mathsf{P}(D)} = \frac{p_3 p_2(1-p_1) + p_3 p_1(1-p_2)}{p_1 p_2(1-p_3) + p_1(1-p_2)p_3 + (1-p_1)p_2 p_3} = \frac{10}{19} > \frac{1}{2}.$$

I.7.71. ANSWER:

$$\frac{\frac{n}{1+n}p_2}{\frac{1}{1+n}p_1 + \frac{n}{n+1}p_2} = \frac{np_2}{p_1 + np_2}.$$

I.7.72. Let z be the number of white balls in the first urn, x the number of white balls in the second urn, and y the number of black balls in the second urn. The probability that n balls taken from the first urn are all white is equal to $(z/(x+y))^n$. The probability that n balls drawn from the second urn are either all white or all black is equal to $(x/(x+y))^n + (y/(x+y))^n$. Then

$$\left(\frac{z}{x+y}\right)^n = \left(\frac{x}{x+y}\right)^n + \left(\frac{y}{x+y}\right)^n,$$

whence $x^n + y^n = z^n$. Thus the problem is equivalent to the famous Fermat Last Theorem.

I.7.74. Denote by x and y the probabilities that a consecutive "successes" will occur earlier than b consecutive "failures", provided that the first trial resulted in a "success" or in a "failure", respectively. Evidently, the required probability is $px + qy$. Consider the case where the first trial resulted in a "success". Then the event of interest to us can happen in the following a mutually exclusive ways: 1) we may obtain $a-1$ "successes" in consecutive $a-1$ trials after the first trial; 2) after $r-2$ $(2 \le r \le a)$ "successes", we obtain a "failure" in the rth trial and then the required event. Then, by the law of total probability, we have

$$x = p^{a-1} + p^{a-2}qy + p^{a-3}qy + \dots + qy = p^{a-1} + \frac{1-p^{a-1}}{1-p}qy = p^{a-1} + (1-p^{a-1})y.$$

Similarly,

$$y = x\left(1 - q^{b-1}\right).$$

Hence

$$px + qy = \frac{p^{a-1}(1-q^b)}{p^{a-1} + q^{b-1} - p^{a-1}q^{b-1}}.$$

I.7.75. ANSWER: a) $\frac{21}{216}$; b) $\frac{7}{216}$.

I.7.76. The probability that there will be ν hits at shots with numbers $n_1, n_2, \dots, n_\nu$ is equal to

$$\prod_{i=1}^{\nu} \frac{3}{4n_i^2} \prod_{n \neq n_1, n_2, \dots} \left(1 - \frac{3}{4n^2}\right) = \prod_{n=1}^{\infty} \left(1 - \frac{3}{4n^2}\right) \prod_{i=1}^{\nu} \frac{\frac{3}{4n_i^2}}{1 - \frac{3}{4n_i^2}}.$$

The probability that the cruiser will not go down at ν hits with numbers $n_1, n_2, \dots, n_\nu$ is equal to

$$\frac{1}{4^\nu} \prod_{n=1}^{\infty} \left(1 - \frac{3}{4n^2}\right) \prod_{i=1}^{\nu} \frac{\frac{3}{4n_i^2}}{1 - \frac{3}{4n_i^2}} = \prod_{n=1}^{\infty} \left(1 - \frac{3}{4n^2}\right) \prod_{i=1}^{\nu} \frac{\frac{3}{16n_i^2}}{1 - \frac{3}{4n_i^2}}.$$

Thus, the desired probability is equal to

$$\sum_{\nu=0}^{\infty} \sum_{n_1, \dots, n_\nu} \prod_{n=1}^{\infty} \left(1 - \frac{3}{4n^2}\right) \prod_{i=1}^{\nu} \frac{\frac{3}{16n_i^2}}{1 - \frac{3}{4n_i^2}} = \prod_{n=1}^{\infty} \left(1 - \frac{3}{4n^2}\right) \sum_{\nu=0}^{\infty} \sum_{n_1, \dots, n_\nu} \prod_{i=1}^{\nu} \frac{\frac{3}{16n_i^2}}{1 - \frac{3}{4n_i^2}}$$

$$= \prod_{n=1}^{\infty} \left(1 - \frac{3}{4n^2}\right) \prod \left(1 + \frac{\frac{3}{16n^2}}{1 - \frac{3}{4n^2}}\right) = \prod_{n=1}^{\infty} \left(1 - \frac{9}{16n^2}\right) = \frac{\sin \frac{3\pi}{4}}{3\pi} = \frac{2\sqrt{2}}{3\pi}.$$

Solutions to Chapter II

§1.

II.1.1. ANSWER: $\Omega = \{\text{HH, HT, TH, TT}\}$;

$$\mathsf{E}\,\xi = \frac{1}{4} \cdot 2 + \frac{1}{2} \cdot 1 = 1; \qquad \mathsf{E}\,\xi^2 = 4 \cdot \frac{1}{4} + 1 \cdot \frac{1}{2} = \frac{3}{2}; \qquad \operatorname{Var} \xi = \mathsf{E}\,\xi^2 - (\mathsf{E}\,\xi)^2 = \frac{1}{2}.$$

II.1.3. ANSWER: $\Omega = \{\omega_1, \omega_2, \dots, \omega_n, \dots\}$, where

$$\omega_1 = \{\text{H}\},\ \omega_2 = \{\text{TH}\},\ \dots,\ \omega_n = \{\underbrace{\text{TT} \dots \text{T}}_{n-1 \text{ times}}, \text{H}\},\ \dots;$$

$\xi(\omega_n) = n$, $\mathsf{P}(\xi(\omega_n) = n) = 2^{-n}$;

$$\mathsf{P}(\xi > 1) = \sum_{n=2}^{\infty} \frac{1}{2^n} = \frac{1}{2}.$$

II.1.6. ANSWER: $\mathsf{E}\,\xi = 0$, $\operatorname{Var} \xi = \frac{n(n+1)}{3}$.

II.1.7. ANSWER:

$$\mathsf{E}\,\xi^n = \begin{cases} 1 & \text{if } n \text{ is an even number,} \\ 0 & \text{if } n \text{ is an odd number.} \end{cases}$$

II.1.16. a) The joint distribution is given by a 6×6 matrix. The main diagonal consists of the elements $q, 2q, \dots, 6q$, where $q = \frac{1}{36}$. On one side of the diagonal all the elements are zeros and on the other side all the elements equal q.

b) ANSWER: $\mathsf{E}\,\xi = \frac{7}{2}$, $\operatorname{Var} \xi = \frac{35}{12}$, $\mathsf{E}\,\eta = \frac{161}{36}$, $\operatorname{Var} \eta = \frac{2555}{1296}$, $\operatorname{cov}(\xi, \eta) = \frac{105}{72}$.

II.1.22. Let $\mathsf{P}(\xi = x_i) = p_i$, $\mathsf{P}(\eta = y_j) = q_j$, and $\mathsf{P}(\xi = x_i, \eta = y_j) = p_{ij}$. According to the condition of the problem,

$$\sum_{i=1}^{m}\sum_{j=1}^{n} \delta_{ij} x_i^h y_j^k = 0, \qquad 0 \le h \le m-1,\ 0 \le k \le n-1,$$

where $\delta_{ij} = p_{ij} - p_i q_j$. Put

$$\alpha_{ik} = \sum_{j=1}^{n} \delta_{ij} y_j^k.$$

Then the unknowns α_{ik}, $1 \le i \le m$, satisfy the following system of linear homogeneous equations:

$$\sum_{i=1}^{m} \alpha_{ik} x_i^h = 0, \qquad 0 \le h \le m-1.$$

The determinant of this system, the so-called *Vandermonde determinant*, is nonzero. Therefore,

$$\sum_{j=1}^{n} \delta_{ij} y_j^k = 0, \qquad 1 \le i \le m.$$

Since the last equality is true for every k, we have a system of linear homogeneous equations for δ_{ij}:

$$\sum_{j=1}^{n} \delta_{ij} y_j^k = 0, \qquad 0 \le k \le n-1,$$

whence $\delta_{ij} = 0$, i.e., $p_{ij} = p_i q_j$.

II.1.23. Put

$$\eta_k = \begin{cases} 1 & \text{if the } k\text{th letter is placed correctly,} \\ 0 & \text{otherwise.} \end{cases}$$

Then $\xi_n = \eta_1 + \cdots + \eta_n$. Since $\mathsf{P}(\eta_k = 1) = n^{-1}$, we have

$$\mathsf{E}\,\eta_k = \frac{1}{n}, \qquad \operatorname{Var}\eta_k = \frac{1}{n} - \frac{1}{n^2} = \frac{n-1}{n^2}, \qquad \mathsf{E}\,\xi_n = \sum_{k=1}^{n} \mathsf{E}\,\eta_k = 1.$$

To calculate $\operatorname{Var}\xi_n$, it is necessary to calculate $\mathsf{E}\,\eta_k\eta_i$, $k \ne i$. Since $\eta_k\eta_i$ assumes only the values 1 or 0 and $\mathsf{P}(\eta_k\eta_i = 1) = n^{-1}(n-1)^{-1}$, we have

$$\mathsf{E}\,\eta_k\eta_i = \frac{1}{n(n-1)},$$

$$\operatorname{cov}(\eta_k, \eta_i) = \mathsf{E}\,\eta_k\eta_i - \mathsf{E}\,\eta_k\,\mathsf{E}\,\eta_i = \frac{1}{n(n-1)} - \frac{1}{n^2} = \frac{1}{n^2(n-1)}.$$

Thus

$$\operatorname{Var}\xi_n = \sum_{k=1}^{n} \operatorname{Var}\eta_k + 2\sum_{k<i} \operatorname{cov}(\eta_k, \eta_i) \frac{n-1}{n} + 2\binom{n}{2}\frac{1}{n^2}\frac{1}{n-1} = 1.$$

II.1.26. ANSWER:

$$\mathsf{P}(\xi+\eta=k)=\begin{cases}\dfrac{k+1}{(n+1)^2}, & 0\le k\le n;\\[2mm] \dfrac{2n+1-k}{(n+1)^2}, & n+1\le k\le 2n.\end{cases}$$

II.1.27. Consider a sequence of independent random variables $\eta_1,\dots,\eta_n$, where

$$\eta_k=\begin{cases}1 & \text{if the } k\text{th trial was a success},\\ 0 & \text{if the } k\text{th trial was a failure}.\end{cases}$$

Then

$$\mathsf{E}\,\eta_k=1\cdot p_k+0\cdot q_k=p_k,\qquad \operatorname{Var}\eta_k=p_k-p_k^2=p_kq_k.$$

Since $\xi_n=\eta_1+\cdots+\eta_n$, we have

$$\mathsf{E}\,\xi_n=\sum_{k=1}^n p_k,\qquad \operatorname{Var}\xi_n=\sum_{k=1}^n p_kq_k.$$

II.1.29. ANSWER: $n=18$, $p=\frac{2}{3}$.

II.1.32. ANSWER:

$$\binom{5}{2}\left(\frac{1}{3}\right)^2\left(\frac{2}{3}\right)^3.$$

II.1.34. ANSWER: a) either $m_0=2$ or $m_0=3$; $p_{14}(2)=p_{14}(3)\approx 0.25$; b) $\mathsf{P}(\xi\ge 4)\approx 0.302$.

II.1.35. ANSWER: a) 0.665; b) 0.618; c) 0.597. The probability of getting at least n 6's at a throw of $6n$ dice is

$$1-\sum_{k=0}^{n-1}\binom{6n}{k}\left(\frac{1}{6}\right)^k\left(\frac{5}{6}\right)^{n-k}.$$

II.1.37. ANSWER:

$$\text{a) } 1-(0.8)^{10}-2\cdot(0.8)^9\approx 0.6242;\qquad \text{b) } \frac{1-(0.8)^{10}-2\cdot(0.8)^9}{1-(0.8)^{10}}\approx 0.6993.$$

II.1.38. ANSWER:

$$\sum_{k=0}^n\binom{n}{k}^2\frac{1}{2^{2n}}=\binom{2n}{n}\frac{1}{2^{2n}}\sim\frac{1}{\sqrt{\pi n}}$$

for large values of n. (Use the Stirling formula.)

II.1.39. a) If a selected box is empty while the second box contains r matches, then the matches were taken total $2N-r$ times, with N times from the box that is empty. If boxes are selected at random, then we are dealing with $2N-r$ successive independent trials, and in each of them the box that became then empty appears with probability $\frac{1}{2}$. Thus the required probability equals the probability of obtaining N successes in the Bernoulli scheme, i.e., $\binom{2N-r}{N}2^{-(2N-r)}$.

b) $x_r = \binom{2N-1-r}{N-1}2^{-(2N-1-r)}$,

c) $\sum_{r=1}^{N} x_r 2^{-(r+1)} = \frac{1}{2^{2N}}\sum_{r=1}^{N}\binom{2N-1-r}{N-1}$.

II.1.40. ANSWER:

$$\binom{5}{3}\frac{1}{2^3}\frac{1}{2^{5-3}} = \frac{40}{128} > \binom{8}{4}\frac{1}{2^4}\frac{1}{2^{8-4}} = \frac{35}{128}.$$

II.1.41. ANSWER: n^{-1}.

II.1.43. ANSWER:

$$\mathsf{E}\,\gamma = 2npq, \qquad \operatorname{Var}\gamma = 2npq(1-2pq) + 2(n-1)pq(p-q)^2.$$

II.1.44. ANSWER:

$$\frac{1}{2^n}\left(\binom{n}{0} + \binom{n}{1} + \cdots + \binom{n}{2r} + \cdots\right) = \frac{1}{r}\sum_{k=0}^{r-1}\cos\frac{nk\pi}{r}\cos^n\frac{k\pi}{r}.$$

II.1.49. ANSWER:

$$\mathsf{P}(\eta = i) = 2q^ip - q^{2i}p - q^{2i+1}p; \qquad \mathsf{P}(\eta = i, \xi_1 = j) = \begin{cases} q^{i+j}p^2, & i > j, \\ (1-q^{i+1})q^ip, & i = j. \end{cases}$$

II.1.50. If $\xi = n$, then the following two events must occur simultaneously:

a) the $(n+1)$th trial results in a success; the probability of this event is p;

b) there were exactly $r-1$ successes in the previous n trials; the probability of this event is $\binom{n}{r-1}p^{r-1}q^{n-r-1}$.

Therefore,

$$\mathsf{P}(\xi = n) = p\binom{n}{r-1}p^{r-1}q^{n-r+1} = \binom{n}{r-1}p^rq^{n-r+1}.$$

II.1.55. ANSWER:

a) $\mathsf{P}(\xi = k) = p^kq + q^kp,\quad \mathsf{E}\,\xi = \frac{p}{q} + \frac{q}{p},\quad \operatorname{Var}\xi = \frac{p}{q^2} + \frac{q}{p^2} - 2;$

b) $\mathsf{P}(\eta = k) = p^2q^{k-1} + q^2p^{k-1},\quad \mathsf{P}(\xi = m, \eta = n) = p^{m+1}q^n + q^{m+1}p^n.$

II.1.58. ANSWER: $\dfrac{1-e^{-\lambda}}{\lambda}$.

II.1.61. ANSWER: $\mathsf{E}\,\xi = a$, $\operatorname{Var}\xi = a(a+1)$.

II.1.62. ANSWER:

$$\mathsf{P}(\xi = k) = \frac{k^n - (k-1)^n}{N^n}; \qquad \mathsf{E}\,\xi = \frac{N^{n+1} - \sum_{k=1}^{N}(k-1)^n}{N^n}.$$

II.1.65. ANSWER: $\dfrac{12!}{2^6 6^{12}} \approx 0.0034.$

II.1.66. First prove that the probability is at its maximum if and only if $p_i k_j \le (k_i+1)p_j$ for any pair (i,j).

II.1.70. ANSWER: $\min\{0, m+n-N\} \le k \le m$,

a) $\mathsf{P}(\xi = k) = \binom{n}{k}\binom{N-n}{m-k}\Big/\binom{N}{m}$;

b) $\mathsf{E}\xi = m\frac{n}{N}$, $\quad \operatorname{Var}\xi = m\frac{n}{N}\frac{N-n}{N}\frac{1-\frac{n}{N}}{1-\frac{1}{N}}$.

§**2.**

II.2.2. $\Omega = \{(u,v)\colon u^2+v^2 \le R^2\}$. If $\xi(\omega)$ is the distance from the point $\omega = (u,v)$ to the origin, then $\xi(\omega) = \sqrt{u^2+v^2}$ and

$$\{\omega\colon \xi(\omega) < x\} = \begin{cases} \varnothing, & x \le 0, \\ \{(u,v)\colon u^2+v^2 < x^2\}, & 0 < x \le R, \\ \Omega, & x > R. \end{cases}$$

Thus $\{\omega\colon \xi(\omega) < x\} \in \mathfrak{A}$. The distribution of ξ is of the form

$$F(x) = \mathsf{P}(\omega\colon \xi(\omega) < x) = \begin{cases} 0, & x \le 0, \\ \dfrac{x^2}{R^2}, & 0 < x \le R, \\ 1, & x > R. \end{cases}$$

II.2.4. Let $x_n \downarrow x$. Then

$$\{\omega\colon \xi(\omega) \le x\} = \bigcap_{n=1}^{\infty} \{\omega\colon \xi(\omega) < x_n\} \in \mathfrak{A}$$

and

$$\{\omega\colon \xi(\omega) < x_{n+1}\} \subset \{\omega\colon \xi(\omega) < x_n\}.$$

In view of the continuity property of probability,

$$\mathsf{P}\{\omega\colon \xi(\omega) \le x\} = \lim_{n\to\infty} \mathsf{P}\{\omega\colon \xi(\omega) < x_n\} = \lim_{n\to\infty} F(x_n) = F(x+0).$$

II.2.7. ANSWER: No.

II.2.9. ANSWER: a) No; b) Yes.

II.2.11. ANSWER: No.

II.2.16. Let A be an atom. Consider the set $A \cap \{\omega\colon \xi(\omega) < c\}$ for every c.

II.2.18. Let $I = \{A\colon A \in B(F_0), \chi_A(\omega) \in H\}$. Using the conditions of the problem, establish that I is a monotone class. According to Problem I.1.36, we have $B(F_0) \subset I$. For an arbitrary function $\xi(\omega)$ measurable with respect to $B(F_0)$ consider the functions $\xi_{j_n}(\omega)$ and $\xi_n(\omega)$:

$$\xi_{j_n}(\omega) = \begin{cases} 1, & \dfrac{j}{2^n} \le \xi(\omega) < \dfrac{j+1}{2^n}, \\ 0, & \text{otherwise}; \end{cases}$$

$$\xi_n(\omega) = \sum_{j=-n}^{n} \frac{\xi_{j_n}(\omega)}{2^n}.$$

Since $\xi_{j_n}(\omega) \in H$ (because $B(F_0) \subset I$), we have $\xi_n(\omega) \in H$ (because of condition 2)). Then $\xi(\omega) = \lim_{n\to\infty} \xi_n(\omega) \in H$ (because of condition 3)).

II.2.20. Let H be the class of those Borel functions of n variables for which the assertion of the problem holds, and let F_0 be the class of half-closed parallelepipeds $\{\vec{x}: a_i \le x_i < b_i,\ 1 \le i \le n\}$ in $\mathbf{R}^n$. Use the assertion of the previous problem to prove that the class H contains all Borel functions.

II.2.21. Consider the indicator of the set A.

§3.

II.3.1. ANSWER: c), d), and e).

II.3.5. ANSWER: e^{-1}.

II.3.6. For $x \le 0$ we have $\mathsf{P}\left(e^{\xi} < x\right) = 0$; for $x > 0$ we have $F_\eta(x) = \mathsf{P}\left(e^{\xi} < x\right) = \mathsf{P}(\xi < \ln x) = F(\ln x)$. Therefore,

$$F_\eta(x) = \begin{cases} 0, & x \le 0, \\ F(\ln x), & x > 0. \end{cases}$$

II.3.7. ANSWER: $F_\eta(x) = 1 - F(-x+0)$.

II.3.8. ANSWER: a) $F(x) = 1 - F(-x+0)$; b) $p(x) = p(-x)$.

II.3.9. ANSWER: $\mathsf{E}\,\eta = 1 - [F(+0) - F(0)]$.

II.3.10. ANSWER:

a) $\begin{cases} F\left(\frac{x-b}{a}\right), & \text{for } a > 0, \\ 1 - F\left(\frac{x-b}{a} + 0\right), & \text{for } a < 0; \end{cases}$

b) $\begin{cases} F\left(\sqrt{x}\right) - F\left(-\sqrt{x} + 0\right), & \text{for } x > 0, \\ 0, & \text{for } x \le 0; \end{cases}$

c) if $g(x)$ is monotone increasing, then the answer is $F(g^{-1}(x))$, where g^{-1} is the inverse of $g(x)$; if $g(x)$ is monotone decreasing, then $1 - F(g^{-1}(x) + 0)$;

f) $\displaystyle\sum_{k=-\infty}^{\infty} \left|F(\arctan x + k\pi) - F\left(-\frac{\pi}{2} + k\pi\right)\right|.$

II.3.11. ANSWER:

a) $\dfrac{x-b}{a}\dfrac{1}{|a|^p}$; b) $p(x) + p(-x)$;

c) $\dfrac{1}{2\sqrt{x}}\left[p\left(\sqrt{x}\right) + p\left(-\sqrt{x}\right)\right]$; d) $p\left(g^{-1}(x)\right)\left|\left[g^{-1}(x)\right]'\right|$.

II.3.14. The uniform distribution on $\left(-\frac{\pi}{2}, \frac{\pi}{2}\right)$.

II.3.15. The uniform distribution on $[0, 1]$.

II.3.18. The normal distribution $N(0, 1)$.

II.3.20. ANSWER:

$$p_\eta(x) = \frac{1}{\pi\sqrt{1-x^2}}, \qquad x \in (-1,1).$$

II.3.21. ANSWER:

$$F_\eta(x) = \begin{cases} 0, & x \le -1, \\ \dfrac{\pi + 2\arcsin x}{2\pi}, & -1 < x < 1, \\ 1, & x \ge 1, \end{cases}$$

$$p_\eta(x) = \frac{1}{\pi\sqrt{1-x^2}}, \qquad x \in (-1,1).$$

II.3.25. Let $0 < x < 1$. Then

$$\mathsf{P}(\eta < x) = \mathsf{P}(F(\xi) < x) = \mathsf{P}\left(\xi < F^{-1}(x)\right) = F\left(F^{-1}(x)\right) = x.$$

For $x \le 0$ we have $\mathsf{P}(\eta < x) = 0$, and for $x > 1$ we have $\mathsf{P}(\eta < x) = 1$. Therefore, η has the uniform distribution on $(0,1)$.

II.3.26. Let ξ be a positive random variable with distribution function $F(x)$. Consider the random variable $\eta = \max(\xi, 1/\xi)$.

II.3.28. ANSWER:

$$F_\eta(x) = \begin{cases} 0, & x \le 0, \\ \dfrac{x^2}{R^2}, & 0 < x \le R, \\ 1, & x > R. \end{cases}$$

II.3.29. Let $\angle OQP = y$; then $\eta = OP = \tan y$ and

$$F_\eta(x) = \frac{1}{\pi}\arctan x + \frac{1}{2}, \qquad p_\eta(x) = \frac{1}{\pi}\frac{1}{1+x^2}$$

(the Cauchy distribution).

II.3.31. ANSWER:

$$F_\xi(x) = \frac{1}{2} + \frac{1}{\pi}\arcsin(x-1), \qquad 0 < x \le 2.$$

II.3.32. ANSWER:

$$F_\gamma(x) = \begin{cases} 0, & x \le 0, \\ \dfrac{2}{\pi}\arcsin\dfrac{x}{2R}, & 0 < x \le 2R, \\ 1, & x > 2R. \end{cases}$$

II.3.33. ANSWER:

$$p_\gamma(x) = \begin{cases} \dfrac{2}{\pi}\dfrac{R}{x^2+R^2}, & x \ge 0, \\ 0, & x < 0. \end{cases}$$

II.3.35. ANSWER: $F(x) = 2x/l$ for $0 < x \le l/2$.

II.3.39. Let us "paste together" the ends of the segment by curving it so as to obtain a circle of length T. Then there will be $n+1$ points on the circle, which divide it into $n+1$ parts. By symmetry, the length of every part has the same distribution. Let $\eta_1, \eta_2, \ldots, \eta_{n+1}$ be the lengths of these parts, $\sum_{k=1}^{n+1} \eta_k = T$. It is enough to find the distribution of any of these parts, say, of η_1. Note that the event $\{\omega\colon \eta_1 > x\}$ occurs if and only if each of the remaining n points belongs to the segment $[x, T]$. Therefore,

$$\mathsf{P}(\eta_1 > x) = \frac{(T-x)^n}{T^n},$$

and

$$F_{\eta_1}(x) = 1 - \mathsf{P}(\eta_1 > x) = 1 - \frac{(T-x)^n}{T^n}.$$

II.3.40. By assumption,

$$\mathsf{P}(\xi \geq x + \Delta x) = \mathsf{P}(\xi \geq x)\mathsf{P}\left(\xi \geq x + \Delta x / \xi \geq x\right) = \mathsf{P}(\xi \geq x)\left(1 - \lambda \Delta x + o(\Delta x)\right).$$

Put $Q(x) = \mathsf{P}(\xi \geq x)$. Then

$$Q(x + \Delta x) - Q(x) = -(\lambda \Delta x + O(\Delta x))Q(x),$$

whence

$$\frac{dQ(x)}{dx} = \lambda Q(x).$$

Taking into account that $Q(0) = 1$ (this is a quite natural assumption meaning that at the initial moment the device gets out of order with probability 0), we get $Q(x) = e^{-\lambda x}$ for $x \geq 0$.

II.3.43. By the definition of conditional probability, we have

$$\mathsf{P}\left(\xi - t < x / \xi \geq t\right) = \frac{\mathsf{P}(\xi - t < x, \xi \geq t)}{\mathsf{P}(\xi \geq t)} = \frac{\int_t^{t+x} \lambda e^{-\lambda u}\, du}{\int_t^{\infty} \lambda e^{-\lambda u}\, du} = \frac{e^{-\lambda t} - e^{-\lambda t} e^{-\lambda x}}{e^{-\lambda t}}.$$

Hence

$$\mathsf{P}\left(\xi - t < x / \xi \geq t\right) = 1 - e^{-\lambda x},$$

i.e., the distribution of $\xi - t$, given $\xi \geq t$, is the same as the distribution of ξ. This is a very important property of an exponential distribution, and among continuous distributions, only the exponential distribution possesses this property (see the next problem).

II.3.44. Let $Q(x) = \mathsf{P}(\xi \geq x) = 1 - F(x)$. Then

$$\mathsf{P}\left(\xi < t + x / \xi \geq t\right) = \frac{\mathsf{P}(t \leq \xi < t + x)}{\mathsf{P}(\xi \geq t)} = \frac{Q(t) - Q(t+x)}{Q(t)} = 1 - Q(x),$$

whence $Q(t+x) = Q(t)Q(x)$. The function $Q(x)$ is monotone. It is easy to demonstrate that every monotone function satisfying the functional equation (1) is of the form $Q(x) = e^{-\lambda x}$. Since $Q(x)$ is bounded ($Q(x) \leq 1$), we have $Q(x) = e^{-\lambda x}$ for $\lambda \geq 0$. Therefore, the random variable ξ has an exponential distribution. Among continuous distributions, only the exponential distribution possesses the absence-of-aftereffect property.

II.3.48. The exponential distribution with parameter λ.

II.3.49. The geometric distribution with parameter $p = 1 - e^{-\lambda}$.

II.3.52. ANSWER:

$$\text{a) } x\int_x^{+\infty}\frac{1}{u}\,dF(u) = x\int_x^{\sqrt{x}}\frac{1}{u}\,dF(u) + x\int_{\sqrt{x}}^{+\infty}\frac{1}{u}\,dF(u)$$
$$\le \int_x^{\sqrt{x}} dF(u) + \sqrt{x}\int_{\sqrt{x}}^{+\infty} dF(u)$$
$$= F\left(\sqrt{x}\right) - F(x) + \sqrt{x}\left[1 - F\left(\sqrt{x}\right)\right].$$

The right-hand side of this inequality vanishes as $x \to +0$.

$$\text{b) } x\int_x^{+\infty}\frac{1}{u}\,dF(u) \le \int_x^{+\infty} dF(u) = 1 - F(x).$$

II.3.53. e) 0.

II.3.57. a) Let ξ be a random variable with continuous distribution function $F(x)$. Then $F(\xi)$ is a random variable uniformly distributed on $(0,1)$. Therefore,

$$\mathsf{E}\,F(\xi) = \int_{-\infty}^{+\infty} F(x)\,dF(x) = \frac{1}{2}.$$

II.3.58. ANSWER: a) $\sigma\sqrt{2/\pi}$.

II.3.59. ANSWER:

$$\frac{1}{\pi}\ln 2 + \frac{1}{2}.$$

II.3.60. Let ξ be the length of a cotton fiber. Then for $x > a$ we have

$$\mathsf{P}\left(\xi < x/\xi > a\right) = \frac{\mathsf{P}(a < \xi < x)}{\mathsf{P}(\xi > a)} = \frac{2}{\sqrt{2\pi}\sigma}\int_x^a \exp\left\{-\frac{(u-a)^2}{2\sigma^2}\right\} du$$

and

$$a'' = \frac{2}{\sqrt{2\pi}\sigma}\int_a^{+\infty} z\exp\left\{-\frac{(z-a)^2}{2\sigma^2}\right\} dz = a + \frac{2\sigma}{\sqrt{2\pi}}.$$

Similarly, $a' = a - 2\sigma/\sqrt{2\pi}$, so that

$$t = \sqrt{\frac{2}{\pi}}\frac{\sigma}{a}.$$

II.3.62. Let $F(x) = \mathsf{P}(|\xi| < x)$ and $\mathsf{E}\,|\xi| = \int_0^\infty x\,dF(x) < \infty$. Then

$$\int_0^\infty x\,dF(x) = \sum_{k=0}^\infty \int_k^{k+1} x\,dF(x) \ge \sum_{k=1}^\infty k\mathsf{P}(k \le |\xi| < k+1) = \sum_{k=1}^\infty \mathsf{P}(|\xi| \ge k).$$

If the series $\sum_{k=1}^\infty \mathsf{P}(|\xi| \ge k)$ converges, then

$$\sum_{k=1}^\infty \mathsf{P}(|\xi| \ge k) = \sum_{k=1}^\infty k\mathsf{P}(k \le |\xi| < k+1) = \sum_{k=0}^\infty (k+1)\mathsf{P}(k \le |\xi| < k+1) - 1$$
$$\ge \sum_{k=0}^\infty \int_k^{k+1} x\,dF(x) - 1 = \int_0^\infty x\,dF(x) - 1.$$

II.3.63. Since $\mathsf{E}\xi$ exists, the integral $\int_{-\infty}^{+\infty} |x|\,dF(x)$ converges. Taking into account the inequality $x(1-F(x)) \le \int_x^{+\infty} y\,dF(y)$, we get

$$\lim_{x\to+\infty}(1-F(x)) \le \lim_{x\to+\infty}\int_x^{\infty} y\,dF(y) = 0.$$

Similarly, from the inequality

$$\int_{-\infty}^{x} |y|\,dF(y) \ge |x|F(x), \qquad x<0,$$

we get $\lim_{x\to-\infty} xF(x) = 0$.

II.3.64. Integrating by parts, we obtain the equalities

$$\text{(I)}\qquad \int_0^x y\,dF(y) = \int_0^x y\,d[F(y)-1] = -x(1-F(x)) + \int_0^x (1-F(y))\,dy;$$

$$\text{(II)}\qquad \int_{-x}^{0} y\,dF(y) = xF(-x) - \int_{-x}^{0} F(y)\,dy.$$

Let $\mathsf{E}\xi$ exist. Then, passing to the limit as $x\to\infty$ and using Problem II.3.63, we get

$$\int_0^{+\infty} y\,dF(y) = \int_0^{+\infty}(1-F(y))\,dy, \qquad \int_{-\infty}^{0} y\,dF(y) = -\int_{-\infty}^{0} F(y)\,dy.$$

Summing up these equalities, we get

$$\text{(III)}\qquad \mathsf{E}\xi = \int_{-\infty}^{\infty} y\,dF(y) = \int_0^{+\infty}(1-F(y))\,dy - \int_{-\infty}^{0} F(y)\,dy.$$

Let the integrals $\int_0^{\infty}(1-F(y))\,dy$ and $\int_{-\infty}^{0} F(y)\,dy$ converge. Then, taking into account the inequalities

$$\frac{2}{x}(1-F(x)) \le \int_{x/2}^{x}[1-F(x)]\,dy \quad \text{and} \quad -\frac{1}{2}xF(x) \le \int_x^{x/2} F(y)\,dy,$$

we have

$$\lim_{x\to+\infty} x[1-F(x)] = 0, \qquad \lim_{x\to-\infty} xF(x) = 0.$$

Now, passing to the limit in the equalities (I) and (II), we see that $\mathsf{E}\xi$ exists and (III) holds.

II.3.69. Answer:

$$p(x) = \frac{x}{2R^2}, \quad x\in[0,2R]; \qquad \frac{3}{4}R.$$

II.3.70. Answer:

$$p(x) = \frac{1}{\sqrt{\pi}}\frac{\Gamma(n/2)}{\Gamma((n-1)/2)}\frac{x^{n-2}}{R^{n-1}}\left(1-\frac{x^2}{4R^2}\right)^{(n-3)/2}, \qquad 0<x<2R.$$

II.3.71. Answer:

$$p_\eta(x) = \frac{2}{\pi\sqrt{1-x^2}}, \quad 0<x<1; \qquad \mathsf{E}\eta = \frac{2}{\pi}.$$

II.3.72. Since the area of the surface of a spherical belt is directly proportional to the height of the belt, we have

$$\mathsf{P}(\eta < x) = \frac{2x}{2} = x, \qquad 0 < x < 1.$$

Therefore, the random variable η is uniformly distributed on the segment $[0, 1]$.

II.3.73. ANSWER:

$$p_\gamma(x) = \frac{x}{\sqrt{1-x^2}}, \quad 0 < x < 1; \qquad \mathsf{E}\,\gamma = \frac{\pi}{4}.$$

II.3.74. Let NS be the diameter of the sphere, and $\angle SNP = \Theta$. Then $NP = 2R\cos\Theta$. But $\cos\Theta$ is the length of projection of the unit vector oriented along NP to the diameter NS. In view of Problem II.3.72, it has the uniform distribution on $[0, 1]$.

II.3.75. ANSWER:

a) $\dfrac{2}{\pi}\arcsin\dfrac{x}{2}, \qquad 0 < x \le 2;$

b) $\dfrac{1}{4}x^2, \qquad 0 < x \le 2.$

II.3.77. ANSWER:

a) $p_\xi(x) = p_\eta(x) = e^{-x}, \qquad x > 0;$

b) $1 - e^{-x} - e^{-y} + e^{-x-y-axy}.$

II.3.78. ANSWER:

$$p_1(x) = 12x^2(1-x), \quad x \in (0,1); \qquad p_2(y) = 2y, \quad y \in (0,1);$$
$$p(x,y) = p_1(x)p_2(y).$$

II.3.80. Since ξ_1 and ξ_2 are independent, the distribution density of the random vector (ξ_1, ξ_2) is

$$p_{(\xi_1,\xi_2)}(u,v) = \frac{1}{\sqrt{2\pi}}e^{-u^2/2}\frac{1}{\sqrt{2\pi}}e^{-v^2/2} = \frac{1}{2\pi}e^{-(u^2+v^2)/2}.$$

Let $D = \{(u,v) : u^2 + v^2 \le R^2\}$. Then

$$\begin{aligned}
\mathsf{P}((\xi_1,\xi_2) \in D) &= \iint_D p_{(\xi_1,\xi_2)}(u,v)\,du\,dv \\
&= \frac{1}{2\pi}\iint_{u^2+v^2\le R^2} \exp\left\{-\frac{u^2+v^2}{2}\right\} du\,dv \\
&= \frac{1}{2\pi}\int_0^R\int_0^{2\pi} e^{-r^2/2} r\,dr\,dy = 1 - e^{-R^2/2}.
\end{aligned}$$

II.3.82. ANSWER:

$$p_\xi(x) = e^{-x}, \quad x > 0; \qquad p_\eta(y) = \frac{1}{(1+y)^2}, \quad y > 0.$$

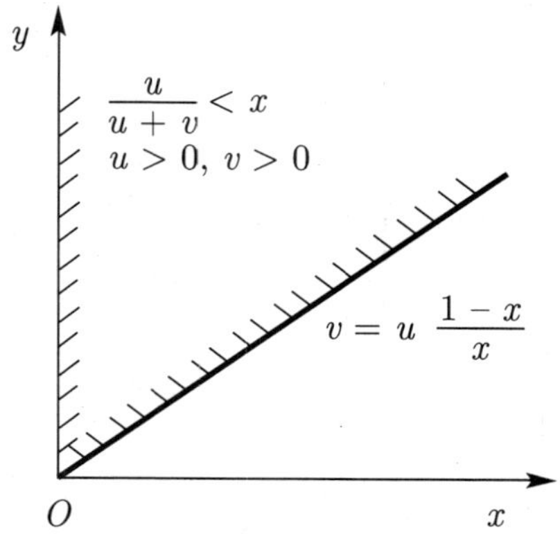

FIGURE 15

II.3.83. ANSWER:

$$p_\xi(x) = \frac{1}{\Gamma(p)} x^{p-1} e^{-x}, \quad x > 0; \qquad p_\eta(x) = \frac{1}{\Gamma(q)} x^{q-1} e^{-x}, \quad x > 0.$$

II.3.84. Since ξ and η are positive random variables, we have

$$0 < \frac{\xi}{\xi+\eta} < 1$$

with probability 1. Therefore,

$$\mathsf{P}\left(\frac{\xi}{\xi+\eta} < x\right) = \begin{cases} 1, & x \geq 1, \\ 0, & x \leq 0. \end{cases}$$

Let $0 < x < 1$. Then

$$\mathsf{P}\left(\frac{\xi}{\xi+\eta} < x\right) = \mathsf{P}((\xi, \eta) \in D),$$

where the domain D (Figure 15) is defined as

$$D = \left\{(u, v) \colon \frac{u}{u+v} < x,\ u > 0,\ v > 0\right\}.$$

Integrating the distribution density of the random vector (ξ, η) over the domain D, we get

$$\mathsf{P}\left(\frac{\xi}{\xi+\eta} < x\right) = \lambda^2 \iint_D e^{-\lambda u - \lambda v}\, du\, dv = \lambda^2 \int_0^{+\infty} du \int_{u(1-x)/x}^{+\infty} e^{-\lambda u - \lambda v}\, dv = x.$$

Thus the random variable $\xi/(\xi+\eta)$ has the uniform distribution on the segment $[0, 1]$.

II.3.85. ANSWER: $1 - x^{-1}$ for $x > 1$.

II.3.86. ANSWER:

$$F_\eta = \begin{cases} 0, & x \leq 0, \\ \dfrac{\lambda_1 x}{(\lambda_1 - \lambda_2)x + \lambda_2}, & 0 < x \leq 1, \\ 1, & x > 1. \end{cases}$$

II.3.87. ANSWER:

$$F_\eta(x) = \begin{cases} 0, & x \le 0, \\ \dfrac{1}{2}\dfrac{x}{1-x}, & 0 < x \le \dfrac{1}{2}, \\ \dfrac{3x-1}{2x}, & \dfrac{1}{2} < x \le 1, \\ 1, & x > 1. \end{cases}$$

II.3.90. Since the random variables ξ_1, ξ_2, and θ are independent, their joint distribution density is equal to

$$\frac{1}{\sqrt{2\pi}} \exp\left\{-\frac{x_1^2}{2}\right\} \frac{1}{\sqrt{2\pi}} \exp\left\{-\frac{x_2^2}{2}\right\},$$

so that

$$\mathsf{P}(\xi_1 \cos\theta + \xi_2 \sin\theta < z) = \frac{1}{(2\pi)^2} \iiint\limits_{\substack{x_1 \cos x_3 + x_2 \sin x_3 < z \\ 0 \le x_3 \le 2\pi}} \exp\left\{-\frac{x_1^2 + x_2^2}{2}\right\} dx_1\, dx_2\, dx_3$$

$$= \frac{1}{(2\pi)^2} \int_0^{2\pi} dx_3 \left(\iint_D \exp\left\{-\frac{x_1^2 + x_2^2}{2}\right\} dx_1\, dx_2 \right),$$

where $D = \{(x_1, x_2)\colon x_1 \cos x_3 + x_2 \sin x_3 < z\}$. After the change of coordinates $u_1 = x_1 \cos x_3 + x_2 \sin x_3$ and $u_2 = x_1 \sin x_3 - x_2 \cos x_3$ in the double integral, we have

$$\mathsf{P}(\xi_1 \cos\theta + \xi_2 \sin\theta < z) = \frac{1}{(2\pi)^2} \int_0^{2\pi} dx_3 \iint_{u_1 < z} \exp\left\{-\frac{u_1^2 + u_2^2}{2}\right\} du_1\, du_2$$

$$= \frac{1}{\sqrt{2\pi}} \int_{-\infty}^{z} e^{-u_1^2/2}\, du_1.$$

Thus the random variable $\xi_1 \cos\theta + \xi_2 \sin\theta$ has the normal distribution with parameters 0 and 1.

II.3.91. a) Let $D = \{(u, v)\colon u + v < x\}$ and

$$\chi_D(u, v) = \begin{cases} 1, & (u, v) \in D, \\ 0, & (u, v) \notin D. \end{cases}$$

Then $\mathsf{E}\,\chi_D(u, v) = \mathsf{P}((\xi_1, \xi_2) \in D)$. On the other hand, by the formula for the expectation of a function of a random vector, we have

$$\mathsf{E}\,\chi_D(\xi_1, \xi_2) = \iint \chi_D(u, v)\, dF_{(\xi_1, \xi_2)}(u, v).$$

Since the random variables ξ_1 and ξ_2 are independent, $F_{(\xi_1,\xi_2)}(u, v) = F_{\xi_1}(u) F_{\xi_2}(v)$ and finally

$$\mathsf{P}(\xi_1 + \xi_2 < x) = \mathsf{P}((\xi_1, \xi_2) \in D) = \iint_D dF_{\xi_1}(u) F_{\xi_2}(v) = \int_{-\infty}^{+\infty} F_{\xi_2}(x - u)\, dF_{\xi_1}(u)$$

$$= \int_{-\infty}^{+\infty} F_{\xi_1}(x - v)\, dF_{\xi_2}(v).$$

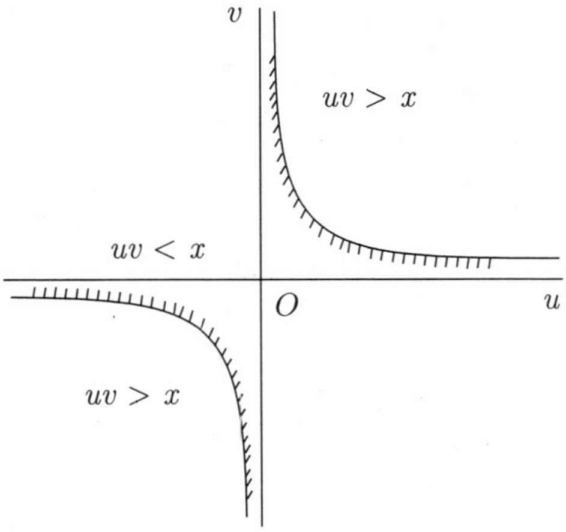

FIGURE 16

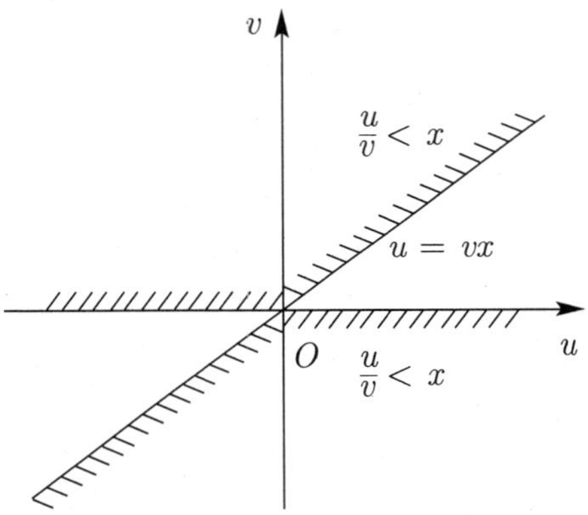

FIGURE 17

II.3.93. ANSWER:

$$\mathsf{P}(\xi_1\xi_2 < x) = \iint_D dF_{\xi_1}(u)\, dF_{\xi_2}(v),$$

where $D = \{(u,v): uv < x\}$ (Figure 16).

II.3.94. ANSWER:

$$\mathsf{P}\left(\frac{\xi_1}{\xi_2} < x\right) = \iint_D dF_{\xi_1}(u)\, dF_{\xi_2}(v),$$

where $D = \left\{(u,v): \frac{u}{v} < x\right\}$ (Figure 17).

II.3.95. ANSWER:

$$p_\eta(x) = \begin{cases} 0, & |x| > 1, \\ 1 - |x|, & |x| \le 1. \end{cases}$$

II.3.96. ANSWER:

a) $p_\eta(x) = \begin{cases} \ln x, & 0 < x < 1, \\ 0, & x \notin (0,1), \end{cases}$ b) $p_\eta(x) = \begin{cases} 1 - |x|, & |x| \le 1, \\ 0, & |x| > 1. \end{cases}$

II.3.97. ANSWER:

$$p_\eta(x) = \frac{1}{a}\left(1 - \frac{|x|}{a}\right), \qquad |x| < a.$$

II.3.99. ANSWER:

$$p_\eta(x) = \frac{|x| + 1}{4} e^{-|x|}.$$

II.3.101. ANSWER: $p_{\xi\eta} = 12x(1-x)^2$, $x \in (0,1)$.

II.3.104. γ has the Cauchy distribution.

II.3.105. ANSWER:

a) $\dfrac{x}{1+x}$, $x > 0$; b) $\dfrac{1}{1+x^2}$, $x > 0$;

c) the expectation does not exist.

II.3.107. Use induction and Problem II.3.106.

II.3.110. ANSWER:

$$p_\eta(z) = \begin{cases} z^2, & 0 < z \le 1, \\ z(2-z), & 1 < z < 2, \\ 0, & \text{otherwise}. \end{cases}$$

II.3.115. ANSWER: $f(x) = 12x(4x - x^2 - 2\ln x - 3)$, $0 < x < 1$.

II.3.117. ANSWER:

$$\text{a) } \frac{\alpha}{\beta}; \qquad \text{b) } \frac{\alpha}{\beta^2}.$$

II.3.118. For $x \le 0$ we have $F_\eta(x) = \mathsf{P}(\xi^2 < x) = 0$ and for $x > 0$ we have

$$F_\eta(x) = \mathsf{P}(\eta < x) = \mathsf{P}\left(-\sqrt{x} < \xi < \sqrt{x}\right) = \frac{1}{\sqrt{2\pi}} \int_{-\sqrt{x}}^{\sqrt{x}} e^{-z^2/2}\, dz$$

$$= \frac{2}{\sqrt{2\pi}} \int_0^{\sqrt{x}} e^{-z^2/2}\, dz.$$

The distribution density of the random variable η is

$$f_\eta(x) = F'_\eta(x) = \frac{2}{\sqrt{2\pi}} e^{-x/2} \frac{1}{2\sqrt{x}} = \frac{1}{\sqrt{2\pi}} \frac{1}{\sqrt{x}} e^{-x/2},$$

which is the density of the gamma distribution with parameters $\left(\frac{1}{2}, \frac{1}{2}\right)$.

II.3.119. We have

$$\begin{aligned}f_\eta(x) &= \int_{-\infty}^{+\infty} f_{\xi_1}(x-y) f_{\xi_2}(y)\,dy \\ &= \frac{\beta^{\alpha_1+\alpha_2}}{\Gamma(\alpha_1)\Gamma(\alpha_2)} \int_0^x (x-y)^{\alpha_1-1} e^{-\beta(x-y)} y^{\alpha_2-1} e^{-\beta y}\,dy \\ &= \frac{\beta^{\alpha_1+\alpha_2}}{\Gamma(\alpha_1)\Gamma(\alpha_2)} e^{-\beta x} \int_0^x (x-y)^{\alpha_1-1} y^{\alpha_2-1}\,dy.\end{aligned}$$

Now make the change of variable $y = tx$ to obtain

$$\begin{aligned}f_\eta(x) &= \frac{\beta^{\alpha_1+\alpha_2}}{\Gamma(\alpha_1)\Gamma(\alpha_2)} B(\alpha_1,\alpha_2) x^{\alpha_1+\alpha_2-1} e^{-\beta x} \\ &= \frac{\beta^{\alpha_1+\alpha_2}}{\Gamma(\alpha_1+\alpha_2)} x^{\alpha_1+\alpha_2-1} e^{-\beta x}, \qquad x > 0.\end{aligned}$$

II.3.122. ANSWER:

$$\frac{\Gamma\left(\frac{m+n}{2}\right)}{\Gamma\left(\frac{n}{2}\right)\Gamma\left(\frac{m}{2}\right)} \frac{x^{(m/2)-1}}{(1+x)^{(m+n)/2}}, \qquad x > 0.$$

II.3.124. Consider $\ln\eta$ and use the fact that $\ln\xi_k^{-1}$ has the exponential distribution.

II.3.128. It is necessary to prove the equality

$$\mathsf{P}(\xi_1-\xi_2 < x, \xi_1+\xi_2 < y) = \mathsf{P}(\xi_1-\xi_2 < x)\mathsf{P}(\xi_1+\xi_2 < y).$$

To this end we make the change of variables $u - v = t$, $u + v = s$ in the integral

$$\iint\limits_{\substack{u-v<x\\u+v<y}} \exp\left\{-\frac{u^2+v^2}{2}\right\} du\,dv$$

and use the fact that each of the random variables $\xi_1-\xi_2$ and $\xi_1+\xi_2$ has the normal distribution $N(0,2)$. It is known that the assertion of the problem is valid only in the case when both random variables ξ_1 and ξ_2 are normally distributed.

II.3.129. For arbitrary $x \ge 0$ and $y \ge 0$ consider the probability

$$\mathsf{P}(\xi_1-\xi_2 > x, \min(\xi_1,\xi_2) > y) = I.$$

Since ξ_1 and ξ_2 are independent, we have

$$\begin{aligned}I &= \iint\limits_{\substack{u-x>x\\ \min(u,v)>y}} \lambda_1 e^{-\lambda_1 u}\lambda_2 e^{-\lambda_2 v}\,du\,dv = \lambda_1\lambda_2 \int_{x+y}^{\infty} e^{-\lambda_1 u}\left[\int_y^{u-x} e^{-\lambda_2 v}\,dv\right] du \\ &= \lambda_1\lambda_2 \int_x^{\infty} e^{-\lambda_1(y+u)}\left[\int_y^{u+y-x} e^{-\lambda_2 v}\,dv\right] du \\ &= \mathsf{P}(\xi_1-\xi_2 > x)\mathsf{P}(\min(\xi_1,\xi_2) > y).\end{aligned}$$

From this it is easy to obtain the independence of the events

$$\{\omega\colon \xi_1-\xi_2 < x\} \quad\text{and}\quad \{\omega\colon \min(\xi_1,\xi_2) < y\}$$

in the case $x < 0$ and $y \ge 0$.

If $x < 0$, then it is necessary to consider the probability

$$\mathsf{P}(\xi_1 - \xi_2 < x, \min(\xi_1, \xi_2) > y) = \mathsf{P}(\xi_2 - \xi_1 > -x, \min(\xi_1, \xi_2) > y)$$

and to use the equality just proved.

REMARK. The following assertion holds: *if random variables ξ_1 and ξ_2 have distribution densities, and the random variables $\xi_1 - \xi_2$ and $\min(\xi_1, \xi_2)$ are independent, then the random variables ξ_1 and ξ_2 have the exponential distributions.*

II.3.134. Consider the case $n = 1$ and use the induction on n to obtain

$$\sum_{k=1}^{n+1} \frac{\xi_k}{2^k} + \frac{\eta}{2^{n+1}} = \frac{\xi_1}{2} + \frac{1}{2}\left(\sum_{k=2}^{n+1} \frac{\xi_k}{2^{k-1}} + \frac{\eta}{2^n}\right).$$

II.3.135. Since $\xi = \tan\varphi$, where φ is a random variable uniformly distributed on $\left(-\frac{\pi}{2}, \frac{\pi}{2}\right)$, we have

$$\frac{1}{\xi} = \tan\left(\frac{\pi}{2} - \varphi\right), \qquad \frac{2\xi}{1-\xi^2} = \tan 2\varphi, \qquad \frac{3\xi - \xi^3}{1 - 3\xi^2} = \tan 3\varphi.$$

II.3.136. If $\xi_1 = \tan\varphi_1$ and $\xi_2 = \tan\varphi_2$, where φ_1 and φ_2 are independent identically distributed random variables on $\left(-\frac{\pi}{2}, \frac{\pi}{2}\right)$, then $\eta = \tan(\varphi_1 + \varphi_2)$.

II.3.137. ANSWER:

$$\frac{2}{\pi^2} \frac{\ln|x|}{x^2 - 1}.$$

II.3.138. ANSWER:

$$\text{a) } \rho = 0; \quad \text{b) } \rho = \frac{\sqrt{21}}{5}; \quad \text{c) } \rho = \frac{4 \cdot \sqrt{6}}{\pi^2}; \quad \text{d) } \rho = 0.$$

II.3.139. ANSWER: $-\frac{1}{2}$.

II.3.143. ANSWER: 2.

II.3.144. ANSWER:

$$p_\xi(x) = \frac{2\sqrt{1 - \frac{x^2}{a^2}}}{\pi a} \quad \text{for} \quad |x| < a.$$

II.3.146. Let $f(x)$ have expectation μ and variance σ^2. Then

$$\mathsf{E}\,\xi_1 = \mathsf{E}\,\xi_2 = \frac{\mu}{2}; \qquad \operatorname{Var}\xi_1 = \operatorname{Var}\xi_2 = \frac{\sigma^2}{3} + \frac{\mu^2}{12}; \qquad \operatorname{cov}(\xi_1, \xi_2) = \frac{\sigma^2}{6} - \frac{\mu^2}{12}.$$

II.3.149. Let A be an arbitrary Borel set in $\mathbf{R}^n$, and B its image under the transformation (2). Then

$$\mathsf{P}\left((\eta_1, \dots, \eta_n) \in A\right) = \mathsf{P}\left((\xi_1, \dots, \xi_n) \in B\right)$$

and

$$\mathsf{P}\left((\eta_1, \dots, \eta_n) \in A\right) = \int_A p_{\eta_1, \dots, \eta_n}(y_1, \dots, y_n)\, dy_1 \cdots dy_n,$$

$$\mathsf{P}\left((\xi_1, \dots, \xi_n) \in B\right) = \int_B p_{\xi_1, \dots, \xi_n}(x_1, \dots, x_n)\, dx_1 \cdots dx_n.$$

Passing to the new variables $y_1, \dots, y_n$ in the last integral by formulas (2), we get

$$\begin{aligned}
&\mathsf{P}((\xi_1, \dots, \xi_n) \in B) \\
&\quad = \int_A p_{\xi_1,\dots,\xi_n}(g_1(y_1, \dots, y_n), \dots, g_n(y_1, \dots, y_n)) \left| \frac{D(x_1, \dots, x_n)}{D(y_1, \dots, y_n)} \right| dy_1 \cdots dy_n \\
&\quad = \mathsf{P}((\eta_1, \dots, \eta_n) \in A) = \int_A p_{\eta_1,\dots,\eta_n}(y_1, \dots, y_n)\, dy_1 \cdots dy_n,
\end{aligned}$$

and

$$\mathsf{P}_{\eta_1,\dots,\eta_n}(y_1, \dots, y_n) = p_{\xi_1,\dots,\xi_n}(g_1(y_1, \dots, y_n), \dots, g_n(y_1, \dots, y_n)) \left| \frac{D(x_1, \dots, x_n)}{D(y_1, \dots, y_n)} \right|$$

since the set A is arbitrary.

II.3.150. According to Problem II.3.149, we have

$$p_{(R,\Phi)}(r, \varphi) = \frac{1}{2\pi\sigma^2} \exp\left\{-\frac{r^2}{2\sigma^2}\right\} r, \qquad 0 \le r \le +\infty, \quad 0 \le \varphi \le 2\pi.$$

Therefore, R and Φ are independent, the distribution density of R being

$$p_R(r) = \frac{1}{\sigma^2} \exp\left\{-\frac{r^2}{2\sigma^2}\right\} r,$$

and the distribution density of Φ being $(2\pi)^{-1}$, $0 \le \varphi \le 2\pi$.

II.3.155. The distribution density of (R, Φ) is

$$g(r, \varphi) = g(r)r = 2\pi g(r) r \frac{1}{2\pi}.$$

Therefore, R and Φ are independent random variables, Φ being uniformly distributed on $[0, 2\pi]$, and R having the distribution density $2\pi g(r)r$.

II.3.157. The distribution density of (R, Θ, Φ) is

$$g(r, \theta, \varphi) = g(r)r^2 \sin\theta = \frac{4\pi g(r) r^2 \sin\theta}{4\pi},$$
$$0 < r < +\infty, \qquad 0 < \theta < \pi, \qquad 0 < \varphi < 2\pi.$$

Therefore R, Θ, and Φ are independent random variables, R having the distribution density $4\pi g(r)r^2$, Θ having the distribution density $\frac{1}{2}\sin\theta$, $0 \le \theta \le \pi$, and Φ having the distribution density $(2\pi)^{-1}$, $0 \le \varphi \le 2\pi$.

II.3.160. Let η be the length of the vector, and let γ be the projection of the vector on the Ox axis. Then $\gamma = \eta\xi$ is the length of the projection of the vector uniformly distributed on the unit sphere. By the condition of the problem, η and ξ are independent random variables. According to Problem II.3.72, the random variable ξ has the uniform distribution on $[0, 1]$. Applying the formula for the distribution function of a product of independent random variables, we obtain a). Assertion b) can be obtained from a) by differentiation.

II.3.161. The distribution density of the length of the projection is

$$f(t) = \sqrt{\frac{2}{\pi}} e^{-t^2/2}, \qquad t > 0.$$

To conclude the proof, use the result of Problem II.3.160.

II.3.163. ANSWER:

$$\begin{aligned}\mathsf{P}(\xi_{(1)} < x) &= 1 - \mathsf{P}(\min(\xi_1, \dots, \xi_n) \geq x) = 1 - \mathsf{P}(\xi_1 \geq x, \dots, \xi_n \geq x) \\ &= 1 - \prod_{k=1}^{n} \mathsf{P}(\xi_k \geq x) = 1 - e^{-\lambda n x}.\end{aligned}$$

II.3.167. a) We calculate the probability of the random event $\{\omega: \xi_{(1)} \geq x\}$. Since

$$\{\omega: \xi_{(1)} \geq x\} = \{\omega: \xi_1 \geq x, \dots, \xi_n \geq x\}$$

and the random variables $\xi_1, \dots, \xi_n$ are independent, we have

$$\mathsf{P}(\xi_{(1)} \geq x) = \prod_{k=1}^{n} \mathsf{P}(\xi_k \geq x) = [1 - F(x)]^n.$$

Hence

$$\mathsf{P}(\xi_{(1)} < x) = 1 - \mathsf{P}(\xi_{(1)} \geq x) = 1 - [1 - F(x)]^n.$$

b) Since $\{\omega: \xi_{(n)} < x\} = \{\omega: \xi_1 < x, \dots, \xi_n < x\}$, we have

$$\mathsf{P}(\xi_{(n)} < x) = \prod_{k=1}^{n} \mathsf{P}(\xi_k < x) = F^n(x).$$

c) The random event $\{\omega: \xi_{(m)} < x\}$ happens if and only if at least m of the random variables $\xi_1, \dots, \xi_n$ do not exceed x. Therefore,

$$\mathsf{P}(\xi_{(m)} < x) = \sum_{k=m}^{n} \binom{n}{k} F^k(x)[1 - F(x)]^{n-k}.$$

II.3.168. Let us calculate the probability $\mathsf{P}(x \leq \xi_{(m)} < x + \Delta x)$ to within $O(\Delta x)$. Each of the random variables $\xi_1, \dots, \xi_n$ falls into one of the intervals $(-\infty, x)$, $[x, x + \Delta x)$, or $[x + \Delta x, \infty)$. Using a polynomial distribution, we get

$$\begin{aligned}&\mathsf{P}(x \leq \xi_{(m)} < x + \Delta x) \\ &\quad = \frac{n!}{(m-1)!\,1!\,(n-m)!} F^{m-1}(x) p(x) \Delta x [1 - F(x)]^{n-m} + o(\Delta x).\end{aligned}$$

Therefore,

$$p_{\xi_{(m)}}(x) = n\binom{n-1}{m} F^{m-1}(x)[1 - F(x)]^{n-m} p(x) \Delta x + o(\Delta x).$$

II.3.169. If $x < y$, then

$$\mathsf{P}(\xi_{(k)} < x, \xi_{(m)} < y) = \sum_{r=k}^{n} \sum_{s=m-r}^{n} \frac{n!}{r!\,s!\,(n-r-s)!} F^s(x)[F(y)-F(x)]^s[1-F(y)]^{n-r-s}.$$

If $y < x$, then

$$\mathsf{P}(\xi_{(k)} < x, \xi_{(m)} < y) = \mathsf{P}(\xi_{m)} < y) = \sum_{k=m}^{n} \binom{n}{k} F^k(y)[1-F(y)]^{n-k}.$$

If we assume that the random variables have distribution density $p(x)$, then we can find the joint distribution density of $\xi_{(k)}$ and $\xi_{(m)}$ by computing

$$\mathsf{P}(x < \xi_{(k)} < x + \Delta x, y < \xi_{(m)} < y + \Delta y)$$

to within $o(\Delta x \Delta y)$. Using the polynomial distribution, we get

$$p_{\xi_{(k)},\xi_{(m)}}(x,y) = \begin{cases} \dfrac{n!}{(k-1)!\,(m-k-1)!\,(n-m)!} G_{k,m}(x,y)p(x)p(y), & x < y, \\ 0, & x > y, \end{cases}$$

where $G_{k,m}(x,y) = F^{k-1}(x)[F(y)-F(x)]^{n-k-1}[1-F(y)]^{n-m}$.

II.3.173. a) The vector $\xi_{(1)}, \dots, \xi_{(n)}$ is uniformly distributed in the domain $D = \{\vec{x}: x_1 \le x_2 \le \dots \le x_n < T\}$. The distribution density is equal to $n!/T^n$ on D and 0 outside D.

b) The random event $\{\omega: \xi_{(k)} < x\}$ happens if and only if at least k of the random variables $\xi_1, \dots, \xi_n$ belong to the interval $[0, x]$. The probability of this event is

$$\mathsf{P}(\xi_{(k)} < x) = \sum_{r=k}^{n} \binom{n}{r} \left(\frac{x}{T}\right)^r \left(1 - \frac{x}{T}\right)^{n-r}.$$

In order to find the distribution density, it is necessary to compute the probability $\mathsf{P}(x < \xi_k < x + h)$ to within $o(h)$. The distribution density can be obtained also by integrating the joint distribution density of $(\xi_1, \dots, \xi_n)$.

II.3.174. Use Problem II.3.119.

II.3.175. a) We have

$$\begin{aligned} \eta_1 &= \xi_{(1)}, \\ \eta_2 &= \xi_{(2)} - \xi_{(1)}, \\ &\dots\dots\dots\dots \\ \eta_n &= \xi_{(n)} - \xi_{(n-1)}. \end{aligned} \tag{1}$$

The transformation (1) is a linear transformation with determinant 1. It maps

$$D = \{\vec{x}: x_1 \le x_2 \le \dots \le x_n \le T,\ x_i \ge 0\}$$

to

$$D^* = \{\vec{u}: u_i \ge 0, u_1 + \dots + u_n \le T\}.$$

Therefore, the distribution density of $(\eta_1, \dots, \eta_n)$ is

$$p_{(\eta_1,\dots,\eta_n)}(u_1, \dots, u_n) = \begin{cases} \dfrac{n!}{T^n}, & u \in D^*, \\ 0, & u \notin D^*, \end{cases}$$

and the random vector $(\eta_1, \dots, \eta_n)$ is uniformly distributed in D^*.

e) Use the result of Problem II.3.174.

II.3.176. ANSWER:

$$p_{\eta_n}(x) = n(n-1)x^{n-2}(1-x), \qquad 0 < x < 1,$$
$$\mathsf{E}\,\eta_n = \frac{n-1}{n+1}.$$

II.3.179. a) Let P and Q be two points taken independently in D according to the uniform distribution, and let ρ_{PQ} be the distance between these points. Then

$$\mathsf{E}\,f(\rho_{PQ}) = \frac{1}{V^2(D)} \int_D \int_D f(|x-y|)\,dx\,dy.$$

On the other hand,

$$\mathsf{E}\,f(\rho_{PQ}) = \int_{-\infty}^{+\infty} f(z)\,dF_D(z).$$

b) Use the result of Problem II.3.36.

c) Use the assertion of Problem II.3.70.

II.3.180. The conditions of the problem imply that

$$I(y) = \mathsf{P}(\eta \ge y) = \int_y^\infty \frac{z-y}{z}\,dF_\xi(z) = \int_y^\infty \left(1 - \frac{y}{z}\right) dF_\xi(z); \tag{1}$$

$$\mathsf{P}(\xi - \eta \ge u) = \int_u^\infty \frac{z-u}{u}\,dF_\xi(z) = I(u);$$

$$\mathsf{P}(\eta \ge y, \xi - \eta > u) = \int_{u+y}^\infty \left(1 - \frac{u+y}{z}\right) dF_\xi(z) = I(u+y).$$

If γ and η are independent, then $I(u+y) = I(u)I(y)$, so that $I(y) = e^{-ay}$, for some constant $a > 0$. Differentiating twice the equality (1), we find $p_\xi(x)$.

II.3.181. Put

$$\eta_{11} = \frac{\xi_{11} + \xi_{12}}{2}, \qquad \eta_{22} = \frac{\xi_{11} - \xi_{22}}{2}, \qquad \eta_{12} = \frac{\xi_{12} + \xi_{21}}{2}, \qquad \eta_{21} = \frac{\xi_{12} - \xi_{21}}{2}.$$

Then

$$\Delta = \left(\eta_{11}^2 + \eta_{21}^2\right) - \left(\eta_{22}^2 + \eta_{12}^2\right).$$

Since random variables η_{ij} are independent and have normal distribution $N(0, \frac{1}{2})$, the random variables $\eta_{11}^2 + \eta_{21}^2$ and $\eta_{22}^2 + \eta_{12}^2$ have exponential distribution with parameter $\lambda = 1$ (the χ^2-distribution with two degrees of freedom). Hence

$$p_\Delta(x) = \frac{1}{2} e^{-|x|}.$$

II.3.182. Since the random variables $\xi_k/(\xi_1+\cdots+\xi_n)$ are identically distributed, we have $\mathsf{E}\,\xi_k/(\xi_1+\cdots+\xi_n)=c$ (i.e., c does not depend on k). Therefore,

$$\mathsf{E}\,\frac{\xi_1+\cdots+\xi_k}{\xi_1+\cdots+\xi_n}=\sum_{i=1}^{k}\mathsf{E}\,\frac{\xi_i}{\xi_1+\cdots+\xi_n}=kc.$$

The constant c can be found from the condition

$$\mathsf{E}\,\frac{\xi_1+\cdots+\xi_n}{\xi_1+\cdots+\xi_n}=1=nc.$$

Thus

$$\mathsf{E}\,\frac{\xi_1+\cdots+\xi_k}{\xi_1+\cdots+\xi_n}=\frac{k}{n}.$$

II.3.183. a) Obviously, $Q_\xi(+\infty)=\mathsf{P}(-\infty<\xi<+\infty)=1$. On the other hand, $Q_\xi(0)=\sup_x\mathsf{P}(\xi=x)$. Thus $Q_\xi(0)$ is equal to the greatest jump of the distribution function of ξ.

b) Let $l_1<l_2$. For any $\varepsilon>0$ there exists an interval $[x_1,x_1+l_1]$ of length l_1 such that $\mathsf{P}(x_1\le\xi\le x_1+l_1)>Q_\xi(l_1)-\varepsilon$. But

$$\mathsf{P}(x_1\le\xi\le x_1+l_2)\ge\mathsf{P}(x_1\le\xi\le x_1+l_1)>Q_\xi(l_1)-\varepsilon,$$

whence $Q_\xi(l_2)\ge Q_\xi(l_1)$.

c) $Q_\xi(l)=\sup_{x\in\mathbf{R}}[F(x+l-0)-F(x)]$.

d) As was shown earlier, $Q_{\xi+\eta}(l)=\sup_{x\in\mathbf{R}}[F_{\xi+\eta}(x+l+0)-F_{\xi+\eta}(x)]$. But

$$F_{\xi+\eta}(x)=\int_{-\infty}^{+\infty}F_\xi(x-y)\,dF_\eta(y).$$

Therefore,

$$Q_{\xi+\eta}(x)=\sup_{x\in\mathbf{R}}\left[\int_{-\infty}^{+\infty}[F_\xi(x+l+0-y)-F_\xi(x-y)]\,dF_\eta(y)\right].$$

By the definition of the function $Q(l)$, we have

$$F_\xi(x+l+0-y)-F_\xi(x-y)\le\sup_{u\in\mathbf{R}}[F_\xi(u+l+0)-F_\xi(u)]=Q_\xi(l).$$

Thus

$$Q_{\xi+\eta}(l)\le\sup_{x\in\mathbf{R}}\int_{-\infty}^{+\infty}Q_\xi(l)\,dF_\eta(x)=Q_\xi(l).$$

The inequality $Q_{\xi+\eta}(l)\le Q_\eta(l)$ can be proved similarly.

e) If the distribution function of ξ is continuous, then $Q_\xi(0)=0$. Therefore, the inequality $Q_{\xi+\eta}(0)\le Q_\xi(0)$ implies that $Q_{\xi+\eta}(0)=0$.

II.3.184. We have

$$\mathsf{P}(x\le\xi\le x+l)=\frac{1}{\pi}\left[\arctan\frac{x+l}{a}-\arctan\frac{x}{a}\right].$$

Equating the derivative of this expression to zero, we find that it takes the maximum value at $x=-l/2$. Hence

$$Q_\xi(l)=\frac{2}{\pi}\arctan\frac{l}{2a}.$$

§4.

II.4.10. ANSWER: $\mathsf{P}(Q(\xi_1, \xi_2) > y) = e^{-y}$, $y > 0$.

II.4.12. a) Using the result of Problem II.3.149, we obtain that the joint distribution density of (R, Φ) is equal to

$$p_{R,\Phi}(r, \varphi) = \frac{r}{2\pi\sigma_1\sigma_2\sqrt{1-\rho^2}} \exp\left\{-\frac{r^2}{2(1-\rho^2)} G(\varphi)\right\},$$

where

$$G(\varphi) = \frac{\cos^2\varphi}{\sigma_1^2} - 2\rho\frac{\cos\varphi\sin\varphi}{\sigma_1\sigma_2} + \frac{\sin^2\varphi}{\sigma_2^2}.$$

b) Integrating $p_{R,\Phi}(r, \varphi)$ with respect to r from 0 to ∞, we get the distribution density of Φ:

$$p_\Phi(\varphi) = \frac{1}{2\pi\sigma_1\sigma_2} \frac{\sqrt{1-\rho^2}}{G(\varphi)}, \qquad 0 \le \varphi \le 2\pi,$$

where $G(\varphi)$ is given above. In particular, Φ has a uniform distribution for $\rho = 0$, $\sigma_1^2 = \sigma_2^2$.

c) To find the distribution density of R, we integrate $p_{R,\Phi}(r, \varphi)$ with respect to φ from 0 to 2π. In doing so, we use the integral

$$I_0(z) = \frac{1}{2\pi} \int_0^{2\pi} e^{-z\sin\theta}\, d\theta,$$

where $I_0(z)$ is the modified Bessel function of the first kind. Then, using the above notation $G(\varphi)$, we get:

$$\begin{aligned} p_R(r) &= \frac{r}{2\pi\sigma_1\sigma_2\sqrt{1-\rho^2}} \int_0^{2\pi} \exp\left\{-\frac{r^2}{2(1-\rho^2)} G(\varphi)\right\} d\varphi \\ &= \frac{r}{2\pi\sigma_1\sigma_2\sqrt{1-\rho^2}} \exp\left\{-\frac{c^2}{4}s_1\right\} \int_0^{2\pi} \exp\left\{-\frac{c^2}{2}\left[s_2\frac{\cos 2\varphi}{2} - \frac{\rho\sin 2\varphi}{\sigma_1\sigma_2}\right]\right\} d\varphi, \end{aligned}$$

where

$$c = \frac{r}{\sqrt{1-\rho^2}}, \qquad s_1 = \frac{1}{\sigma_1^2} + \frac{1}{\sigma_2^2}, \qquad s_2 = \frac{1}{\sigma_1^2} - \frac{1}{\sigma_2^2}.$$

Therefore,

$$\begin{aligned} p_R(r) &= \frac{c}{\sigma_1\sigma_2} \exp\left\{-\frac{c^2}{4}s_1\right\} \frac{1}{2\pi} \int_0^{2\pi} \exp\left\{-\frac{c^2}{2}\sqrt{\frac{s_2^2}{4} + \frac{\rho^2}{\sigma_1^2\sigma_2^2}} \sin(2\varphi + \gamma)\right\} d\varphi \\ &= \frac{c}{\sigma_1\sigma_2} \exp\left\{-\frac{c^2}{4}s_1\right\} I_0\left(\frac{c^2}{2}\sqrt{\frac{s_2^2}{4} + \frac{\rho^2}{\sigma_1^2\sigma_2^2}}\right). \end{aligned}$$

In particular, if $\rho = 0$ and $\sigma_1^2 = \sigma_2^2 = \sigma^2$, then $c = r$, $s_2 = 0$, and

$$p_R(r) = \frac{r}{\sigma^2} \exp\left\{-\frac{r^2}{2\sigma^2}\right\},$$

since $I_0(0) = 1$.

II.4.13. Use the result of the previous problem to obtain

$$\begin{aligned}\mathsf{P}(\xi\eta > 0) &= \mathsf{P}\left(\left\{\omega\colon 0 \le \Phi \le \frac{\pi}{2}\right\} \cup \left\{\omega\colon \pi \le \Phi \le \frac{3\pi}{2}\right\}\right)\\ &= \frac{\sqrt{1-\rho^2}}{\pi}\int_0^{2\pi}\frac{d\varphi}{1-\rho\sin 2\varphi} = \frac{\sqrt{1-\rho^2}}{2\pi}\int_0^{2\pi}\frac{du}{1-\rho\sin u}.\end{aligned}$$

Calculate the last integral by introducing the new variable $t = \tan\frac{\varphi}{2}$. Then

$$\mathsf{P}(\xi\eta > 0) = \frac{1}{2} + \frac{1}{\pi}\arcsin\rho.$$

II.4.14. Using the symmetry, we get

$$\mathsf{E}\max(\xi,\eta) = \frac{1}{\pi\sqrt{1-\rho^2}}\iint_{\{x>y\}} x\exp\left\{\frac{x^2-2\rho xy+y^2}{2(1-\rho^2)}\right\}dx\,dy.$$

We make the change of variables $x = v$, $y - \rho x = u\sqrt{1-\rho^2}$. Then the integral takes the form

$$\begin{aligned}\frac{1}{\pi}\iint\limits_{\left\{u<\frac{1-\rho}{\sqrt{1-\rho^2}}v\right\}} ve^{-(v^2+u^2)/2}\,dv\,du &= \frac{1}{\pi}\int_{-\infty}^{+\infty}e^{-u^2/2}\left[\int_{\sqrt{(1-\rho^2)/(1-\rho)}} ve^{-v^2/2}\,dv\right]du\\ &= \frac{1}{\pi}\int_{-\infty}^{+\infty}e^{-u^2/(1-\rho)}\,du = \sqrt{\frac{1-\rho}{\pi}}.\end{aligned}$$

§5.

II.5.1. ANSWER:

$$\begin{aligned}\mathsf{E}\,\xi^2 &= \int_\Omega \xi^2(\omega)\,d\mathsf{P} = \int_{\{\omega:|\xi(\omega)|\ge\varepsilon\}}\xi^2(\omega)\,d\mathsf{P} + \int_{\{\omega:|\xi(\omega)|<\varepsilon\}}\xi^2(\omega)\,d\mathsf{P}\\ &\ge \int_{\{\omega:|\xi(\omega)|<\varepsilon\}}\xi^2(\omega)\,d\mathsf{P} \ge \varepsilon^2\mathsf{P}(\{\omega\colon|\xi(\omega)| > \varepsilon\}).\end{aligned}$$

II.5.5. ANSWER:

$$\mathsf{E}\,f(\xi) = \int_{-\infty}^{+\infty}f(x)\,dF(x) = \int_{-\infty}^{\varepsilon}f(x)\,dF(x) + \int_{\varepsilon}^{+\infty}f(x)\,dF(x) \ge f(\varepsilon)\mathsf{P}(\xi > \varepsilon).$$

II.5.12. a) Let $\varepsilon > 0$. Consider the function $f(x) = (x+c)^2$, $c > 0$. Since $f(x) \ge (c+\varepsilon)^2$ for $x > \varepsilon$, in accordance with Problem II.5.6 we have

$$\mathsf{P}(\xi > \varepsilon) \le \frac{\mathsf{E}(\xi+c)^2}{(\varepsilon+c)^2} = \frac{c^2+\sigma^2}{(\varepsilon+c)^2}.$$

The right-hand side of the inequality attains its minimum value $\sigma^2/(\sigma^2+\varepsilon^2)$ at $c = \sigma^2/\varepsilon$.

b) Let $\varepsilon < 0$. Put $f(x) = (x-c)^2$. Since $f(x) \ge (\varepsilon-c)^2$ for $x < \varepsilon$, in accordance with Problem II.5.6, we have

$$\mathsf{P}(\xi < \varepsilon) \le \frac{\mathsf{E}(\xi-c)^2}{(\varepsilon-c)^2} = \frac{\sigma^2+c^2}{(\varepsilon-c)^2}.$$

The right-hand side of this inequality attains its minimum value $\sigma^2/(\sigma^2+\varepsilon^2)$ at $c=-\sigma^2/\varepsilon$.

II.5.19. Replacing ξ and η with $a\xi$ and $b\eta$, respectively, we get an equivalent inequality; thus, it suffices to prove the inequality for random variables that satisfy the conditions $\mathsf{E}\,\xi^2=1$ and $\mathsf{E}\,\eta^2=1$. But in this case the inequality follows from the inequality $|\xi\eta|\le\xi^2+\eta^2$ (to see this, it is enough to take the expectations of both sides of this inequality).

II.5.20. The function $u(x)=\ln x$ is convex on $(0,+\infty)$. Therefore,

$$(1-t)\ln x_1+t\ln x_2\le\ln[(1-t)x_1+tx_2]$$

for any $x_1>0$, $x_2>0$, and $0<t<1$. Hence

$$x_1^{1-t}x_2^t\le(1-t)x_1+tx_2. \tag{1}$$

It is sufficient to prove the inequality for random variables ξ and η satisfying the condition $\mathsf{E}\,|\xi|^p=\mathsf{E}\,|\eta|^q=1$. In the inequality (1), put

$$t=1/q,\qquad 1-t=1/p,$$

substitute $x_1=|\xi|^p$, $x_2=|\eta|^q$, and take the expectations of both sides of the obtained inequality. Then $\mathsf{E}\,|\xi\eta|\le 1$.

II.5.21. According to the Cauchy inequality,

$$\left(\mathsf{E}\,|\xi|^t\right)^2=\left(\mathsf{E}\,|\xi|^{t-h}|\xi|^{t+h}\right)^2\le\mathsf{E}\,|\xi|^{t-h}\,\mathsf{E}\,|\xi|^{t+h}.$$

Taking the logarithm of this inequality, we get $2u(t)\le u(t-h)+u(t+h)$, which means that $u(t)$ is convex.

II.5.22. a) If the random variable ξ is symmetric, then its distribution function coincides with the distribution function of the random variable $-\xi$. Therefore,

$$\mathsf{E}\cosh(2y\xi)=\mathsf{E}\,\frac{e^{2y\xi}+e^{-2y\xi}}{2}=\mathsf{E}\,e^{2y\xi},$$

and the assertion follows.

b) We have

$$\tau(\xi)=\sup_{y>0}\sqrt{G(y)}=\sigma,$$

where

$$G(y)=\frac{1}{2y^2}\ln\frac{1}{\sqrt{2\pi\sigma^2}}\int_{-\infty}^{+\infty}\exp\left\{2yx-\frac{x^2}{2\sigma^2}\right\}dx.$$

d) Put $y_0=x/2\tau^2(\xi)$. Then

$$\tau^2(\xi)\ge\frac{1}{2y_0^2}\ln\mathsf{E}\exp\{2y_0\xi\}\quad\text{or}\quad\exp\{2y_0^2\tau^2(\xi)\}\le\int_{-\infty}^{+\infty}\exp\{2y_0z\}\,dF(z),$$

where $F(z)$ is the distribution function of ξ. Furthermore,

$$\int_{-\infty}^{+\infty}\exp\{2y_0z\}\,dF(z)\ge\exp\{2y_0x\}\int_x^{\infty}dF(x)=\exp\{2y_0x\}\mathsf{P}(\xi>x),$$

so that $\mathsf{P}(\xi>x)\le\exp\{2y_0^2\tau^2(\xi)-2y_0x\}=\exp\{-x^2/2\tau^2(\xi)\}$, since $y_0=x/2\tau^2(\xi)$.

II.5.23. Note that

$$\mathsf{E}\exp\{2y\xi\} \le \exp\{2\tau^2(\xi)y^2\}$$

for $y > 0$. Using the Cauchy inequality, we have

$$\begin{aligned}\mathsf{E}\exp\{2y(\xi_1+\xi_2)\} &\le (\mathsf{E}\exp\{4y\xi_1\}\,\mathsf{E}\exp\{4y\xi_2\})^{1/2}\\ &\le \left(\exp\{8y^2\tau^2(\xi_1)\}\exp\{8y^2\tau^2(\xi_2)\}\right)^{1/2}\\ &\le \exp\{2y^2\}2\left[\tau^2(\xi_1)+\tau^2(\xi_2)\right],\end{aligned}$$

whence $\tau^2(\eta) \le \left[\tau^2(\xi_1)+\tau^2(\xi_2)\right]$. Similarly, $\tau^2(-\eta) \le \left[\tau^2(-\xi_1)+\tau^2(-\xi_2)\right]$, and since $\tau^*(\eta) \le \max(\tau(\eta), \tau(-\eta))$, the assertion follows.

§6.

II.6.1. ANSWER:
a) $\mathsf{E}\left(\xi/\mathfrak{A}\right) = \xi$ with probability one.
b) $\mathsf{P}\left(\mathsf{E}\left(\xi/\mathfrak{B}\right) = \mathsf{E}\,\xi\right) = 1$.
c) With probability one,

$$\mathsf{E}\left(\xi/\mathfrak{B}\right) = \frac{1}{\mathsf{P}(A)}\int_A \xi\,\mathsf{P}(d\omega)\chi_A + \frac{1}{\mathsf{P}(\overline{A})}\int_{\overline{A}} \xi\,d\mathsf{P}(d\omega)\chi_{\overline{A}}.$$

II.6.2. With probability one,

$$\mathsf{P}\left(A/\eta\right) = \sum_{k=1}^{\infty}\frac{1}{\mathsf{P}(\eta = x_k)}\mathsf{P}(A\cap\{\omega\colon \eta = x_k\})\chi_{\{\omega:\eta=x_k\}},$$

$$\mathsf{E}\left(\xi/\eta\right) = \sum_{k=1}^{\infty}\frac{1}{\mathsf{P}(\eta = x_k)}\int_{\{\omega:\eta=x_k\}}\xi\,\mathsf{P}(d\omega)\chi_{\{\omega:\eta=x_k\}}.$$

II.6.3. For any k, $1 \le k \le n$,

$$\mathsf{P}\left(A/B_k\right) = \frac{1}{a_k - a_{k-1}}\mathsf{P}(A\cap(a_{k-1}, a_k]), \qquad a_0 = 0,\ a_n = 1.$$

With probability one,

$$\mathsf{P}\left(A/\mathfrak{B}\right) = \sum_{k=1}^{n}\frac{1}{a_k - a_{k-1}}\mathsf{P}(A\cap(a_{k-1}, a_k])\chi_{(a_{k-1},a_k]},$$

$$\mathsf{E}\left(\xi/\eta\right) = \sum_{k=1}^{n}\frac{1}{a_k - a_{k-1}}\int_{a_{k-1}}^{a_k}\xi(\omega)\,d\omega\chi_{(a_{k-1},a_k]}.$$

II.6.4. With probability one,

$$\mathsf{E}\left(\xi/\mathfrak{B}_1\right) = \left[\sum_{k=-\infty}^{\infty} f(\xi+k)\right]^{-1}\sum_{k=-\infty}^{\infty}(\xi+k)f(\xi+k),$$

$$\mathsf{E}\left(\xi/\mathfrak{B}_2\right) = \frac{\xi f(\xi) - \xi f(-\xi)}{f(\xi) + f(-\xi)}.$$

II.6.5. Assertion a) is a consequence of the following integral property: *if h is a $\mathfrak{B}$-measurable function on Ω and*

$$\int_B h(\omega)\,\mathsf{P}(d\omega) \geq 0,$$

for any $B \in \mathfrak{B}$, then $\mathsf{P}(h \geq 0) = 1$. The other assertions follow from a) and the definition of conditional expectation.

II.6.6. This is a consequence of equality (2) in the definition of conditional expectation for $B = \Omega$ (see the beginning of §II.6).

II.6.7. Indeed, for $B \in \mathfrak{B}$,

$$\int_B \mathsf{E}\left(\xi/\mathfrak{B}\right)\,\mathsf{P}(d\omega) = \int_B \xi\,\mathsf{P}(d\omega) = \int_\Omega \xi\chi_B\,\mathsf{P}(d\omega) = \mathsf{E}(\xi\chi_B).$$

The independence of ξ and χ_B implies the equality $\mathsf{E}(\xi\chi_B) = \mathsf{E}\,\xi\,\mathsf{E}\,\chi_B$, so that

$$\int_B \mathsf{E}\left(\xi/\mathfrak{B}\right)\,\mathsf{P}(d\omega) = \int_B \mathsf{E}\,\xi\,\mathsf{P}(d\omega).$$

II.6.8. If $\mathfrak{B}_1$ and $\mathfrak{B}_2$ are independent, then the required inequality holds by the result of the previous problem. On the other hand, if $A \in \mathfrak{B}_1$, then, by the condition of the problem, for $\xi = \chi_A$ we have

$$\mathsf{P}\left(\mathsf{E}\left(\chi_A/\mathfrak{B}_2\right) = \mathsf{P}(A)\right) = 1.$$

Therefore, by the definition of the conditional expectation, for any $B \in \mathfrak{B}_2$,

$$\int_B \mathsf{E}\left(\chi_A/\mathfrak{B}_2\right)\,\mathsf{P}(d\omega) = \int_B \mathsf{P}(A)\,\mathsf{P}(d\omega) = \int_B \chi_A\,\mathsf{P}(d\omega),$$

i.e.,

$$\mathsf{P}(A)\mathsf{P}(B) = \mathsf{P}(A \cap B).$$

II.6.9. By the definition of the conditional expectation,

$$\int_B \mathsf{E}\left(\mathsf{E}\left(\xi/\mathfrak{B}_2\right)/\mathfrak{B}_1\right)\,\mathsf{P}(d\omega) = \int_B \mathsf{E}\left(\xi/\mathfrak{B}_2\right)\,\mathsf{P}(d\omega)$$

for $B \in \mathfrak{B}_1$. Since $B \in \mathfrak{B}_2$, we have

$$\int_B \mathsf{E}\left(\xi/\mathfrak{B}_2\right)\,\mathsf{P}(d\omega) = \int_B \xi\,\mathsf{P}(d\omega)$$

and

$$\int_B \xi\,\mathsf{P}(d\omega) = \int_B \mathsf{E}\left(\xi/\mathfrak{B}_1\right)\,\mathsf{P}(d\omega).$$

Therefore, for $B \in \mathfrak{B}_1$,

$$\int_B \mathsf{E}\left(\mathsf{E}\left(\xi/\mathfrak{B}_2\right)/\mathfrak{B}_1\right)\,\mathsf{P}(d\omega) = \int_B \mathsf{E}\left(\xi/\mathfrak{B}_1\right)\,\mathsf{P}(d\omega).$$

II.6.10. By the definition of $\mathsf{E}\left(\eta/\mathfrak{B}\right)$, for $B \in \mathfrak{B}$ we have

$$\int_B \eta \,\mathsf{P}(d\omega) = \int_B \mathsf{E}\left(\eta/\mathfrak{B}\right) \,\mathsf{P}(d\omega),$$

that is,

$$\int_\Omega \eta\chi_B \,\mathsf{P}(d\omega) = \int_\Omega \mathsf{E}\left(\eta/\mathfrak{B}\right) \chi_B \,\mathsf{P}(d\omega),$$

whence

$$\int_\Omega \eta\zeta \,\mathsf{P}(d\omega) = \int_\Omega \mathsf{E}\left(\eta/\mathfrak{B}\right) \zeta \,\mathsf{P}(d\omega)$$

for any random variable ζ of the form

$$\zeta = \sum_{k=1}^{n} c_k \chi_{B_k},$$

where $c_k \in \mathbf{R}$, $B_k \in \mathfrak{B}$ for all $1 \le k \le n$. Now it is necessary to use the theorem on approximation of $\mathfrak{B}$-measurable functions by functions of the above form and the theorem on the passage to the limit under the sign of Lebesgue integral.

II.6.11. A consequence of the previous two problems.

II.6.12. The desired equality holds for $\mathsf{E}\left(\xi/\mathfrak{B}\right)$ by the result of Problem II.6.11. If for a $\mathfrak{B}$-measurable random variable ζ,

$$\mathsf{E}(\xi\eta) = \mathsf{E}(\zeta\eta),$$

then for $\eta = \chi_B$, $B \in \mathfrak{B}$, we get

$$\int_B \xi \,\mathsf{P}(d\omega) = \int_B \zeta \,\mathsf{P}(d\omega),$$

i.e., ζ is a variant of conditional expectation.

II.6.13. According to Problem II.6.5, under the conditions of the problem,

$$\mathsf{P}\left(\mathsf{E}\left(\xi_{n+1}/\mathfrak{B}\right) \ge \mathsf{E}\left(\xi_n/\mathfrak{B}\right) \ge 0\right) = 1, \qquad n \ge 1.$$

Now use the Lebesgue theorem on the passage to the limit in the equality

$$\int_B \mathsf{E}\left(\xi_n/\mathfrak{B}\right) \,\mathsf{P}(d\omega) = \int_B \xi_n \,\mathsf{P}(d\omega), \qquad B \in \mathfrak{B},\ n \ge 1.$$

II.6.14. According to Problem II.6.5,

$$\left|\mathsf{E}\left(\xi_n/\mathfrak{B}\right) - \mathsf{E}\left(\xi/\mathfrak{B}\right)\right| \le \mathsf{E}\left(|\xi_n - \xi|/\mathfrak{B}\right), \qquad n \ge 1.$$

Hence it suffices to demonstrate that

$$\mathsf{P}\left(\lim_{n\to\infty} \mathsf{E}\left(|\xi_n - \xi|/\mathfrak{B}\right) = 0\right) = 1.$$

To do this, consider the sequence

$$\left\{2\eta - \sup_{k\ge n} |\xi_k - \xi|,\ n \ge 1\right\}$$

and use the result of Problem II.6.13.

II.6.15. Apply the result of Problem II.6.13 to the sequences

$$\left\{\xi - \sup_{k\geq n} \xi_k,\ n \geq 1\right\}, \quad \text{and} \quad \left\{\inf_{k\geq n} \xi_k - \eta,\ n \geq 1\right\}.$$

II.6.16. Apply the result of Problem II.6.13 to the sequence

$$\left\{\sum_{k=1}^{n} \xi_k,\ n \geq 1\right\}.$$

II.6.17. According to Problems II.6.5 and II.6.10, the desired equality holds for a function of the form

$$(*) \qquad g(x, y) = \sum_{k=1}^{n} a_k(x) b_k(x).$$

For a function g satisfying the condition of the problem there exists a sequence $\{g_n\}$ of functions of the form $(*)$ such that $\mathsf{P}\left(g_n(\xi, \eta) \to g(\xi, \eta) \text{ as } n \to \infty\right) = 1$. Now use the result of Problem II.6.14.

II.6.18. A consequence of Problems II.6.17 and II.6.7.

II.6.19. Put $B = \Omega$ in formula (1) at the beginning of §II.6.

II.6.20. The properties of conditional probabilities listed in this problem are consequences of the corresponding properties of conditional expectation, since $\mathsf{P}\left(A/\mathfrak{B}\right) = \mathsf{E}\left(\chi_A/\mathfrak{B}\right)$. For instance, the equality d) is a consequence of Problem II.6.13 and e) is a consequence of Problem II.6.16.

II.6.21. A consequence of Problem II.6.16.

II.6.22. ANSWER:

$$\mathsf{P}\left(A/\mathfrak{B}_1\right) = \left\{ \sum_{n=-\infty}^{\infty} [f(x + 2n\pi) + f(\pi - x + 2n\pi)] \right\}^{-1}$$
$$\times \sum_{n=-\infty}^{\infty} [\chi_A(x + 2n\pi) f(x + 2n\pi) + \chi_A(\pi - x + 2n\pi) f(\pi - x + 2n\pi)].$$

II.6.23. A consequence of Problem II.6.9.

II.6.25. ANSWER:

$$\begin{aligned}
\int_0^{\infty} x\, dF_B(x) &= \sum_{n=0}^{\infty} \int_n^{n+1} x\, dF_B(x) \leq \sum_{n=0}^{\infty} (n+1)\, [F_B(n+1) - F_B(n)] \\
&= \sum_{n=0}^{\infty} (n+1) \mathsf{P}\left(n \leq \xi < n + 1/B\right) \\
&= \frac{1}{\mathsf{P}(B)} \sum_{n=0}^{\infty} (n+1) \mathsf{P}\left(\{n \leq \xi < n+1\} \cap B\right) \\
&\leq \frac{1}{\mathsf{P}(B)} \sum_{n=0}^{\infty} (n+1) \mathsf{P}(n \leq \xi < n+1) \\
&\leq \frac{1}{\mathsf{P}(B)} \int_0^{\infty} (x+1)\, d\mathsf{P}(\xi < x).
\end{aligned}$$

Similarly,

$$\int_{-\infty}^{0} |x|\, dF_B(x) \le \frac{1}{\mathsf{P}(B)} \int_{-\infty}^{0} (|x|+1)\, d\mathsf{P}(\xi < x).$$

II.6.26. For a real Borel function f such that $\mathsf{E}\,|f(\xi)| < \infty$, a more general equality

$$\mathsf{E}\left(f(\xi)/B\right) = \frac{1}{\mathsf{P}(B)} \int_B f(\xi)\, \mathsf{P}(d\omega),$$

or the equality

$$\int_{-\infty}^{\infty} f(x)\, dF_B(x) = \frac{1}{\mathsf{P}(B)} \int_B f(\xi(\omega))\, \mathsf{P}(d\omega),$$

is true. This equality holds for $f = \chi_A$, where A is a Borel set on $\mathbf{R}$:

$$\int_{-\infty}^{\infty} \chi_A(x)\, dF_B(x) = F_B(A) = \widetilde{\mathsf{P}}(\xi \in A) = \frac{\mathsf{P}(\{\omega : \xi \in A\} \cap B)}{\mathsf{P}(B)}$$
$$= \frac{1}{\mathsf{P}(B)} \int_B \chi_A(\xi(\omega))\, \mathsf{P}(d\omega).$$

Obviously, this equality is valid also for linear combinations of indicators. Now use the theorem on the passage to the limit.

II.6.27. A consequence of the previous problem.

II.6.28. For $B = \{\omega : \xi > 0\}$ establish that

$$F_B(x) = \frac{\mathsf{P}(\xi < x, \xi > 0)}{\mathsf{P}(\xi > 0)} = \left[\int_0^{\infty} f(u)\, du\right]^{-1} \int_0^x f(u)\, du.$$

Then

$$\mathsf{E}\left(\xi/\xi > 0\right) = \left[\int_0^{\infty} f(u)\, du\right]^{-1} \int_0^{\infty} u f(u)\, du.$$

Similarly,

$$\mathsf{E}\left(\xi/a \le \xi \le b\right) = \left[\int_a^{b} f(u)\, du\right]^{-1} \int_a^{b} u f(u)\, du.$$

II.6.29. The function $\sum_{n=1}^{\infty} \mathsf{E}\left(\xi/B_n\right) \chi_{B_n}$ is $\mathfrak{B}$-measurable. Calculate the integral

$$\int_B \left\{ \sum_{n=1}^{\infty} \mathsf{E}\left(\xi/B_n\right) \chi_{B_n} \right\} \mathsf{P}(d\omega).$$

See Problem II.6.26.

II.6.30. The sequence of events $\{\{\omega : \nu = n\}, n \ge 1\}$ is a complete group of events. Therefore, according to Problem II.6.29, we have

$$\mathsf{E}\left(\sum_{k=1}^{\nu} \xi_k\right) = \sum_{n=1}^{\infty} \mathsf{E}\left(\sum_{k=1}^{\nu} \xi_k/\{\nu = n\}\right) \mathsf{P}(\nu = n)$$
$$= \sum_{n=1}^{\infty} \left(\sum_{k=1}^{n} \xi_k/\{\nu = n\}\right) \mathsf{P}(\nu = n).$$

Since the random variable $\sum_{k=1}^{n}\xi_k$ and the event $\{\omega\colon \nu = n\}$ are independent, we get (Problem II.6.24)

$$\mathsf{E}\left(\sum_{k=1}^{n}\xi_k/\{\nu=n\}\right)=\mathsf{E}\sum_{k=1}^{n}\xi_k=n\,\mathsf{E}\,\xi_1.$$

Therefore,

$$\mathsf{E}\sum_{k=1}^{\nu}\xi_k=\sum_{n=1}^{\infty}n\,\mathsf{E}\,\xi_1\mathsf{P}(\nu=n)=\mathsf{E}\,\nu\,\mathsf{E}\,\xi_1.$$

II.6.31. For a fixed $t \geq 0$, let

$$B_k=B_k(t)=\left\{\omega\colon\sum_{j=0}^{k-1}\tau_j\leq t<\sum_{j=0}^{k}\tau_j\right\}.$$

The sequence $\{B_n, n\geq 1\}$ is a complete group of events. According to Problems II.6.29 and II.6.24, we have

$$m(t)=\sum_{k=1}^{\infty}\mathsf{E}\left(\xi(t)/B_k\right)\mathsf{P}(B_k)=\sum_{k=1}^{\infty}\mathsf{E}\left(\xi_k/B_k\right)\mathsf{P}(B_k)=\sum_{k=1}^{\infty}\mathsf{E}\,\xi_k\mathsf{P}(B_k)=m.$$

To calculate $r(s,t)$, assume that $s<t$ and consider the complete group of events

$$B_{jk}=B_j(s)\cap B_k(t),\qquad k,j\geq 1.$$

As before, we get

$$\begin{aligned}
r(s,t)&=\sum_{j,k=1}^{\infty}\mathsf{E}\left((\xi(s)-m)(\xi(t)-m)/B_{jk}\right)\mathsf{P}(B_{jk})\\
&=\sum_{j,k=1}^{\infty}\mathsf{E}\left((\xi_j-m)(\xi_k-m)/B_{jk}\right)\mathsf{P}(B_{jk})=\sum_{j,k=1}^{\infty}\mathsf{E}(\xi_j-m)(\xi_k-m)\mathsf{P}(B_{jk})\\
&=\sum_{j=1}^{\infty}\mathsf{E}(\xi_j-m)^2\mathsf{P}(B_{jj})=\sigma^2\exp\{-\lambda(t-s)\}.
\end{aligned}$$

Thus

$$m(t)=m,\qquad r(s,t)=\sigma^2\{-\lambda|t-s|\},\qquad s,t\in\mathbf{R}.$$

II.6.32. The σ-algebra generated by the random variable η coincides with the σ-algebra generated by the complete group of events $\{\omega\colon \eta=y_n\}$, $n\geq 1$. According to Problem II.6.29, the equality

$$\mathsf{E}\left(\xi/\eta\right)=\sum_{n=1}^{\infty}\mathsf{E}\left(\xi/\eta=y_n\right)\chi_{\{\eta=y_n\}}=\sum_{n=1}^{\infty}\frac{1}{\mathsf{P}(\eta=y_n)}\int_{\{\eta=y_n\}}\xi\chi_{\{\eta=y_n\}}\,\mathsf{P}(d\omega)$$

holds with probability one.

II.6.33. A consequence of Problem II.6.27.

II.6.34. For a convex function f and a point x_0 there exists a straight line with slope λ such that

$$f(x) \geq f(x_0) + \lambda(x - x_0), \qquad x \in \mathbf{R}.$$

Put $x_0 = \mathsf{E}\left(\xi/\mathfrak{B}\right)$ and $x = \xi$. Then

$$f(\xi) \geq f\left(\mathsf{E}\left(\xi/\mathfrak{B}\right)\right) + \lambda\left(\xi - \mathsf{E}\left(\xi/\mathfrak{B}\right)\right).$$

Now take the conditional expectation.

II.6.35. A consequence of Problems II.6.6 and II.6.34:

$$\mathsf{E}\,|\xi|^r = \mathsf{E}\left[\mathsf{E}\left(|\xi|^r/\mathfrak{B}\right)\right] \geq \mathsf{E}\left[\left|\mathsf{E}\left(\xi/\mathfrak{B}\right)\right|^r\right].$$

II.6.36. According to Problem II.6.34, we have

$$\mathsf{P}\left(\left|\mathsf{E}\left(\xi/\mathfrak{B}\right)\right|^r \leq \mathsf{E}\left(|\xi|^r/\mathfrak{B}\right)\right) = 1.$$

Therefore,

$$\begin{aligned}\int_{\{|\mathsf{E}(\xi/\mathfrak{B})|^r > a\}} \left|\mathsf{E}\left(\xi/\mathfrak{B}\right)\right|^r \mathsf{P}(d\omega) &\leq \int_{\{|\,\mathsf{E}(\xi/\mathfrak{B})|^r > a\}} \mathsf{E}\left(|\xi|^r/\mathfrak{B}\right) \mathsf{P}(d\omega) \\ &= \int_{\{|\mathsf{E}(\xi/\mathfrak{B})|^r > a\}} |\xi|^r \, \mathsf{P}(d\omega).\end{aligned}$$

The last equality holds by the definition of conditional expectation of the random variable $|\xi|^r$, since $\left\{\omega\colon \left|\mathsf{E}\left(\xi/\mathfrak{B}\right)\right|^r > a\right\} \in \mathfrak{B}$. Furthermore, according to the Chebyshev inequality,

$$\mathsf{P}\left(\left|\mathsf{E}\left(\xi/\mathfrak{B}\right)\right|^r > a\right) \leq \frac{1}{a}\,\mathsf{E}\left|\mathsf{E}\left(\xi/\mathfrak{B}\right)\right|^r \leq \frac{1}{a}\,\mathsf{E}\left[\mathsf{E}\left(|\xi|^r/\mathfrak{B}\right)\right] = \frac{1}{a}\,\mathsf{E}\,|\xi|^r.$$

II.6.38. Indeed,

$$\mathsf{E}\left[\xi\,\mathsf{E}\left(\eta/\mathfrak{B}\right)\right] = \mathsf{E}\left\{\mathsf{E}\left(\xi\,\mathsf{E}\left(\eta/\mathfrak{B}\right)/\mathfrak{B}\right)\right\} = \mathsf{E}\left\{\mathsf{E}\left(\xi/\mathfrak{B}\right)\mathsf{E}\left(\eta/\mathfrak{B}\right)\right\}.$$

II.6.39. Indeed, according to Problem II.6.35, $\mathsf{E}\left[\mathsf{E}\left(\xi/\mathfrak{B}\right)\right]^2$ exists and

$$\begin{aligned}\mathsf{E}(\xi - \eta)^2 &= \mathsf{E}\left[\xi - \mathsf{E}\left(\xi/\mathfrak{B}\right) + \mathsf{E}\left(\xi/\mathfrak{B}\right) - \eta\right]^2 \\ &= \mathsf{E}\left[\xi - \mathsf{E}\left(\xi/\mathfrak{B}\right)\right]^2 + \mathsf{E}\left[\mathsf{E}\left(\xi/\mathfrak{B}\right) - \eta\right]^2,\end{aligned}$$

since, in view of Problem II.6.10, we have

$$\mathsf{E}\left\{\left[\xi - \mathsf{E}\left(\xi/\mathfrak{B}\right)\right]\left[\mathsf{E}\left(\xi/\mathfrak{B}\right) - \eta\right]\right\} = \mathsf{E}\left\{\left[\mathsf{E}\left(\xi/\mathfrak{B}\right) - \eta\right]\mathsf{E}\left(\left[\xi - \mathsf{E}\left(\xi/\mathfrak{B}\right)\right]/\mathfrak{B}\right)\right\} = 0.$$

II.6.40. The solution of this problem is similar to that of the previous one. The random variable $\mathsf{E}\left(\xi/\eta\right)$ is always a random function of the form $f(\eta)$ (prove this fact). Then the required assertion is a consequence of the previous problem.

II.6.41. Indeed, according to Problem II.6.35, we have

$$\mathsf{E}\left|\mathsf{E}\left(\xi/\mathfrak{B}\right)\right|^r < \infty, \qquad \mathsf{E}\left|\mathsf{E}\left(\xi_n/\mathfrak{B}\right)\right|^r < \infty, \quad n \geq 1.$$

Then

$$\mathsf{E}\left|\mathsf{E}\left(\xi_n/\mathfrak{B}\right) - \mathsf{E}\left(\xi/\mathfrak{B}\right)\right|^r = \mathsf{E}\left|\mathsf{E}\left((\xi_n - \xi)/\mathfrak{B}\right)\right|^r \leq \mathsf{E}\,|\xi_n - \xi|^r.$$

II.6.42. We have

$$\mathsf{E}\left(\varphi/\mathfrak{B}\right) = \frac{1}{n!}\sum \varphi(x_{k_1}, x_{k_2}, \ldots, x_{k_n}),$$

where the summation is over all $n!$ permutations $k_1, \ldots, k_n$ of the set $1, 2, \ldots, n$.

$$\text{a) } \bar{x} = \frac{1}{n}\sum_{i=1}^{n} x_i; \qquad \text{b) } \frac{1}{n-1}\sum_{i=1}^{n}(x_i - \bar{x})^2.$$

II.6.43. A consequence of Problem II.6.34. A simpler proof can be obtained if one uses an explicit form of $\mathsf{E}\left(\varphi/\mathfrak{B}\right)$ and the Cauchy inequality.

II.6.44. A consequence of Problem II.6.10.

II.6.45. A consequence of Problem II.6.7.

II.6.46. A consequence of Problem II.6.9.

II.6.47. Prove that

$$\mathsf{P}\left(\mathsf{E}\left(\xi_k/\bar{\xi}\right) = \mathsf{E}\left(\xi_1/\bar{\xi}\right)\right) = 1, \qquad 1 \le k \le n.$$

Then

$$\mathsf{E}\left(\xi_k/\bar{\xi}\right) = \frac{1}{n}\sum_{i=1}^{n} \mathsf{E}\left(\xi_i/\bar{\xi}\right) = \mathsf{E}\left(\bar{\xi}/\bar{\xi}\right) = \bar{\xi}$$

with probability one. The independence condition can be generalized as follows: $F(x_1, \ldots, x_n) = \mathsf{P}(\xi_1 < x_1, \ldots, \xi_n < x_n)$ *is a symmetric function of its arguments.*

II.6.49. The first equality is equivalent to the equality $\varphi(R) = \mathsf{E}\,\xi$. Let $\mathfrak{B}$ be the σ-algebra generated by the random variable η. To prove the second equality, it is necessary to demonstrate that

$$\int_B g(\eta)\,\mathsf{P}(d\omega) = \int_B \mathsf{E}\left(\xi/\eta\right)\,\mathsf{P}(d\omega), \qquad B \in \mathfrak{B},$$

or

$$\int_B g(\eta)\,\mathsf{P}(d\omega) = \int_B \xi\,\mathsf{P}(d\omega).$$

Let $C \in S$ be such that $\eta^{-1}(C) \in B$. Then

$$\begin{aligned}\int_B g(\eta)\,\mathsf{P}(d\omega) &= \int_{\{\eta \in C\}} g(\eta)\,\mathsf{P}(d\omega) \\ &= \int_C g(y)\,dF(y) = \varphi(C) = \int_{\{\eta\in C\}} \xi\,\mathsf{P}(d\omega) = \int_B \xi\,\mathsf{P}(d\omega).\end{aligned}$$

II.6.50. Indeed, in the notation of the previous problem, for $C \in S$ we have

$$\begin{aligned}\varphi(C) &= \int_{\{\eta\in C\}} \xi\,\mathsf{P}(d\omega) = \iint_{\{y \in C\}} x f(x,y)\,dx\,dy \\ &= \int_C \left[\int_{-\infty}^{\infty} x f(x,y)\,dx\right] dy = \int_C \left[\frac{1}{f_2(y)}\int_{-\infty}^{\infty} x f(x,y)\,dx\right] f_2(y)\,dy.\end{aligned}$$

The proof of the second equality is similar to that of the first with ξ replaced by $\chi_{\{\xi < x\}}$.

II.6.51. ANSWER:

$$\mathsf{P}\left(\xi < x/y - \Delta y \le \eta < y + \Delta y\right) = \frac{\mathsf{P}(\xi < x, y - \Delta y \le \eta < y + \Delta y)}{\mathsf{P}(y - \Delta y \le \eta < y + \Delta y)} = \frac{\int_{-\infty}^{x}\left[\int_{y-\Delta y}^{y+\Delta y} f(u,v)\,dv\right] du}{\int_{y-\Delta y}^{y+\Delta y} f_2(v)\,dv};$$

$$\mathsf{E}\left(\xi/y - \Delta y \le \eta < y + \Delta y\right) = \frac{\int_{-\infty}^{\infty} u\left[\int_{y-\Delta y}^{y+\Delta y} f(u,v)\,dv\right] du}{\int_{y-\Delta y}^{y+\Delta y} f_2(v)\,dv}.$$

II.6.53. ANSWER:

$$\mathsf{E}\left(\xi/\xi + \eta \in B\right) = \frac{\int_B\left[\int_{-\infty}^{\infty} uf(u, v-u)\,du\right] dv}{\int_B\left[\int_{-\infty}^{\infty} f(u, v-u)\,du\right] dv}.$$

II.6.55. By the condition of the problem, the conditional distribution density of the random variable ξ_n, given $\xi_{n-1} = y$, equals 1 for $y \le x \le y+1$ and 0 for all other values of x. Therefore,

$$\mathsf{E}\left(\xi_n/\xi_{n-1} = y\right) = \int_y^{y+1} x\,dx = y + \frac{1}{2};$$

$$\mathsf{E}\left(\xi_n/\xi_{n-1}\right) = \xi_{n-1} + \frac{1}{2}$$

(see Problem II.6.50). Hence,

$$\mathsf{E}\,\xi_n = \mathsf{E}\left[\mathsf{E}\left(\xi_n/\xi_{n-1}\right)\right] = \mathsf{E}\left(\xi_{n-1} + \frac{1}{2}\right) = \mathsf{E}\,\xi_{n-1} + \frac{1}{2},$$

and since $\mathsf{E}\,\xi_n = \frac{n}{2}$, we have $\mathsf{E}\,\xi_1 = \frac{1}{2}$.

II.6.56. ANSWER:

$$\mathsf{E}\left(\xi_n/\xi_{n-1}\right) = \frac{1}{2}\xi_{n-1} + \frac{1}{2},$$

and, similarly to Problem II.6.55,

$$\mathsf{E}\,\xi_n = \frac{1}{2}\,\mathsf{E}\,\xi_{n-1} + \frac{1}{2}, \qquad \mathsf{E}\,\xi_1 = \frac{1}{2},$$

so that

$$\mathsf{E}\,\xi_n = \frac{1}{2} + \frac{1}{4} + \cdots + \frac{1}{2^n} = 1 - \frac{1}{2^n}.$$

II.6.57. To verify b), prove first that

$$\mathsf{E}\left(\overline{\xi}/\max_{1\le k\le n}\xi_k\right) = \mathsf{E}\left(\xi_1/\max_{1\le k\le n}\xi_k\right).$$

Then calculate the joint distribution function of the random variables ξ_1 and $\max_{1\le k\le n}\xi_k$ and use the result of Problem II.6.50.

II.6.60. a) Consider the corresponding conditional distribution densities.
b) Use a), the result of Problem II.6.59, and the equality

$$\mathsf{E}\left(\xi^2/\eta = y\right) = \int_{-\infty}^{\infty} x^2 f(x/y)\,dx.$$

II.6.61. If the random variables ξ_n, $n \geq 1$, have positive densities, the assertion can be easily proved by using conditional densities. Otherwise, an alternative approach can be used. Let g and h be real Borel functions on $\mathbf{R}$. By the definition of $\mathsf{E}\left(g(\eta_k)h(\eta_m)/\eta_n = y\right)$ (see Problem II.6.48), for any Borel set $C \subset \mathbf{R}$ we have

$$\int_{\{\eta_n \in C\}} g(\eta_k)h(\eta_m)\,\mathsf{P}(d\omega) = \int_C \mathsf{E}\left(g(\eta_k)h(\eta_m)/\eta_n = y\right)\,dF(y),$$

where F is the distribution function of the random variable η_n. Let $\mathfrak{B}$ be the σ-algebra generated by the random variables η_k and η_n. According to Problems II.6.18 and II.6.10, we have

$$\begin{aligned}\int_{\{\eta_n \in C\}} g(\eta_k)h(\eta_m)\,\mathsf{P}(d\omega) &= \mathsf{E}\left[\chi_{\{\eta\in C\}}g(\eta_k)h(\eta_m)\right]\\ &= \mathsf{E}\left\{\mathsf{E}\left(\chi_{\{\eta\in C\}}g(\eta_k)h(\eta_m)/\mathfrak{B}\right)\right\}\\ &= \mathsf{E}\left\{\chi_{\{\eta\in C\}}g(\eta_k)\left[\mathsf{E}\,h(y+\xi_{n+1}+\cdots+\xi_m)\right]\Big|_{y=\eta_n}\right\}\\ &= \int_C \mathsf{E}\left(g(\eta_k)/\eta_n = y\right)\mathsf{E}\left(h(\eta_m)/\eta_n = y\right)\,dF(y).\end{aligned}$$

II.6.62. A consequence of the definition of conditional expectation.

II.6.63. The events

$$A_k = \{\omega\colon \mathsf{E}\left(|\xi|/\mathfrak{B}_j\right) < \alpha, 1 \leq j < k, \mathsf{E}\left(|\xi|/\mathfrak{B}_k\right) \geq \alpha\}, \qquad 1 \leq k \leq m,$$

are pairwise disjoint and $A_k \in \mathfrak{B}_k$. Therefore,

$$\begin{aligned}\mathsf{P}(A) &= \sum_{k=1}^m \mathsf{P}(A_k) = \sum_{k=1}^m \int_{A_k} \mathsf{P}(d\omega) \leq \frac{1}{\alpha}\sum_{k=1}^m \int_{A_k} \mathsf{E}\left(|\xi|/\mathfrak{B}_k\right)\,\mathsf{P}(d\omega)\\ &= \frac{1}{\alpha}\sum_{k=1}^m \int_{A_k} |\xi|\,\mathsf{P}(d\omega) \leq \frac{1}{\alpha}\,\mathsf{E}\,|\xi|,\end{aligned}$$

where $A = \bigcup_{k=1}^m A_k$.

II.6.64. c) Prove that η is $\mathfrak{B}$-measurable and verify the equality

$$\int_B \eta\,\mathsf{P}(d\omega) = \int_B \mathsf{E}\left(\xi/\mathfrak{B}\right)\,\mathsf{P}(d\omega) = \int_B \xi\,\mathsf{P}(d\omega), \qquad B \in \mathfrak{B}.$$

II.6.65. Use the following assertions:
1) if ξ is $\mathfrak{B}_k$-measurable for some k, then the equality

$$\lim_{n\to\infty} \mathsf{E}\left(\xi/\mathfrak{B}_n\right) = \xi$$

holds with probability 1;

2) if ξ is $\mathfrak{B}$-measurable, then for any $\varepsilon > 0$ there exist a $k \geq 1$ and a $\mathfrak{B}_k$-measurable random variable η such that

$$\mathsf{E}\,|\xi - \eta| < \varepsilon;$$

3) the inequality of Problem II.6.63.

It suffices to demonstrate that for a $\mathfrak{B}$-measurable random variable ξ such that $\mathsf{E}\,|\xi| < \infty$ the relation

$$\mathsf{P}\left(\lim_{n\to\infty} \mathsf{E}\left(\xi/\mathfrak{B}_n\right) = \xi\right) = 1$$

holds, since, according to Problem II.6.9,

$$\mathsf{E}\left(\xi/\mathfrak{B}_n\right) = \mathsf{E}\left(\mathsf{E}\left(\xi/\mathfrak{B}\right)/\mathfrak{B}_n\right), \qquad n \geq 1.$$

For a given $\varepsilon > 0$, let k be such that there exists a $\mathfrak{B}$-measurable random variable ξ_ε with $\mathsf{E}\,|\xi - \xi_\varepsilon| < \varepsilon$. Since $\mathsf{P}\left(\mathsf{E}\left(\xi_\varepsilon/\mathfrak{B}_n\right) = \xi_\varepsilon\right) = 1$ for $n \geq k$, we have

$$\left|\mathsf{E}\left(\xi/\mathfrak{B}_n\right) - \xi\right| \leq \left|\mathsf{E}\left((\xi - \xi_\varepsilon)/\mathfrak{B}_n\right)\right| + |\xi_\varepsilon - \xi|$$

with probability 1. Therefore, for any $\alpha > 0$ we have

$$\mathsf{P}\left(\limsup_{n\to\infty}\left|\mathsf{E}\left(\xi/\mathfrak{B}_n\right) - \xi\right| \geq \alpha\right) \leq \mathsf{P}\left(\sup_{n\geq k}\left|\mathsf{E}\left((\xi - \xi_\varepsilon)/\mathfrak{B}_n\right)\right| + |\xi_\varepsilon - \xi| \geq \alpha\right)$$
$$\leq \frac{2}{\alpha}\,\mathsf{E}\,|\xi - \xi_\varepsilon| \leq \frac{2\varepsilon}{\alpha}.$$

Solutions to Chapter III

§**1.**

III.1.1. a) Either $\mathsf{P}(A) = 0$ or $\mathsf{P}(B) = 1$;
b) either $\mathsf{P}(A) = 0$ or $\mathsf{P}(B) = 0$;
c) either $\mathsf{P}(A) = 0$ or $\mathsf{P}(A) = 1$.

III.1.2. a) The events are independent.
b) The events are independent for $a < \frac{1}{2}$ and $b = 1 - a$.
c) The events are pairwise independent, but this family is not a family of independent events.

III.1.3. Every A_n is the union of 2^{n-1} pairwise disjoint events $I(n,k)$, $k = 0, 1, \ldots, 2^{n-1} - 1$. The length of $I(n,k)$ is equal to 2^{-n}. If $m < n$, then the interval

$$I(m,k) = \left[\frac{2k}{2^m}, \frac{2k+1}{2^m}\right] = \left[\frac{2 \cdot 2^{n-m} \cdot k}{2^n}, \frac{(2k+1) \cdot 2^{n-m}}{2^n}\right]$$

intersects 2^{n-m-1} intervals $I(n,i)$. Therefore, the length of $A_m \cap A_n$ is equal to

$$2^{m-1} \cdot \frac{1}{2^n} \cdot 2^{n-m-1} = \frac{1}{4}.$$

Hence

$$\mathsf{P}(A_m \cap A_n) = \frac{1}{4} = \mathsf{P}(A_m)\mathsf{P}(A_n).$$

It may be proved analogously that

$$\mathsf{P}(A_{n_1} \cap A_{n_2} \cap \cdots \cap A_{n_s}) = \frac{1}{2^s}$$

for all $n_1 < n_2 < \cdots < n_s$, $s \geq 1$.

III.1.4. a) $\prod_{i=1}^n (1 - p_i)$.

b) $\sum' \prod_{k=1}^m p_{i(k)} \prod'(1-p_j)$, where $\sum'$ denotes the summation over all m-tuples $(i(1), i(2) \dots, i(m))$ such that $1 \leq i(1) < i(2) < \cdots < i(m) \leq n$ and $\prod'$ denotes the product over all j such that $1 \leq j \leq n$ and $j \neq i(k), 1 \leq k \leq m$.

III.1.5. Let A_k be the event that a success has occurred in a sequence of trials with numbers k, $k+1$, $k+2$, $k+3$, $k+4$. It is easy to see that $\mathsf{P}(A_k) = p^5$ for every $k \geq 1$, and that $A_1, A_6, A_{11}, \dots$ are independent events. The Borel–Cantelli lemma implies, for $p > 0$, that the probability that we are looking for, is equal to 1. Analogously, if $0 < p < 1$, then every fixed sequence of successes and failures occurs infinitely often with probability 1.

III.1.6. a) The following proposition is a corollary of the Borel–Cantelli lemma: *If the series of the problem is convergent, then*

$$\mathsf{P}\left(\limsup_{k\to\infty} B_k^{(m)}\right) = 0, \qquad m \geq 1.$$

If the series of the problem is divergent, then it is necessary to apply the second part of the Borel–Cantelli lemma to the following sequences of events:

$$\left\{B_{2i(m+1)+1}^{(m)}, i \geq 0\right\}, \qquad \left\{B_{2i(m+1)-m}^{(m)}, i \geq 1\right\}.$$

b) The following proposition is a corollary of a): *If the series of part* a) *is divergent for all m, then*

$$\mathsf{P}\left(\limsup_{k\to\infty} B_k^{(m)}\right) = 1, \qquad m \geq 1,$$

and the first case is proved because the intersection of events of probability one also has probability one. In the second case, it is necessary to use the additional condition of the problem and the identity

$$\mathsf{P}\left(\limsup_{k\to\infty} B_k^{(m)}\right) = \lim_{n\to\infty} \mathsf{P}\left(\bigcup_{k=n}^{\infty} B_k^{(m)}\right) = 0.$$

III.1.7. $\prod_{n=1}^\infty (1 - p_n) = 0$. The variable ν is an integer-valued random variable, and

$$\mathsf{P}(\nu = 1) = p_1, \qquad \mathsf{P}(\nu = n) = p_n \prod_{k=1}^{n-1} (1 - p_k), \quad n \geq 2.$$

The distribution of ν is called the generalized geometric distribution.

III.1.8. The Borel–Cantelli lemma implies that the events A_n, $n \geq 1$, occur infinitely often with probability one. Therefore, the values of the random variable ν are finite with probability one.

a) The probability $\mathsf{P}(\nu_{k+1} - \nu_k = m/\nu_k = n) = \mathsf{P}(\nu_1 = m)$ does not depend on n.

b) Use a) and prove similarly that

$$\mathsf{P}\Big(\bigcap_{k=1}^{N} B_k^{(m_k)}\Big) = \mathsf{P}\Big(\bigcap_{k=1}^{N-1} B_k^{(m_k)}\Big)\mathsf{P}(\nu_1 = m_N).$$

III.1.9. b) Put

$$B = \bigcap_{k=m}^{\infty} \bar{A}_k, \qquad m \geq 1,$$

and use de Morgan's rules.

III.1.10. This is a corollary of the Borel–Cantelli lemma: *If A occurs, then only a finite number of A_n, $n \geq 1$, occur.*

III.1.11. $\mathsf{P}(A) = 0$.

III.1.12. Let $\alpha_n = \chi_{A_n}$. It is easy to see that

$$\mathsf{E}\,\alpha_n = \mathsf{P}(A_n), \qquad \mathsf{E}\,\alpha_m\alpha_n = \mathsf{P}(A_m A_n),$$

$$\operatorname{Var}\Big(\sum_{k=1}^{n}\alpha_k\Big) = \mathsf{E}\Big(\sum_{k=1}^{n}\alpha_k\Big)^2 - \Big[\mathsf{E}\Big(\sum_{k=1}^{n}\alpha_k\Big)\Big]^2 = \sum_{j,k=1}^{n}\mathsf{P}(A_jA_k) - \Big[\sum_{k=1}^{n}\mathsf{P}(A_k)\Big]^2.$$

According to the Chebyshev inequality, we have

$$\mathsf{P}\Big(\Big|\sum_{k=1}^{n}\alpha_k - \sum_{k=1}^{n}\mathsf{P}(A_k)\Big| \geq \frac{1}{2}\sum_{k=1}^{n}\mathsf{P}(A_k)\Big) \leq 4\left[\frac{\sum_{j,k=1}^{n}\mathsf{P}(A_jA_k)}{\big(\sum_{k=1}^{n}\mathsf{P}(A_k)\big)^2} - 1\right].$$

Put

$$d_n = \mathsf{P}\Big(\sum_{k=1}^{n}\alpha_k \geq \frac{1}{2}\sum_{k=1}^{n}\mathsf{P}(A_k)\Big).$$

It follows from the last inequality that $\liminf_{n\to\infty} d_n = 0$. Therefore, there exists an increasing sequence of natural numbers such that

$$\sum_{k=1}^{\infty} d_{n(k)} < \infty.$$

By the Borel–Cantelli lemma, with probability 1 either only finitely many events

$$\Big\{\sum_{k=1}^{n(m)}\alpha_k \leq \frac{1}{2}\sum_{k=1}^{n(m)}\mathsf{P}(A_k)\Big\}, \qquad m \geq 1,$$

occur or all but a finite number of the events

$$\Big\{\sum_{k=1}^{n(m)}\alpha_k > \frac{1}{2}\sum_{k=1}^{n(m)}\mathsf{P}(A_k)\Big\}, \qquad m \geq 1,$$

occur. Together with the conditions of the problem, this implies the convergence of the series $\sum_{n=1}^{\infty}\alpha_n$ with probability one.

III.1.13. This is a corollary of the previous problem. The pairwise independence distinguishes between this problem and the Borel–Cantelli lemma.

§2.

III.2.2. To prove that $\mathfrak{A}_1$ and $\mathfrak{A}_2$ are independent, consider the collection $\mathfrak{B}_1$ of events A in $\mathfrak{A}_1$ such that

$$\mathsf{P}(A \cap \{\omega: \xi_2 < c\}) = \mathsf{P}(A)\mathsf{P}(\xi_2 < c), \qquad A \in \mathfrak{B}_1,\ c \in \mathbf{R}.$$

Prove that $\mathfrak{A}_1 \subset \mathfrak{B}_1$. Proceed in the same way in the case of n σ-algebras.

III.2.3. Let $\mathfrak{B}_n$ be the Borel σ-algebra in $\mathbf{R}^n$. The σ-algebra of events

$$\{\{\omega: (\xi_1, \xi_2, \dots, \xi_n) \in B_n\}, B_n \in \mathfrak{B}_n\}$$

is generated by the collection of events

$$\left\{\bigcap_{k=1}^{n} \{\omega: \xi_k \in C_k\}, C_k \in \mathfrak{B}_1, 1 \le k \le n\right\}.$$

III.2.4. This is a consequence of the zero-one law. For example, in the case of the random variable $\limsup_{n\to\infty} \xi_n$, the event

$$\left\{\omega: \limsup_{n\to\infty} \xi_n \ge c\right\} = \bigcap_{n=1}^{\infty} \bigcup_{k=n}^{\infty} \{\omega: \xi_k \ge c\}$$

for all $c \in \mathbf{R}$ belongs to the tail σ-algebra of the sequence $\{\mathfrak{A}_n, n \ge 1\}$, where $\mathfrak{A}_n$ is the σ-algebra generated by the random variable ξ_n, $n \ge 1$. Now we use the following remark. Let $(\Omega, \mathfrak{A}, \mathsf{P})$ be a probability space such that for every $A \in \mathfrak{A}$ either $\mathsf{P}(A) = 0$ or $\mathsf{P}(A) = 1$. Let ξ be an $\mathfrak{A}$-measurable random variable. Then ξ is a constant with probability one. To prove this statement, consider events $\{\omega: \xi < r\}$ for rational numbers r.

III.2.5–7. These are corollaries of the zero-one law. First prove that corresponding random variables are measurable with respect to the tail σ-algebra, then proceed similarly to the solution of Problem III.2.4.

III.2.9. Use the Borel–Cantelli lemma and the divergence of the series

$$\sum_{n=1}^{\infty} \mathsf{P}(|\xi_n| > cn).$$

The latter can be be shown as follows. Assume that the series converges. Then

$$\begin{aligned}\sum_{n=1}^{\infty} \mathsf{P}(|\xi_n| > cn) &= \sum_{n=1}^{\infty} \mathsf{P}(|\xi_1| > cn) = \sum_{n=1}^{\infty}\sum_{m=n}^{\infty} \mathsf{P}\left(m+1 \ge \frac{|\xi_1|}{c} > m\right) \\ &= \sum_{n=1}^{\infty} n\mathsf{P}\left(n+1 \ge \frac{|\xi_1|}{c} > n\right) \ge \frac{1}{c}\int_{|x|\ge c} (|x| - c)\, dF(x),\end{aligned}$$

where F is the distribution of ξ_1. This contradicts the assumption $\mathsf{E}\,|\xi_1| = +\infty$.

III.2.10. Apply the Borel–Cantelli lemma to the sequence $\{\omega: |\xi_n|^\alpha > cn\}$, $n \ge 1$, and use the fact that the series $\sum_{n=1}^{\infty} \mathsf{P}(|\xi| > n)$ converges if and only if the series $\mathsf{E}\,|\xi|$ does.

III.2.11. The solution is similar to that of Problem III.2.10.

III.2.12. We must prove that there exists a number $c > 0$ such that the inequality $\xi_n > c\ln n$ is satisfied only for a finite set of numbers n. In other words, it is necessary to prove that

$$\mathsf{P}\left(\limsup_{n\to\infty} A_n\right) = 0,$$

where $A_n = \{\omega\colon \xi_n > c\ln n\}$, $n \geq 1$. We have

$$\mathsf{P}(A_n) = \mathsf{P}(\xi_n > c\ln n) = \lambda \int_{c\ln n}^{\infty} e^{-\lambda u}\,du = n^{-\lambda c},$$

where $\lambda > 0$ is the parameter of the distribution. Choose c such that $\lambda c > 1$. Then

$$\sum_{n=1}^{\infty} \mathsf{P}(A_n) < \infty.$$

III.2.13. See the solution of the preceding problem.

III.2.14. To prove the equality $\mathsf{P}(\eta < x) = x$, $x \in [0,1]$, consider the binary expansion of x:

$$x = \sum_{n=1}^{\infty} \frac{a_n}{2^n},$$

where the a_n take only values 0 or 1. We consider only those recurring expansions that do not have 1 as their repetend. The event $\{\omega\colon \eta < x\}$ is the union of disjoint events

$$\{\omega\colon \xi_1 < a_1\},\ \{\omega\colon \xi_1 = a_1, \xi_2 < a_2\},\ \{\omega\colon \xi_1 = a_1, \xi_2 = a_2, \xi_3 < a_3\}, \dots .$$

Thus

$$\mathsf{P}(\eta < x) = \sum_{n=1}^{\infty} \mathsf{P}(\xi_1 = a_1, \dots, \xi_{n-1} = a_{n-1}, \xi_n = a_n).$$

Since random variables ξ_n, $n \geq 1$, are independent,

$$\begin{aligned}&\mathsf{P}(\xi_1 = a_1, \dots, \xi_{n-1} = a_{n-1}, \xi_n < a_n)\\ &\quad = \mathsf{P}(\xi_1 = a_1)\mathsf{P}(\xi_2 = a_2)\cdots \mathsf{P}(\xi_{n-1} = a_{n-1})\mathsf{P}(\xi_n < a_n) = \frac{a_n}{2^n}.\end{aligned}$$

Therefore

$$\mathsf{P}(\eta < x) = \sum_{n=1}^{\infty} \frac{a_n}{2^n} = x.$$

III.2.15. ANSWER: 0.

The first solution. By the Borel–Cantelli lemma, any finite sequence of experiment results occurs infinitely many times with probability 1 (see Problem III.1.5).

The second solution. Prove that η is uniformly distributed on the interval $[0,1]$ (see Problem III.2.4).

III.2.16. The variable x is uniformly distributed on the interval $[0,1]$. Since $\mathsf{P}(x = 1) = 0$, without loss of generality we may assume that it takes values in $[0,1)$. We have $\mathsf{P}(\xi_1 = i) = \frac{1}{10}$ for all i, $1 \leq i \leq 9$, since $\xi_1 = i$ if and only if x is within the interval $\left[\frac{i}{10}, \frac{i+1}{10}\right)$ of length $\frac{1}{10}$. Similarly, $\mathsf{P}(\xi_k = i) = \frac{1}{10}$, since $\xi_k = i$ if and only if x is in the interval

$$I(m_1, m_2, \dots, m_{k-1}, i) = \left[\frac{m_1}{10} + \frac{m_2}{10^2} + \dots + \frac{m_{k-1}}{10^{k-1}} + \frac{i}{10^k}, \frac{m_1}{10} + \dots + \frac{m_{k-1}}{10^{k-1}} + \frac{i+1}{10^k}\right).$$

The numbers $m_1, m_2, \dots, m_{k-1}$ take only integer values $0, 1, \dots, 9$. The intervals $I(m_1, m_2, \dots, m_{k-1}, i)$ are disjoint and each has the length 10^{-k}. The total number of such intervals is 10^{k-1}. To prove that ξ_k and ξ_r are independent it is necessary to show that

$$\mathsf{P}(\xi_k = i, \xi_r = j) = \mathsf{P}(\xi_k = i)\mathsf{P}(\xi_r = j) = 1/10^2.$$

As before, $\mathsf{P}(\xi_k = i, \xi_r = j)$ denotes the probability for x to be in

$$I(m_1, m_2, \dots, m_{k-1}, i) \cap I(\overline{m}_1, \overline{m}_2, \dots, \overline{m}_{k-1}, i).$$

For $k < r$ these intersections are either of length 10^{-r} or empty. The total number of nonempty intersections is 10^{r-2}, since only 10^{-k-1+r} intervals of the second kind $I(\overline{m}_1, \overline{m}_2, \dots, \overline{m}_{k-1}, i)$ intersect any interval $I(m_1, m_2, \dots, m_{k-1}, i)$. Therefore, $\mathsf{P}(\xi_k = i, \xi_r = j) = 10^{-2}$. Analogously, for $k_1 < k_2 < \dots < k_s$,

$$\mathsf{P}(\xi_{k_1} = i_1, \xi_{k_2} = i_2, \dots, \xi_{k_r} = i_r) = 1/10^r.$$

III.2.17. Using the independence we have

$$\mathsf{P}\left(\frac{1}{n} \max_{1 \le k \le n} \xi_k < x\right) = \mathsf{P}(\xi_1 < nx, \dots, \xi_n < nx) = \prod_{k=1}^{n} \mathsf{P}(\xi_k < kx) = \left(\frac{1}{\pi}\int_{-\infty}^{nx} \frac{du}{1+u^2}\right)^n.$$

III.2.18. Use the property that $\mathsf{E}\,|\xi| < \infty$ for a random variable ξ if and only if

$$\sum_{n=1}^{\infty} \mathsf{P}(|\xi| \ge n) < \infty.$$

The expectation of the random variables $\eta_n = \max_{1 \le k \le n} |\xi_k|$ is finite for all $n \ge 1$. Further,

$$\mathsf{P}(\eta_n \ge x) = \mathsf{P}\left(\bigcup_{k=1}^{n} \{\omega \colon |\xi_k| \ge x\}\right) \le n[1 - F(x)],$$

where $F(x) = \mathsf{P}(|\xi_n| < x)$, $x > 0$. Hence

$$\frac{1}{n}\mathsf{E}\,\eta_n \le \frac{1}{n}\sum_{k=1}^{\infty} \mathsf{P}(\eta_k \ge k) \le \sum_{n=1}^{\infty} [1 - F(n)] < \infty,$$

since $\mathsf{E}\,|\xi_n| < \infty$. Therefore, the series

$$\sum_{k=1}^{\infty} \frac{1}{n}\mathsf{P}(\eta_n \ge k)$$

converges uniformly with respect to n, and we get the desired result as $n \to \infty$.

III.2.19. Use the representation

$$\mathsf{E}\,\xi = \int_0^\infty \mathsf{P}(\xi \geq x)\,dx$$

for a random variable ξ with $\mathsf{P}(\xi \geq 0) = 1$. It is sufficient to prove that for some $a > 0$,

$$\int_a^\infty [1 - \mathsf{P}(\eta < x)]\,dx < \infty.$$

From

$$\prod_{n=1}^N \mathsf{P}(|\xi_n| < x) = \prod_{n=1}^N [1 - \mathsf{P}(|\xi_n| \geq x)] \geq 1 - \sum_{n=1}^N \mathsf{P}(|\xi_n| \geq x)$$

and the Chebyshev inequality we obtain

$$\prod_{n=1}^\infty \mathsf{P}(|\xi_n| < x) \geq 1 - \frac{1}{x^2}\sum_{n=1}^\infty \mathsf{E}\,\xi_n^2.$$

Since the variables ξ_n, $n \geq 1$, are independent and the probability is continuous,

$$\mathsf{P}(\eta < x) \geq 1 - \frac{1}{x^2}\sum_{n=1}^\infty \mathsf{E}\,\xi_n^2,$$

and we get the desired statement.

III.2.20. Use the formula

$$\mathsf{E}\,\eta_\nu = \sum_{n=1}^\infty \mathsf{E}\,(\eta_\nu/\nu = n)\,\mathsf{P}(\nu = n)$$

and an analogous representation for $\mathsf{E}\,\eta_\nu^2$. ANSWER: $\mathsf{E}\,\eta_\nu = \mathsf{E}\,\nu \cdot \mathsf{E}\,\xi_1$, $\operatorname{Var}\eta_\nu = (\mathsf{E}\,\xi_1)^2 \operatorname{Var}\nu + \mathsf{E}\,\nu \cdot \operatorname{Var}\xi_1$.

III.2.21. The variable ν can be defined as

$$\nu = \begin{cases} 2, & \xi_2 > \xi_1, \\ k, & \xi_k > \xi_1, \xi_j \leq \xi_1, 2 \leq j \leq k-1, k \geq 3, \\ +\infty, & \xi_n \leq \xi_1, n \geq 2. \end{cases}$$

At first glance this definition looks complicated, although it is not needed for computations. The probability of the event that the first random variable is maximal among the n independent random variables equals

$$\mathsf{P}(\nu > n) = \mathsf{P}(\xi_2 \leq \xi_1, \xi_3 \leq \xi_1, \dots, \xi_n \leq \xi_1).$$

Since the random variables are independent and identically distributed and the probability is continuous, we get

$$\mathsf{P}(\xi_2 \leq \xi_1, \xi_3 \leq \xi_1, \dots, \xi_n \leq \xi_1) = \mathsf{P}(\xi_1 \leq \xi_2, \xi_3 \leq \xi_2, \dots, \xi_n \leq \xi_2),$$

because the joint distribution of $\xi_1, \xi_2, \dots, \xi_n$ coincides with that of $\xi_2, \xi_1, \dots, \xi_n$. Therefore, the n numbers

$$\mathsf{P}(\xi_i \leq \xi_k, i \neq k, 1 \leq i \leq n), \qquad k = 1, 2, \dots, n,$$

are identical. It is clear that at least one of the variables is maximal, and thus $\mathsf{P}(\nu > n) = n^{-1}$. Therefore,

$$\mathsf{P}(\nu = n) = \mathsf{P}(\nu > n-1) - \mathsf{P}(\nu > n) = \frac{1}{(n-1)n}, \qquad n \geq 2.$$

Note that $\mathsf{P}(\nu = +\infty) = 0$, since $\sum_{n=2}^{\infty} \mathsf{P}(\nu = n) = 1$, and that $\mathsf{E}\,\nu = +\infty$.

III.2.22. The solution is similar to that of Problem III.2.21.

III.2.23. The random variable ν can be defined as

$$\nu = \begin{cases} n & \text{if } \xi_1 + \xi_2 + \dots + \xi_{n-1} \leq 1, \text{ but } \xi_1 + \xi_2 + \dots + \xi_n > 1, \\ +\infty & \text{if } \xi_1 + \xi_2 + \dots + \xi_n \leq 1 \text{ for all } n \geq 1. \end{cases}$$

Note that

$$\{\omega\colon \nu = n\} = \{\omega\colon \xi_1 + \xi_2 + \dots + \xi_{n-1} \leq 1\} \setminus \{\omega\colon \xi_1 + \xi_2 + \dots + \xi_n \leq 1\}$$

and

$$\{\omega\colon \xi_1 + \xi_2 + \dots + \xi_n \leq 1\} \subset \{\omega\colon \xi_1 + \xi_2 + \dots + \xi_{n-1} \leq 1\}.$$

Consequently, for $n \geq 2$,

$$\mathsf{P}(\nu = n) = \mathsf{P}(\xi_1 + \xi_2 + \dots + \xi_{n-1} \leq 1) - \mathsf{P}(\xi_1 + \xi_2 + \dots + \xi_n \leq 1).$$

The joint probability density of the random variables $\xi_1, \dots, \xi_n$ is

$$\lambda^n \exp\{-\lambda(x_1 + \dots + x_n)\}, \quad \text{for } x_i \geq 0,\ 1 \leq i \leq n.$$

Thus

$$\begin{aligned}\mathsf{P}(\xi_1 + \xi_2 + \dots + \xi_n \leq 1) &= \lambda^n \underset{\substack{x_i \geq 0, 1 \leq i \leq n, \\ x_1 + \dots + x_n \leq 1}}{\int \overset{n}{\cdots} \int} \exp\{-\lambda(x_1 + \dots + x_n)\}\, dx_1 \cdots dx_n \\ &= \frac{\lambda^n}{(n-1)!} \int_0^1 x^{n-1} e^{-\lambda x}\, dx, \qquad n \geq 1,\end{aligned}$$

whence

$$\begin{aligned}\mathsf{P}(\nu = n) &= \frac{\lambda^{n-1}}{(n-2)!} \int_0^1 x^{n-2} e^{-\lambda x}\, dx - \frac{\lambda^n}{(n-1)!} \int_0^1 x^{n-1} e^{-\lambda x}\, dx \\ &= \frac{\lambda^{n-1}}{(n-2)!} \int_0^1 x^{n-2} e^{-\lambda x}\, dx \\ &\quad - \frac{\lambda^n}{(n-1)!} \left[-x^{n-1} \frac{e^{-\lambda x}}{\lambda} \Big|_0^1 + \int_0^1 \frac{n-1}{\lambda} x^{n-2} e^{-\lambda x}\, dx \right] \\ &= \frac{\lambda^{n-1}}{(n-1)!} e^{-\lambda}\end{aligned}$$

and

$$\mathsf{P}(\nu = 1) = \mathsf{P}(\xi_1 > 1) = e^{-\lambda}.$$

Hence ν has the Poisson distribution with parameter λ.

III.2.24. The random variable ξ_ν is defined as

$$\xi_\nu = \begin{cases} \xi_1 & \text{if } \xi_1 > 1, \\ \xi_2 & \text{if } \xi_1 \le 1 \text{ and } \xi_1 + \xi_2 > 1, \\ \xi_n & \text{if } n > 2 \text{ and } \xi_1 + \dots + \xi_{n-1} \le 1,\ \xi_1 + \dots + \xi_n > 1, \\ +\infty, & \xi_1 + \dots + \xi_n \le 1,\ n \ge 1. \end{cases}$$

Therefore,

$$\begin{aligned} \mathsf{P}(\xi_\nu < x) &= \sum_{n=1}^{\infty} \mathsf{P}(\xi_\nu < x, \nu = n) \\ &= \mathsf{P}(1 < \xi_1 < x) + \sum_{n=2}^{\infty} \mathsf{P}\Big(\xi_n < x, \sum_{i=1}^{n-1} \xi_i \le 1 < \sum_{i=1}^{n} \xi_i\Big). \end{aligned}$$

Let

$$g_{n-1}(x) = \frac{(\lambda x)^{n-2}}{(n-2)!} \lambda e^{-\lambda x}, \quad \text{for } x \ge 0,\ n > 1,$$

be the distribution density of the sum $\xi_1 + \dots + \xi_{n-1}$. Taking into account that the variables $\eta_n = \xi_1 + \dots + \xi_{n-1}$ and ξ_n are independent for $n > 1$, we have

$$\begin{aligned} \mathsf{P}(\xi_n < x, \eta \le 1, \eta + \xi_n > 1) &= \int_{1-x}^{1} g_{n-1}(u) \int_{1-u}^{x} \lambda e^{-\lambda v}\, dv\, du \\ &= \int_{1-x}^{1} g_{n-1}(u) \left[e^{\lambda(1-u)} - e^{-\lambda x}\right] du \end{aligned}$$

for $x < 1$, if $n > 1$ and $\mathsf{P}(1 < \xi_1 < x) = 0$. Note that

$$\sum_{n=2}^{\infty} g_{n-1}(x) = \lambda, \qquad x > 0,$$

$$\mathsf{P}(\xi_\nu < x) = \int_{1-x}^{1} \lambda \left[e^{-\lambda(1-u)} - e^{-\lambda x}\right] du = 1 - e^{-\lambda x} - \lambda x e^{-\lambda x}.$$

Now let $x > 1$. In this case

$$\mathsf{P}(1 < \xi_1 < x) = e^{-\lambda} - e^{-\lambda x},$$

$$\mathsf{P}(\xi_n < x, \eta \le 1 < \eta + \xi_n) = \int_0^1 g_{n-1}(u) \left[e^{-\lambda(1-u)} - e^{-\lambda x}\right] du,$$

and

$$\mathsf{P}(\xi_\nu < x) = e^{-\lambda} - e^{-\lambda x} + \int_0^1 \lambda \left[e^{-\lambda(1-u)} - e^{-\lambda x}\right] du = 1 - e^{-\lambda x} - \lambda e^{-\lambda x}.$$

Finally we have

$$\mathsf{P}(\xi_\nu < x) = \begin{cases} 1 - e^{-\lambda x} - \lambda x e^{-\lambda x}, & x \le 1, \\ 1 - e^{-\lambda x} - \lambda e^{-\lambda x}, & x > 1. \end{cases}$$

Therefore, the distributions of the random variables ξ_ν and ξ_n, $n \ge 1$, are different. It is easy to check that $\mathsf{E}\,\xi_\nu = 2\,\mathsf{E}\,\xi_1$.

III.2.25. The random variable ν takes only natural values, i.e.,

$$\sum_{n=1}^{\infty} \mathsf{P}(\nu = n) = 1.$$

Indeed,

$$\{\omega : \nu = n\} = \{\omega : \xi_1 + \cdots + \xi_{n-1} < 1, \xi_1 + \cdots + \xi_n \geq 1\}$$

and, moreover, these events are disjoint. Thus

$$1 - \sum_{n=1}^{\infty} \mathsf{P}(\nu = n) = 1 - \mathsf{P}\Big(\bigcup_{n=1}^{\infty} \{\omega : \nu = n\}\Big) = \mathsf{P}(\xi_1 + \cdots + \xi_n < 1, n \geq 1).$$

Let n_0 be such that $n_0 \mathsf{E}\,\xi_1 > 1$, and let $S_n = \xi_1 + \cdots + \xi_n$. It is easily seen that

$$\begin{aligned} \mathsf{P}(S_n < 1, n \geq 1) &= \mathsf{P}(S_n - \mathsf{E}\,S_n < 1 - \mathsf{E}\,S_n, n \geq 1) \\ &\leq \mathsf{P}(|S_{n^2} - \mathsf{E}\,S_{n^2}| > \mathsf{E}\,S_{n^2} - 1, n \geq n_0). \end{aligned}$$

The Chebyshev inequality implies

$$\mathsf{P}\left(|S_{n^2} - \mathsf{E}\,S_{n^2}| > \mathsf{E}\,S_{n^2} - 1\right) \leq \frac{n^2 \operatorname{Var} \xi_1}{\left(n^2 \mathsf{E}\,\xi_1 - 1\right)^2}.$$

Therefore, according to the Borel–Cantelli lemma, infinitely many events

$$\{\omega : |S_{n^2} - \mathsf{E}\,S_{n^2}| > \mathsf{E}\,S_{n^2} - 1\}, \qquad n \geq n_0,$$

occur with probability 0, and moreover,

$$\mathsf{P}(S_n < 1, n \geq 1) = 0.$$

For arbitrary x and y and for fixed $k \geq 1$, consider

$$\begin{aligned} \mathsf{P}(\xi_\nu < x, \xi_{\nu+k} < y) &= \sum_{n=1}^{\infty} \mathsf{P}(\xi_\nu < x, \xi_{\nu+k} < y, \nu = n) \\ &= \sum_{n=1}^{\infty} \mathsf{P}(\xi_n < x, \xi_{n+k} < y, S_{n-1} < 1, S_n \geq 1). \end{aligned}$$

Since the random variables ξ_k, $k \geq 1$, are independent and identically distributed,

$$\begin{aligned} \mathsf{P}(\xi_\nu < x, \xi_{\nu+k} < y) &= \sum_{n=1}^{\infty} \mathsf{P}(\xi_n < x, S_{n-1} < 1, S_n \geq 1)\mathsf{P}(\xi_{n+k} < y) \\ &= \sum_{n=1}^{\infty} \mathsf{P}(\xi_\nu < x, \nu = n)\mathsf{P}(\xi_1 < y) = \mathsf{P}(\xi_\nu < x)\mathsf{P}(\xi_1 < y). \end{aligned}$$

Letting $x \to \infty$, we get $\mathsf{P}(\xi_{\nu+k} < y) = \mathsf{P}(\xi_1 < y)$. Thus the random variables ξ_ν and $\xi_{\nu+k}$ are independent and $\xi_{\nu+k}$ has the same distribution as ξ_1. It easy to show that $\{\xi_{\nu+k}, k \geq 1\}$ is a sequence of independent identically distributed random variables. Note that the distribution of ξ_ν does not coincide with that of ξ_1.

III.2.26. Let $F_n(x) = \mathsf{P}(S_n < x)$. Prove by induction that

$$F_n(x) = \frac{x^n}{n!}, \qquad 0 \le x \le 1, \ n \ge 1.$$

Moreover,

$$\mathsf{P}(\nu = n) = \mathsf{P}(S_{n-1} \le 1, S_n > 1) = \int_0^1 y\, dF_{n-1}(y) = \frac{1}{n}\frac{1}{(n-2)!}.$$

III.2.27. The expectation to be estimated is the sum of n^{2r} terms of the form $\mathsf{E}\,(\xi_{i_1}\xi_{i_2}\cdots\xi_{i_{2r}})$. If a term has a subscript, say i_1, that is different from all the others, then the variable ξ_{i_1} does not depend on the remaining factors and this term is equal to zero. Thus, there are at most $n^r\binom{2r}{r}$ nonzero terms of the form $\mathsf{E}\,(\xi_{i_1}^2\xi_{i_2}^2\cdots\xi_{i_r}^2)$, and none of them exceeds $\left(1+\mathsf{E}\,\xi_1^{2r}\right)^r$.

III.2.28. First check that

$$\sum_{n=1}^{\infty}\int_{|x|>2a_n} dF(x) = +\infty, \qquad \mathsf{P}\left(\limsup_{n\to\infty}\{|\xi_n| > 2a_n\}\right) = 1.$$

III.2.29. Let $S_n = \sum_{k-1}^n \xi_k$. It is necessary to prove that

$$\mathsf{P}\bigg(\limsup_{n\to\infty}\frac{|S_n|}{n} = +\infty\bigg) = 1.$$

Problem III.2.9 implies that for any $c > 0$, infinitely many events $\{\omega\colon |\xi_n| > cn\}$, $n \ge 1$, occur with probability one. On the other hand, if for some n the event

$$\{\omega\colon |\xi_n| > cn\} = \{\omega\colon |S_n - S_{n-1}| > cn\}$$

occurs, then at least one of either the events

$$\left\{\omega\colon |S_n| > \frac{1}{2}cn\right\},$$

$$\left\{\omega\colon |S_{n-1}| > \frac{1}{2}cn\right\} = \left\{\omega\colon |S_{n-1}| > \frac{c}{2}\frac{n}{n-1}(n-1)\right\}$$

or the events

$$\left\{\omega\colon |S_n| > \frac{1}{2}cn\right\}, \qquad \left\{\omega\colon |S_{n-1}| > \frac{c}{2}(n-1)\right\}$$

occurs. Thus for any $c > 0$ infinitely many events $\{\omega\colon |S_n| > cn\}$, $n \ge 1$, occur with probability one and

$$\mathsf{P}\left(\limsup_{n\to\infty}\frac{|S_n|}{n} \ge c\right) = 1.$$

III.2.30. Determine first the distribution of $\frac{1}{n}\sum_{k=1}^n \xi_k$. Note that $\mathsf{E}\,|\xi_1| = +\infty$. For any $n \ge 1$ the random variable $\frac{1}{n}\sum_{k=1}^n \xi_k$ has the same distribution as ξ_1. Thus

$$\mathsf{P}(A_n) = \frac{1}{\pi}\int_c^{\infty}\frac{dx}{1+x^2} = \delta > 0, \qquad A_n = \left\{\omega\colon \frac{1}{n}\sum_{k=1}^n \xi_k > c\right\}, \quad c > 0,$$

whence

$$\mathsf{P}\Big(\bigcup_{m=n}^{\infty} A_m\Big) \geq \delta, \qquad n \geq 1,$$

and

$$\mathsf{P}\left(\limsup_{n\to\infty} A_n\right) \geq \delta.$$

By the zero-one law, the random variable

$$\limsup_{n\to\infty} \frac{1}{n}\sum_{k=1}^{n} \xi_k$$

is a constant with probability 1. Therefore,

$$\mathsf{P}\left(\limsup_{n\to\infty} \frac{S_n}{n} = +\infty\right) = 1,$$

and similarly

$$\mathsf{P}\left(\liminf_{n\to\infty} \frac{S_n}{n} = -\infty\right) = 1.$$

III.2.31. For the events A_n (see the solution to Problem III.2.30) we have

$$\mathsf{P}(A_n) = \frac{1}{\pi}\int_c^{\infty} \frac{dx}{a+x^2} \leq \frac{1}{\pi c},$$

whence

$$\mathsf{P}\left(\frac{1}{n^{1+\alpha}}\sum_{k=1}^{n} \xi_k > c\right) \leq \frac{1}{\pi c n^{\alpha}}.$$

By the Borel–Cantelli lemma, for any fixed $c > 0$ only a finite number of events

$$\left\{\omega\colon \frac{1}{n^{1+\alpha}}\sum_{k=1}^{n} \xi_k > c\right\}, \qquad n \geq 1,$$

occur with probability one. Therefore,

$$\mathsf{P}\left(\limsup_{n\to\infty} \frac{1}{n^{1+\alpha}}\sum_{k=1}^{n} \xi_k \leq c\right) = 1.$$

Since $c > 0$ is arbitrary,

$$\mathsf{P}\left(\limsup_{n\to\infty} \frac{1}{n^{1+\alpha}}\sum_{k=1}^{n} \xi_k \leq 0\right) = 1.$$

It is obvious that

$$\mathsf{P}\left(\liminf_{n\to\infty} \frac{1}{n^{1+\alpha}}\sum_{k=1}^{n} \xi_k \geq 0\right) = 1.$$

III.2.32. A set is everywhere dense if its intersection with any open interval is nonempty. For the purposes of this problem, it is sufficient to consider intervals with rational endpoints. The probability of the event that no term of the sequence lies within such an interval is zero.

§3.

III.3.1. The set of points of convergence of a sequence $\{\xi_n, n \geq 1\}$ to a random variable ξ is defined as follows:

$$\bigcap_{k=1}^{\infty} \bigcup_{m=1}^{\infty} \bigcap_{n=m}^{\infty} \left\{\omega\colon |\xi_n(\omega) - \xi(\omega)| < \frac{1}{k}\right\}.$$

III.3.2. According to the Cauchy criterion, the set of points of convergence coincides with the set of points where $\{\xi_n, n \geq 1\}$ is a Cauchy sequence:

$$\bigcap_{k=1}^{\infty} \bigcup_{N=1}^{\infty} \bigcap_{m=N}^{\infty} \bigcap_{n=N}^{\infty} \left\{\omega\colon |\xi_m(\omega) - \xi_n(\omega)| < \frac{1}{k}\right\}$$
$$= \bigcap_{k=1}^{\infty} \bigcup_{n=1}^{\infty} \bigcap_{m=1}^{\infty} \left\{\omega\colon |\xi_{m+n}(\omega) - \xi_n(\omega)| < \frac{1}{k}\right\}.$$

III.3.3. Use Problem III.3.1 and the fact that the events

$$\limsup_{n\to\infty} \left\{\omega\colon |\xi_n(\omega) - \xi(\omega)| \geq \frac{1}{k}\right\}, \qquad k \geq 1,$$

monotonically increase.

III.3.4. If $\mathsf{P}(\xi_n \to \xi \text{ as } n \to \infty) = 1$, then the desired statement follows from the equality

$$(*) \qquad \mathsf{P}\Bigl(\bigcup_{k=1}^{\infty} \bigcap_{N=1}^{\infty} \bigcup_{n=N}^{\infty} \left\{\omega\colon |\xi_n(\omega) - \xi(\omega)| \geq \frac{1}{k}\right\}\Bigr) = 0$$

and the continuity of the probability. This equality holds under the conditions of the problem, since the probability is self-additive.

III.3.5. Note that

$$\bigcap_{k=1}^{\infty} \bigcup_{N=1}^{\infty} \bigcap_{n=N}^{\infty} \left\{\omega\colon |\xi_n(\omega) - \xi(\omega)| < \frac{1}{k}\right\}$$
$$= \bigcap_{k=1}^{\infty} \bigcup_{N=1}^{\infty} \bigcap_{n=N}^{\infty} \left[\left\{\omega\colon |\eta_n(\omega) - \xi(\omega)| < \frac{1}{k}\right\} \cup \left(A_n \cap \left\{\omega\colon |\xi_n(\omega) - \xi(\omega)| < \frac{1}{k}\right\}\right)\right],$$

where $A_n = \{\omega\colon \eta_n \neq \xi_n\}$, $\mathsf{P}(A_n) = 0$, $n \geq 1$. Thus

$$\mathsf{P}(\xi_n \to \xi \text{ as } n \to \infty) = \mathsf{P}(\xi_n \to \eta \text{ as } n \to \infty) = 1.$$

Moreover, the intersection of two sets of probability one has the same probability.

III.3.6. The intersection of any finite number of sets of probability one has probability one.

III.3.7. For $\varepsilon > 0$ we have

a) $\mathsf{P}(\xi_n > \varepsilon) \leq 2^{-[\log n]} \to 0, \qquad n \to \infty,$

b) $\displaystyle\bigcup_{n=N}^{\infty} \{\omega\colon \xi_n > \varepsilon\} \supset \bigcup_{n=2^m}^{2^{m+1}-1} A_n = [0,1], \qquad 2^m > N.$

(See Problem III.3.4.) For the subsequence $\{\xi_{2^n}, n \geq 1\}$ we have

$$\mathsf{P}\Big(\bigcup_{n=N}^{\infty}\{\omega: \xi_{2^n} > \varepsilon\}\Big) \leq \frac{1}{2^N} \to 0, \qquad N \to \infty.$$

III.3.8. This problem follows from Problems III.3.3 and III.3.4.

III.3.9. See Problem III.3.4.

III.3.10. It follows from the condition of the problem that

$$\mathsf{P}\Big(\bigcap_{m=1}^{\infty}\bigcup_{n=m}^{\infty}\{\omega: |\xi_{n+1}(\omega) - \xi_n(\omega)| > \varepsilon_n\}\Big) = 0,$$

so that the event

$$\bigcup_{m=1}^{\infty}\bigcap_{n=m}^{\infty}\{\omega: |\xi_{n+1}(\omega) - \xi_n(\omega)| \leq \varepsilon_n\}$$

has probability one. Check that this event is the set of points ω where $\{\xi_n(\omega), n \geq 1\}$ is a Cauchy sequence.

III.3.11. By Problem III.3.9, it is sufficient to prove that

$$\sum_{n=1}^{\infty}\mathsf{P}(|\xi_n| > \varepsilon) < +\infty, \qquad \varepsilon > 0.$$

This series is convergent in view of the conditions of the problem and the Chebyshev inequality

$$\mathsf{P}(|\xi_n| > \varepsilon) \leq \frac{\mathsf{E}\,|\xi_n|^r}{\varepsilon^r}.$$

III.3.12. This is a different formulation of Problem III.3.4.

III.3.13. It is necessary to prove that $\mathsf{P}(|\xi_n - \xi| > \varepsilon) \to 0$ as $n \to \infty$. According to Problem II.3.4, we have a much stronger property

$$\mathsf{P}\Big(\bigcup_{m=n}^{\infty}\{\omega: |\xi_n(\omega) - \xi(\omega)| > \varepsilon\}\Big) \to 0, \qquad n \to \infty.$$

Also see Problem III.3.12.

III.3.14. Example. Let $\{\xi_n, n \geq 1\}$ be a sequence of independent random variables, where the ξ_n take values 0 or n^{-1} with probabilities $1 - n^{-1}$ or n^{-1}, respectively, $n \geq 1$. Then for $\varepsilon > 0$ we have $\mathsf{P}(|\xi_n| > \varepsilon) = 1/n$ starting with some n. Thus $\xi_n \to 0$ in probability as $n \to \infty$. By the Borel–Cantelli lemma,

$$\mathsf{P}\Big(\bigcap_{n=1}^{\infty}\bigcup_{m=n}^{\infty}\{\omega: |\xi_n(\omega)| > \varepsilon\}\Big) = 1, \qquad \varepsilon > 0,$$

whence

$$\mathsf{P}\left(\lim_{n\to\infty}|\xi_n| = +\infty\right) = 1.$$

Also see Problem III.3.7.

III.3.15. a) This is a direct consequence of the definition of an atom.

b) This is a consequence of σ-additivity of the probability.

c) Let $\{A_n, n \geq 1\}$ be a sequence of all the disjoint atoms of the space $(\Omega, \mathfrak{A}, \mathsf{P})$. Then the set

$$A_0 = \Omega \setminus \bigcup_{n=1}^{\infty} A_n$$

does not contain atoms. Prove that for any fixed p, $0 < p \leq \mathsf{P}(A_0)$, there exists $B \subset A_0$, $B \in \mathfrak{A}$, with $\mathsf{P}(B) \leq p$. Then construct a set of probability p.

III.3.16. a) Assume the contrary.

b) In this case

$$\Omega = \bigcup_{n=1}^{\infty} A_n, \qquad \mathsf{P}(A_n) > 0, \quad n \geq 1.$$

Any random variable ξ may be represented in the form

$$\xi(\omega) = \sum_{n=1}^{\infty} c_n I_{A_n}(\omega),$$

where $\{c_n, n \geq 1\} \subset \mathbf{R}$, and I_A is the indicator of A. A sequence

$$\xi_N = \sum_{n=1}^{\infty} c_n^{(N)} I_{A_n}(\omega), \qquad N \geq 1,$$

converges to zero in probability if and only if $c_n^{(N)} \to 0$ for any n as $N \to \infty$, since

$$\mathsf{P}(|\xi_N| > \varepsilon) = \sum_{n=1}^{\infty} \mathsf{P}(\{\omega: |\xi_N| > \varepsilon\} \cap A_n) = \sum_{n=1}^{\infty} I_{(\varepsilon,\infty)}\big(|c_n^{(N)}|\big)\mathsf{P}(A_n).$$

In this case

$$\mathsf{P}(\xi_N \to 0 \text{ as } N \to \infty) = 1.$$

c) If $\mathsf{P}(A_0) > 0$, then for any n consider a representation of A_0

$$A_0 = \bigcup_{k=1}^{n} B_{nk},$$

where B_{nk}, $1 \leq k \leq n$, are disjoint and $\mathsf{P}(B_{nk}) = n^{-1}\mathsf{P}(A_0)$. Set $I_{nk} = I_{B_{nk}}$. Check that the sequence $I_{11}, I_{21}, I_{22}, I_{31}, I_{32}, I_{33}, \ldots$ converges to zero in probability and does not have the property of convergence with probability one.

Also see Problem III.3.7.

III.3.17. The set where at least one of the conditions $\xi_n \geq 0$, $\xi_n \geq \xi_{n+1}$, $n \geq 1$, is not satisfied has probability 0. Therefore, we can assume that these conditions are satisfied everywhere. According to Problem III.3.4, it is necessary to prove that for all $\varepsilon > 0$,

$$\lim_{N\to\infty} \mathsf{P}\bigg(\bigcup_{n=N}^{\infty} \{\omega: |\xi_n(\omega)| \geq \varepsilon\}\bigg) = 0$$

or

$$\lim_{N\to\infty} \mathsf{P}\Big(\bigcap_{n=N}^{\infty}\{\omega\colon |\xi_n(\omega)|<\varepsilon\}\Big)=1.$$

But

$$\mathsf{P}\Big(\bigcap_{n=N}^{\infty}\{\omega\colon |\xi_n(\omega)|<\varepsilon\}\Big)=\mathsf{P}(|\xi_N|<\varepsilon).$$

III.3.18. See Problem III.3.5.

III.3.19. If $\xi_n \to \xi$ in probability as $n\to\infty$, then, evidently, any subsequence converges in probability to the variable ξ. Therefore, any subsequence contains a part converging to ξ with probability one. Now, if the conditions of the problem are satisfied, but $\{\xi_n, n\ge 1\}$ is not converging to ξ with probability one, then there exist $\varepsilon>0$ and $\delta>0$ such that

$$\mathsf{P}(|\xi_{n(k)}-\xi|\ge\varepsilon)\ge\delta$$

for some subsequence $\{\xi_{n(k)}, k\ge 1\}$. This subsequence does not contain a part converging to ξ with probability one.

III.3.20. Use Problem III.3.19.

III.3.21. This is a corollary of Problem III.3.20.

III.3.22. Take the indicator of the set

$$\{(y_1,y_2,\dots,y_k)\colon y_1<x_1, y_2<x_2,\dots,y_n<x_n\}$$

as the function g in Problem III.3.20. Then

$$\mathsf{P}\left(g\left(\xi_n^{(1)},\dots,\xi_n^{(k)}\right)=1\right)=\mathsf{P}\left(\xi_n^{(i)}<x_i, 1\le i\le k\right),$$
$$\mathsf{P}\left(g\left(\xi^{(1)},\dots,\xi^{(k)}\right)=1\right)=\mathsf{P}\left(\xi^{(i)}<x_i, 1\le i\le k\right),$$

and

$$\mathsf{P}\left(\left|g\left(\xi_n^{(1)},\dots,\xi_n^{(k)}\right)-g\left(\xi^{(1)},\dots,\xi^{(k)}\right)\right|>\frac{1}{2}\right)\to 0,\qquad n\to\infty.$$

Now consider the difference

$$\mathsf{P}\left(g\left(\xi_n^{(1)},\dots,\xi_n^{(k)}\right)=1\right)-\mathsf{P}\left(g\left(\xi^{(1)},\dots,\xi^{(k)}\right)=1\right).$$

III.3.23. This is a corollary of Problem III.3.22

III.3.24. *The necessity* follows from Problem III.3.23. To prove *the sufficiency,* take $\varepsilon>0$ and consider the probability

$$\mathsf{P}(|\xi_n-c|>\varepsilon)=\mathsf{P}(\xi_n<c-\varepsilon)+\mathsf{P}(\xi_n>c+\varepsilon)\to F(c-\varepsilon)+1-F(c+\varepsilon),\qquad n\to\infty.$$

The sufficiency is not valid if c is not a constant.

III.3.25. Consider a sequence of independent identically distributed random variables with the distribution function F and prove that it is not true that this sequence is a Cauchy sequence with probability one.

III.3.26. See Problem III.3.22.

III.3.27. Consider a relation

$$\begin{aligned}\mathsf{P}(|\xi_n\eta_n| > \varepsilon) &= \mathsf{P}(|\xi_n\eta_n| > \varepsilon, |\xi_n| > a) + \mathsf{P}(|\xi_n\eta_n| > \varepsilon, |\xi_n| \le a) \\ &\le \mathsf{P}(|\xi_n| > a) + \mathsf{P}\left(|\eta_n| > \frac{\varepsilon}{a}\right).\end{aligned}$$

III.3.29. It is clear that

$$\begin{aligned}\mathsf{P}(|\xi_n - \xi_m| > \varepsilon) &\le \mathsf{P}(|\xi_n - \xi| + |\xi - \xi_m| > \varepsilon) \\ &\le \mathsf{P}\left(\left\{\omega\colon |\xi_n - \xi| > \frac{\varepsilon}{2}\right\} \cup \left\{\omega\colon |\xi - \xi_m| > \frac{\varepsilon}{2}\right\}\right) \\ &\le \mathsf{P}\left(|\xi_n - \xi| > \frac{\varepsilon}{2}\right) + \mathsf{P}\left(|\xi - \xi_m| > \frac{\varepsilon}{2}\right),\end{aligned}$$

where ξ is the limit in probability of the sequence $\{\xi_n, n \ge 1\}$.

III.3.30. First, using the Cauchy property, construct a subsequence $\{\xi_{n(k)}, k \ge 1\}$ such that

$$\sum_{k=1}^{\infty} \mathsf{P}\left(|\xi_{n(k+1)} - \xi_{n(k)}| > \frac{1}{2^k}\right) < \infty.$$

According to Problem III.3.10, this subsequence converges to ξ with probability one. Then use the Cauchy property again.

III.3.31. See Problem III.3.23.

III.3.34. For example,

$$F_n(x) = \begin{cases} 0, & x \le n, \\ 1, & x > n, \end{cases} \qquad n \ge 1.$$

Even a more general assertion is valid: *For any nondecreasing left-continuous function F such that $0 \le F(x) \le 1$, $x \in \mathbf{R}$, there exists a sequence of probability distribution functions converging to F at points where F is continuous.*

III.3.35. The limit F is a nondecreasing function. So, we only need to check that

$$F(-a) \to 0, \qquad F(a) \to 1, \qquad a \to +\infty.$$

III.3.36. This is a corollary of the Chebyshev inequality and Problem III.3.35.

III.3.37. This is a corollary of the Chebyshev inequality.

III.3.38. The first assertion follows from the inequality

$$(*) \qquad |a+b|^r \le c_r\left(|a|^r + |b|^r\right), \qquad c_r = \begin{cases} 1, & 0 < r \le 1, \\ 2^{r-1}, & r > 1, \end{cases} \qquad a, b \in \mathbf{R}.$$

A sequence $\{\xi_n, n \ge 1\}$ is a Cauchy sequence in mean of order r if for any fixed $\varepsilon > 0$ there exists N such that $\mathsf{E}\,|\xi_m - \xi_n|^r < \varepsilon$ for all $m, n > N$. As in the solution of Problem III.3.37, we prove that $\{\xi_n\}$ is a Cauchy sequence in probability. Using Problem III.3.30, conclude that it converges in probability to a random variable ξ. By the Fatou lemma we get

$$\mathsf{E}\,|\xi_m - \xi|^r \le \varepsilon, \qquad m \ge N,$$

whence, in view of $(*)$, $\mathsf{E}\,|\xi|^r < \infty$ and the sequence $\{\xi_n, n \ge 1\}$ converges in mean of order r to the random variable ξ.

III.3.39. Let $\{\xi_n, n \geq 1\}$ be a sequence of independent random variables with the distributions

$$\mathsf{P}(\xi_n = 0) = 1 - \frac{1}{n}, \qquad \mathsf{P}(\xi_n = 1) = \frac{1}{n}, \qquad n \geq 1.$$

For this sequence, $\mathsf{E}\,|\xi_n|^r = \frac{1}{n}$, for $n \geq 1$, that is, if $r > 0$, then $\{\xi_n\}$ converges to zero in mean of order r, but

$$\mathsf{P}\left(\limsup_{n\to\infty} \xi_n = 1\right) = 1.$$

III.3.40. The statement is evident if the random variables have no moments. For example, if ξ is a random variable with the distribution density

$$f(x) = \frac{\ln 2}{x \ln^2 x}, \qquad x \geq 2,$$

then $\mathsf{E}\,\xi^r = +\infty$ for all $r > 0$, but the sequence $\{\frac{\xi}{n}, n \geq 1\}$ converges to 0 with probability one. Moreover, the statement is also valid in the case where the moments exist. For example, if $\{\xi_n, n \geq 1\}$ is a sequence of random variables such that

$$\mathsf{P}(\xi_n = 0) = 1 - \frac{1}{n^2}, \qquad \mathsf{P}(\xi_n = n) = \frac{1}{n^2}, \qquad n \geq 1,$$

then $\mathsf{E}\,\xi_n^2 = 1$ for all $n \geq 1$ and $\mathsf{P}(\xi_n \to 0 \text{ as } n \to \infty) = 1$.

III.3.41. According to Problem III.3.20, $g(\xi_n) \to g(\xi)$ in probability as $n \to \infty$. Therefore, the desired assertion follows from the Lebesgue dominated convergence theorem.

III.3.42. This is a corollary of the following statement: *If* $\mathsf{E}\,|\xi|^r < \infty$ *for some* $r > 0$, *then* $(\mathsf{E}\,|\xi|^s)^{1/s} \leq (\mathsf{E}\,|\xi|^r)^{1/r}$ *for* $0 < s < r$.

III.3.43. It is sufficient to consider the case $s = r$ (see Problem III.3.42). For $r > 1$ the desired statement follows from the Minkowski inequality

$$(\mathsf{E}\,|\alpha + \beta|^r)^{1/r} \leq (\mathsf{E}\,|\alpha|^r)^{1/r} + (\mathsf{E}\,|\beta|^r)^{1/r},$$

which leads to the estimate

$$\left|(\mathsf{E}\,|\alpha|^r)^{1/r} - (\mathsf{E}\,|\beta|^r)^{1/r}\right| \leq (\mathsf{E}\,|\alpha - \beta|^r)^{1/r}.$$

For $r \leq 1$ use inequality $(*)$ of Problem III.3.38, which is simpler.

III.3.44. Let independent random variables ξ_n, $n \geq 1$, have the distributions

$$\mathsf{P}(\xi_n = 0) = 1 - \frac{1}{n^2}, \qquad \mathsf{P}(\xi_n = -n^2) = \mathsf{P}(\xi_n = n^2) = \frac{1}{2n^2}, \qquad n \geq 1.$$

Then $\xi_n \to 0$ with probability one as $n \to \infty$, and

$$\mathsf{E}\,\xi_n = 0, \qquad n \geq 1,$$
$$\mathsf{E}\,|\xi_n - 0| = \mathsf{E}\,|\xi_n| = 1, \qquad n \geq 1.$$

III.3.45. First note that $\mathsf{P}(|\xi_n| > c) = 0$ and $\xi_n - \xi \to 0$ in probability as $n \to \infty$. Then use Problem III.3.41 with

$$g(x) = \begin{cases} |x|^r, & |x| \le 2c, \\ (2c)^r, & |x| > 2c. \end{cases}$$

III.3.46. Conditions of the problem imply that for any polynomial Q,

$$\lim_{n\to\infty} \int_{-c}^{c} Q(x)\, dF_n(x) = \int_{-c}^{c} Q(x)\, dF(x),$$

where $F_n(x) = \mathsf{P}(\xi_n < x)$, $n \ge 1$, $F(x) = \mathsf{P}(\xi < x)$. Use the Weierstrass theorem on approximation of a given function by polynomials and Theorem 1 stated at the beginning of §III.3.

III.3.47. This is a corollary of the Lebesgue dominated convergence theorem.

III.3.48. By the Fatou lemma, $\mathsf{E}\,|\xi|^r$ is finite, since for a subsequence $\{\xi_{n(k)}, k \ge 1\}$ such that $\mathsf{P}(\xi_{n(k)} \to \xi \text{ as } k \to \infty) = 1$, we have

$$\liminf_{n\to\infty} \mathsf{E}\,|\xi_{n(k)}|^r \ge \mathsf{E} \liminf_{n\to\infty} |\xi_{n(k)}|^r = \mathsf{E}\,|\xi|^r.$$

Also, for all ε, $0 < \varepsilon < a$, we have

$$\begin{aligned} \mathsf{E}\,|\xi_n - \xi|^s &= \int_\Omega |\xi_n(\omega) - \xi(\omega)|^s \, d\mathsf{P} \\ &= \left(\int_{\{|\xi_n - \xi| \le \varepsilon\}} + \int_{\{\varepsilon < |\xi_n - \xi| \le a\}} + \int_{\{|\xi_n - \xi| > a\}} \right) |\xi_n(\omega) - \xi(\omega)|\, d\mathsf{P} \\ &\le \varepsilon^s + a^s \mathsf{P}(|\xi_n - \xi| > \varepsilon) + a^{s-r}\, \mathsf{E}\,|\xi_n - \xi|^r. \end{aligned}$$

This estimate yields $\mathsf{E}\,|\xi_n - \xi|^s \to 0$ as $n \to \infty$. The example to Problem III.3.40 shows that in general we cannot put $s = r$.

III.3.49. This is a corollary of Problems III.3.48 and III.3.42.

III.3.50. If $\xi_n \to \xi$ in probability as $n \to \infty$, then $d(\xi_n, \xi) \to 0$ as $n \to \infty$ by the Lebesgue dominated convergence theorem. On the other hand, if $d(\xi_n, \xi) \to 0$ as $n \to \infty$, then the Chebyshev inequality implies the convergence $\xi_n \to \xi$ in probability as $n \to \infty$. See Problem III.3.42.

III.3.51. The solution is similar to that of Problem III.3.50.

III.3.52. a) Use the inequality of Problem III.3.38.
b) Use the Minkowski inequality.
c), d) See the solution to Problem III.3.38.
e) This is a corollary of the inequality $|\xi|^s \le 1 + |\xi|^r$.

III.3.53. For a given $\varepsilon > 0$, consider points $-a = x_0 < x_1 < \cdots < x_m = a$ such that

$$F(-a) < \frac{\varepsilon}{3}, \qquad 1 - F(a) < \frac{\varepsilon}{3}, \qquad F(x_{k+1}) - F(x_k) < \frac{\varepsilon}{3}$$

for $0 \le k \le m-1$. We have

$$\sup_{x<-a} |F_n(x) - F(x)| \le F_n(-a) + F(-a);$$

$$\sup_{x>a} |F_n(x) - F(x)| \le 1 - F_n(a) + 1 - F(a);$$

$$\sup_{x_k \le x \le x_{k+1}} |F_n(x) - F(x)| \le \sup_{0\le k\le m-1} \{|F_n(x_{k+1}) - F(x_k)|, |F_n(x_k) - F(x_{k+1})|\}.$$

Let n_0 be such that

$$|F_n(x_k) - F(x_k)| < \frac{\varepsilon}{3}, \qquad 0 \le k \le m, \quad n \ge n_0.$$

Then

$$\sup_{-\infty<x<\infty} |F_n(x) - F(x)| < \varepsilon, \qquad n \ge n_0.$$

III.3.54. Consider the inequality

$$\begin{aligned} |\mathsf{E}(\xi_n\eta_n) - c\,\mathsf{E}\,\eta_n| &\le \mathsf{E}(|\xi_n - c|\cdot|\eta_n|) \\ &= \int_{\{|\eta_n|>A\}} |\xi_n - c|\cdot|\eta_n|\,d\mathsf{P} + \int_{\{|\eta_n|\le A\}} |\xi_n - c|\cdot|\eta_n|\,d\mathsf{P} \\ &\le (a+c)\sup_{n\ge 1}\int_{\{|\eta_n|>A\}} |\eta_n|\,d\mathsf{P} + A\int_\Omega |\xi_n - c|\,d\mathsf{P}. \end{aligned}$$

III.3.56. Consider the inequalities

$$\int_{\{|\xi_n|>a\}} |\xi_n|\,d\mathsf{P} \le \int_\Omega |\xi_n - \xi|\,d\mathsf{P} + \int_{\{|\xi_n|>a\}} |\xi|\,d\mathsf{P},$$

$$\mathsf{P}(|\xi_n| > a) \le \mathsf{P}\left(|\xi_n - \xi| > \frac{a}{2}\right) + \mathsf{P}\left(|\xi| > \frac{a}{2}\right).$$

Then use the convergence $\xi_n \to \xi$ in probability as $n \to \infty$ and the absolute continuity of an integral.

III.3.59. Use assertion c) of the preceding problem to show that $\mathsf{E}\,|\xi| < \infty$.

III.3.60. For a number c such that $a(t)/t \ge \delta$ for $t \ge c$, consider the inequality

$$\int_{\{|\xi_n|>c\}} |\xi_n|\,d\mathsf{P} \le \frac{1}{\delta}\int_{\{|\xi_n|>c\}} a(|\xi_n|)\,d\mathsf{P} \le \frac{1}{\delta}\,\mathsf{E}\,a(|\xi_n|).$$

III.3.61. This is a corollary of Problem III.3.60.

REMARK. Definition III.3.56 and Problem III.3.60 may be applied to every collection of random variables (not necessarily denumerable).

III.3.62. Indeed,

$$\int_{\{|S_n|>A\sqrt{n}\}} \frac{1}{\sqrt{n}}|S_n|\,d\mathsf{P} \le \frac{1}{\sqrt{n}}\left(\int_\Omega S_n^2\,d\mathsf{P}\cdot\mathsf{P}(|S_n| > A\sqrt{n})\right)^{1/2} \le \frac{\sigma^2}{A}.$$

III.3.63. This is a corollary of the Cauchy inequality.

III.3.64. Since $\mathsf{E}\,\eta_n^2 \to \mathsf{E}\,\eta^2$ as $n \to \infty$, this is a corollary of the Cauchy inequality:

$$|\,\mathsf{E}(\xi_n\eta_n) - \mathsf{E}(\xi\eta)| \le |\,\mathsf{E}(\xi_n - \xi)\eta_n| + |\,\mathsf{E}\,\xi(\eta_n - \eta)|$$
$$\le \left(\mathsf{E}(\xi_n - \xi)^2\,\mathsf{E}\,\eta_n^2\right)^{1/2} + \left(\mathsf{E}\,\xi^2\,\mathsf{E}(\eta_n - \eta)^2\right)^{1/2}.$$

III.3.65. The necessity follows from the Lebesgue dominated convergence theorem, and the sufficiency from the Chebyshev inequality.

III.3.66. See Problem III.3.20.

III.3.69. Consider the relation

$$\mathsf{P}(|\zeta_n| > \varepsilon) = \mathsf{P}(|\zeta_n| > \varepsilon, |\xi_n| \le a, |\eta_n| \le \delta)$$
$$+\,\mathsf{P}(|\zeta_n| > \varepsilon, \{\omega\colon |\xi_n| > a\} \cup \{\omega\colon |\eta_n| > \delta\})$$
$$\le \mathsf{P}(|\zeta_n| > \varepsilon, |\xi_n| \le a, |\eta_n| \le \delta) + \mathsf{P}(|\xi_n| > a) + \mathsf{P}(|\eta_n| > \delta),$$

where $\zeta_n = g(\xi_n+\eta_n)-g(\xi_n)$. For a given $\gamma > 0$, take $a > 0$ such that $\mathsf{P}(|\xi_n| > a) < \gamma/2$ for all $n \ge 1$. Then choose $\delta > 0$ such that

$$|g(x + y) - g(x)| < \varepsilon, \qquad |y| < \delta, \quad |x| \le a.$$

Also take n_0 such that $\mathsf{P}(|\eta_n| > \delta) < \frac{\gamma}{2}$ for $n \ge n_0$. Then $\mathsf{P}(|\zeta_n| > \varepsilon) < \gamma$ for $n \ge n_0$.

III.3.70. See Problem III.3.69.

III.3.71. See the beginning of §III.3.

III.3.72. Use Problems III.3.68, III.3.70, III.3.71, and the Lebesgue dominated convergence theorem. To prove that $\xi_n+\eta_n$ converges in law to $\xi+c$, it is necessary to show that

$$\lim_{n\to\infty} \mathsf{E}\,g(\xi_n + \eta_n) = \mathsf{E}\,g(\xi + c).$$

We have

$$\mathsf{E}\,g(\xi_n + \eta_n) = \mathsf{E}\left[g(\xi_n + \eta_n) - g(\xi_n + c)\right] + \mathsf{E}\,g(\xi_n + c),$$

and moreover, the sequence $\{g(\xi_n+\eta_n)-g(\xi_n+c), n \ge 1\}$ is bounded and converges to zero in probability.

III.3.73. Put $\eta_n = \xi_n + \delta_n$, $n \ge 1$, and use Problems III.3.68, III.3.27, and III.3.72.

III.3.74. If $0 < \varepsilon_n < \frac{1}{2}$, then $|\xi_n - \eta_n| \le 2\varepsilon_n|\xi_n|$. See Problem III.3.71.

III.3.75. See Problem III.3.71.

III.3.76. See Problem III.3.4.

III.3.78. By the condition of the problem,

$$\lim_{m\to\infty} \mathsf{P}\left(\sup_{n\ge m} |\xi_n - \xi| \ge \varepsilon\right) = 0$$

for each $\varepsilon > 0$, and

$$\lim_{m\to\infty} \mathsf{P}\left(\inf_{n\ge m} \nu(n) \ge c\right) = 1$$

for all $c > 0$. Therefore, for any N_1,

$$\mathsf{P}\left(\sup_{n\ge N}|\xi_{\nu(n)}-\xi|\ge\varepsilon\right)=\mathsf{P}\left(\sup_{n\ge N}|\xi_{\nu(n)}-\xi|\ge\varepsilon,\ \inf_{n\ge N}\nu(n)\ge N_1\right)$$
$$+\mathsf{P}\left(\sup_{n\ge N}|\xi_{\nu(n)}-\xi|\ge\varepsilon,\ \inf_{n\ge N}\nu(n)<N_1\right)$$
$$\le\mathsf{P}\left(\sup_{n\ge N_1}|\xi_n-\xi|\ge\varepsilon\right)+\mathsf{P}\left(\inf_{n\ge N}\nu(n)<N_1\right).$$

III.3.79. By the Borel–Cantelli lemma, infinitely many events $\{\omega\colon \xi_n(\omega)=1\}$, $n\ge 1$, occur with probability one.

III.3.80. For all $\varepsilon>0$ and N we have

$$\mathsf{P}(|\xi_{\nu(n)}-\xi|\ge\varepsilon)=\mathsf{P}(|\xi_{\nu(n)}-\xi|\ge\varepsilon,\nu(n)<N)+\sum_{k=N}^{\infty}\mathsf{P}(|\xi_{\nu(n)}-\xi|\ge\varepsilon,\nu(n)=k)$$
$$\le\mathsf{P}(\nu(n)<N)+\sum_{k=N}^{\infty}\mathsf{P}(|\xi_k-\xi|\ge\varepsilon,\nu(n)=k)$$
$$=\mathsf{P}(\nu(n)<N)+\sum_{k=N}^{\infty}\mathsf{P}(|\xi_k-\xi|\ge\varepsilon)\mathsf{P}(\nu(n)=k)$$
$$\le\mathsf{P}(\nu(n)<N)+\mathsf{P}(\nu(n)\ge N)\sup_{k\ge N}\mathsf{P}(|\xi_k-\xi|\ge\varepsilon)$$
$$\le\mathsf{P}(\nu(n)<N)+\sup_{k\ge N}\mathsf{P}(|\xi_k-\xi|\ge\varepsilon).$$

III.3.81. Indeed,

$$\mathsf{P}(\xi_{\nu(n)}<x)=\mathsf{P}(\xi_{\nu(n)}<x,\nu(n)<N)+\sum_{k=N}^{\infty}\mathsf{P}(\xi_{\nu(n)}<x,\nu(n)=k)$$
$$\le\mathsf{P}(\nu(n)<N)+\sup_{n\ge N}\mathsf{P}(\xi_n<x).$$

III.3.82. Consider the relation

$$\mathsf{E}\,|\xi_{\nu(n)}-\xi|^r=\sum_{k=1}^{\infty}\mathsf{E}\left[|\xi_{\nu(n)}-\xi|^r I(\nu(n)=k)\right]$$
$$\le\sup_{n\ge1}\mathsf{E}\,|\xi_n-\xi|^r\mathsf{P}(\nu(n)<N)+\sup_{n\ge N}\mathsf{E}\,|\xi_n-\xi|^r$$

for natural N, where $I(A)$ is the indicator of the event A.

III.3.83. It is sufficient to consider the case $|f_n|<1$, $n\ge1$. In the general case consider the functions $\arctan f_n$, $n\ge 1$. Then

$$\mathsf{E}[f_{2n}(\xi_1,\dots,\xi_{2n})-f_n(\xi_1,\dots,\xi_n)]^2=\mathsf{E}[f_{2n}(\xi_1,\dots,\xi_{2n})-f_n(\xi_{n+1},\dots,\xi_{2n})]^2,$$

whence

$$\lim_{n\to\infty}\mathsf{E}[f_{2n}(\xi_1,\dots,\xi_{2n})-f_n(\xi_{n+1},\dots,\xi_{2n})]^2=0.$$

Furthermore, consider

$$\mathsf{E}[f_n(\xi_1,\dots,\xi_n)-f_n(\xi_{n+1},\dots,\xi_{2n})]^2$$

and prove that

$$\lim_{n\to\infty} \operatorname{Var} f_n(\xi_1, \ldots, \xi_n) = 0.$$

III.3.84. Let $\Omega = [0,1]$, $\mathfrak{A}$ the σ-algebra of Lebesgue measurable sets, and P Lebesgue measure on $\mathfrak{A}$. For the function F_n define a function H_n such that $H_n(y) = x$ if $F_n(x) = y$ and there exists only one such a number x; $H_n(y) = x_{np}$ if $F_n(x) = y$ holds for all $x \in (x_0, x_{np}]$ (we consider the maximal interval with this property), and $H_n(y) = x_p$ if $F_n(x_p) < y < F_n(x_p + 0)$, $0 \le y \le 1$. The variable $H_n(\omega)$, $\omega \in \Omega$, is random and has the distribution function F_n. Check that $\mathsf{P}(H_n \to H \text{ as } n \to \infty) = 1$.

III.3.86. Use Problems III.3.84 and III.3.49.

III.3.89. This is a corollary of the Lebesgue dominated convergence theorem, since $g(\vec{\xi}^{(n)}) \to g(\vec{\xi})$ in probability as $n \to \infty$. See Problem III.3.20.

III.3.90. See Problem III.3.89.

III.3.91. See Problem III.3.19.

III.3.92. See the case $m = 1$ in Problem III.3.53.

III.3.93. This is a corollary of Theorem 1 at the beginning of §III.3.

III.3.94. The solution is similar to that of Problem III.3.69.

III.3.95. Use Problem III.3.94, the Lebesgue dominated convergence theorem, and the theorem on weak convergence.

§**4.**

III.4.1. This is a corollary of the Chebyshev inequality.

III.4.2. This is a corollary of the Lebesgue dominated convergence theorem.

III.4.3. For all $x \in \mathbf{R}$ the random variables $\{f(x+\xi_n), n \ge 1\}$ are independent and identically distributed with mean value

$$\int_0^1 f(u)\,du.$$

III.4.4. Let $\{\xi_n, n \ge 1\}$ be a sequence of independent random variables uniformly distributed on the interval $[0,1]$. According to the law of large numbers,

$$S_n = \frac{1}{n}\sum_{k=1}^{n} \xi_k \to \mathsf{E}\,\xi_1 = \frac{1}{2} \quad \text{as } n \to \infty$$

in probability. The continuity of f implies

$$f(S_n) \to f\left(\frac{1}{2}\right) \quad \text{as } n \to \infty$$

in probability, so that the boundedness of f and the Lebesgue dominated convergence theorem yield the relation

$$\mathsf{E}\,f(S_n) \to \mathsf{E}\,f\left(\frac{1}{2}\right) = f\left(\frac{1}{2}\right) \quad \text{as } n \to \infty.$$

Now write an expression for $\mathsf{E} f(S_n)$ by using the independence of ξ_n, $n \geq 1$.

III.4.5. ANSWER: 2^{-m}.

III.4.6. The solution is similar to that of Problem III.4.4.

III.4.7. Let $\{\xi_n, n \geq 1\}$ be a sequence of independent random variables uniformly distributed on $[0,1]$ and

$$\eta_n = \frac{1}{n}\sum_{k=1}^{n} f(\xi_k), \qquad \zeta_n = \frac{1}{n}\sum_{k=1}^{n} g(\xi_k), \quad n \geq 1.$$

Then

$$\eta_n \to \int_0^1 f(x)\,dx, \qquad \zeta_n \to \int_0^1 g(x)\,dx$$

in probability as $n \to \infty$, and therefore the sequence $\{\eta_n/\zeta_n, n \geq 1\}$ converges in probability. Moreover, $0 \leq \eta_n/\zeta_n \leq c$, $n \geq 1$. Apply the Lebesgue dominated convergence theorem to $\mathsf{E}\,(\eta_n/\zeta_n)$.

III.4.8. See Problem III.4.7.

III.4.9. Note that

$$n\int_0^1 \overset{n}{\cdots} \int_0^1 \left[f\left(\frac{x_1+x_2+\cdots+x_n}{n}\right) - f\left(\frac{1}{2}\right)\right] dx_1\,dx_2\cdots dx_n$$
$$= n\,\mathsf{E}\left[f\left(\frac{\xi_1+\xi_2+\cdots+\xi_n}{n}\right) - f\left(\frac{1}{2}\right)\right]$$

for a sequence of independent random variables $\{\xi_n, n \geq 1\}$ uniformly distributed on $[0,1]$. Using the Taylor formula for some $0 < \theta < 1$,

$$f(x) = f\left(\frac{1}{2}\right) + f'\left(\frac{1}{2}\right)\left(x - \frac{1}{2}\right) + \frac{1}{2}f''\left(\frac{1}{2}\right)\left(\frac{1}{2} + \theta\left(x - \frac{1}{2}\right)\right)\left(x - \frac{1}{2}\right)^2,$$

and the evident equality

$$\mathsf{E}\left[\frac{\xi_1+\xi_2+\cdots+\xi_n}{n} - \frac{1}{2}\right] = 0,$$

we get

$$\mathsf{E}\left[f\left(\frac{\xi_1+\xi_2+\cdots+\xi_n}{n}\right) - f\left(\frac{1}{2}\right)\right]$$
$$= \frac{1}{2}\mathsf{E}\left\{f''\left(\frac{1}{2} + \theta\left(\frac{\xi_1+\xi_2+\cdots+\xi_n}{n} - \frac{1}{2}\right)\right)\left[\frac{\xi_1+\xi_2+\cdots+\xi_n}{n} - \frac{1}{2}\right]^2\right\}.$$

By the law of large numbers,

$$\frac{\xi_1+\xi_2+\cdots+\xi_n}{n} \to \frac{1}{2}$$

in probability as $n \to \infty$. It is evident that

$$n\,\mathsf{E}\left[\frac{\xi_1+\xi_2+\cdots+\xi_n}{n} - \frac{1}{2}\right]^2 \to \frac{1}{12}.$$

Now it is easy to check that

$$\lim_{n\to\infty} \mathsf{E}\left[f\left(\frac{\xi_1+\xi_2+\cdots+\xi_n}{n}\right)-f\left(\frac{1}{2}\right)\right]=\frac{1}{24}f''\left(\frac{1}{2}\right).$$

III.4.10. Let $\{\xi_n, n\geq 1\}$ be the sequence defined in the solution of Problem III.4.9. The integrals are equal to

a) $\mathsf{P}\left(\xi_1^2+\xi_2^2+\cdots+\xi_n^2\leq\sqrt{n}\right)=\mathsf{P}\left(\dfrac{\xi_1^2+\xi_2^2+\cdots+\xi_n^2}{n}\leq\dfrac{1}{\sqrt{n}}\right)$,

b) $\mathsf{P}\left(\dfrac{\xi_1^2+\xi_2^2+\cdots+\xi_n^2}{n}\leq\dfrac{1}{4}\right)$,

c) $\mathsf{P}\left(\dfrac{\xi_1^2+\xi_2^2+\cdots+\xi_n^2}{n}\leq\dfrac{1}{2}\right)$,

respectively. According to the weak law of large numbers,

$$\frac{\xi_1^2+\xi_2^2+\cdots+\xi_n^2}{n}\to\mathsf{E}\,\xi_1^2$$

in probability as $n\to\infty$. ANSWER: a) 0; b) 0; c) 1.

III.4.11. The solution is similar to that of Problem III.4.10.

III.4.12. This is a corollary of the Lebesgue dominated convergence theorem and the law of large numbers, which gives us

$$\frac{1}{n}\sum_{k=1}^{n}\xi_k\to x$$

in probability as $n\to\infty$.

III.4.13. This is a corollary of Khintchine's theorem.

III.4.14. See Problem III.4.13.

III.4.16. First prove that for all $c>0$,

$$\sup_{n\geq 1}\mathsf{P}(|\xi_n|>c)\to 0\quad\text{as } n\to\infty.$$

To do this we use several times the equality

$$\xi_n=-\alpha\xi_{n-1}+\varepsilon_n=-\alpha(-\alpha\xi_{n-2}+\varepsilon_{n-1})+\varepsilon_n$$

and the Chebyshev inequality. Then consider the expression

$$\frac{\xi_{n+1}}{n}+\frac{\alpha}{n}\xi_n+\frac{1+\alpha}{n}\sum_{k=1}^{n}\xi_k=\frac{1}{n}\sum_{k=1}^{n+1}\varepsilon_k.$$

III.4.17. It is clear that

$$\xi_n^2+\alpha\xi_n\xi_{n-1}=\xi_n\varepsilon_n,\qquad \xi_n\xi_{n-1}+\alpha\xi_{n-1}^2=\xi_{n-1}\varepsilon_n,\qquad n\geq 1.$$

Sum up these equalities and consider the right-hand side.

III.4.18. This is a corollary of the law of large numbers for a sequence of identically distributed random variables with a finite variance. Consider, for instance, m sequences separately.

III.4.19. Check that

$$S^2 = \frac{1}{n}\sum_{k=1}^{n}\xi_k^2 - (\bar{x})^2$$

and apply the law of large numbers.

III.4.20. For $i \geq 1$ put

$$\chi_i = \begin{cases} 1, & \xi_i < x, \\ 0, & \xi_i \geq x, \end{cases} \qquad i \geq 1.$$

The random variables χ_i, $1 \leq i \leq n$, are independent and

$$F_n(x) = \frac{1}{n}\sum_{k=1}^{n}\chi_k, \qquad \mathsf{E}\,F_n(x) = F(x), \qquad \operatorname{Var} F_n(x) = \frac{F(x)[1-F(x)]}{n}.$$

III.4.21. According to the Chebyshev inequality, for $\varepsilon > 0$ we have

$$\mathsf{P}\left(\left|\frac{1}{n}\sum_{k=1}^{n}\xi_k\right| \geq \varepsilon\right) \leq \frac{1}{\varepsilon^{r/2}}\frac{1}{n^{r/2}}\,\mathsf{E}\left|\sum_{k=1}^{n}\xi_k\right|^{r/2} \leq \frac{1}{\varepsilon^{r/2}}\frac{1}{n^{r/2}}\sum_{k=1}^{n}\mathsf{E}\left(|\xi_k|^{r/2}\right)$$
$$\leq \frac{1}{\varepsilon^{r/2}}\frac{1}{n^{r/2}}\sum_{k=1}^{n}\sqrt{\mathsf{E}\,|\xi_k|^r}.$$

III.4.24. Consider the inequalities

$$\begin{aligned}|\,\mathsf{E}\,f(\eta_n) - f(x)| &\leq \mathsf{E}\,|f(\eta_n) - f(x)| \\ &= \mathsf{E}\left(|f(\eta_n) - f(x)| / |\eta_n - x| < \delta\right)\mathsf{P}(|\eta - x| < \delta) \\ &\quad + \mathsf{E}\left(|f(\eta_n) - f(x)| / |\eta_n - x| \geq \delta\right)\mathsf{P}(|\eta - x| \geq \delta)\end{aligned}$$

and use the Chebyshev inequality.

III.4.25. The Weierstrass theorem on the uniform approximation of a continuous function by polynomials with a given accuracy is a consequence of this problem. To solve the problem, represent $B_n(f;x)$ as the expectation and check the conditions of the preceding problem.

III.4.26. See the hint to the preceding problem.

III.4.29. a) For $c > 0$ we have

$$\begin{aligned}\frac{1}{n}\int_{\{|S_n|>cn\}}|S_n|\,d\mathsf{P} &= \frac{1}{n}\int_{\{S_n>cn\}}S_n\,d\mathsf{P} + \frac{1}{n}\int_{\{S_n<-cn\}}(-S_n)\,d\mathsf{P} \\ &= \int_{\{|S_n|>cn\}}|\xi_1|\,d\mathsf{P},\end{aligned}$$

where $S_n = \xi_1 + \xi_2 + \cdots + \xi_n$, $n \geq 1$. Next apply the Khintchine law of large numbers.

b) This is a corollary of a) and Problem III.3.59.

III.4.30. By Theorem 1 at the beginning of §III.4, $S_n n \to 0$ in probability as $n \to \infty$. By the law of total probability,

$$\mathsf{P}\left(\frac{|S_{\nu+1}|}{\nu+1} > \varepsilon\right) = \sum_{n=0}^{\infty} \mathsf{P}\left(\frac{|S_{\nu+1}|}{\nu+1} > \varepsilon/\{\nu = n\}\right) \mathsf{P}(\nu = n)$$
$$= \sum_{n=0}^{\infty} \mathsf{P}\left(\frac{|S_n|}{n} > \varepsilon\right) \frac{\lambda^n}{n!} e^{-n}.$$

These two relations imply the desired result.

III.4.31. See Problem III.3.80.

§**5.**

III.5.1. This is a corollary of the Kolmogorov inequality.

III.5.2. This is a corollary of the Kolmogorov inequality.

III.5.3. To prove a) use the following inclusion:

$$\left\{\max_{m \le l \le n} \left|\sum_{k=m}^{l} \xi_k\right| \ge 2a + c\right\}$$
$$\subset \bigcup_{u=m}^{n-1} \left\{\left|\sum_{k=m}^{v} \xi_k < a\right|, v \le u-1; \left|\sum_{k=m}^{u} \xi_k\right| \ge a\right\} \cap \left\{\max_{u+1 \le i \le n} \left|\sum_{k=u+1}^{i} \xi_k\right| \ge a\right\}.$$

III.5.4. The proof of this generalization of the Kolmogorov inequality is completely analogous to that of Theorem 1 at the beginning of §III.5. To complete the proof use the Jensen inequality $\mathsf{E}\, g(\xi) \ge g(\mathsf{E}\, \xi)$ for a convex function g.

III.5.5. Consider the inequality

$$2\mathsf{P}(S_n > l) + \mathsf{P}(S_n = l) = 2\sum_{k=1}^{n} \mathsf{P}(S_1 < l, \dots, S_{k-1} < l, S_k = l, S_n - S_k > 0)$$
$$+ \sum_{k=1}^{n} \mathsf{P}(S_1 < l, \dots, S_{k-1} < l, S_k = l, S_n - S_k = 0)$$

and use

$$2\mathsf{P}(S_n - S_k > 0) + \mathsf{P}(S_n - S_k = 0) = 1.$$

III.5.7. The proof is similar to that of the Kolmogorov inequality if we use the basic properties of the conditional expectation.

III.5.8. See solution to Problem III.5.7.

III.5.9. See solution to Problem III.5.7.

III.5.10. Use the Kolmogorov inequality and the formula

$$\int_0^{\infty} x\, dF(x) = \int_0^{\infty} [1 - F(x)]\, dx,$$

which is valid for any distribution function F on $[0, \infty)$.

§**6.**

III.6.1. ANSWER:

a) $\bigcap_{k=1}^{\infty}\bigcup_{N=1}^{\infty}\bigcap_{n=N}^{\infty}\left\{\omega: |\xi_1+\cdots+\xi_n-\xi|<\frac{1}{k}\right\}$,

b) $\bigcap_{k=1}^{\infty}\bigcup_{N=1}^{\infty}\bigcap_{n=0}^{\infty}\left\{\omega: |\xi_N+\xi_{N+1}\cdots+\xi_{N+n}|<\frac{1}{k}\right\}$.

III.6.2. a) This is, in fact, a condition for $\mathsf{P}(\xi_n\to 0 \text{ as } n\to\infty)=1$.
b) See Problem III.6.1 and use the zero-one law.
c) and d) follow from the zero-one law.

III.6.3. According to the Borel–Cantelli lemma only a finite number of events $\{\omega: |\xi_n|\geq\varepsilon_n\}$, $n\geq 1$, occur with probability one. Therefore, the series $\sum_{n=1}^{\infty}\xi_n$ converges absolutely with probability one.

III.6.4. Apply Beppo Levy's lemma to the sequence

$$\left\{\sum_{k=1}^{n}\xi_k, n\geq 1\right\}.$$

III.6.5. See Problem III.6.4.

III.6.7. Use the fact that a generic term of a series that converges with probability one, tends to zero with probability one.

III.6.8. The series $\sum_{n=1}^{\infty}\xi_n$ converges in mean square if and only if

$$S_n=\xi_1+\xi_2+\cdots+\xi_n, \qquad n\geq 1,$$

is a Cauchy sequence in mean square. Consider the equality

$$\mathsf{E}(S_n-S_m)^2=\mathsf{E}\left(\sum_{k=m+1}^{n}\xi_k\xi_k\right)^2=\sum_{k=m+1}^{n}(\mathsf{E}\,\xi_k)^2, \qquad m<n.$$

III.6.9. Check the Cauchy property for the sequence

$$\left\{\sum_{k=1}^{n}\xi_k, n\geq 1\right\}.$$

III.6.10. Let, for instance,

$$\xi_n=\frac{(-1)^n}{\sqrt{n}}+\eta_n, \qquad n\geq 1,$$

where $\{\eta_n, n\geq 1\}$ is a sequence of uncorrelated random variables with the distributions

$$\mathsf{P}\left(\eta_n=-\frac{1}{n}\right)=\mathsf{P}\left(\eta_n=\frac{1}{n}\right)=\frac{1}{2}, \qquad n\geq 1.$$

III.6.11. This is a corollary of the Cauchy inequality.

III.6.12. See Problem III.6.11.

III.6.13. See Problem III.6.11.

III.6.14. Check that the corresponding sequences are Cauchy sequences. To do this use the representation

$$S_n = \sum_{k=1}^{n} \eta_k, \qquad \eta_k = 2\xi_k \sum_{j=1}^{k-1} a_{jk}\xi_j + a_{kk}\xi_k^2,$$

the orthogonality of the variables

$$\eta_k - \mathsf{E}\,\eta_k, \qquad \eta_j - \mathsf{E}\,\eta_j, \qquad k \neq j,$$

and the equality

$$\operatorname{Var} \eta_k^2 = 4\sum_{j=l}^{k-1} a_{jk}^2 + a_{kk}^2 \left(\mathsf{E}\,\xi_k^4 - 1\right).$$

III.6.15. Prove that the series $\sum_{n=1}^{\infty} \operatorname{Var} \xi_n$ converges.

III.6.16. Use the sufficient condition for an infinite product to be convergent and Problem III.6.4.

III.6.17. The convergence of the series $\sum_{n=1}^{\infty} \mathsf{E}\,\xi_n^2$ implies the convergence with probability one of the series $\eta = \sum_{n=1}^{\infty} \xi_n^2$ (see Problem III.6.4). On the other hand, if the last series converges with probability one, then

$$\mathsf{P}(\eta < \infty) = 1, \qquad \mathsf{P}\left(e^{-\eta} > 0\right) = 1.$$

Therefore $\mathsf{E}\,e^{-\eta} > 0$. It is easy to check that

$$\mathsf{E}\,e^{-\eta} = \prod_{n=1}^{\infty} \mathsf{E}\,e^{-\xi_n^2}.$$

After simple computations we get

$$\mathsf{E}\,e^{-\xi_n^2} = \frac{1}{\sqrt{(1 + \mathsf{E}\,\xi_n^2)}}, \qquad n \geq 1.$$

Therefore, the product

$$\prod_{n=1}^{\infty} \frac{1}{\sqrt{(1 + \mathsf{E}\,\xi_n^2)}}$$

converges. The last property holds if and only if the series $\sum_{n=1}^{\infty} \mathsf{E}\,\xi_n^2$ converges.

III.6.18. See Problem III.6.17.

III.6.19. According to Theorem 1 of §III.5, the series $\sum_{n=1}^{\infty} \xi_n$ converges with probability one. Thus the expectation $\mathsf{E}\exp\{i\sum_{n=1}^{\infty} \xi_n\}$ exists. Since random variables ξ_n, $n \geq 1$, are independent,

$$\mathsf{E}\exp\left\{i \sum_{n=1}^{N} \xi_n\right\} = \prod_{n=1}^{N} \mathsf{E}\exp\{i\xi_n\}, \qquad N \geq 1.$$

Using the Lebesgue dominated convergence theorem we obtain

$$\lim_{N\to\infty} \mathsf{E}\exp\left\{i\sum_{n=1}^{N} \xi_n\right\} = \mathsf{E}\exp\left\{i\sum_{n=1}^{\infty} \xi_n\right\}.$$

This and the preceding equality imply the result.

III.6.20. Use the equality

$$\operatorname{Var}\xi = \sum_{k=1}^{n} \operatorname{Var}\xi_k + \operatorname{Var}(\xi - \xi_1 - \xi_2 - \cdots - \xi_n)$$

to prove the convergence of the series $\sum_{n=1}^{\infty} \operatorname{Var}\xi$. Then apply Theorem 1 at the beginning of §III.6.

III.6.21. Use Problem III.6.9 and Theorem 1 at the beginning of §III.6.

III.6.22. This is a corollary of Theorem 1 at the beginning of §III.6.

III.6.25. a) A series converges in probability if and only if the sequence of its partial sums is a Cauchy sequence in probability. Therefore, for every $\varepsilon > 0$ we have

$$\mathsf{P}\left(\left|\sum_{k=m}^{\infty}(a_k\xi_k + b_k)\right| \geq \varepsilon\right) \to 0$$

as $m, n \to \infty$. The random variable

$$\eta_{mn} = \sum_{k=m}^{n}(a_k\xi_k + b_k)$$

is Gaussian and

$$\mathsf{E}\,\eta_{mn} = \mu_{mn} = \sum_{k=m}^{n} b_k, \qquad \operatorname{Var}\eta_{mn} = \sigma_{mn}^2 = \sum_{k=m}^{n} a_k^2.$$

Thus

$$\begin{aligned}
\mathsf{P}\left(\left|\sum_{k=m}^{\infty}(a_k\xi_k + b_k)\right| \geq \varepsilon\right) &= \frac{1}{\sqrt{2\pi}\sigma_{mn}} \int_{|x|\geq\varepsilon} \exp\left\{-\frac{(x-\mu_{mn})^2}{2\sigma_{mn}^2}\right\} dx \\
&= \frac{1}{\sqrt{2\pi}} \int_{x\sigma_{mn}+\mu_{mn}\geq\varepsilon} \exp\left\{-\frac{x^2}{2}\right\} dx \\
&= \frac{1}{\sqrt{2\pi}} \int_{\frac{\varepsilon-\mu_{mn}}{\sigma_{mn}}}^{\infty} \exp\left\{-\frac{x^2}{2}\right\} dx + \frac{1}{\sqrt{2\pi}} \int_{-\infty}^{-\frac{\varepsilon+\mu_{mn}}{\sigma_{mn}}} \exp\left\{-\frac{x^2}{2}\right\} dx.
\end{aligned}$$

In order that this expression tend to 0, it is necessary and sufficient that $\sigma_{mn} \to 0$ and $\mu_{mn} \to 0$ as $m, n \to \infty$.

b) See Problem III.6.24.

III.6.26. Use Problem III.6.25.

III.6.27. According to Problem III.2.14, the random variable $(\eta + 1)/2$ is uniformly distributed on $[0, 1]$.

III.6.29. Use the Borel–Cantelli lemma.

III.6.30. Use the Kolmogorov three series theorem.

III.6.31. Use the Kolmogorov three series theorem. It is sufficient to prove that the sequences $\{\xi_n, n \geq 1\}$ and $\{\xi_n^c, n \geq 1\}$ are Khintchine equivalent, that is,

$$\sum_{n=1}^{\infty} \mathsf{P}(|\xi_n| > c) < \infty.$$

Under the conditions of the problem we have

$$|\mathsf{E}\,\xi_n^c| < \frac{c}{2}, \qquad n \geq N,$$

and

$$\operatorname{Var}\xi_n^c \geq \int_{\{|\xi_n|>c\}} (\xi_n^c - \mathsf{E}\,\xi_n^c)^2 \, d\mathsf{P} \geq \frac{c^2}{4}\mathsf{P}(|\xi_n| > c), \qquad n \geq N.$$

III.6.32. Use the Kolmogorov three series theorem. While proving the necessity, first prove that the sequence $\{a_n, n \geq 1\}$ is bounded. If a subsequence $\{a_{n(k)}, k \geq 1\}$ tends to infinity as $k \to \infty$, then $\xi_{n(k)} \to 0$ with probability one as $k \to \infty$.

III.6.34. Use the inequality

$$\mathsf{P}\left(\max_{1\leq k\leq n} \left| \sum_{i=1}^{k} \xi_i \right| \leq a \right) \leq \frac{(a+c)^2}{\sum_{i=1}^{n} \mathsf{E}\,\xi_i^2}.$$

III.6.35. This is a corollary of the zero-one law and the theorems of §III.6.

III.6.36. ANSWER: $\sum_{n=1}^{\infty} a_n < \infty$.

III.6.37. For positive numbers a_n, $n \geq 1$, a desired necessary and sufficient condition can be formulated as follows: $a_n \to 0$ as $n \to \infty$ and the series

$$\sum_{n=1}^{\infty} a_n \ln \frac{1}{a_n}$$

converges.

III.6.38. This is a corollary of the Kolmogorov three series theorem.

III.6.39. This is a corollary of the Kolmogorov three series theorem.

III.6.40. Consider corresponding partial sums and use the theorems of §III.6.

III.6.41. To prove the statement of the problem use Theorem 3 at the beginning of §III.5; the inequality

$$\mathsf{P}\left(\max_{1\leq k\leq n} |\xi_k| > c \right) \leq \mathsf{P}\left(\max_{1\leq k\leq n} \left| \sum_{j=k}^{n} \xi_j \right| > c \right) \leq \mathsf{P}\left(\max_{1\leq k\leq n} \left| \sum_{j=1}^{k} \xi_j \right| > \frac{c}{2} \right);$$

the convergence with probability one of the series

$$\sum_{n=1}^{\infty} \left(\xi_n - \xi_n^{(c)} \right), \qquad \xi^{(c)} = \begin{cases} \xi, & |\xi| \leq c, \\ 0, & |\xi| > c; \end{cases}$$

the boundedness in probability of the sums $\sum_{k=1}^{n} \xi_k^{(c)}$, $n \geq 1$; and the three series theorem.

III.6.42. Check that the convergence of the series

$$\sum_{n=1}^{\infty} \mathsf{P}(|\xi_n + \eta_n| > c)$$

implies the existence of numbers $\tilde{a}_n$, $n \geq 1$, such that the series

$$\sum_{n=1}^{\infty} \mathsf{P}(|\xi_n + \tilde{a}_n| > c), \qquad \sum_{n=1}^{\infty} \mathsf{P}(|\eta_n - \tilde{a}|_n > c)$$

converge.

III.6.43. Use the symmetrization. Consider a sequence of random variables $\{\xi_n', n \geq 1\}$ such that $\xi_1, \xi_1', \xi_2, \xi_2', \dots, \xi_n, \xi_n', \dots$ are independent and for all $n \geq 1$ the variables ξ_n and ξ_n' have the same distribution. Then $\{\xi_n - \xi_n', n \geq 1\}$ is a sequence of independent symmetric random variables such that the series

$$\sum_{n=1}^{\infty} (\xi_n - \xi_n')$$

converges with probability one. Then use the results of the preceding two problems.

III.6.44. Observe that the sequence of partial sums is bounded in probability. Then use Problem III.6.25 or Problem III.6.41.

III.6.45. The sequence of partial sums is bounded in probability, hence the series $\sum_{n=1}^{\infty} \mathsf{E}\,\xi_n$ converges.

III.6.46. First prove that the sequence of partial sums is bounded in probability. Then use Problem III.6.41.

III.6.47. Use the Kolmogorov three series theorem.

III.6.48. See Problem III.6.19.

III.6.49. Use Theorem 2 at the beginning of §III.5.

§**7.**

III.7.1. This is a corollary of the strong law of large numbers.

III.7.4. Use the equalities

$$S_n = \sum_{k=1}^{n} \frac{\xi_k}{n}, \qquad \xi_m = m(S_m - S_{m-1}), \quad S_0 = 0.$$

III.7.8. By the Borel–Cantelli lemma, it is not true that the sequence $\{\frac{\xi_n}{n}, n \geq 1\}$ converges to zero with probability one (see Problem III.7.6).

III.7.9. See Problem III.7.2.

III.7.10. See Problem III.7.7 and Theorem 1 at the beginning of §III.6.

III.7.11. Take into account that

$$\mathsf{P}\left(\xi_n/n \to 0 \text{ as } n \to \infty\right) = 1$$

and use the Borel–Cantelli lemma to prove that the series

$$\sum_{n=1}^{\infty} \mathsf{P}(|\xi_n| > n)$$

converges.

III.7.12. See Problems III.7.4 and III.6.5.

III.7.13. Prove that the series $\sum_{n=1}^{\infty} \frac{\xi_n}{n}$ converges with probability one (see Problem III.7.9).

III.7.14. Prove that

$$\sum_{n=1}^{\infty} \mathsf{E}\left(\frac{1}{n}\sum_{k=1}^{n} \xi_k\right)^4 < \infty.$$

III.7.15. See Problems III.6.46 and III.7.4.

III.7.16. Use Problem III.2.16 and the strong law of large numbers.

III.7.19. Under the conditions of the problem,

$$S_{2^n}/2^n \to 0, \qquad n \to \infty,$$

with probability one (see Problem III.3.9). Using the inequality of Theorem 3 at the beginning of §III.5, prove that

$$\max_{2^n \le m \le 2^{n+1}} \left| \frac{S_m}{m} - \frac{S_{2^n}}{2^n} \right| \le \frac{1}{2^n} \max_{2^n \le m \le 2^{n+1}} |S_m - S_{2^n}| + \frac{S_{2^n}}{2^n} \to 0$$

with probability one as $n \to \infty$.

III.7.20. Use the following properties of Gaussian distributions:

$$\mathsf{P}(|\xi| \ge A) \sim \frac{2}{A\sqrt{2\pi}} e^{-A^2/2},$$

where ξ is a Gaussian random variable with $\mathsf{E}\,\xi = 0$, $\mathsf{E}\,\xi^2 = 1$, and a sum of independent Gaussian random variables is also Gaussian.

III.7.21. Check that it is not true that the random variable

$$\frac{1}{2^k} \sum_{i=2^k+1}^{2^{k+1}} (\xi_i - \mathsf{E}\,\xi_i)$$

tends to 0 with probability one.

III.7.22. The limit of the sequence under consideration exists and equals zero with probability one if and only if

$$\lim_{N\to\infty} \mathsf{P}\left(\sup_{m \ge N} \sup_{n \ge 1} \frac{1}{m} |\xi_{n+1} + \cdots + \xi_{n+m}| < \varepsilon \right) = 0.$$

III.7.23. The solution of this problem is similar to that of Problem III.7.4, where the following assertion must be taken into account: *If $x_n \to x$ as $n \to \infty$, then*

$$\frac{1}{a_n} \sum_{k=1}^{n} (a_k - a_{k-1}) x_k \to x \qquad \textit{as } n \to \infty.$$

III.7.24. See Problem III.7.23.

III.7.25. Prove that $\mathsf{P}\left(\frac{\xi_n}{a_n} \to 0 \text{ as } n \to \infty\right) = 1$.

III.7.26. This is a corollary of Problem III.7.24.

III.7.27. See Problem III.7.26.

III.7.28. It is necessary to prove that for every $\varepsilon > 0$ there exists c such that the inequality

$$|\xi_1 + \xi_2 + \cdots + \xi_n| > cn^{\frac{1}{2}+\varepsilon}$$

is valid with probability one only for a finite number of indices n. See Problem III.7.27.

III.7.29. See Problems III.2.30 and III.2.31. First we note that

$$\eta_n = \frac{1}{n^{1+\alpha}} \sum_{k=1}^{n} \xi_k \to 0$$

in probability as $n \to \infty$, since

$$\mathsf{P}(|\eta_n| \geq \varepsilon) \leq \frac{1}{\pi\varepsilon n^{\alpha}}$$

(see the solution of Problem III.2.30). By the Borel–Cantelli lemma the subsequence $\{\eta_{2^n}, n \geq 1\}$ converges to zero with probability one as $n \to \infty$. To prove that $\eta_n \to 0$ with probability one as $n \to \infty$, it is necessary to verify that for every $\varepsilon > 0$ only a finite number of events $A_n = \{\omega\colon |\eta_n| \geq \varepsilon\}$, $n \geq 1$, occur with probability one, that is,

$$\mathsf{P}\left(\limsup_{n\to\infty} A_n\right) = \mathsf{P}\left(\bigcap_{n=1}^{\infty} \bigcup_{i=n}^{\infty} A_i\right) = 0.$$

To prove this equality, it is sufficient to show that

$$\lim_{n\to\infty} \mathsf{P}\left(\bigcup_{i=n}^{\infty} A_i\right) = 0.$$

For an integer $n \geq 1$ define another integer $m(n)$ such that

$$2^{m(n)} \leq n < 2^{m(n)+1}.$$

If $\eta_n \to 0$ in probability, then for all $\varepsilon > 0$ there exists N such that

$$\mathsf{P}\left(|S_{2^{m(n)+1}} - S_n| < \frac{\varepsilon}{2} n^{1+\alpha}\right) \geq \frac{1}{2}$$

for $n \geq N$, where

$$S_n = \sum_{k=1}^{n} \xi_k.$$

Indeed,

$$\frac{S_n}{n^{1+\alpha}} \to 0, \qquad \frac{S_{2^{m(n)+1}}}{n^{1+\alpha}} = \frac{2^{(m(n)+1)(1+\alpha)}}{n^{1+\alpha}} \cdot \frac{S_{2^{m(n)+1}}}{2^{(m(n)+1)(1+\alpha)}} \to 0$$

in probability as $n \to \infty$, since

$$\frac{2^{(m(n)+1)(1+\alpha)}}{n^{1+\alpha}} \leq 2^{1+\alpha}.$$

For $n \geq 1$ put

$$B_n = \left\{\omega \colon |S_{2^{m(n)+1}} - S_n| < \frac{\varepsilon}{2} n^{1+\alpha}\right\}, \qquad p_n = \mathsf{P}(B_n), \qquad D_n = A_n \cap B_n,$$
$$C_n = \left\{\omega \colon |S_{2^{m(n)+1}}| \geq \frac{\varepsilon}{2} n^{1+\alpha}\right\}.$$

It is easy to show that $A_n \cap B_n \subset C_n$, $n \geq 1$, and A_k and B_k are independent events for $k \leq n$. Now for $n \geq N$ we get

$$\begin{aligned} \mathsf{P}\Big(\bigcup_{i=n}^{\infty} A_i\Big) &= \mathsf{P}(A_n) + \mathsf{P}\left(A_{n+1} \cap \bar{A}_n\right) + \mathsf{P}\left(A_{n+2} \cap \bar{A}_{n+1} \cap \bar{A}_n\right) + \cdots \\ &\leq 2\left[\mathsf{P}(A_n)p_n + \mathsf{P}\left(A_{n+1}\bar{A}_n\right)p_{n+1} + \mathsf{P}\left(A_{n+2}A_n\bar{A}_{n+1}\right)p_{n+2} + \cdots\right] \\ &= 2\left[\mathsf{P}(A_n \cap B_n) + \mathsf{P}\left(A_{n+1} \cap B_{n+1} \cap \bar{A}_n\right) + \cdots\right] \\ &\leq 2\left[\mathsf{P}(D_n) + \mathsf{P}\left(D_{n+1}\overline{D_n}\right) + \mathsf{P}\left(D_{n+2}\overline{D_n D_{n+1}}\right) + \cdots\right] \\ &= 2\mathsf{P}\Big(\bigcup_{i=n}^{\infty} A_i B_i\Big) \leq 2\mathsf{P}\Big(\bigcup_{i=n}^{\infty} C_i\Big). \end{aligned}$$

Therefore, $\lim_{n\to\infty} \mathsf{P}\left(\bigcup_{i=n}^{\infty} C_i\right) = 0$ by virtue of the convergence with probability one of the sequence $\{\eta_{2^n}, n \geq 1\}$ to zero.

III.7.30. The necessity of the condition is obvious. To prove the sufficiency we assume that

$$\mathsf{E}\left(\eta_{2^n} - \mathsf{E}\,\eta_{2^n}\right)^2 = \frac{1}{2^{2n}} \sum_{k=1}^{2^n} \operatorname{Var} \xi_k \to 0$$

as $n \to \infty$. Then

$$\frac{1}{n^2} \sum_{k=1}^{n} \operatorname{Var} \xi_k \to 0.$$

Indeed, let $m(n)$ be such that

$$2^{m(n)} \leq n < 2^{m(n)+1}.$$

We have

$$\frac{1}{n^2} \sum_{k=1}^{n} \operatorname{Var} \xi_k \leq \frac{1}{2^{m(n)+2}} \sum_{k=1}^{2^{m(n)+1}} \operatorname{Var} \xi_k.$$

Then

$$\mathsf{E}\left(\eta_n - \mathsf{E}\,\eta_n\right)^2 = \frac{1}{n^2} \sum_{k=1}^{n} \operatorname{Var} \xi_k \to 0$$

as $n \to \infty$.

REMARK. It is clear that the statement of the problem remains true if we consider orthogonal, instead of independent, random variables ξ_k, $k \geq 1$:

$$\mathsf{E}(\xi_i - \mathsf{E}\,\xi_i)(\xi_k - \mathsf{E}\,\xi_k) = 0, \qquad i \neq k.$$

III.7.31. Use the property that the correspondence between characteristic functions and distribution functions is continuous.

§**8.**

III.8.3. First obtain the representation

$$\zeta_\nu = \zeta_1 + \sum_{k=1}^{N-1} \xi_k \chi_k^*,$$

where $\xi_k = \zeta_{k+1} - \zeta_k$ and $\chi_k^* = 1$ for $k \le \nu$, and $\chi_k^* = 0$ otherwise.

III.8.4. First obtain the representation

$$\eta_n = \zeta_1 + \sum_{k=1}^{n-1} \xi_k \chi_k^*,$$

where $\xi_k = \zeta_{k+1} - \zeta_k$ and $\chi_k^* = 1$ for $k \le \nu$, and $\chi_k^* = -1$ otherwise.

III.8.5. Use the equality

$$\mathsf{E}\left(\max_{k\le n} \zeta_k^+\right)^\alpha = \alpha \int_0^\infty a^{\alpha-1} \mathsf{P}\left(\max_{k\le n} \zeta_k \ge a\right) da$$

and the Hölder inequality.

III.8.6. Put

$$\eta_n = \sum_{k=1}^{n} \mathsf{E}\left(\zeta_k - \zeta_{k-1}/\zeta_1, \dots, \zeta_{k-1}\right), \qquad n \ge 1.$$

III.8.8. The sequence $\{\eta_n = \mathsf{E}\left(\zeta/F_n\right), n \ge 1\}$ is a martingale.

III.8.9. See Problems II.6.35 and II.6.9. Note that in proving the convergence we may consider only F-measurable random variables, since

$$\mathsf{E}\left(\xi/F_n\right) = \mathsf{E}\left[\mathsf{E}\left(\xi/F\right)/F_n\right]$$

(see Problem II.6.9). It is necessary to prove that

$$\mathsf{P}\left(\lim_{n\to\infty} \mathsf{E}\left(\xi/F_n\right) = \xi\right) = 1$$

for an F-measurable random variable ξ with $\mathsf{E}\,|\xi| < \infty$. Then use the representation

$$\xi = \xi^+ - \xi^-, \qquad \xi^+ = \max(0, \xi), \qquad \xi^- = \min(0, \xi)$$

and Problem III.8.8.

III.8.10. First compute

$$\mathsf{E}\left(\zeta_n/\xi_1, \xi_2, \dots, \xi_{n-1}\right) = \zeta_{n-1}\,\mathsf{E}\,\frac{\xi_n}{m_n} = \zeta_{n-1}$$

(see Problems II.6.10 and II.6.7). Then in view of Problem II.6.9, we have

$$\mathsf{E}\left(\zeta_n/\zeta_1, \zeta_2, \dots, \zeta_{n-1}\right) = \mathsf{E}\left[\mathsf{E}\left(\zeta_n/\xi_1, \xi_2, \dots, \xi_{n-1}\right)/\zeta_1, \zeta_2, \dots, \zeta_{n-1}\right] = \zeta_{n-1}.$$

III.8.11. The solution is similar to that of Problem III.8.10.

Solutions to Chapter IV

§1.

IV.1.1. ANSWER:

$$\text{a) } \frac{p}{1-z(1-p)}; \qquad \text{b) } (pr+1-p)^m;$$

$$\text{c) } e^{\lambda(z-1)}; \qquad \text{d) } \frac{1-z^{N+1}}{(N+1)(1-z)}.$$

IV.1.5. ANSWER:

$$p_n = \begin{cases} 0, & \text{for } n = 2k,\ k \geq 0, \\ \dfrac{1}{2q}\binom{k+1}{\frac{1}{2}} 4pq^{k+1}(-1)^{k+1}, & \text{for } n = 2k+1,\ k \geq 0. \end{cases}$$

IV.1.6. Find the moment generating functions of the random variables $n + \xi + \eta - \zeta$, $2n + \xi + \eta - 2\zeta$, $\xi + \eta + \zeta$ and the coefficients of z^n, z^{2n}, and z^{n+1} in the expressions for these moment generating functions.

ANSWER:

a) $\mathsf{P}(\xi + \eta = \zeta) = (n-1)(2n)^{-2}$;

b) $\mathsf{P}(\xi + \eta = 2\zeta) = (2n)^{-1}$ if n is even and $(n^2+1)(2n^3)^{-1}$ if n is odd;

c) $\mathsf{P}(\xi + \eta + \zeta = n+1) = (n-1)(2n)^{-2}$.

IV.1.7. ANSWER:

$$\varphi_{\nu_N^+}(z) = z^N\left(1-(1-p)^{N+1}\right) + \frac{p(z(1-p))^{N+1}}{1-z(1-p)},$$

$$\varphi_{\nu_N^-}(z) = \frac{p\left(1-((1-p)z)^{N+1}\right)}{1-(1-p)z} + z^N(1-p)^{N+1}.$$

IV.1.11. ANSWER: $e^{\lambda(pz_1z_2+(1-p)z_1-1)}$.

IV.1.12. ANSWER:

$$\frac{p}{1-(1-p)z_1}\left(\frac{u}{1-(1-u)z_2}\left(1-(1-p)z_1^{N+1}\right) + \frac{v}{1-(1-v)z_2}((1-p)z_1)^{N+1}\right).$$

IV.1.17. ANSWER:

$$\frac{1}{(N+1)^2(1-z_2)}\left((1+z_2)\frac{1-z_1^{N+1}}{1-z_1} - \frac{1-(z_1z_2)^{N+1}}{1-z_1z_2} - z_2\frac{z_2^{N+1}-z_1^{N+1}}{z_2-z_1}\right).$$

IV.1.25. b) Represent $P(s)$ in the form

$$P(s) = \frac{A_1}{1-z_1^{-1}s} + \frac{A_2}{1-z_2^{-1}s}$$

and establish the formula

$$P_n = \frac{A_1}{z_1^n} + \frac{A_2}{z_2^n}.$$

IV.1.28. ANSWER:

$$u(z) = \frac{z^2(3u_2 - u_1 - u_0) + z(3u_1 - u_0) + 3u_0}{3 - z - z^2 - z^3},$$

$$u_n \to \frac{3u_2 + 2u_1 + u_0}{6} \quad \text{as } n \to \infty.$$

IV.1.30. Use the property that the number of successes N_n satisfies the relation

$$N_n = \sum_{r=0}^{n-1} N_r N_{n-r-1}.$$

IV.1.31. ANSWER: $(p_1 z_1 + \cdots + p_l z_l)^n$.

IV.1.32. c) Use the following recurrence relations for the probabilities $p(r,k) = \mathsf{P}(\tau_1 = k, \tau_2 = r)$ and $p_i(r) = \mathsf{P}(\tau_i = r)$, $i = 1, 2$:

$$p(r,k) = \begin{cases} (1 - p_1 - p_2)p(r-1, k-1), & \text{for } r > 1,\ k > 1, \\ p_1 p_2 (k-1), & \text{for } r = 1,\ k > 1, \\ p_2 p_1 (r-1), & \text{for } r > 1,\ k = 1, \\ 0, & \text{for } r = 1,\ k = 1; \end{cases}$$

$$\varphi(z_1, z_2) = \frac{p_1 p_2 z_1 z_2}{1 - (1 - p_1 - p_2) z_1 z_2} \left(\frac{z_2}{1 - z_2(1 - p_2)} + \frac{z_1}{1 - z_1(1 - p_1)} \right).$$

IV.1.34. a) Let τ_i be the time when the ith result occurs. Find the joint moment generating function of the random variables μ_{ij} and τ_i,

$$\varphi(z) = \frac{p_i z_2}{1 - z_2(1 - p_i - p_j + z_2 p_j)};$$

b) $\mathsf{E}\, \mu_{li}\mu_{lj} = \dfrac{2p_i p_j}{p_l^2}$.

IV.1.35. Establish the recurrence formula

$$u_n = p_i(1 - u_{n-1}) + (1 - p_i)u_{n-1}, \quad n \geq 1, \qquad u_0 = 1.$$

IV.1.36. The system of recurrence relations for the probabilities u_n and v_n is of the form

$$\begin{cases} u_n = (1 - p_i)u_{n-1} + p_i(1 - u_{n-1} - v_{n-1}), \\ v_n = (1 - p_i)v_{n-1} + p_i u_{n-1}, \\ n \geq 1, \end{cases}$$

where $u_0 = 1$, $v_0 = 0$.

IV.1.37. Denote by v_n the probability that the pattern $a_i a_j$ appears for the first time in $(2n-1)$st and $(2n)$th trials and a_i is the result of the first trial. For the probabilities u_n and v_n we get the system of recurrence relations

$$u_n = (1 - p_i)u_{n-1} + p_i v_{n-1},$$
$$v_n = (1 - p_i - p_j)u_{n-1} + p_i v_{n-1},$$
$$n \geq 1,$$

where $u_0 = v_0 = 0$, $v_1 = p_j$. The moment generating function of the sequences $\{u_n\}$ and $\{v_n\}$ is given by

$$u_n(z) = \frac{p_i p_j z^2}{(1-(1-p_i)z)(1-p_i z) - p_i(1-p_i-p_j)z^2}.$$

IV.1.40. Use the relation

$$\mathsf{P}(\gamma_{in} \le r) = \mathsf{P}(\nu_{ir} > n).$$

IV.1.42. a) Use the independence of the random variable ν and the random vector $(\nu_1(n), \dots, \nu_l(n))$ to prove that

$$\mathsf{E}\, z_1^{\alpha_1} \cdots z_l^{\alpha_l} = \sum_{n=0}^{\infty} \mathsf{E}\, z_1^{\nu_1(n)} \cdots z_l^{\nu_l(n)} \mathsf{P}(\nu = n).$$

IV.1.43. Denote the number of balls in the ith cell by α_i. Find the joint moment generating function of the random variables $\alpha_1, \dots, \alpha_m$.

IV.1.44. Use the scheme described in Problem IV.1.42 for a geometric random variable ν with parameter α and geometric random variables ξ_n with parameter p.

IV.1.46. a) For the probabilities $f_n = \mathsf{P}(\nu = n)$, $n \ge 1$, obtain the recurrence relation

$$f_{n+2} = (1-p')(1-p'')f_n, \qquad n \ge 1,$$

where $f_0 = 0$, $f_1 = p''$.

b) Use the scheme described in Problem IV.1.42 for a geometric random variable ν with a parameter p'.

IV.1.47. Use the recurrence relations

$$u_{2k} = (1-p')u_{2k-1} + p'(1-u_{2k-1}), \qquad k \ge 1,$$
$$u_{2k+1} = (1-p'')u_{2k} - p''(1-u_{2k}), \qquad k \ge 0,$$

where $u_0 = 1$, $u_1 = 1 - p''$.

IV.1.48. Let v_n be the probability that the pattern SF appears for the first time in $(2n-1)$st and $(2n)$th trials and the result of the first trial was a "success". For the probabilities u_n and v_n obtain the system of recurrence relations

$$\begin{cases} u_n = (1-p')(1-p'')u_{n-1} + p'v_{n-1}, \\ v_n = p''p'v_{n-1}, \end{cases}$$

for $n \ge 1$, where $u_1 = v_1 = p''(1-p')$.

§**2.**

IV.2.2. First prove that

$$L(s,z) = \sum_{n=0}^{\infty} z^n \psi_n(s),$$

where

$$\psi_n(s) = \int_0^{\infty} e^{-st} \mathsf{P}(\nu(t) = n)\, dt.$$

Then, using the relation given in Problem IV.2.1, establish that

$$\psi_n(s) = \frac{1}{s}\psi^n(s)(1-\psi(s)).$$

IV.2.3. ANSWER:

$$\mathsf{P}(\nu(t)=n) = \frac{(\lambda t)^n}{n!}e^{-\lambda t}, \qquad n \ge 0,$$
$$N(t) = \lambda t, \qquad t \ge 0.$$

IV.2.4. ANSWER:

$$\mathsf{P}(\nu(t)\ge n) = \mathsf{P}(\xi_1+\dots+\xi_n \le t) = u_n(t) = \frac{1}{T^n n!}\sum_{k=0}^{n}(-1)^k\binom{n}{k}((t-ka)_+)^n,$$

where

$$(x-c)_+ = \begin{cases} x-c, & \text{for } x \ge c,\\ 0, & \text{for } x < c.\end{cases}$$

IV.2.5. Use the property that the Erlang distribution of order r with parameter λ coincides with the distribution of the sum of r independent exponential random variables with parameter λ. ANSWER:

$$\mathsf{P}(\nu(t)=n) = \sum_{k=nr}^{nr+r-1}\frac{(\lambda t)^k}{k!}e^{-\lambda t}, \qquad n \ge 0.$$

IV.2.6. Use the relation given in Problem IV.2.1.

IV.2.9. Use the relation given in Problem IV.2.1 to obtain

$$\mathsf{P}\left(\frac{\nu(t)-\frac{t}{\mu}}{b\mu^{-3/2}\sqrt{t}} > x\right) = \mathsf{P}\left(\sum_{k=1}^{m+1}\xi_k - \mu \le t - \mu(m+1)\right),$$

where $m = \left[b\mu^{-3/2}x\sqrt{t}+\frac{t}{\mu}\right]$.

IV.2.12. First prove that

$$\mathsf{E}\,e^{-s\tau} = \int_0^\infty e^{-st}\,d\mathsf{P}(\tau<t) = s\int_0^\infty \mathsf{P}(\tau\le t)e^{-st}\,dt$$

for any nonnegative random variable τ.

IV.2.25. ANSWER:

$$\mathsf{E}\,z^{\nu(t)} = \frac{1}{z}\left(\frac{zp}{1-z(1-p)}\right)^{[t/\tau]+1},$$
$$\mathsf{E}\,\nu(t) = \frac{1}{1-p}\left(\left[\frac{t}{\tau}\right]+1\right)-1.$$

IV.2.27. Let a renewal scheme be constructed with respect to random variables $\{\xi_n, n \geq 1\}$. Introduce auxiliary random variables $\{\xi_n^\tau, n \geq 1\}$ as follows:

$$\xi_k^\tau = \begin{cases} 0 & \text{if } \xi_k < \tau, \\ \tau & \text{if } \xi_k \geq \tau, \end{cases}$$

where τ is such that $\mathsf{P}(\xi_1 < \tau) = F(\tau) = 1$. Use Problem IV.2.25 to show that $\mathsf{E}\,\nu_\tau^2(t) < \infty$, where $\nu_\tau(t)$ is the number of renewals over time t in a renewal scheme constructed by the random variables ξ_k^τ. Then use Problem IV.2.26.

IV.2.28. Show that $t^{-1}\nu(t) \leq \tau^{-1}$ and use Problem IV.26.

IV.2.29. Let $\nu_\tau(t)$ be defined as in the solution to Problem IV.2.27. Establish corresponding estimates for $\mathsf{E}(\nu_\tau(t)/t)^2$. Then use Problem IV.2.26.

IV.2.30. Use an appropriate case of the Lebesgue dominated convergence theorem.

IV.2.31. The function $g(s) = \mathsf{E}\, e^{is\xi_1}$ is integrable and

$$N(t) = \sum_{n=1}^{\infty} \int_0^t \frac{1}{2\pi} \int_{-\infty}^{\infty} g^n(s) e^{-isu}\, ds\, du = \frac{1}{2\pi} \sum_{n=1}^{\infty} \int_{-\infty}^{\infty} g^n(s) \frac{1 - e^{-ist}}{is}\, ds.$$

IV.2.32. See the preceding problem.

IV.2.33. The set T is a subset of $\widehat{T} = \bigl\{t \colon \sum_{k=1}^{\infty} \mathsf{P}(\xi_1 + \cdots + \xi_k = t) > 0\bigr\}$.

IV.2.34. Use Problem IV.2.12.

IV.2.35. To prove the uniqueness, show that the difference of two arbitrary solutions satisfies the equation

$$v(x) = \int_0^{x+0} v(x-y)\, F^{*n}(dy), \qquad x \geq 0,$$

for all $n \geq 1$ and use $F^{*n}(x) \to 0$ as $x \to +\infty$. Here $F^{*n}(y)$ is the n-fold convolution of the distribution function $F(x)$.

IV.2.40. c) Use the equality

$$k\left(q_k - \frac{1}{\mu}\right) = \frac{1}{\pi i} \int_{-\pi}^{\pi} \left(\frac{1}{1 - \psi(s)} + \frac{1}{ias}\right) d\cos ks.$$

IV.2.41. a) Prove that the function $N(t) = \int_0^t u(s)\, ds$ satisfies the renewal equation with $q(x) = F(x)$, $x \geq 0$, and use Problems IV.2.35 and IV.2.36.

c) Let $R = \sup_t f(t) < \infty$. First show that for the density

$$f_2(t) = \int_0^t f(t-y) f(y)\, dy$$

the following estimate is valid:

$$f_2(t) \leq 2R\left(1 - F\left(\frac{t}{2}\right)\right).$$

Then use the property that the difference $u(t) - f(t)$ satisfies the renewal equation with $q(x) = f_2(x)$, $x \geq 0$.

IV.2.44. ANSWER: $\mathsf{P}(\gamma_t^+ > u) = e^{-\lambda u}$.

IV.2.45. ANSWER: $\mathsf{P}(\gamma_n^+ \geq k) = (1-p)^{k-1}$, $k \geq 1$.

IV.2.46. ANSWER:

$$\mathsf{P}(\gamma_r^- = k, \gamma_r^+ = l) = \sum_{n=0}^{\infty} e^{-\lambda n} \frac{(\lambda n)^{r-k} e^{-\lambda} \lambda^{k+l}}{(r-k)!\,(k+l)!}, \qquad 0 \leq k \leq r < l + r.$$

Here $(\lambda n)^{r-k} = 1$ for $n = 0$, $r = k$.

IV.2.47. ANSWER: $\mathsf{E}\,\gamma_t = (2 - e^{-\lambda t})/\lambda$.

IV.2.49. ANSWER:

$$\mathsf{P}(\gamma_n^+ = k) = \sum_{l=r-m-k}^{r} \sum_{n=1}^{\infty} \binom{mn}{l} p^l (1 - p^{mn-l}) \binom{m}{k+r-l} p^{k+r-l} (1-p)^{m-k-r+l}.$$

§3.

IV.3.1. First prove that $\nu(t)$, $t \geq 0$, is a homogeneous process with independent increments and

$$\mathsf{P}(\nu(t) = n) = e^{-\lambda t} \frac{(\lambda t)^n}{n!}, \qquad n \geq 0.$$

To solve the problem use the independence of random variables $\nu(t), \xi_1, \xi_2, \dots$.

IV.3.8. First prove that under the conditions of the problem the characteristic function of the random variable ξ_1 is given by

$$\mathsf{E}\, e^{iz\xi_1} = 1 + iaz - \frac{1}{2} z^2 \delta^2 + o(z^2).$$

Then show that

$$\mathsf{E} \exp\left\{ \frac{iz(\xi(t) - \lambda a t)}{\sqrt{\delta^2 \lambda t}} \right\} \to e^{-z^2/2} \quad \text{as } t \to \infty.$$

IV.3.13. The stochastic process $\xi(s + \tau_1) - \xi_1$, $s \geq 0$, is independent of the random variables ξ_1 and τ and has the same distribution as $\xi(s)$, $s \geq 0$. To derive the desired equation use the representation $\eta_{t,x} = tg(x)$ if $\tau_1 > t$, and

$$\begin{aligned} \eta_{t,x} &= \tau_1 g(x) + \int_{\tau_1}^{t} g(x + \xi_1 + (\xi(s) - \xi_1))\, ds \\ &= \tau_1 g(x) + \int_{0}^{t-\tau_1} g(x + \xi_1 + (\xi(s+\tau_1) - \xi_1))\, ds \end{aligned}$$

if $\tau_1 \leq t$.

IV.3.14. ANSWER: b) $\mathsf{E}\, e^{i(z_1\xi_1(t) + z_2\xi_2(t))} = e^{-\lambda t(1 - \varphi_1(z_1)\varphi_2(z_2))}$, where

$$\varphi_j(z) = \mathsf{E}\, e^{iz\xi_j(t)}, \qquad j = 1, 2.$$

IV.3.16. ANSWER:

$$\mathsf{E}\, e^{iz\xi(t)} = \exp\left\{ -(\lambda_1+\lambda_2)\int_0^\infty \left(1-e^{itx}\right)\, dG(x)\right\},$$

where

$$G(x) = \frac{\lambda_1}{\lambda_1+\lambda_2} F_1\left(\frac{x}{a_1}\right) + \frac{\lambda_2}{\lambda_1+\lambda_2} F_2\left(\frac{x}{a_2}\right).$$

IV.3.17. ANSWER: a) $\mathsf{E}\, e^{iz\xi(t,A_j)} = \exp\{-\lambda t \mathsf{P}(\xi_1 \in A_j)(1-e^{iz})\}$.

IV.3.18. ANSWER:

$$\text{a) } \mathsf{P}(\xi(t)=m) = e^{-\lambda t}\sum_{k=0}^{\infty} \frac{(\lambda t)^{2k+m}}{(2k+m)!}\binom{2k+m}{k} q^k p^{k+m}.$$

b) We have $\tau_1(x) = \sum_{k=1}^{\widetilde{\tau}(x)} \tau_k$, where the random variables $\widetilde{\tau}(x)$, τ_k, $k\ge 1$, are jointly independent, $\widetilde{\tau}(x)$ has the same distribution as the first passage time of the point x for a Bernoulli walk (see Problem IV.4.46), and τ_k, $k \ge 1$, are exponential random variables with a parameter λ.

IV.3.19. Use the relation

$$\sup_{t\ge 0} \zeta(t) = \gamma\tau_1 + \sup_{n\ge 1}\sum_{k=1}^{n} (\gamma\tau_{k+1} + \xi_k).$$

IV.3.21. a) Let τ_n, $n\ge 1$, and ξ_n, $n \ge 1$, be sequences of random variables that are used to construct the Poisson process $\xi(t)$, $t\ge 0$. Put

$$\begin{aligned}
A_{m,n} &= \{\omega\colon \Gamma_{1,n+1} \le x, \Gamma_{1,n-1} + \gamma\tau_n > x\},\\
B_{m,n} &= \left\{\omega\colon S_{n-1} < x+u-\frac{1}{\gamma}(x-\Gamma_{1,n-1})\right\},\\
C_{m,n} &= \{\omega\colon \Gamma_{n+1,n+m-1} \le y; \Gamma_{n+1,n+m} + \gamma\tau_{n+m} > y\},\\
D_{m,n} &= \left\{\omega\colon S_{n+m-1} - S_n < y+v-\frac{1}{\gamma}(y-\Gamma_{n+1,n+m-1})\right\},\\
\Gamma_{m,n} &= \sum_{k=m}^{n} (\gamma\tau_k - \xi_k),\\
S_n &= \tau_1 + \cdots + \tau_n.
\end{aligned}$$

Then use the following relation:

$$\mathsf{P}\big(Q(x) > u+x, Q(x+y) - Q(x) \le y+v\big) = \sum_{m,n=1}^{\infty} \mathsf{P}(A_{m,n}\cap B_{m,n}\cap C_{m,n}\cap D_{m,n}).$$

b) Use the estimate

$$\mathsf{P}(\Theta(x) \le \varepsilon) \ge \mathsf{P}(\tau_1 > x), \qquad x \le \varepsilon.$$

c) Using problems a) and b) show that $\varphi(s,x) = \mathsf{E}\, e^{-s\Theta(x)}$, as a function of x, is continuous for all s and satisfies the functional equation

$$\varphi(s,x+y) = \varphi(s,x)\varphi(s,y).$$

d) The process $\xi'(t) = \xi(t+\tau_1) - \xi_1$, $t \geq 0$, does not depend on the random variables τ_1 and ξ_1 and is a compound Poisson process with the same characteristic function as $\xi(t)$, $t \geq 0$.

e) Taking the Laplace transform in equation d), show that

$$\mathsf{E}\, e^{-s\Theta(x)} = \exp\left\{-\frac{x}{\gamma}(s + \Phi(\omega(s))\right\}.$$

f) Use the property that for any nonnegative random variable Θ,

$$\lim_{t\to\infty} \mathsf{P}(\Theta \leq t) = \lim_{s\to 0} \mathsf{E}\, e^{-s\Theta}.$$

IV.3.22. e) $\frac{1}{n}\sum_{k=1}^{n} \xi_k \to a$ and $\nu(t) \to +\infty$ with probability one as $n \to \infty$ and $t \to \infty$, respectively. Use this property.

IV.3.24. a) Put $U(t,x) = \mathsf{P}(\xi(t) < x)$. For $U(t,x)$ derive the following integral equation:

$$U(t,x) = \chi_{(0,\infty)}(x)\mathsf{P}(\tau_1 > t) + \int_0^{t+0}\int_{-\infty}^{\infty} U(t-s, x-y)\, d\mathsf{P}(\tau_1 < s, \xi_1 < y).$$

IV.3.25. a) With probability one,

$$\frac{1}{n}\sum_{k=1}^{n} \xi_k \to c \quad \text{and} \quad \frac{\nu(t)}{t} \to \frac{1}{a}$$

as $n \to \infty$ and $t \to \infty$, respectively (see Problem IV.3.22 e)). Use this property.

b) Use the estimate

$$\mathsf{P}\left(\left|\sum_{k=1}^{\nu(t)} \xi_k - \sum_{k=1}^{[t/a]} \xi_k\right| > \Delta\sqrt{t}\right) \leq \mathsf{P}(|\nu(t) - [t/a]| > \delta t) + 2\mathsf{P}\left(\max_{0\leq r\leq \delta t}\left|\sum_{k=1}^{r} \xi_k\right| > \Delta\sqrt{t}\right)$$

and the Kolmogorov inequality.

c) Use part b) to show that the random variables

$$\frac{\xi(t)}{\sqrt{\delta^2 t/c}} \quad \text{and} \quad \sum_{k=1}^{[t/a]} \frac{\xi_k}{\sqrt{\delta^2 t/c}}$$

have the same limit distribution.

d) Use the representation

$$\xi(t) - \frac{c}{a}t = \sum_{k=1}^{\nu(t)}\left(\xi_k - \frac{c}{a}\tau_k\right) - \frac{c}{a}\varkappa(t),$$

where $\varkappa(t) = t - \sum_{k=1}^{\nu(t)} \tau_k$, to show that

$$\frac{\varkappa(t)}{\sqrt{t}} \xrightarrow{\mathsf{P}} 0$$

as $t \to \infty$ (see part b)).

§4.

IV.4.1. Compute the total number of different trajectories of a random walk from a to y.

IV.4.2. Show that there exists a one-to-one correspondence between the trajectories of a random walk of length n that are described, respectively, on the left- and right-hand sides of the desired equality.

IV.4.10. Show that for the random variable τ_1 the following representation is valid:

(a) $$\tau_1 \cong 1 + \begin{cases} 0, & \text{with probability } p, \\ \tau_2, & \text{with probability } q, \end{cases}$$

where $\xi \cong \eta$ means that the random variables ξ and η are identically distributed.

Then apply the first statement of the problem and derive from (a) a quadratic equation to find $f(s) = \mathsf{E}\, s^{\tau_1}$. To choose a root use that $f(s) \to 0$ as $s \to 0$.

IV.4.14. Prove that the probability $P_x(0,z)$, $x = 1, 2, \ldots, z-1$, satisfies the system of linear equations

$$P_x(0,z) = qP_{x-1}(0,z) + pP_{x+1}(0,z), \qquad x = 1, 2, \ldots, z-1,$$
$$P_0(0,z) = 1, \qquad P_z(0,z) = 0.$$

Note that this system has a unique solution.

IV.4.16. ANSWER: iii) $\mathsf{E}\, \mu_z = \frac{1}{q}(p/q)^{z-1}$,

$$\mathsf{E}\, \mu_y \mu_z = \left.\frac{\partial^2 \varphi_y(u\varphi_{z-y}(v))}{\partial u \partial v}\right|_{u,v=1} = \varphi_y'(1)\varphi_{z-y}'(1) + \varphi_{z-y}''(1)\varphi_y'(1)$$
$$= \frac{1}{q^2}\left(\frac{p}{q}\right)^{z+y-2}\left[2\left(1 - \frac{q}{p}y\right) - \frac{q}{p}y(p-q)\right]\frac{1}{p-q}.$$

IV.4.17. a) For the probabilities $P_x^{(n)}(0,z)$, derive the recurrence relations

$$P_x^{(n)}(0,z) = qP_{x-1}^{(n-1)}(0,z) + pP_{x+1}^{(n-1)}(0,z), \qquad k = 1, \ldots, z-1, \quad n \geq 1,$$
$$P_0^{(n)}(0,z) = \begin{cases} 1, & \text{for } n = 0, \\ 0, & \text{for } n \geq 1, \end{cases}$$
$$P_z^{(n)}(0,z) = 0 \quad \text{for } n \geq 0.$$

b) Taking into account the relation $\lambda_1(s)\lambda_2(s) = q/p$ we have

$$\varphi_x(s,z) = \left(\frac{q}{p}\right)^z \frac{\lambda_2^{x-z}(s) - \lambda_1^{x-z}(s)}{\lambda_1^z(s) - \lambda_2^z(s)},$$

where

$$\lambda_1(s) = \frac{1 + \sqrt{1 - 4pqs^2}}{2ps}, \qquad \lambda_2(s) = \frac{1 - \sqrt{1 - 4pqs^2}}{2ps}.$$

IV.4.28. To solve Problems IV.4.27 and IV.4.28, show that the random variables γ_k^+, $k \geq 0$, ν^+, and η^+ satisfy $\eta^+ = \gamma_{\nu^+}^+$.

IV.4.31. Let A_n be the event that a moment n is the weak ladder epoch for a random walk. Prove that

$$\mathsf{E}\, \bar{\nu}^+ = \sum_{n=1}^{\infty} \mathsf{P}(A_n)$$

and use Problems IV.4.25 and IV.4.30.

IV.4.32. Prove that a random walk is bounded from above under the condition of the problem (use Problems IV.4.27, IV.4.28, IV.4.31).

IV.4.43. In this case the distribution of the ladder height γ^+ is exponential, $\mathsf{P}(\gamma^+ < x) = 1 - e^{-\alpha x}$, and $\psi([0,x)]$ is the expectation of renewals over time x in a renewal scheme with an exponential distribution of the renewal time (see Problem IV.2.3).

IV.4.49. Let ${}_y f_{zx}^{(n)} = \mathsf{P}\left(\tau_x = n < \tau_y / S_0 = z\right)$. First prove that ${}_x f_{00}^{(n)} = {}_0 f_{xx}^{(n)}$.

IV.4.51. Use Problems IV.4.49 and IV.4.50.

IV.4.53. Introduce the moment generating functions for the sequences $\mathsf{P}(\tau = n)$ and $\mathsf{P}(S_n = 0)$ and use Problem IV.4.52 to get a relation between these functions.

IV.4.55. Use the property

$$\mathsf{P}(\xi = k) = \frac{1}{2\pi} \int_{-\pi}^{\pi} e^{-izk} \varphi(z)\, dz,$$

which is valid for any integer-valued random variable ξ with the characteristic function $\varphi(z)$.

IV.4.62. Use Problems IV.4.61 and IV.4.59. ANSWER: $0 < \alpha \le 1$.

IV.4.63. Prove first that the probabilities u_r satisfy the following recurrence relation:

$$u_r = p_r + q(u_{r+1} + u_1 u_r), \qquad r \ge 1.$$

IV.4.64. b) $\mathsf{P}(\eta^+ \ge n) = (1 - \mathsf{P}(\tau_1 = +\infty))^n$, $n \ge 0$.

d) To find $\varphi(s)$ derive a quadratic equation and use that $\varphi(s) \to 0$ as $s \to 0$ to choose a desired root.

§5.

IV.5.1. Use the relation $p_{ij}^{(n)} = \sum_{k \in H} p_{ik} p_{kj}^{(n-1)}$, $n \ge 1$.

IV.5.4. ANSWER:

$$p_{11}(z) = \frac{1 - z + z\beta}{(1-z)^2 + z(1-z)(\alpha+\beta)},$$

$$p_{11}^{(n)} = \frac{\beta}{2-\alpha-\beta} - \left(\frac{\beta}{2-\alpha-\beta} - 1\right)(1-\alpha-\beta)^n,$$

$$p_{21}(z) = \frac{z\beta}{(1-z)^2 + z(1-z)(\alpha+\beta)},$$

$$p_{21}^{(n)} = \frac{\beta}{2-\alpha-\beta} - \frac{\beta}{2-\alpha-\beta}(1-\alpha-\beta)^n.$$

IV.5.5. ANSWER:

$$\mathsf{P}\left(\eta_0 = i / \eta_n = j\right) = \frac{p_{ij}^{(n)} p_i}{\sum_{k \in H} p_{kj}^{(n)} p_k}.$$

IV.5.6. ANSWER:

$$\mathsf{P}\left(\eta_r = l/\eta_0 = i, \eta_n = j\right) = \frac{p_{lj}^{(n-r)} p_{il}^{(r)}}{p_{ij}^{(n)}}.$$

IV.5.9. The random variables ξ_m and ξ_n are independent for $m < n-1$, while for $m = n-1$ we have

$$p_{-1,k}^{(n-1)} = \begin{cases} 0, & k = 1, \\ \frac{1}{2}, & k = -1, 0; \end{cases} \quad p_{0,k}^{(n-1)} = \begin{cases} \frac{1}{2}, & k = 0, \\ \frac{1}{4}, & k = -1, 1; \end{cases} \quad p_{+1,k}^{(n-1)} = \begin{cases} 0, & k = -1, \\ 1, & k = 0, 1. \end{cases}$$

IV.5.10. The probability $\mathsf{P}(\eta_n^- = -1)$ equals $2p(1-p)$ for all $n \geq 0$. The sequence η_n, $n \geq 0$, is a Markov chain if and only if either $p = 0$, or $p = \frac{1}{2}$, or $p = 1$.

IV.5.12. Prove first that

$$\mathsf{P}\left(\eta_{n+k} = j_k, 1 \leq k \leq r/\eta_n = i, \eta_k = i_k, 1 \leq k \leq n-1\right) = P_1 P_2,$$

where

$$P_1 = \mathsf{P}\left(\eta_{n+r} = j_r/\eta_{n+r-1} = j_{r-1}, \ldots, \eta_{n+1} = j_1, \eta_n = i, \eta_k = i_k, 1 \leq k \leq n-1\right),$$
$$P_2 = \mathsf{P}\left(\eta_{n+k} = j_k, 1 \leq k \leq r-1/\eta_n = i, \eta_k = i_k, 1 \leq k \leq n-1\right).$$

IV.5.15. ANSWER: $\alpha = \beta$.

IV.5.17. First prove that

$$\mathsf{P}(\eta_{\tau+k} = j_k, 1 \leq k \leq r, \eta_\tau = i, \tau = n, \eta_k = i_k, 1 \leq k \leq \tau - 1)$$
$$= p_{ii_1} \cdots p_{i_{r-1}i_r} \mathsf{P}(\eta_\tau = i, \tau = n, \eta_k = i_k, 1 \leq k \leq \tau - 1).$$

Then, summing up these relations over n, obtain an equality that is equivalent to the desired one.

IV.5.24. If a state i is not essential, then there exists a chain of states $i = i_0, i_1, \ldots, i_n = j$ such that $p_{i_0 i_1} \cdots p_{i_{n-1} i_n} > 0$ and $p_{ji}^{(n)} = 0$ for all $n \geq 1$. Show that

$$1 - f_{ii} \geq p_{i_0 i_1} \cdots p_{i_{n-1} i_n}.$$

IV.5.25. Use relations given in Problems IV.5.22 and IV.5.23 to show that the series $\sum_{n=0}^{\infty} p_{ij}^{(n)}$ is convergent.

IV.5.32. First prove that $p_{ii}^{(rk)} > 0$ for all $r \geq 1$ if $p_{ii}^{(k)} > 0$. Then show that $p_{jj}^{(n+kr+m)} > 0$ for all $i \geq 1$ if $p_{ij}^{(n)} > 0$, $p_{ji}^{(m)} > 0$, and $p_{ii}^{(k)} > 0$. Derive from this that the period d_j of a state j is a divisor of the number k.

IV.5.33. Let $\eta_0 = i \in C_r$ and η_k, $k \geq 1$, be the numbers such that $p_{ii}^{(nk)} > 0$. By the definition of the period these numbers are of the form $\eta_k = dl_k$, $k \geq 1$. Then the set of numbers $\{l_k, k \geq 1\}$ coincides with the set of numbers l such that $p_{ii}^{(l)} > 0$. Check this property and show that the greatest common divisor of numbers l_k, $k \geq 1$, equals 1.

IV.5.38. If $i \to j$, then there exists a chain of states $i = i_0, i_1, \ldots, i_n = j$ such that $i \neq i_1, \ldots, i_{n-1}$ and $p_{i_0 i_1} \cdots p_{i_{n-1} i_n} > 0$. Prove that ${}_k f_{ij} > 0$ if $k \neq i_1, \ldots, i_{n-1}$, and ${}_i f_{kj} > 0$ otherwise.

IV.5.39. Prove that the probabilities ${}_1f_{iN}$, $1 \le i \le N$, satisfy the system of linear equations

$$\begin{cases} {}_1f_{iN} = p_1 f_{i+1,N} + q_1 f_{i-1,N}, & 1 < i < N, \\ {}_1f_{1N} = 0, \qquad {}_1f_{NN} = 1. \end{cases}$$

Note that we may search for a solution of this system of the form

$${}_1f_{iN} = c_1 \left(\frac{q}{p}\right)^{N-i} + c_2.$$

IV.5.40. The random variables $\tau_i^{(N)}$ are Markov times. First show that

$$\begin{aligned} &\mathsf{P}\left(\tau_i^{(N)} = \tau_i^{(N-1)} + n < \tau_j / \tau_i^{(N-1)} < \tau_j\right) \\ &\quad = \mathsf{P}\left(\eta_k \notin \{i, j, k, n-1\}, \eta_n = i / \eta_0 = i\right), \qquad N > 1. \end{aligned}$$

Then derive from this equality the following recurrence formula:

$$\mathsf{P}\left(\tau_i^{(N)} < \tau_j / \eta_0 = i\right) = {}_j f_{ii} \mathsf{P}\left(\tau_i^{(N-1)} / \eta_0 = i\right), \qquad N > 1.$$

IV.5.41. a) Check that $\mathsf{P}(\mu_{ik} > n) = \mathsf{P}(\tau_k^{(n)} < \tau_j)$.

IV.5.42. Use the definition of the random variable τ_j to prove that

$$\tau_j = \sum_{k \ne j} \mu_{jk}.$$

IV.5.43. Use the equality

$$\begin{aligned} \mathsf{P}\left(\tau_j = +\infty / \eta_0 = i\right) = \mathsf{P}&\left(\tau_i^{(N)} = \tau_j = +\infty / \eta_0 = i\right) \\ &+ \mathsf{P}\left(\tau_i^{(N)} < \tau_j = +\infty / \eta_0 = i\right) \end{aligned}$$

to get the estimate

$$\mathsf{P}\left(\tau_j = +\infty / \eta_0 = i\right) \le \mathsf{P}\left(\tau_i^{(N)} < \tau_j / \eta_0 = i\right) = {}_j f_{ii}^N.$$

IV.5.45. If a state is recurrent, then so is any other state j. To prove this statement use the estimate

$$f_{jj} \ge (1 - {}_i f_j)(1 - {}_j f_i).$$

IV.5.52. a) and b). In both cases, to solve the problem it is necessary to solve the system of recurrence equations

$$v_n = a_n + p_n v_{n+1} + q_n v_{n-1} \tag{1}$$

with the corresponding boundary conditions and free terms a_n. Putting $u_n = v_n - v_{n-1}$ we obtain from (1) that

$$q_n u_n = a_n + p_n u_{n+1}.$$

Then using the corresponding boundary conditions find an explicit expression for u_n.

IV.5.54. The random variables $\tau_i^{(n)}$, $n \geq 0$, are Markov times. Use the relations

$$\tau_i^{(n)} = \sum_{k=0}^{n-1} \varkappa_i^{(k)},$$

$$\mathsf{P}\left(\varkappa_i^{(n+1)} \geq r / \varkappa_i^{(n)} = r_n, \ldots, \varkappa_i^{(0)} = r_0\right)$$
$$= \mathsf{P}\left(\eta_{\tau_i+k} \neq i, 1 \leq k \leq r / \tau_i^{(n+1)} = r_0 + \cdots + r_n, \tau_i^{(n)} = r_1\right).$$

IV.5.56. First prove that

$$\{\omega\colon \nu_i(n) \geq r\} = \{\omega\colon \tau_i^{(r)} \leq n\}.$$

IV.5.57. The random variables $\tau_i^{(n)}$, $n \geq 1$, are Markov times. Use the relations

$$\mathsf{P}\left(\varkappa_i^{(n+1)} > r, \mu_{ij}^{(n+1)} = l / \varkappa_i^{(k)} = r_k, \mu_{ij}^{(k)} = l_k, 0 \leq k \leq n\right)$$
$$= \mathsf{P}\left(\eta_{\tau_i+m} \neq i, 1 \leq m \leq r, \sum_{m=1}^{r} \delta_j^{\eta_{\tau_i+m}} = l / A\right),$$

where

$$A = \left\{\omega\colon \tau_i^{(n+1)} = r_0 + \cdots + r_n, \varkappa_i^{(k)} = r_k, \mu_{ij}^{(k)} = l_k, 0 \leq k \leq n\right\}.$$

Show that the random vectors $(\varkappa_i^{(n)}, \mu_{ij}^{(n)})$, $n \geq 0$, are independent and identically distributed for $n \geq 1$.

IV.5.58. According to Problem IV.5.41 b), $\mathsf{E}\,\mu_{ij} < \infty$ for a recurrent Markov chain. Then use Problems IV.5.56 and IV.5.57.

IV.5.59. Use the estimate

$$\sum_{k=1}^{\nu_i(n)} \mu_{ij}^{(k-1)} \leq \nu_i(n) \leq \sum_{k=1}^{\nu_i(n)+1} \mu_{ij}^{(k-1)}$$

to show that with probability one

$$\frac{\nu_j(n)}{n} \to \frac{\mathsf{E}\,\mu_{ij}}{m_{ii}}, \qquad n \to \infty.$$

IV.5.60. a) Use Problem IV.5.59.

IV.5.66. ANSWER:

$$q_j = \frac{1 - pq^{-1}}{1 - (pq^{-1})^N} (pq^{-1})^{j-1}, \qquad 1 \leq j \leq N.$$

IV.5.68. Let $q_i = \lim_{n\to\infty} p_{ij}^{(n)}$, $j \geq 0$, be the stationary probabilities. Write an equation for q_j and obtain for the moment generating functions

$$Q(z) = \sum_{j=0}^{\infty} q_j z^j \quad \text{and} \quad Q_k(z) = \sum_{j=0}^{k-1} q_j z^j$$

the relation

$$Q(z) = (pz+q)^m Q_k(z) \frac{z^k - 1}{z^k + (pz+q)^m}.$$

IV.5.73. Let $Q_j = p_1 \cdots p_j$. Then the recurrence and ergodicity conditions are written as follows:

$$\text{Recurrence:} \qquad Q_j \to 0 \quad \text{as } j \to \infty,$$

$$\text{Ergodicity:} \qquad \sum_{j=1}^{\infty} Q_j < \infty.$$

The stationary probabilities q_j are given by

$$q_j = Q_j \left(\sum_{i=1}^{\infty} Q_i \right)^{-1}.$$

IV.5.74. Let

$$\varphi(z) = \frac{p_0}{z} + \sum_{k=1}^{\infty} p_k z^{k-1}.$$

Then the recurrence and ergodicity conditions are written as follows:

$$\text{Recurrence:} \qquad \varphi'(1) \leq 0,$$
$$\text{Ergodicity:} \qquad \varphi'(1) < 0.$$

The moment generating function of the stationary distribution is given by

$$Q(z) = \frac{(1+\varphi'(1))\varphi(z)(1-z)}{\varphi(z) - z}.$$

IV.5.75. a) First prove that the probabilities ${}_{i-k}f_i = q_k$ do not depend on $i \geq k$. Then obtain the following recurrence relations:

$$1 - q_k = (1-q_1)(1-q_{k-1}), \qquad k > 1,$$

and use it to get the equation

$$1 - q_1 = q + p(1-q_1)^2 + r(1-q_1).$$

Solutions to Chapter V

§1.

V.1.14. Represent the integrand in the form of a series where the coefficient of p_k equals z^{-1}.

V.1.15. Integrate the equality

$$e^{-izt}\varphi(z) = \int_{-\infty}^{\infty} e^{iz(x-t)}\,dF(x)$$

with respect to the distribution $G(z)$.

V.1.16. Let

$$p_a(x) = \frac{1}{\sqrt{2\pi a^2}} \exp\left\{-\frac{x^2}{2a^2}\right\}$$

be the distribution density of a Gaussian random variable with expectation 0 and variance a^2. First prove that

$$\text{(a)} \qquad \frac{1}{2\pi}\int_{-\infty}^{\infty} e^{izt}\varphi(z)\exp\left\{-\frac{1}{2}a^2z^2\right\}dz = \int_{-\infty}^{\infty} p_a(x-t)\,dF(x).$$

Use the remark that the right-hand side of (a) is the probability density of the convolution of the distribution function $F(x)$ and the Gaussian distribution with zero mean and variance a^2. Let $F_a(x)$ be the corresponding distribution function, and $f_a(x)$ its probability density. Prove that $F_a(x) \to F(x)$ as $a \to 0$ at any point x where F is continuous. Then show that $p(x)$ is a continuous bounded function and $p_a(x) \to p(x)$ as $a \to 0$ for all $x \in \mathbf{R}$, so that for any interval $[c, d]$ we have

$$F_a(d) - F_a(c) = \int_c^d p_a(x)\,dx \to \int_c^d p(x)\,dx$$

as $a \to 0$.

V.1.17. Let $\Phi(\cdot)$ be the convolution of the distribution function $F(x)$ and the uniform distribution on $[0, h]$. Use the inversion formula for the probability density of the distribution function $\Phi(\cdot)$.

V.1.18. Let $F_a(x)$ be the convolution of the distribution function $F(x)$ and the Gaussian distribution with zero mean and variance a^2. Note that it is a distribution function. Write equality (a) of Problem V.1.16 for the distribution $F_a(x)$ and use the convergence $F_a(x) \to F(x)$ as $a \to 0$, which is valid at any point x where F is continuous.

V.1.19. The integral on the right under the limit sign is the distribution function of the sum of a random variable ξ with the distribution function $F(x)$ and a Gaussian random variable that is independent of ξ and has zero mean and variance a^2.

V.1.22. a) Calculate the characteristic function of the random variable $\xi = \xi_1 - \xi_2$, where ξ_1 and ξ_2 are independent exponential random variables with parameter 1 (the random variable ξ has two-sided exponential distribution).

b) Use the inversion formula for densities and part a).

V.1.23. a) Use integration by parts.

b) Use the inversion formula for densities and part a).

V.1.25. Let $\varphi_1(z)$ be a periodic function with period 2 such that $\varphi_1(z) = 1-|z|$ for $|z| \le 1$, and

$$\varphi_2(z) = 2\left(\varphi_1\left(\frac{z}{2}\right) - \frac{1}{2}\right).$$

Check that $\varphi_1(z)$ and $\varphi_2(z)$ are characteristic functions, and show that $\varphi_1^2(z) = \varphi_2^2(z)$.

V.1.29. a) Use the change of variables $y = x(1 - iz)$.

V.1.30. ANSWER: $\varphi(z) = (1 - 2iz)^{-n/2}$.

V.1.34. ANSWER: $(1 + z^2)^{-1/2}$.

V.1.35. To construct an appropriate example use Problem V.1.22 b).

V.1.39. ANSWER: a)

$$G(x) = \frac{1}{\mu_2} \int_{-\infty}^{x} y^2 \, dF(y).$$

V.1.41. Use the inequality

$$\left| e^{i\xi x} \left(e^{itx} - 1 - \frac{itx}{1!} - \cdots - \frac{(itx)^{n-1}}{(n-1)!} \right) \right| \leq \frac{(tx)^n}{n!}.$$

V.1.44. Use the formula

$$\frac{1 - u(z)}{z^2} = \int_0^\infty \frac{1 - \cos zx}{z^2 x^2} x^2 \, dF(x),$$

where $u(z) = \operatorname{Re} \varphi(z)$. Prove that the expression on the right tends to $u''(0)$ as $z \downarrow 0$, provided $u''(0)$ exists. Then show that the integrand converges to $\frac{1}{2}$ as $z \to 0$.

V.1.53. Show that any such function can be represented as a limit of linear combinations of characteristic functions that satisfy the conditions given in the problem.

V.1.54. Use Problem V.1.16.

V.1.67. ANSWER:

$$\varphi(z) = \left(1 - \frac{a}{\lambda} iz + \frac{1}{2} \frac{z^2 \sigma^2}{\lambda} \right)^{-1}.$$

V.1.71. ANSWER:

$$G(x) = \int_0^x \frac{e^{-hy}}{\varphi(h)} \, dF(y), \qquad x \geq 0.$$

V.1.72. ANSWER:

$$G(x) = \frac{1}{\mu_n} \int_0^x y^n \, dF(y), \qquad x \geq 0.$$

V.1.76. Write $g(\bar{z}_k - \bar{z}_j)\varphi(\bar{z}_j - \bar{z}_j)$ in the form

$$\int_{\mathbf{R}} g(\bar{z}_k - \bar{z}_j)^{i(\bar{z}_k, \bar{x})} e^{-i(\bar{z}_j, \bar{x})} \, dF(x_1, \ldots, x_n),$$

where F is the distribution function that corresponds to $\varphi(\bar{z})$.

V.1.79. To calculate the integral that represents the characteristic function use polar coordinates:

$$\varphi(u,v)=\frac{\sin u}{u}\frac{\sin v}{v}+\left(\frac{\cos v}{v}-\frac{\sin v}{v^2}\right)\left(-\frac{\cos u}{u}+\frac{3\sin u}{u^3}+\frac{6\cos u}{u^3}-6\frac{\sin u}{u^3}\right)$$
$$-\left(\frac{\cos u}{u}-\frac{\sin u}{u^2}\right)\left(-\frac{\cos v}{v}+\frac{3\sin v}{v^2}+\frac{6\cos v}{v^3}-\frac{6\sin v}{v^4}\right).$$

V.1.80. ANSWER:

$$\psi(z_1,\dots,z_l)=\exp\left\{i\sum_{j=1}^{l}b_jz_j\right\}\varphi\left(\sum_{j=1}^{l}a_{j1}z_j,\dots,\sum_{j=1}^{l}a_{jm}z_j\right).$$

V.1.82. Using the formula for density transformations (see Problem II.3.149) prove that for any $\bar z=(z_1,\dots,z_m)$ the random variable $z_1\xi_1+\dots+z_m\xi_m$ is Gaussian and find its mean and variance. Then use Problem V.1.81 to find $\varphi(z_1,\dots,z_m)$.

V.1.85. ANSWER:
a) $r_{12}r_{34}+r_{13}r_{24}+r_{14}r_{23}$;
b) 0;
c) $8r_{12}r_{13}r_{23}+2\left(r_{12}^2+r_{13}^2+r_{23}^2\right)+1$.

§2.

V.2.4. First show that condition (1) implies

$$\lim_{t\to\infty}\frac{\int_{|x|\le t}x^2\,dF(x)}{\int_{|x|\le\varepsilon t}x^2\,dF(x)}=1$$

for all $\varepsilon>0$. Let B_n be a solution of the equation

$$n\int_{|x|\le B_n}x^2\,dF(x)=B_n^2.$$

Prove that

$$\lim_{n\to\infty}\left(1-\varphi\left(\frac{z}{B_n}\right)\right)=-\frac{z^2}{2},$$

where $\varphi(z)$ is the characteristic function of the random variable ξ_k.

V.2.5. Find the characteristic function of

$$\frac{1}{\sqrt{k_n}}\sum_{k=1}^{\nu_n}\xi_k$$

and apply the continuity theorem. To justify the passage to the limit, show that

$$\psi\left(\frac{z}{\sqrt{k_n}}\right)^{k_ny}\to\exp\left\{-\frac{z^2}{2}\sigma^2y\right\}$$

as $n\to\infty$, uniformly in $y\in[a,b]$ for any finite interval $[a,b]$ that does not contain 0. Here $\psi(z)$ is the characteristic function of ξ_1.

V.2.6. a) Use the estimate

$$\left| e^{iy} - 1 - iy - \frac{y^2}{2} \right| \le \frac{|y|^3}{6}.$$

V.2.15. First show that the distributions of the random variables $S_{\varepsilon n} - \mathsf{E}\, S_{\varepsilon n}$ defined in Problem V.2.14 weakly converge to a corresponding Gaussian distribution. To do this obtain the following representation for the characteristic function $\varphi_{kn}(s)$ of the random variable $\xi_{kn}^{(\varepsilon_n)} - \mathsf{E}\, \xi_{kn}^{(\varepsilon_n)}$:

$$\varphi_{kn}(s) = 1 - \frac{s^2}{2} \operatorname{Var} \xi_{kn}^{(\varepsilon_n)} + \varepsilon_n \frac{s^3}{6} \operatorname{Var} \xi_{kn}^{(\varepsilon_n)} \Theta_{kn}(s),$$

where $|\Theta_{kn}(s)| \le 1$. Then use Problem V.2.14 b) and c).

V.2.18. First prove that for any sequence of characteristic functions $\varphi_k(z)$, $k \ge 1$, such that $|\varphi_k(z)|^2 \to 1$ there exists a sequence of real numbers $\{\alpha_n, n \ge 1\}$ such that $\varphi_n(z) e^{-i\alpha_n z} \to 1$ as $n \to \infty$.

V.2.20. To prove the necessity of the condition a), use the relation

$$\prod_{k=1}^{k_n} \mathsf{P}(\xi_{kn} \le \varepsilon) = \mathsf{P}\Big(\max_{k=1,\dots,k_n} \xi_{kn} \le \varepsilon \Big) > \mathsf{P}\Big(\sum_{k=1}^{k_n} \xi_{kn} \le \varepsilon \Big).$$

On the other hand, to prove the necessity of the condition b), apply the Chebyshev inequality to the random variables

$$\xi'_{kn} = \begin{cases} \xi_{kn} & \text{if } \xi_{kn} \le \varepsilon, \\ 0 & \text{otherwise.} \end{cases}$$

V.2.21. First show that $|\varphi(z)| < 1$ for $z \ne 0$ and $|\varphi(z)| \to 0$ as $z \to \pm\infty$. Then use the relation $|1 - \varphi(z)|^2 \sim b|z|^2$ as $z \to 0$.

V.2.22. a) Using Problem V.2.21, first prove that

$$(1) \qquad \int_{|z| > \sqrt{n \ln n}} \left| \varphi\left(\frac{z}{\sqrt{n}} \right) \right|^n dz \le \sqrt{n} e^{-(u-\nu)\delta \ln n} \int_{-\infty}^{\infty} |\varphi(z)|^\nu \, dz.$$

Then use the inequality

$$(2) \qquad \left| \varphi\left(\frac{z}{\sqrt{n}} \right) \right|^n \le e^{-\delta z^2} \quad \text{for } |z| < \sqrt{n \ln n}$$

and the central limit theorem.

b) Show that

$$\int_{-\infty}^{\infty} |\varphi(z)|^2 \, dz < \infty$$

and apply estimates (1) and (2).

V.2.24. Use the inequality

$$\left| \varphi(t) e^{-iat} - 1 + \frac{z^2}{2} \right| \le \int_{-\infty}^{\infty} \left| e^{izx} - izx + \frac{x^2 z^2}{2} \right| dF(x) \le c \int_{-\infty}^{\infty} |z|^{2+\delta} |x|^{2+\delta} \, dF(x),$$

which is valid since the function

$$\frac{e^{ix} - 1 - ix + \frac{x^2}{2}}{|x|^{2+\delta}}$$

is bounded for $0 < \delta \leq 1$.

V.2.28. To prove the assertion use the estimate

$$\left|pe^{iz} + q\right| = \sqrt{1 - 2pq\sin^2\frac{z}{2}} \leq \exp\left\{-pq\sin^2\frac{z}{2}\right\} \leq \exp\{-\delta pqz^2\},$$

where $\delta > 0$ is a constant.

V.2.30. Find an explicit expression for the distribution density $p_n(x)$ of the random variable

$$\frac{\xi_1 + \cdots + \xi_n}{\sqrt{2n}}$$

and use it to determine an asymptotic expansion for $p_n(x)$. In so doing, use the representation $\xi_k = \eta_k + \zeta_k$, where η_k and ζ_k are independent, η_k is Gaussian, and $\mathsf{P}(\zeta_k = -1) = \mathsf{P}(\zeta_k = 1) = \frac{1}{2}$.

§3.

V.3.1. Use the inequality of Problem V.1.61.

V.3.8. A sum of independent random variables is nonnegative with probability one if and only if all summands are nonnegative with probability one. To prove that a random variable with the characteristic function (d) is nonnegative use the property that for all $\varepsilon > 0$ the distribution function corresponding to the characteristic function

$$\exp\left\{\int_{\varepsilon}^{\infty}\left(e^{ix} - 1\right)\,dG(x)\right\},$$

is of the form

$$\sum_{k=0}^{\infty}\frac{a_\varepsilon^k}{k!}e^{-a_\varepsilon}\Phi^{*k}(x),$$

where

$$\Phi(x) = \begin{cases} 0, & \text{for } x \leq \varepsilon, \\ \frac{G(x)-G(\varepsilon+0)}{G(+\infty)-G(\varepsilon+0)}, & \text{for } x > \varepsilon, \end{cases}$$

$$a_\varepsilon = G(+\infty) - G(\varepsilon + 0), \qquad \Phi^{*0}(x) = \begin{cases} 0, & \text{for } x \leq 0, \\ 1, & \text{for } x > 0. \end{cases}$$

V.3.15. A sum of independent random variables has an arithmetic distribution if and only if every summand has an arithmetic distribution.

§4.

V.4.15. c) Show that the eigenfunctions of equation (2) satisfy the condition

$$g_k(0) = 0, \qquad g_k'(T) = 0, \qquad g_k'(t) = -\lambda_k g_k(t).$$

V.4.21. Let B be a nonsingular matrix, that is, $\det(B) \neq 0$, and let $\bar{l}_i$, $i = 1, \ldots, k$, be an orthonormal system of eigenvectors of the matrix B. Prove that $w_i(t) - \left(\bar{l}_i, \bar{w}(t)\right)$, $i = 1, \ldots, k$, are independent real-valued Wiener processes.

§**5.**

V.5.1. Introduce the random variables

$$\tau_a^{(h)} = \min\{kh: w(kh) > a\}, \qquad h = 0, 1, \dots,$$

and use the independence of the events $\{\omega: \tau_a^{(h)} = kh\}$ and $\{\omega: w(kh + t_i) < x_i\}$, $i = 1, \dots, l$. Also take into account that $\tau_a^{(h)} \to \tau_a$ as $h \to 0$ and that $w(t)$ is continuous with probability one. Then show that for each $s \le +\infty$,

$$\mathsf{P}(w(\tau_a + t_i) < x_i, i = 1, \dots, l, \tau_a < s) = \mathsf{P}(w(t_i) < x_i, i = 1, \dots, l)\mathsf{P}(\tau_a < s).$$

V.5.2. Use the relation

$$\widetilde{w}(t) = a + w(t + \tau_a) - w(\tau_a)$$

for $t \ge \tau_a$ and Problem V.5.1.

V.5.3. Apply the relation

$$\mathsf{P}\Big(\sup_{t \le T} w(t) > a, w(T) < x\Big) = \int_0^T \mathsf{P}\left(w(T) - w(\tau_a) < x - a/\tau_a = s\right) d\mathsf{P}(\tau_a < s)$$

to prove by Problem V.5.1 that

$$\mathsf{P}\Big(\sup_{t \le T} w(t) > a, w(T) < x\Big) = \mathsf{P}(w(t) > 2a - x).$$

V.5.8. Prove that for all intervals (α, β),

$$\mathsf{P}\Big(\sup_{\alpha < t < \beta} w(t) = a\Big) = 0$$

and use the following assertion. *For $\tau_a' < \tau_a$, there exist rational α and β, $\alpha < \beta$, such that $\sup_{\alpha < t < \beta} w(t) = a$ if $\alpha < \tau_a' < \beta < \tau_a$.*

V.5.9. Use the symmetry of the process $w(t)$ to show that

$$P_{t_1,t_2} = 2 \int_{-\infty}^0 \mathsf{P}(\tau_{-a} \le t_2 - t_1)\, d\mathsf{P}(w(t_1) < a).$$

ANSWER: $P_{t_1,t_2} = \frac{2}{\pi} \arccos \sqrt{t_1/t_2}$.

V.5.10. Use Problem V.5.9.

V.5.13. Find the joint probability density of the random variables $w^+(T)$ and $w(T)$.

V.5.14. For $t' < t$ we have

$$X(t) = \max\Big\{ \sup_{t' \le u \le t} (w(u) - w(t')), X(t') - w(t) - w(t')\Big\}.$$

V.5.16. Since $|w(t)|$ and $X(t)$ are Markov processes, it is sufficient to check that the distribution functions

$$\mathsf{P}\left(X(t) < y/X(t_0) = y_0\right) \quad \text{and} \quad \mathsf{P}\left(|w(t)| < y/w(t_0) = y_0\right)$$

coincide for $t_0 < t$.

V.5.17. ANSWER:

$$\frac{\arcsin\sqrt{t_0/t_2}}{\arcsin\sqrt{t_0/t_1}}.$$

V.5.18. Find

$$\mathsf{P}\left(w(t) \neq 0, 0 < t_0 \leq t \leq t_2 / w(t) \neq 0, 0 < t_0 \leq t \leq t_1\right)$$

and pass to the limit as $t_0 \to 0$.

V.5.20. First prove that

$$\mathsf{P}\left(w(1) \leq x / w(u) \geq 0, 0 \leq u \leq 1\right) = \mathsf{P}\left(w(1) \leq x / w(u) \leq w(1), 0 \leq u \leq 1\right)$$

and use Problem V.5.13.

V.5.22. ANSWER:

$$u(x) = \frac{x-c}{b-c}.$$

V.5.24. To solve the problem use the function $q_{a,b}(t,x)$,

$$q_{a,b}(t,x) = \mathsf{P}\Big(\sup_{0\leq s\leq t} w(s) < b-x, \inf_{0\leq s\leq t} w(s) \geq a-x\Big), \qquad a < x < b, \quad t > 0$$

(see Problem V.5.23). This function is related to the distribution function of the random variable $\zeta_{a-x,b-x}$ as follows:

$$q_{a,b}(t,x) = \mathsf{P}(\zeta_{a-x,b-x} > t).$$

Selected Titles in This Series

(Continued from the front of this publication)

125 **Masayoshi Nagata,** Theory of commutative fields, 1993
124 **Masahisa Adachi,** Embeddings and immersions, 1993
123 **M. A. Akivis and B. A. Rosenfeld,** Élie Cartan (1869–1951), 1993
122 **Zhang Guan-Hou,** Theory of entire and meromorphic functions: deficient and asymptotic values and singular directions, 1993
121 **I. B. Fesenko and S. V. Vostokov,** Local fields and their extensions: A constructive approach, 1993
120 **Takeyuki Hida and Masuyuki Hitsuda,** Gaussian processes, 1993
119 **M. V. Karasev and V. P. Maslov,** Nonlinear Poisson brackets. Geometry and quantization, 1993
118 **Kenkichi Iwasawa,** Algebraic functions, 1993
117 **Boris Zilber,** Uncountably categorical theories, 1993
116 **G. M. Fel′dman,** Arithmetic of probability distributions, and characterization problems on abelian groups, 1993
115 **Nikolai V. Ivanov,** Subgroups of Teichmüller modular groups, 1992
114 **Seizô Itô,** Diffusion equations, 1992
113 **Michail Zhitomirskiĭ,** Typical singularities of differential 1-forms and Pfaffian equations, 1992
112 **S. A. Lomov,** Introduction to the general theory of singular perturbations, 1992
111 **Simon Gindikin,** Tube domains and the Cauchy problem, 1992
110 **B. V. Shabat,** Introduction to complex analysis Part II. Functions of several variables, 1992
109 **Isao Miyadera,** Nonlinear semigroups, 1992
108 **Takeo Yokonuma,** Tensor spaces and exterior algebra, 1992
107 **B. M. Makarov, M. G. Goluzina, A. A. Lodkin, and A. N. Podkorytov,** Selected problems in real analysis, 1992
106 **G.-C. Wen,** Conformal mappings and boundary value problems, 1992
105 **D. R. Yafaev,** Mathematical scattering theory: General theory, 1992
104 **R. L. Dobrushin, R. Kotecký, and S. Shlosman,** Wulff construction: A global shape from local interaction, 1992
103 **A. K. Tsikh,** Multidimensional residues and their applications, 1992
102 **A. M. Il′in,** Matching of asymptotic expansions of solutions of boundary value problems, 1992
101 **Zhang Zhi-fen, Ding Tong-ren, Huang Wen-zao, and Dong Zhen-xi,** Qualitative theory of differential equations, 1992
100 **V. L. Popov,** Groups, generators, syzygies, and orbits in invariant theory, 1992
99 **Norio Shimakura,** Partial differential operators of elliptic type, 1992
98 **V. A. Vassiliev,** Complements of discriminants of smooth maps: Topology and applications, 1992 (revised edition, 1994)
97 **Itiro Tamura,** Topology of foliations: An introduction, 1992
96 **A. I. Markushevich,** Introduction to the classical theory of Abelian functions, 1992
95 **Guangchang Dong,** Nonlinear partial differential equations of second order, 1991
94 **Yu. S. Il′yashenko,** Finiteness theorems for limit cycles, 1991
93 **A. T. Fomenko and A. A. Tuzhilin,** Elements of the geometry and topology of minimal surfaces in three-dimensional space, 1991
92 **E. M. Nikishin and V. N. Sorokin,** Rational approximations and orthogonality, 1991
91 **Mamoru Mimura and Hirosi Toda,** Topology of Lie groups, I and II, 1991
90 **S. L. Sobolev,** Some applications of functional analysis in mathematical physics, third edition, 1991
89 **Valeriĭ V. Kozlov and Dmitriĭ V. Treshchëv,** Billiards: A genetic introduction to the dynamics of systems with impacts, 1991

(See the AMS catalog for earlier titles)